SIXTH EDITION

Basic Statistics for the Behavioral Sciences

For my wife, Karen, the love of my life

BRIEF CONTENTS

PREFACE   xxii

1   **Introduction to Statistics**   1

2   **Statistics and the Research Process**   12

3   **Frequency Distributions and Percentiles**   36

4   **Measures of Central Tendency: The Mean, Median, and Mode**   60

5   **Measures of Variability: Range, Variance, and Standard Deviation**   84

6   **z-Scores and the Normal Curve Model**   109

7   **Correlation Coefficients**   135

8   **Linear Regression**   160

9   **Using Probability to Make Decisions about Data**   185

10 **Introduction to Hypothesis Testing**   207

11 **Performing the One-Sample *t*-Test and Testing Correlation Coefficients**   234

12 **The Two-Sample *t*-Test**   260

13 **The One-Way Analysis of Variance**   290

14 **The Two-Way Analysis of Variance**   318

15 **Chi Square and Other Nonparametric Procedures**   351

APPENDIXES

A   **Additional Statistical Formulas**   379

B   **Using SPSS**   392

C   **Statistical Tables**   418

D   **Answers to Odd-Numbered Questions**   438

GLOSSARY   453

INDEX   461

vii

CONTENTS

PREFACE   xxii

1 Introduction to Statistics   1

GETTING STARTED   1

WHY IS IT IMPORTANT TO LEARN STATISTICS
(AND HOW DO YOU DO THAT?)   1

*What Are Statistics? 1   Why Must I Learn Statistics? 1   What Do
Researchers Do with Statistics? 1   But I'm Not Interested in Research;
I Just Want to Help People! 2   But I Don't Know Anything about Research! 2
What If I'm Not Very Good at Math? 2   What If I'm Not Very Good at
Statistics? 2   I Looked through the Book: Statistics Aren't Written in English! 3
But, Those Formulas! 3   So All I Have to Do Is Learn How to Compute
the Answers? 3   What about Using a Computer to Do Statistics? 3
All Right, So How Do I Learn Statistics? 3*

REVIEW OF MATHEMATICS USED IN STATISTICS   4

Basic Statistical Notation   4

*Identifying Mathematical Operations 5   Determining the Order of
Mathematical Operations 5   Working with Formulas 5*

Rounding Numbers   6

Transforming Scores   6

Proportions 7   Percents 7

Creating Graphs   7

PUTTING IT ALL TOGETHER   9

CHAPTER SUMMARY   9

KEY TERMS   10

REVIEW QUESTIONS   10

APPLICATION QUESTIONS   10

2 Statistics and the Research Process   12

GETTING STARTED   12

THE LOGIC OF RESEARCH   12

Samples and Populations   13

Obtaining Data by Measuring Variables   14

Understanding Relationships   15

The Strength of a Relationship 16   When No Relationship Is Present 17

A QUICK REVIEW   18

   Graphing Relationships   18

APPLYING DESCRIPTIVE AND INFERENTIAL STATISTICS   20

Descriptive Statistics   20

Inferential Statistics   21

Statistics versus Parameters   21

UNDERSTANDING EXPERIMENTS AND CORRELATIONAL STUDIES   22

Experiments   22

   The Independent Variable   22   *Conditions of the Independent Variable   23*
   The Dependent Variable   23   *Drawing Conclusions from Experiments   24*

Correlational Studies   25

A QUICK REVIEW   26

A Word about Causality   26

THE CHARACTERISTICS OF SCORES   27

The Four Types of Measurement Scales   27

Continuous versus Discrete Scales   28

A QUICK REVIEW   29

STATISTICS IN PUBLISHED RESEARCH: USING STATISTICAL TERMS   30

PUTTING IT ALL TOGETHER   30

CHAPTER SUMMARY   31

KEY TERMS   32

REVIEW QUESTIONS   32

APPLICATION QUESTIONS   33

3   Frequency Distributions and Percentiles   **36**

GETTING STARTED   36

NEW STATISTICAL NOTATION   36

WHY IS IT IMPORTANT TO KNOW ABOUT FREQUENCY DISTRIBUTIONS   37

SIMPLE FREQUENCY DISTRIBUTIONS   37

Presenting Simple Frequency in a Table   37

A QUICK REVIEW   38

Graphing a Simple Frequency Distribution   38

   Bar Graphs   39   *Histograms   40*   *Polygons   40*

A QUICK REVIEW   42

TYPES OF SIMPLE FREQUENCY DISTRIBUTIONS   42

The Normal Distribution   42

Other Common Frequency Polygons   43

Labeling Frequency Distributions   45

A QUICK REVIEW   46

RELATIVE FREQUENCY AND THE NORMAL CURVE   47

Presenting Relative Frequency in a Table or Graph   48

Finding Relative Frequency Using the Normal Curve   49

A QUICK REVIEW   *51*

COMPUTING CUMULATIVE FREQUENCY AND PERCENTILE   51
Computing Cumulative Frequency   52
Computing Percentiles   52
Finding Percentile Using the Area Under the Normal Curve   53
A QUICK REVIEW   *54*

STATISTICS IN PUBLISHED RESEARCH: APA PUBLICATION RULES   54

A WORD ABOUT GROUPED FREQUENCY DISTRIBUTIONS   54

PUTTING IT ALL TOGETHER   55
Using the SPSS Appendix   *55*

CHAPTER SUMMARY   55

KEY TERMS   56

REVIEW QUESTIONS   56

APPLICATION QUESTIONS   57

INTEGRATION QUESTIONS   58

SUMMARY OF FORMULAS   59

4   Measures of Central Tendency: The Mean, Median, and Mode   60

GETTING STARTED   60

NEW STATISTICAL NOTATION   60

WHY IS IT IMPORTANT TO KNOW ABOUT CENTRAL TENDENCY?   61

WHAT IS CENTRAL TENDENCY?   61

THE MODE   62
Uses of the Mode   63

THE MEDIAN   63
Uses of the Median   65
A QUICK REVIEW   *65*

THE MEAN   65
Uses of the Mean   66
A QUICK REVIEW   *67*

Comparing the Mean, Median, and Mode   67

TRANSFORMATIONS AND THE MEAN   69

DEVIATIONS AROUND THE MEAN   69
Using the Mean to Predict Scores   71
A QUICK REVIEW   *72*

DESCRIBING THE POPULATION MEAN   72

SUMMARIZING RESEARCH   73
Summarizing an Experiment   73
Graphing the Results of an Experiment   75
   Line Graphs 75   Bar Graphs 76
A QUICK REVIEW   *77*

Inferring the Relationship in the Population   78

STATISTICS IN PUBLISHED RESEARCH: USING THE MEAN   79

PUTTING IT ALL TOGETHER   79
Using the SPSS Appendix   79

CHAPTER SUMMARY   79

KEY TERMS   80

REVIEW QUESTIONS   80

APPLICATION QUESTIONS   81

INTEGRATION QUESTIONS   82

SUMMARY OF FORMULAS   83

5   Measures of Variability: Range, Variance, and Standard Deviation   84

GETTING STARTED   84

NEW STATISTICAL NOTATION   84

WHY IS IT IMPORTANT TO KNOW ABOUT MEASURES OF VARIABILITY?   85

A QUICK REVIEW   86

THE RANGE   87

UNDERSTANDING THE VARIANCE AND STANDARD DEVIATION   87
The Sample Variance   88
The Sample Standard Deviation   89

A QUICK REVIEW   92

COMPUTING THE SAMPLE VARIANCE AND SAMPLE
STANDARD DEVIATION   93
Computing the Sample Variance   93
Computing the Sample Standard Deviation   94

A QUICK REVIEW   94
Mathematical Constants and the Standard Deviation   95

THE POPULATION VARIANCE AND THE POPULATION
STANDARD DEVIATION   95
Estimating the Population Variance and Population Standard Deviation   96
Computing the Estimated Population Variance and Standard Deviation   98
Interpreting the Estimated Population Variance and Standard Deviation   99

A QUICK REVIEW   99

A SUMMARY OF THE VARIANCE AND STANDARD DEVIATION   100

APPLYING THE VARIANCE AND STANDARD DEVIATION TO RESEARCH   101
Variability and Strength of a Relationship   101
Variability and Errors in Prediction   102
Accounting for Variance   103

STATISTICS IN PUBLISHED RESEARCH: REPORTING VARIABILITY   103

PUTTING IT ALL TOGETHER   104
Using the SPSS Appendix   104

CHAPTER SUMMARY   104

KEY TERMS   105

REVIEW QUESTIONS   105

APPLICATION QUESTIONS   106

INTEGRATION QUESTIONS   107

SUMMARY OF FORMULAS   108

6   **z-Scores and the Normal Curve Model**   **109**

GETTING STARTED   109

NEW STATISTICAL NOTATION   109

WHY IS IT IMPORTANT TO KNOW ABOUT z-SCORES?   110

UNDERSTANDING z-SCORES   110

Computing z-Scores   112

Computing a Raw Score When z Is Known   113

A QUICK REVIEW   114

INTERPRETING z-SCORES USING THE z-DISTRIBUTION   115

USING z-SCORES TO COMPARE DIFFERENT VARIABLES   116

USING z-SCORES TO DETERMINE THE RELATIVE FREQUENCY OF RAW SCORES   117

The Standard Normal Curve   118

Using the z-Table   120

A QUICK REVIEW   123

STATISTICS IN PUBLISHED RESEARCH: USING z-SCORES   123

USING z-SCORES TO DESCRIBE SAMPLE MEANS   124

The Sampling Distribution of Means   124

The Standard Error of the Mean   126

Computing a z-Score for a Sample Mean   127

Another Example Combining All of the Above   128

Describing the Relative Frequency of Sample Means   128

Summary of Describing a Sample Mean with a z-Score   129

A QUICK REVIEW   130

PUTTING IT ALL TOGETHER   130

Using the SPSS Appendix   130

CHAPTER SUMMARY   130

KEY TERMS   131

REVIEW QUESTIONS   131

APPLICATION QUESTIONS   132

INTEGRATION QUESTIONS   133

SUMMARY OF FORMULAS   134

7 **The Correlation Coefficient** **135**

GETTING STARTED   135

NEW STATISTICAL NOTATION   135

WHY IS IT IMPORTANT TO KNOW ABOUT CORRELATION COEFFICIENTS?   136

UNDERSTANDING CORRELATIONAL RESEARCH   136
Drawing Conclusions from Correlational Research   137
Distinguishing Characteristics of Correlational Analysis   138
Plotting Correlational Data: The Scatterplot   138

TYPES OF RELATIONSHIPS   139
Linear Relationships   139
Nonlinear Relationships   141
How the Correlation Coefficient Describes the Type of Relationship   142

STRENGTH OF THE RELATIONSHIP   142
Perfect Association   143
Intermediate Association   144
Zero Association   146
A QUICK REVIEW   *147*

THE PEARSON CORRELATION COEFFICIENT   147
A QUICK REVIEW   *150*

THE SPEARMAN RANK-ORDER CORRELATION COEFFICIENT   151
A QUICK REVIEW   *153*

THE RESTRICTION OF RANGE PROBLEM   153

STATISTICS IN PUBLISHED RESEARCH: CORRELATION COEFFICIENTS   154

PUTTING IT ALL TOGETHER   155
Using the SPSS Appendix   *155*

CHAPTER SUMMARY   155

KEY TERMS   156

REVIEW QUESTIONS   156

APPLICATION QUESTIONS   157

INTEGRATION QUESTIONS   159

SUMMARY OF FORMULAS   159

8 **Linear Regression 160**

GETTING STARTED   160

NEW STATISTICAL NOTATION   160

WHY IS IT IMPORTANT TO KNOW ABOUT LINEAR REGRESSION?   161

UNDERSTANDING LINEAR REGRESSION   161

THE LINEAR REGRESSION EQUATION   163

Computing the Slope   164
Computing the Y Intercept   166
Describing the Linear Regression Equation   166
Plotting the Regression Line 166
Computing Predicted Y Scores 167
A QUICK REVIEW   *168*

DESCRIBING THE ERRORS IN PREDICTION   168
Computing the Variance of the Y Scores around Y' 169
Computing the Standard Error of the Estimate   170
A QUICK REVIEW   *171*

Interpreting the Standard Error of the Estimate   171
The Strength of a Relationship and Prediction Error   173

COMPUTING THE PROPORTION OF VARIANCE ACCOUNTED FOR   174
Using r to Compute the Proportion of Variance Accounted For   176
Applying the Proportion of Variance Accounted For   177
A QUICK REVIEW   *178*

A WORD ABOUT MULTIPLE CORRELATION AND REGRESSION   179

STATISTICS IN PUBLISHED RESEARCH: LINEAR REGRESSION   179

PUTTING IT ALL TOGETHER   179
Using the SPSS Appendix   180

CHAPTER SUMMARY   180

KEY TERMS   181

REVIEW QUESTIONS   181

APPLICATION QUESTIONS   181

INTEGRATION QUESTIONS   183

SUMMARY OF FORMULAS   184

9 Using Probability to Make Decisions about Data   185

GETTING STARTED   185

NEW STATISTICAL NOTATION   185

WHY IS IT IMPORTANT TO KNOW ABOUT PROBABILITY?   186

THE LOGIC OF PROBABILITY   186

COMPUTING PROBABILITY   188
Creating Probability Distributions   188
Factors Affecting the Probability of an Event   188
A QUICK REVIEW   *189*

OBTAINING PROBABILITY FROM THE STANDARD NORMAL CURVE   189
Determining the Probability of Individual Scores   190
Determining the Probability of Sample Means   191
A QUICK REVIEW   *192*

RANDOM SAMPLING AND SAMPLING ERROR   193

DECIDING WHETHER A SAMPLE REPRESENTS A POPULATION   194

A QUICK REVIEW   *196*

Setting Up the Sampling Distribution   197
Identifying the Critical Value   198
Deciding if the Sample Represents the Population   199
Other Ways to Set up the Sampling Distribution   200

A QUICK REVIEW   *202*

PUTTING IT ALL TOGETHER   202

CHAPTER SUMMARY   203

KEY TERMS   203

REVIEW QUESTIONS   204

APPLICATION QUESTIONS   204

INTEGRATION QUESTIONS   205

SUMMARY OF FORMULAS   206

10   Introduction to Hypothesis Testing   207

GETTING STARTED   207

NEW STATISTICAL NOTATION   207

WHY IS IT IMPORTANT TO KNOW ABOUT THE *z*-TEST?   208

THE ROLE OF INFERENTIAL STATISTICS IN RESEARCH   208

SETTING UP INFERENTIAL PROCEDURES   209

Creating the Experimental Hypotheses   209
Designing a One-Sample Experiment   210
Creating the Statistical Hypotheses   211
   The Alternative Hypothesis 211   The Null Hypothesis 212

A QUICK REVIEW   *213*

The Logic of Statistical Hypothesis Testing   213

PERFORMING THE *z*-TEST   215

Setting Up the Sampling Distribution for a Two-Tailed Test   215
Computing *z*   216
Comparing the Obtained *z* to the Critical Value   217

INTERPRETING SIGNIFICANT RESULTS   218

INTERPRETING NONSIGNIFICANT RESULTS   219

SUMMARY OF THE *z*-TEST   220

A QUICK REVIEW   *221*

THE ONE-TAILED TEST   221

The One-Tailed Test for Increasing Scores   221
The One-Tailed Test for Decreasing Scores   223

A QUICK REVIEW   *223*

STATISTICS IN PUBLISHED RESEARCH: REPORTING
SIGNIFICANCE TESTS   224

ERRORS IN STATISTICAL DECISION MAKING   224

Type I Errors: Rejecting H_0 When H_0 Is True   224
Type II Errors: Retaining H_0 When H_0 Is False   226
Comparing Type I and Type II Errors   227
Power   228
A QUICK REVIEW   **229**

PUTTING IT ALL TOGETHER   229

CHAPTER SUMMARY   230

KEY TERMS   231

REVIEW QUESTIONS   231

APPLICATION QUESTIONS   231

INTEGRATION QUESTIONS   233

SUMMARY OF FORMULAS   233

11   Performing the One-Sample *t*-Test and Testing Correlation Coefficients   234

GETTING STARTED   234

WHY IS IT IMPORTANT TO KNOW ABOUT *t*-TESTS?   234

PERFORMING THE ONE-SAMPLE *t*-TEST   235
Computing t_{obt} 236
A QUICK REVIEW   **237**
The *t*-Distribution and Degrees of Freedom   238
Using the *t*-Tables   240
Interpreting the *t*-Test   240
The One-Tailed *t*-Test   241
Some Help When Using the *t*-Tables   242
A QUICK REVIEW   **242**

ESTIMATING μ BY COMPUTING A CONFIDENCE INTERVAL   243
Computing the Confidence Interval   244
A QUICK REVIEW   **246**
Summary of the One-Sample *t*-Test   246

STATISTICS IN PUBLISHED RESEARCH: REPORTING THE *t*-TEST   246

SIGNIFICANCE TESTS FOR CORRELATION COEFFICIENTS   247
Testing the Pearson *r*   247
   The Sampling Distribution of r 249   Interpreting r 250
   One-Tailed Tests of r 251
Testing the Spearman r_S 252
Summary of Testing a Correlation Coefficient   253
A QUICK REVIEW   **253**

MAXIMIZING THE POWER OF STATISTICAL TESTS   253

PUTTING IT ALL TOGETHER   255
Using the SPSS Appendix   255

CHAPTER SUMMARY   255

KEY TERMS   256

REVIEW QUESTIONS   256

APPLICATION QUESTIONS   257

INTEGRATION QUESTIONS   258

SUMMARY OF FORMULAS   259

12 The Two-Sample *t*-Test   260

GETTING STARTED   260

NEW STATISTICAL NOTATION   260

WHY IS IT IMPORTANT TO KNOW ABOUT THE TWO-SAMPLE *t*-TEST?   261

UNDERSTANDING THE TWO-SAMPLE EXPERIMENT   261

THE INDEPENDENT-SAMPLES *t*-TEST   262
Statistical Hypotheses for the Independent-Samples *t*-Test   262
The Sampling Distribution for the Independent-Samples *t*-Test   263
Computing the Independent-Samples *t*-Test   264
   *Estimating the Population Variance 264   Computing the Standard Error of
   the Difference 265   Computing* t_{obt} *266*

A QUICK REVIEW   **267**
Interpreting the Independent-Samples *t*-Test   267
Confidence Interval for the Difference between Two μs   268
Performing One-Tailed Tests with Independent Samples   269

SUMMARY OF THE INDEPENDENT-SAMPLES *t*-TEST   270

A QUICK REVIEW   **270**

THE RELATED-SAMPLES *t*-TEST   271
The Logic of Hypotheses Testing in the Related-Samples *t*-Test   272

STATISTICAL HYPOTHESES FOR THE RELATED-SAMPLES *t*-TEST   273
Computing the Related-Samples *t*-Test   274
Interpreting the Related-Samples *t*-Test   275
Computing the Confidence Interval for μ_D 277
Performing One-Tailed Tests with Related Samples   277

SUMMARY OF THE RELATED-SAMPLES *t*-TEST   278

A QUICK REVIEW   **278**

DESCRIBING THE RELATIONSHIP IN A TWO-SAMPLE EXPERIMENT   279
Graphing the Results of a Two-Sample Experiment   279
Measuring Effect Size in the Two-Sample Experiment   280
   Effect Size Using Cohen's d *280   Effect Size Using Proportion of Variance
   Accounted For 282*

STATISTICS IN PUBLISHED RESEARCH: THE TWO-SAMPLE EXPERIMENT   283

PUTTING IT ALL TOGETHER   284
Using the SPSS Appendix   284

CHAPTER SUMMARY   284

KEY TERMS   285

REVIEW QUESTIONS   285

APPLICATION QUESTIONS   286

INTEGRATION QUESTIONS   288

SUMMARY OF FORMULAS   289

13   The One-Way Analysis of Variance   290

GETTING STARTED   290

NEW STATISTICAL NOTATION   290

WHY IS IT IMPORTANT TO KNOW ABOUT ANOVA?   291

AN OVERVIEW OF ANOVA   291

How ANOVA Controls the Experiment-Wise Error Rate   292

Statistical Hypotheses in ANOVA   293

The Order of Operations in ANOVA: The F Statistic and Post Hoc Comparisons   293

A QUICK REVIEW   *294*

UNDERSTANDING THE ANOVA   294

The Mean Square within Groups   295

The Mean Square between Groups   295

Comparing the Mean Squares: The Logic of the F-Ratio   296

The Theoretical Components of the F-ratio   298

A QUICK REVIEW   *299*

PERFORMING THE ANOVA   299

Computing the F_{obt} 300

 Computing the Sums of Squares 300   *Computing the Degrees of*
 Freedom 302   *Computing the Mean Squares 303*   *Computing the* F *304*

Interpreting F_{obt} 304

A QUICK REVIEW   *306*

PERFORMING POST HOC COMPARISONS   306

Fisher's Protected t-Test   307

Tukey's *HSD* Multiple Comparisons Test   307

A QUICK REVIEW   *309*

SUMMARY OF STEPS IN PERFORMING A ONE-WAY ANOVA   309

ADDITIONAL PROCEDURES IN THE ONE-WAY ANOVA   310

The Confidence Interval for Each Population μ 310

Graphing the Results in ANOVA   311

Describing Effect Size in the ANOVA   311

STATISTICS IN PUBLISHED RESEARCH: REPORTING ANOVA   312

PUTTING IT ALL TOGETHER   312

Using the SPSS Appendix   312

CHAPTER SUMMARY   313

KEY TERMS   314

REVIEW QUESTIONS   314

APPLICATION QUESTIONS   314

INTEGRATION QUESTIONS   316

SUMMARY OF FORMULAS   317

14 The Two-Way Analysis of Variance   318

GETTING STARTED   318

NEW STATISTICAL NOTATION   318

WHY IS IT IMPORTANT TO KNOW ABOUT THE TWO-WAY ANOVA?   319

UNDERSTANDING THE TWO-WAY DESIGN   319

OVERVIEW OF THE TWO-WAY, BETWEEN-SUBJECTS ANOVA   320
The Main Effect of Factor A   321
The Main Effect of Factor B   323
A QUICK REVIEW   324
Interaction Effects   324
A QUICK REVIEW   326

COMPUTING THE TWO-WAY ANOVA   327
Computing the Sums and Means   328
Computing the Sums of Squares   329
Computing the Degrees of Freedom   332
Computing the Mean Squares   332
Computing F 334
Interpreting Each F 335
A QUICK REVIEW   336

INTERPRETING THE TWO-WAY EXPERIMENT   336
Graphing and Post Hoc Comparisons with Main Effects   336
Graphing the Interaction Effect   338
Performing Post Hoc Comparisons on the Interaction Effect   339
A QUICK REVIEW   341
Interpreting the Overall Results of the Experiment   341
Describing the Effect Size: Eta Squared   342

SUMMARY OF THE STEPS IN PERFORMING A TWO-WAY ANOVA   343

PUTTING IT ALL TOGETHER   344
Using the SPSS Appendix   344

CHAPTER SUMMARY   344

KEY TERMS   345

REVIEW QUESTIONS   345

APPLICATION QUESTIONS   346

INTEGRATION QUESTIONS   348

SUMMARY OF FORMULAS   349

15 Chi Square and Other Nonparametric Procedures   351

GETTING STARTED   351

WHY IS IT IMPORTANT TO KNOW ABOUT NONPARAMETRIC PROCEDURES?   351

CHI SQUARE PROCEDURES   352

ONE-WAY CHI SQUARE   352
Hypotheses and Assumptions of the One-Way Chi Square   353
Computing the One-Way Chi Square   354
Interpreting the One-Way Chi Square   355
Testing Other Hypotheses with the One-Way Chi Square   356
A QUICK REVIEW   *357*

THE TWO-WAY CHI SQUARE   357
Computing the Two-Way Chi Square   359
A QUICK REVIEW   *360*
Describing the Relationship in a Two-Way Chi Square   361

STATISTICS IN PUBLISHED RESEARCH: REPORTING CHI SQUARE   362

NONPARAMETRIC PROCEDURES FOR RANKED DATA   363
The Logic of Nonparametric Procedures for Ranked Data   363
Resolving Tied Ranks   364
Choosing a Nonparametric Procedure   364
Tests for Two Independent Samples: The Mann–Whitney *U* Test and the Rank Sums Test   365
   The Mann–Whitney U *Test 365*   *The Rank Sums Test 366*
The Wilcoxon *T* Test for Two Related Samples   368
The Kruskal–Wallis *H* Test   369
The Friedman χ^2 Test   371

PUTTING IT ALL TOGETHER   373
Using the SPSS Appendix   *373*

CHAPTER SUMMARY   373

KEY TERMS   374

REVIEW QUESTIONS   374

APPLICATION QUESTIONS   374

INTEGRATION QUESTIONS   377

SUMMARY OF FORMULAS   378

APPENDIXES

A Additional Statistical Formulas   379

A.1   CREATING GROUPED FREQUENCY DISTRIBUTIONS   379
A.2   PERFORMING LINEAR INTERPOLATION   381
A.3   THE ONE-WAY, WITHIN-SUBJECTS ANALYSIS OF VARIANCE   384

B Using SPSS   392

B.1   ENTERING DATA   392
B.2   FREQUENCY DISTRIBUTIONS AND PERCENTILE   395

B.3   CENTRAL TENDENCY, VARIABILITY, AND z-SCORES   397

B.4   CORRELATION COEFFICIENTS AND THE LINEAR REGRESSION EQUATION   399

B.5   THE ONE-SAMPLE t-TEST AND SIGNIFICANCE TESTING
OF CORRELATION COEFFICIENTS   402

B.6   TWO-SAMPLE t-TESTS   404

B.7   THE ONE-WAY, BETWEEN-SUBJECTS ANOVA   407

B.8   THE TWO-WAY, BETWEEN-SUBJECTS ANOVA   409

B.9   CHI SQUARE PROCEDURES   411

B.10   NONPARAMETRIC TESTS FOR RANKED SCORES   414

B.11   THE ONE-WAY, WITHIN-SUBJECTS ANOVA   416

C Statistical Tables   418

TABLE 1   PROPORTIONS OF AREA UNDER THE STANDARD NORMAL CURVE:
THE z-TABLES   419

TABLE 2   CRITICAL VALUES OF t: THE t-TABLES   423

TABLE 3   CRITICAL VALUES OF THE PEARSON CORRELATION COEFFICIENT:
THE r-TABLES   424

TABLE 4   CRITICAL VALUES OF THE SPEARMAN RANK-ORDER CORRELATION
COEFFICIENT: THE r_S-TABLES   425

TABLE 5   CRITICAL VALUES OF F: THE F-TABLES   426

TABLE 6   VALUES OF STUDENTIZED RANGE STATISTIC, q_K 429

TABLE 7   CRITICAL VALUES OF CHI SQUARE: THE X^2-TABLES   431

TABLE 8   CRITICAL VALUES OF THE MANN–WHITNEY U 432

TABLE 9   CRITICAL VALUES OF THE WILCOXON T 436

D Answers to Odd-Numbered Questions 438

GLOSSARY   453

INDEX   461

PREFACE TO THE INSTRUCTOR

After almost 20 years of writing and rewriting this book, I still obsess over creating the clearest, most understandable explanation of each statistical procedure. My problem with many textbooks is that they take too much of a statistics-for-statistics-sake approach. They produce students who can compute an answer on demand, but who do not understand why researchers would do so or what the answer reveals about data. I am not enamored by the eloquence of formulas. Instead, I concentrate on showing students the eloquence of the *logic* of statistics. When we simplify the jargon and boil them down to concrete ideas, statistics have practical purposes and they really do make sense. I believe that giving students an understanding of this is the most important component of any introductory course.

The premise of this book is that statistics make sense when presented within the context of behavioral research. Therefore, each procedure is introduced using a simple study with readily understandable goals. The focus is that research examines relationships and that statistics are for describing and inferring such relationships. Each discussion ends, however, by returning to an interpretation of the study in terms of behaviors. Although the early examples involve very simple questions taken from everyday life, in later chapters, as students develop their statistical thinking, the examples become more representative of real research.

A textbook should work very hard at explaining concepts and tying them together. Too often books simply offer up a concept and let students and their instructor sort it out. My approach is that if it is important enough to mention, then it is important enough to fully explain. To this end, the narrative attempts to *teach* the material—clearly and patiently—the way a good teacher does. Further, I believe the best teachers are those who can remember what it was like when they were first learning a concept, before they spoke the technical language and could think in such terms. Therefore, I do not forget that, from the student's perspective, everything about this course is new and often very strange and a little scary.

However, this book does not pander to student weaknesses and fears regarding mathematics. On the one hand, the book is geared toward students who are neither proficient in math nor interested in becoming so, and who rather grudgingly learn statistics. On the other hand, I expect that, ultimately, students will be capable of performing and understanding the basic statistical procedures found in modern research—as "junior" researchers. Therefore, the tone is always "At first this may *appear* difficult, but you can do it." Thus, formulas are introduced in terms of what they accomplish, and examples are worked through in a step-by-step manner. The similarities among different procedures are stressed, showing that they answer similar questions. And, the most difficult concepts are presented in small chunks that span different chapters, so that they are less overwhelming when fully revealed.

I have retained the above goals throughout this revision. At the same time, I have tried to keep the material readable and engaging so that students enjoy it as well as learn from it. I include humor, at times I talk directly to students, I point out potential mistakes, and I provide tips on how to get through the course. In addition, several recurring individuals give a little "plot" to the book, providing continuity among topics, and alerting students to particular pitfalls. Throughout, I have tried to dispel the notion that statistics are incomprehensible and boring, and to show that learning statistics can be fun and rewarding.

MAJOR CHANGES IN THE SIXTH EDITION

Although there are numerous changes in this edition, the most notable are:

New *Integration Questions*   Additional end-of-chapter questions address two issues. First, a chapter must often refer to a concept from a previous chapter, necessarily assuming that students remember its discussion. If they do not, new material is usually lost. Second although students know to compute a correlation in the correlation chapter or a *t*-test in the *t*-test chapter, they have difficulty when asked to select the appropriate procedure for a proposed study from the entire set of procedures discussed in the course. Therefore, these new questions (1) force students to revisit previous concepts to ensure their integration with the present chapter, and (2) provide practice at selecting procedures for specific studies from among all procedures discussed to that point.

Revision of Effect Size   The explanations of the very difficult concept of proportion of variance accounted for were combined and reworked, with the major explanations occurring once with linear regression and once with two-sample *t*-tests. (All previous procedures for computing this in the various parametric and nonparametric designs were retained.) In addition, a new discussion of computing and interpreting Cohen's *d* was included.

Revision of SPSS Appendix   A complete, stand-alone guide to using SPSS to compute the procedures discussed in the textbook is provided in Appendix B. It was revised to be compatible with the new SPSS (PASW) version 17, as well as with previous versions. Notably, the instructions include performing the one-way within-subjects ANOVA. A new feature—*Using the SPSS Appendix*—was added to many chapters to better integrate the appendix with each chapter. The book is still organized so that instructors may easily include or exclude discussions of SPSS.

Revision of the Entire Book   I did not merely slap a new cover on the previous edition. I performed a page-by-page revision of the entire book using my recent experiences from teaching statistics every semester, using reviews from other instructors, and using trends in the literature. This resulted in new explanations, new examples, new diagrams and tables, and new pedagogical devices. I also streamlined and modernized the narrative, and I reviewed and revised the end-of chapter problems.

CHAPTER CONTENT AND REVISIONS

Chapter 1 serves as a brief preface for the student and reviews basic math and graphing techniques. Much of this is material that instructors often present at the first class meeting, but having it in a chapter helps reinforce and legitimize the information.

Chapter 2 introduces the terminology, logic, and goals of statistics while integrating them with the purpose and logic of behavioral research. An explanation of using descriptive statistics to predict Y scores by using the relationship with X was added, and the discussion of scales of measurement was revised.

Chapter 3 presents simple, relative, and cumulative frequency, as well as percentile. The introduction to the proportion of the area under the normal curve was revised. Grouped distributions are briefly discussed, with additional information in Appendix A. The formulas for computing percentile were deleted.

Chapter 4 introduces measures of central tendency but focuses on the characteristics of the mean. The discussion of using the mean to predict individual scores was revised, as was the discussion of using the mean to summarize experiments.

Chapter 5 discusses measures of variability. The introduction to variability was revised. Emphasis is first given to interpreting the variance and standard deviation using their defining formulas, and then the computing formulas are introduced. The chapter ends with a new discussion of errors in prediction and an introduction to accounting for variance.

Chapter 6 deals with z-scores while the building blocks of central tendency and variability are still fresh in students' minds. The chapter then makes a rather painless transition to sampling distributions and z-scores for sample means, to set up for later inferential procedures. (Instructions for using linear interpolation with statistical tables are presented in Appendix A.)

Chapter 7 presents correlation coefficients, first explaining *type* and *strength*, and then showing the computations of the Pearson and Spearman coefficients. The concept of a "good" predictor was introduced. The section on correlations in the population was moved to Chapter 11 and a briefer version of resolving tied ranks was moved to Chapter 15.

Chapter 8 presents linear regression, explaining its logic and then showing the computations for the components of the regression equation and the standard error of the estimate. The explanation of errors in prediction, r^2, and the proportion of variance accounted for was revised.

Chapter 9 begins inferential statistics by discussing probability as it is used by behavioral researchers. Then probability is linked to random sampling, representativeness, and sampling error. The focus now quickly moves to computing the probability of sample means. Then the logic of using probability to make decisions about the representativeness of sample means is presented, along with the mechanics of setting up and using a sampling distribution. This is done without the added confusion of the formal hypotheses and terminology of significance testing.

Chapter 10 presents statistical hypothesis testing using the z-test. Here significance testing is presented within the context of experiments, including the terminology and symbols, the interpretation of significant and nonsignificant results, Type I and Type II errors, and an introduction to power.

Chapter 11 presents the one-sample t-test and the confidence interval for a population mean. Because they are similar to t-tests, significance tests of the Pearson and Spearman correlation coefficients are also included, with a new introduction of the population correlation coefficient moved from Chapter 7. The chapter ends with a revised discussion of how to design a powerful study.

Chapter 12 covers the independent- and the dependent-samples t-tests and versions of the confidence interval used with each. The chapter ends with revised discussions of how to interpret two-sample experiments and using the point-biserial correlation to measure effect size. A new discussion of using Cohen's d to measure effect size was added.

Chapter 13 introduces the one-way, between-subjects ANOVA. The discussion of experiment-wise error and the statistical underpinnings of ANOVA were revised and simplified. Post hoc tests for equal and unequal ns, eta squared, and interpreting ANOVA are also discussed. (The one-way, within-subjects ANOVA, with formulas, is described in Appendix A.)

Chapter 14 deals with the two-way, between-subjects ANOVA; post hoc tests for main effects and unconfounded comparisons in an interaction; and graphing and interpreting interactions. The two-way, within-subjects ANOVA and the two-way, mixed-design ANOVA are also introduced.

Chapter 15 first covers the one-way and two-way chi square. The discussion of the general logic of nonparametric procedures was revised and is followed by the Mann–Whitney, rank sums, Wilcoxon, Kruskal–Wallis, and Friedman tests (with appropriate post hoc tests and measures of effect size).

The text is designed to also serve as a reference book for later course work and projects, especially the material in Chapters 14 and 15 and the appendices. Also, the less common procedures tend to occur at the end of a chapter and are presented so that instructors may easily skip them without disrupting the discussion of the major procedures. Likewise, as much as possible, chapters are designed to stand alone so that instructors may reorder or skip topics. This is especially so for correlation and regression, which some instructors prefer covering after t-tests, while others place them after ANOVA.

PEDAGOGICAL FORMAT AND FEATURES

A number of features enhance the book's usefulness as a study tool and as a reference.

- Many mnemonics and analogies are used throughout the text book to promote retention and understanding.
- Each chapter begins with "Getting Started" which lists previously discussed concepts that students should review, followed by the learning goals for the chapter.
- "New Statistical Notation" sections introduce statistical notations at the beginning of the chapter in which they are needed but, to reduce student confusion, are introduced before the conceptual issues.
- An opening section in each chapter titled "WHY IS IT IMPORTANT TO KNOW ABOUT . . . ?" introduces the major topic of the chapter, immediately placing it in a research context.
- Important procedural points are emphasized by a "REMEMBER," a summary reminder set off from the text.
- Computational formulas are labeled and highlighted in color throughout the text.
- Key terms are highlighted in bold, reviewed in the chapter summary, and listed in a "Key Terms" section at the end of the chapter. An end-of-text glossary is included.
- Graphs and diagrams are explained in captions and fully integrated into the discussion.

- "Putting It All Together" sections at the end of each chapter provide advice, cautions, and ways to integrate material from different chapters.

- As part of "Putting It All Together," the section called *Using the SPSS Appendix* identifies the procedures from the chapter that SPSS will perform, and indicates the subsection of the appendix containing the relevant instructions.

- Each "Chapter Summary" section provides a substantive review of the material, not merely a list of the topics covered.

- Approximately 30 multipart questions are provided at the end of each chapter. The questions are separated into "Review Questions," which require students to define terms and outline procedures, and "Application Questions," which require students to perform procedures and interpret results. Then the "Integration Questions," require students to combine information from the previous different chapters. Odd-numbered questions have final and intermediate answers provided in Appendix D. Even-numbered questions have answers in the online Instructor's Resource Manual.

- A Summary of Formulas is provided at the end of each chapter.

- A glossary of symbols appears on the inside back cover. Tables on the inside front cover provide guidelines for selecting descriptive and inferential procedures based on the type of data or research design employed.

SUPPLEMENTARY MATERIALS

Supporting the book are several resources for students and instructors:

- *Student Workbook and Study Guide* Additional review material for students is available in the Student Workbook and Study Guide, revised by Deborah J. Hendricks and Richard T. Walls. Each chapter contains a review of objectives, terms, and formulas; a programmed review; conceptual and computational problems (with answers); and a set of multiple-choice questions similar to those in the Instructor's Resource Manual with Test Bank. A final chapter, called "Getting Ready for the Final Exam," facilitates student integration of the entire course. The workbook can be ordered separately, or may be bundled with the text.

- *Instructor's Resource Manual with Test Bank* This supplement, also revised by Deborah J. Hendricks and Richard T. Walls, contains approximately 750 test items and problems as well as suggestions for classroom activities, discussion, and use of statistical software. It also includes answers to the even-numbered end-of-chapter questions from the book. It is available in print or electronically.

- *SPSS Software* SPSS software is available for sale to students who schools do not license SPSS.

- *The Book Companion Website* offers a variety of study tools and useful resources, such as flashcards, crossword puzzles, a glossary, and web quizzes by chapter. Go to www.cengage.com/psychology/heiman to view all available resources.

- *Examview® CD* Featuring automatic grading, ExamView® allows you to create, deliver, and customize tests and study guides (both print and online) in minutes.

- *Microsoft® PowerPoint® Lecture Outlines* These simple Power Point lecture outlines can easily be adapted to fit your lectures, and include many key figures from the book.

ACKNOWLEDGMENTS

I gratefully acknowledge the help and support of the many professionals associated with Cengage learning. In particular my thanks go to Rebecca Rosenberg, Assistant Editor, Psychology, and to Jane Potter, Senior Sponsoring Editor, Psychology, for their hard work and support. I also want to thank the following reviewers, whose feedback I truly value, even if I cannot always comply with their requests:

Andrew Smiler, SUNY Oswego

Jacki Reihman, SUNY Oswego

Jon Lu, Concordia University

Mike Mangan, University of New Hamphire, Durham

Carloyn J. Mebert, University of New Hamphsire

Maria Cuddy-Casey, Immaculata University

SIXTH EDITION

Basic Statistics for the Behavioral Sciences

1 Introduction to Statistics

GETTING STARTED

Your goals in this chapter are to learn

- Why researchers learn statistics.
- The general purpose of statistical procedures.
- The basic math that's needed.

Okay, so you're taking a course in statistics. You probably are curious—and maybe a little anxious—about what it involves. After all, statistics are math! Well, relax. Statistics do not require that you be a math wizard. Students in the behavioral sciences *throughout the world* take a course like the one you are about to take, and they get through it. In fact, statistics can be fun! They are challenging, there is an elegance to their logic, and you can do nifty things with them. So, keep an open mind, be prepared to do a little work, and you'll be amazed by what happens. You'll find that statistics are interesting and educational, they help you to think logically, and they make behavioral research much easier to understand.

In this chapter we first deal with some common misconceptions that students have about statistics. Then we'll review the basic math that you'll be using.

WHY IS IT IMPORTANT TO LEARN STATISTICS (AND HOW DO YOU DO THAT?)

Here are some frequently asked questions that will teach you something about statistics and your statistics course.

What Are Statistics?   The term *statistics* is often used as a shorthand for *statistical procedures.* These are formulas and calculations developed by statisticians that psychologists and other behavioral researchers employ when "analyzing" the results of their research. Also, some of the answers that we compute are called statistics.

Why Must I Learn Statistics?   Statistics are an integral part of psychology and other behavioral sciences, so statistics and statistical concepts are used every day. Therefore, to understand your chosen field of study, you must understand statistics. You've already experienced this if you've ever read a published research article—you probably skipped the section titled "Results." Now you won't have to.

What Do Researchers Do with Statistics?   Statistics are tools used in research. They are needed because the behavioral sciences are based on empirical research. The word *empirical* means that knowledge is obtained through observation and measurement, and behavioral research measures behaviors. Such measurement results in numbers, or scores. These scores obtained in research are the **data.** (By the way, the word *data* is plural, so say "the data *are* . . .".) For example, to study intelligence, researchers measure the IQ scores of different individuals; to study memory, we examine data that reflect the number of things that people remember or forget; to study social interactions, we measure the distance people stand apart or their anxiety when meeting someone. And so on. Thus, any study typically produces a very large batch of scores that must be made manageable and meaningful. At this point, statistics are applied because *they help us to make sense out of the data*. The procedures we will discuss do this in four ways. First, some procedures *organize* the scores so that we can more clearly see any patterns in the data. Often this simply involves creating a table or graph. Second, other statistics *summarize* the scores. We don't need to examine each of the hundreds of scores that may be obtained in a study. Instead, a summary—such as the average score—allows us to quickly and easily understand the general characteristics of the data. Third, statistics *communicate* the results of a study. Researchers have created techniques and rules for this and, because everyone uses the same rules, it is much easier for us to communicate with each other, especially in published research reports. Finally, statistics are used to *conclude* what the data indicate. All behavioral research is designed to answer a question about a behavior and, ultimately, we must decide what the data tell us about that behavior.

But I'm Not Interested in Research; I Just Want to Help People!   Even if you are not interested in becoming a researcher, statistics are necessary for comprehending other people's research. Let's say that you become a therapist or counselor. You hear of a new therapy that says the way to "cure" people of some psychological problem is to scare the living daylights out of them. This sounds crazy but what is important is the research that does or does not support this therapy. As a responsible professional, you would evaluate the research supporting this therapy before you would use it. You could not do so without understanding statistics.

But I Don't Know Anything about Research!   This book is written for students who have not yet studied how to conduct research. When we discuss each statistic, we also discuss simple studies that employ the procedure, and this will be enough. Later, when you study research methods, you will know the appropriate statistical procedures to use.

What if I'm Not Very Good at Math?   This is *not* a math course. We will discuss some *research tools* that happen to involve mathematical operations. But it is simple math: adding, subtracting, multiplying, dividing, finding square roots, and drawing simple graphs. Also, we will continuously review the math operations as they are needed. Best of all, statisticians have already developed the statistics we'll discuss, so we won't be deriving formulas, performing proofs, or doing other "mystery" math. We will simply learn *when* to use the procedure that statisticians say is appropriate for a given situation, then compute the answer and then determine *what* it tells us about the data. (Eventually you'll understand the tables on the inside of the front cover that summarize which procedures are used in which type of study.)

What if I'm Not Very Good at Statistics?   This course is not a test of whether you should change your college major! First, there are not all that many procedures to learn, and these fancy sounding "procedures" include such simple things as computing an average or drawing a graph. Second, researchers usually do not memorize the formulas. (For quick reference, at the end of each chapter in this book is a list of the formulas discussed.) Finally, statistics are simply a tool used in research, just like a wrench is a tool used to repair automobile engines. A mechanic does not need to be an expert wrencher who loves to wrench, and you do not need be an expert statistician who loves statistics. Rather, in the same way that a mechanic must understand how to correctly use a wrench, your goal is to be able to correctly *use* statistics.

I Looked through the Book: Statistics Aren't Written in English!   Statistics do involve many strange symbols and unfamiliar terms. But these are simply the shorthand "code" for communicating statistical results and for simplifying statistical formulas. A major part of learning statistics is merely learning the code. Think of it this way: To understand research you must speak the language, and you are about to learn the language of statistics. Once you speak this language, much of the mystery surrounding statistics evaporates.

But, Those Formulas!   What makes some formulas *appear* difficult is that they are written in a code that communicates a sequence of operations: You first might square the scores, then add them together, then divide by some other number, and so on. However, most chapters begin with a section called "New Statistical Notation," which explains the symbols that you'll encounter, and then each formula is presented with example data and step-by-step instructions on how to work through it. (Each example involves an unrealistically small number of scores, although real research involves large numbers of scores.) There are also in-chapter "Quick Reviews" where you can practice what you have read and at the end of each chapter are additional practice problems. With practice the formulas become easy, and then the rest of the mystery surrounding statistics will evaporate.

So All I Have to Do Is Learn How to Compute the Answers?   No! Statistical procedures are a *tool* that you must learn to *apply*. Ultimately you want to make sense of data, and to do that, you must compute the appropriate statistic and then correctly interpret it. More than anything else, you need to learn *when* and *why* to use each procedure and how to *interpret* its answer. Be sure to put as much effort into this as you put into the calculations.

What about Using a Computer to Do Statistics?   At first glance, you might think that this book was written before the invention of computers. However, we focus on using formulas to compute answers "by hand" because that is the only way for you to understand statistics. (Having a computer magically produce an answer might sound attractive, but with no idea *how* the answer was produced, you'd be overwhelmed.) However, researchers usually do use computers to compute statistics. Therefore, Appendix B explains the basics of how to use "SPSS," which is one of the leading statistical computer programs. Recently, the new version 17 was released and, for some reason, the name was changed to *PASW*. However, we'll still refer to it as *SPSS*. The instructions in Appendix B are appropriate for version 17, and where needed, different instructions are provided so they are appropriate for earlier versions also. Even if your instructor cannot include this program in your statistics course, eventually you will want to learn it or one like it. Then you can compute the statistics discussed in this book quickly and accurately. (You'll love it—it is *soooo* easy!)

Recognize that, although SPSS is a wizardlike program, it cannot select the appropriate statistical procedure for you nor interpret the answer. You really must learn *when to use each statistic* and *what the answer means.*

All Right, So How Do I Learn Statistics?   Avoid these pitfalls when studying statistics:

- Don't skim the material. You do not speak the language yet, so you must translate the terminology and symbols into things that you understand, and that takes time and effort.

- Don't try to "cram" statistics. You won't learn anything (and your brain will melt). Instead, work on statistics a little every day, digesting the material in bite-sized pieces. This is the most effective—and least painful—way to learn statistics.

- Don't skip something if it seems difficult. Concepts build upon one another, and material in one chapter sets up a concept in a later chapter so that things are bite sized for you. Go back and review when necessary. (The beginning of each chapter lists what you should understand from previous chapters.)

An effective strategy in this course includes

- Learn (memorize), understand, and use the terminology. You are learning a new language, and you can do this only by practicing it. In this book, terminology printed in **boldface** type is important, as is anything labeled REMEMBER. Use the glossary. Making and studying flashcards is a good idea. At the end of a chapter, use the "Chapter Summary" and "Key Terms" to improve your knowledge of the terminology.

- Learn when and why to use each procedure. You'll find this information at the beginning of chapters in the section "Why Is It Important to Know About . . . ?"

- Practice using the formulas. Master the formulas at each step because they often reappear later as part of more complicated formulas. Test yourself with the in-chapter "Quick Reviews." If you cannot understand or perform a review, do not proceed! Go back—you missed something.

- Complete all of the problems at the end of each chapter, because the only way to learn statistics is to *do* statistics. The "Review Questions" are for practicing the terminology, definitions and concepts presented in the chapter. The "Application Questions" give you practice computing and interpreting the answers from each procedure. However, you must not forget about procedures from past chapters. Therefore, starting in Chapter 3, you'll see "Integration Questions," which help you to combine the information from different chapters. Remember, your task is to build a complete mental list of all of the different procedures in this book. Eventually you'll be in faced with picking the one procedure that's appropriate for a study from all those we've discussed. (Often this occurs on your final exam.)

- For all end-of-chapter problems, the answers to the odd-numbered problems are in Appendix D and your instructor has the answers to the even-numbered problems. Make a serious attempt at solving the problem first and only then look at the answer. (This is the practice test before the real test.)

- Pay extra attention to the sections titled "Interpreting the . . ." and "Statistics in Published Research" so that you learn how to interpret statistical results, draw conclusions, and understand what researchers do—and do not—say. These really are the most important aspects of learning statistics.

REVIEW OF MATHEMATICS USED IN STATISTICS

The remainder of this chapter reviews the math used in performing statistical procedures. As you'll see, there are accepted systems for statistical notation, for rounding an answer, for transforming scores, and for creating graphs.

Basic Statistical Notation

Statistical notation refers to the standardized code for symbolizing the mathematical operations performed in the formulas and for symbolizing the answers we obtain.

Identifying Mathematical Operations   We write formulas in statistical notation so that we can apply them to any data. We usually use the symbol X or Y to stand for each individual score obtained in a study. When a formula says to do something to X, it means to do it to all the scores you are calling X. When a formula says to do something to Y, it means to do it to all the scores called X.

The mathematical operations we'll perform are simple ones. Addition is indicated by the plus sign, and subtraction is indicated by the minus sign. We read from left to right, so $X - Y$ is read as "X minus Y." (I *said* this was simple!) This order is important because $10 - 4$, for example, is $+6$, but $4 - 10$ is -6. With subtraction, pay attention to what is subtracted from what and whether the answer is positive or negative.

We indicate division by forming a fraction, such as X/Y. The number above the dividing line is called the numerator, and the number below the line is called the denominator. *Always express fractions as decimals,* dividing the denominator *into* the numerator. (After all, 1/2 equals .5, not 2!)

Multiplication is indicated in one of two ways. We may place two components next to each other: XY means "multiply X times Y." Or we may indicate multiplication using parentheses: $4(2)$ and $(4)(2)$ both mean "multiply 4 times 2."

The symbol X^2 means square the score, so if X is 4, X^2 is 16. Conversely, $\sqrt{X}$ means "find the square root of X," so $\sqrt{4}$ is 2. (The symbol $\sqrt{}$ also means "use your calculator.")

Determining the Order of Mathematical Operations   Statistical formulas often call for a series of mathematical steps. Sometimes the steps are set apart by parentheses. Parentheses mean "the quantity," so always find the quantity inside the parentheses first and then perform the operations outside of the parentheses on that quantity. For example, $(2)(4 + 3)$ indicates to multiply 2 times "the quantity 4 plus 3." So first add, which gives $(2)(7)$, and then multiply to get 14.

A square root sign also operates on "the quantity," so always compute the quantity inside the square root sign first. Thus, $\sqrt{2 + 7}$ means find the square root of the quantity $2 + 7$; so $\sqrt{2 + 7}$ becomes $\sqrt{9}$, which is 3.

Most formulas are giant fractions. Pay attention to how far the dividing line is drawn because the length of a dividing line determines the quantity that is in the numerator and the denominator. For example, you might see a formula that looks like this:

$$\frac{\dfrac{6}{3} + 14}{\sqrt{64}} = \frac{2 + 14}{\sqrt{64}} = \frac{16}{\sqrt{64}} = \frac{16}{8} = 2$$

The longest dividing line means you should divide the square root of 64 into the quantity in the numerator. The dividing line in the fraction in the numerator is under only the 6, so first divide 6 by 3, which is 2. Then add 14, for a total of 16. In the denominator, the square root of 64 is 8. After dividing, the final answer is 2.

If you become confused in reading a formula, remember that there is an order of precedence of mathematical operations. Often this is summarized with PEMDAS meaning that, unless otherwise indicated, first compute inside any **P**arentheses, then compute **E**xponents (squaring and square roots), then **M**ultiply or **D**ivide, and finally, **A**dd or **S**ubtract. Thus, for $(2)(4) + 5$, multiply 2 times 4 first and then add 5. For $2^2 + 3^2$, square first, which gives $4 + 9$, which is then 13. On the other hand, $(2 + 3)^2$ is 5^2, which is 25.

Working with Formulas   We use a formula to find an answer, and we have symbols that stand for that answer. For example, in the formula $B = AX + K$, the B stands for the answer we will obtain. The symbol for the unknown answer is always isolated on one side of the equal sign, but we will know the numbers to substitute for the symbols on the other side of the equal sign. For example, to find B, say that $A = 4$, $X = 11$, and $K = 3$. In working any formula, the first step is to copy the formula and then rewrite it, replacing the symbols with their known values. Thus, start with

$$B = AX + K$$

Filling in the numbers gives

$$B = 4(11) + 3$$

Rewrite the formula after performing each mathematical operation. Above, multiplication takes precedence over addition, so multiply and then rewrite the formula as

$$B = 44 + 3$$

After adding,

$$B = 47$$

For simple procedures, you may have an urge to skip rewriting the formula after each step. Don't! That's a good way to introduce errors.

Rounding Numbers

Close counts in statistics, so you must carry out calculations to the appropriate number of decimal places. Usually, you must "round off" your answer. The rule is this: *Always carry out calculations so that your final answer after rounding has two more decimal places than the original scores.* Usually, we have whole-number scores (e.g., 2 and 11) so the final answer contains two decimal places. But say the original scores contain one decimal place (e.g., 1.4 and 12.3). Here the final answer should contain *three* decimal places.

However, *do not round off at each intermediate step in a formula; round off only at the end!* Thus, if the final answer is to contain two decimal places, round off your intermediate answers to at least three decimal places. Then after you've completed all calculations, round off the final answer to two decimal places.

To round off a calculation use the following rules:

If the number in the next decimal place is 5 or greater, round up. For example, to round to two decimal places, an answer of 2.366 is rounded to 2.370, which becomes 2.37.

If the number in the next decimal place is less than 5, round down: an answer of 3.524 is rounded to 3.520, which becomes 3.52.

We add zeroes to the right of the decimal point to indicate the level of precision we are using. For example, rounding 4.996 to two decimal places produces 5, but to show we used the precision of two decimal places, we report it as 5.00.

> *REMEMBER* Round off your final answer to two more decimal places than are in the original scores.

Transforming Scores

Many statistical procedures are nothing more than elaborate transformations. A **transformation** is a mathematical procedure for systematically converting a set of scores into a different set of scores. Adding 5 to each score is a transformation, or converting "number correct" into "percent correct" is a transformation.

We transform data for one of two reasons. First, transformations make scores easier to work with. For example, if all of the scores contain a decimal, we might multiply every score by 10 to eliminate the decimals. Second, transformations make different kinds of scores comparable. For example, if you obtained 8 out of 10 on a statistics test and 75 out of 100 on an English test, it would be difficult to compare the two scores. However, if you transformed each grade to percent correct, you could then directly compare performance on the two tests.

In statistics, we rely heavily on transformations to proportions and percents.

Proportions   A **proportion** is a decimal number between 0 and 1 that indicates a fraction of the total. To transform a number to a proportion, simply divide the number by the total. If 4 out of 10 people pass an exam, then the proportion of people passing the exam is 4/10, which equals .4. Or, if you score 6 correct on a test out of a possible 12, the proportion you have correct is 6/12, which is .5.

We can also work in the opposite direction from a known proportion to find the number out of the total it represents. Here, multiply the proportion times the total. Thus, to find how many questions out of 12 you must answer correctly to get .5 correct, multiply .5 times 12, and voilà, the answer is 6.

Percents   We can also transform a proportion into a percent. A **percent** (or percentage) is a proportion multiplied by 100. Above, your proportion correct was .5, so you had (.5)(100) or 50% correct. Altogether, to transform the original test score of 6 out of 12 to a percent, first divide the score by the total to find the proportion and then multiply by 100. Thus, (6/12)(100) equals 50%.

To transform a percent back into a proportion, divide the percent by 100 (above, 50/100 equals .5). Altogether, to find the test score that corresponds to a certain percent, transform the percent to a proportion and then multiply the proportion times the total number possible. Thus, to find the score that corresponds to 50% of 12, transform 50% to the proportion, which is .5, and then multiply .5 times 12. So, 50% of 12 is equal to (50/100)(12), which is 6.

Recognize that a percent is a whole unit: Think of 50% as 50 of those things called percents. On the other hand, a decimal in a percent is a proportion of *one* percent. Thus, .2% is .2, or two-tenths, of one percent, which is .002 of the total.

Creating Graphs

One type of statistical procedure is none other than plotting graphs. In case it's been a long time since you've drawn one, recall that the horizontal line across the bottom of a graph is the X axis, and the vertical line at the left-hand side is the Y axis. (Draw the Y axis so that it is about 60 to 75% of the length of the X axis.) Where the two axes intersect is always labeled as a score of zero on X and a score of zero on Y. On the X axis, scores become larger positive scores as you move to the *right*. On the Y axis, scores become larger positive scores as you move *upward*.

Say that we measured the height and weight of several people. We decide to place weight on the X axis and height on the X axis. (How to decide this is discussed later.) We plot the scores as shown in Figure 1.1. Notice that because the lowest height score is 63, the lowest label on the X axis is also 63. The symbol // in the axis indicates that we cut out the part between 0 and 63. We do this with either axis when there is a large gap between 0 and the lowest score we are plotting.

FIGURE 1.1

Plot of height and weight scores

Person	Height	Weight
Jane	63	130
Bob	64	140
Mary	65	155
Tony	66	160
Sue	67	165
Mike	68	170

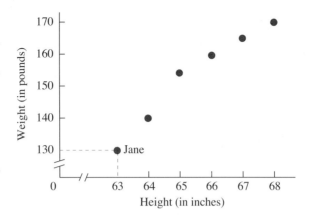

In the body of the graph we plot the scores from the table on the left. Jane is 63 inches tall and weighs 130 pounds, so we place a dot above the height of 63 and opposite the weight of 130. And so on. Each dot on the graph is called a **data point.** Notice that you read the graph by using the scores on one axis and the data points. For example, to find the weight of the person who has a height of 67, travel vertically from 67 to the data point and then horizontally to the Y axis: 165 is the corresponding weight.

In later chapters you will learn when to connect the data points with lines and when to create other types of figures. Regardless of the final form of a graph, always label the X and Y axes to indicate what the scores measure (not just X and Y), and always give your graph a title indicating what it describes.

When creating a graph, make the spacing between the labels for the scores on an axis reflect the spacing between the actual scores. In Figure 1.1 the labels 64, 65, and 66 are equally spaced on the graph because the difference between 64 and 65 is the same as the difference between 65 and 66. However, in other situations, the labels may not be equally spaced. For example, the labels 10, 20, and 40 would not be equally spaced because the distance between these scores is not equal.

Sometimes there are so many different scores that we cannot include a label for each one. Be careful here, because the units used in labeling each axis then determine the impression the graph gives. Say that for the previous weight scores, instead of labeling the Y axis in units of 10 pounds, we labeled it in units of 100 pounds, as shown in Figure 1.2. This graph shows the same data as Figure 1.1, but changing the scale on the Y axis creates a much flatter pattern of data points. This gives the misleading impression

FIGURE 1.2

Plot of height and weight scores using a different scale on the Y axis

Height	Weight
63	130
64	140
65	155
66	160
67	165
68	170

that regardless of their height, the people all have about the same weight. However, looking at the actual scores, you see that this is not the case. Thus, always label the axes in a way that honestly presents the data, without exaggerating or minimizing the pattern formed by the data points.

PUTTING IT ALL TOGETHER

That's all the basic math you'll need to get started. You are now ready to begin learning to use statistics. In fact, you already use statistics. If you compute your grade average or if you ask your instructor to "curve" your grades, you are using statistics. When you understand from the nightly news that Senator Fluster is projected to win the election or when you learn from a television commercial that Brand X "significantly" reduces tooth decay, you are using statistics. You simply do not yet know the formal names for these statistics or the logic behind them. But you will.

CHAPTER SUMMARY

1. All empirical research is based on observation and measurement, resulting in numbers, or scores. These scores are the *data.*

2. Statistical procedures are used to make sense out of data: They are used to *organize, summarize,* and *communicate* data and to draw *conclusions* about what the data indicate.

3. The goal in learning statistics is to know *when* to perform a particular procedure and how to *interpret* the answer.

4 *Statistical notation* refers to the standardized code for symbolizing the mathematical operations performed in the formulas and the answers obtained.

5. Unless otherwise indicated, the order of mathematical operations is to compute inside parentheses first, then square or find square roots, then multiply or divide, and then add or subtract.

6. Round off the final answer in a calculation to two more decimal places than are in the original scores. If the digit in the next decimal place is equal to or greater than 5, round up; if the digit is less than 5, round down.

7. A *transformation* is a procedure for systematically converting one set of scores into a different set of scores. Transformations make scores easier to work with and make different kinds of scores comparable.

8. A *proportion* is a decimal between 0 and 1 that indicates a fraction of the total. To transform a score to a proportion, divide the score by the total. To determine the score that produces a particular proportion, multiply the proportion times the total.

9. To transform a proportion to a *percent,* multiply the proportion times 100. To transform an original score to a percent, find the proportion by dividing the score by the total and then multiplying by 100.

10. To transform a percent to a proportion, divide the percent by 100. To find the original score that corresponds to a particular percent, transform the percent to a proportion and then multiply the proportion times the total.

11. A *data point* is a dot plotted on a graph to represent a pair of X and Y scores.

KEY TERMS

data *1*	proportion *7*
data point *8*	statistical notation *4*
percent *7*	transformation *6*

REVIEW QUESTIONS

(Answers for odd-numbered problems are in Appendix D.)

1. Why do researchers need to learn statistics?
2. What does the term *statistical notation* refer to?
3. (a) To how many places should you round a final answer? (b) If you are rounding to two decimal places, what are the rules for rounding up or down?
4. (a) What is a *transformation*? (b) Why do we transform data?
5. If given no other information, what is the order in which to perform mathematical operations?
6. What is a *percentage*?
7. What is a *data point*?
8. A researcher measures the IQ scores of a group of college students. What four things will the researcher use statistics for?
9. What is a *proportion* and how is it computed?
10. How do you transform a percentage to a proportion?

APPLICATION QUESTIONS

11. (a) What proportion is 5 out of 15? (b) What proportion of 50 is 10? (c) One in a thousand equals what proportion?
12. For each of the following, to how many places will you round off your final answer? (a) When measuring the number of questions students answered correctly on a test. (b) When measuring what proportion of the total possible points students have earned in a course. (c) When counting the number of people having various blood types. (d) When measuring the number of dollar bills possessed by each person in a group.
13. Transform each answer in question 11 to a percent.
14. The intermediate answers from some calculations based on whole-number scores are $X = 4.3467892$ and $Y = 3.3333$. We now want to find $X^2 + Y^2$. After rounding, what values of X and Y do we use?
15. Round off the following numbers to two decimal places: (a) 13.7462, (b) 10.043, (c) 10.047, (d) .079, and (e) 1.004.
16. For $Q = (X + Y)(X^2 + Y^2)$ find the value of Q when $X = 3$ and $Y = 5$.
17. Using the formula in question 16, find Q when $X = 8$ and $Y = -2$.
18. For $X = 14$ and $Y = 4.8$, find D:

$$D = \left(\frac{X - Y}{Y}\right)(\sqrt{X})$$

19. Using the formula in question 18, find D for $X = 9$ and $Y = -4$.
20. Of the 40 students in a gym class, 13 played volleyball, 12 ran track (4 of whom did a push-up), and the remainder were absent. (a) What proportion of the class

ran track? (b) What percentage played volleyball? (c) What percentage of the runners did a push-up? (d) What proportion of the class was absent?

21. In your statistics course, there are three exams: I is worth 40 points, II is worth 35 points, and III is worth 60 points. Your professor defines a passing grade as earning 60% of the points. (a) What is the smallest score you must obtain on each exam to pass it? (b) In total you can earn 135 points in the course. How many points must you earn from the three exams *combined* to pass the course? (c) You actually earn a total of 115 points during the course. What percent of the total did you earn?

22. There are 80 students enrolled in statistics. (a) You and 11 others earned the same score. What percent of the class received your score? (b) Forty percent of the class received a grade of C. How many students received a C? (c) Only 7.5% of the class received a D. How many students is this? (d) A student claims that .5% of the class failed. Why is this impossible?

23. (a) How do you space the labels on the X or Y axis of a graph? (b) Why must you be careful when selecting the amounts used as these labels?

24. Create a graph showing the data points for the following scores.

X Score	Y Score
20	10
25	30
35	20
45	60
25	55
40	70
45	30

2 Statistics and the Research Process

GETTING STARTED

To understand this chapter, recall the following:

- From Chapter 1, (1) that we use statistics to make sense out of data and (2) how to create and interpret graphs.

Your goals in this chapter are to learn

- The logic of *samples* and *populations* in behavioral research.
- How to recognize a *relationship* between scores and what is meant by the *strength* of a relationship.
- What *descriptive* statistics are used for.
- What *inferential* statistics are used for.
- What the difference is between an *experiment* and a *correlational* study and how to recognize the *independent variable,* the *conditions,* and the *dependent variable* in an experiment.
- What the four *scales of measurement* are.

Because statistics are part of the process of conducting research, to understand statistics you need to first understand a little about research. Therefore, this chapter discusses the basics of behavioral research, the general ways that statistics are used in research, and the major aspects of how we conduct a study that influence which statistics are used.

THE LOGIC OF RESEARCH

Behavioral scientists study the "laws of nature" regarding the behavior of living organisms. That is, researchers assume there are specific influences that govern every behavior of all members of a particular group. Although any single study is a very small step in this process, the goal is to understand every factor that influences behavior. Thus, when researchers study such things as the mating behavior of sea lions, social interactions between humans, or neural firing in a rat's brain, they are ultimately studying the laws of nature.

The reason a study is a small step is because nature is very complex. Therefore, research involves a series of translations in which we simplify things so that we can examine a specific influence on a specific behavior in a specific situation. Then, using our findings, we generalize back to the broader behaviors and laws we began with. For example, here's an idea for a simple study. Say that we think a law of nature is that people must study information in order to learn it. We translate this into the more specific hypothesis that "the more you study statistics, the better you'll learn them." Next we will translate the hypothesis into a situation where we can observe and measure specific people who study specific material in different amounts, to see if they *do* learn differently. Based on what we observe, we will have evidence for working back to the general law regarding studying and learning.

Part of this translation process involves *samples* and *populations*.

Samples and Populations

When researchers want to talk of a behavior occurring in some group in nature, they say it occurs in the population. A **population** is the entire group of individuals to which a law of nature applies. The population might be broadly defined (such as all animals, all mammals, or all humans), but it can be more narrowly defined (such as all women, all four-year-old English-speaking children in Canada, or all presidents of the United States). For our studying research, the population might be all college students taking statistics. Notice that, although ultimately researchers discuss the population of individuals, we sometimes talk of the population of *scores,* as if we have already measured the behavior of everyone in the population in a particular situation.

The population contains all past, present, and future members of the group, so we usually consider it to be infinitely large. However, to measure an infinitely large population would take roughly forever! Instead, we measure a sample from the population. A **sample** is a relatively small subset of a population that is intended to represent, or stand in for, the population. Thus, we might study the students in your statistics class as a sample representing the population of all college students enrolled in statistics. The individuals measured in a sample are called the **participants** (or sometimes, the *subjects*) and it is the scores from the sample(s) that constitute our data. As with a population, sometimes we discuss a sample of *scores* as if we have already measured the participants in a particular situation.

Notice that the definitions of a sample and a population depend on your perspective. Say that we study the students in your statistics class. If these are the only individuals we are interested in, then we have measured the population of scores. Or if we are interested in the population of all college students studying statistics, then we have a sample of scores that represent that population. But if we are interested in both the populations of college men and college women who are studying statistics, then the men in the class are one sample and the women in the class are another sample, and each represents its respective population. Finally, scores from one student can be a sample representing the population of all scores that the student might produce. Thus, a population is any complete group of scores that would be found in a particular situation, and a sample is a subset of those scores that we actually measure in that situation.

The logic behind samples and populations is this: We use the scores in a sample to *infer*—to estimate—the scores we would expect to find in the population, if we could measure them. Then, by translating the scores back into the behaviors they reflect, we can infer the behavior of the population. Thus, when the television news uses a survey to predict who will win the presidential election, they are using the scores from a sample (usually containing about 1200 voters) to infer the voting behavior of the population of over 100 million voters. Likewise, if we observe that greater studying leads to better learning for a sample of statistics students, we will infer that similar scores and behaviors would be found in the population of all statistics students. Then, because the population is the entire group to which the law of nature applies, we *are* describing how nature works. Thus, whenever we say a finding applies to the population, we are really describing how a law of nature applies to everyone out there in the world.

> *REMEMBER* The *population* is the entire group of individuals—and scores—to which our conclusions apply, based on our observation of a *sample,* which is a subset of the population.

Recognize that the above logic assumes that our sample is *representative* of the population. We will discuss this issue in detail in Chapter 9, but put simply, a representative sample accurately reflects the individuals, behaviors, and scores found in the population. Essentially, a representative sample is a good example—a miniversion—of the larger population. With such a sample, our inferences about the scores and behaviors found in the population will also be accurate, and so we can *believe* what our data seem to be telling us about nature. Thus, if your class is representative of all statistics students, then the scores in the class are a good example of the scores that the population would produce, and we can believe that everyone would behave as the class does.

Researchers try to create a representative sample by freely allowing the types of individuals found in the population to occur in the sample. To accomplish this, we create a *random sample:* the individuals in our sample are randomly selected from the population. This means that who gets chosen depends simply on the luck of the draw (like drawing names from a hat). Because we don't influence which participants are selected, the different types of individuals are free to occur in our sample as they do in the population, so the sample's characteristics "should" match the population.

However, random sampling is not foolproof because it may *not* produce a representative sample: Just by the luck of the draw, we may select participants whose characteristics do *not* match those of the population. Then the sample will be *unrepresentative,* inaccurately reflecting the behavior of the population. For example, maybe unknown to us, a large number of individuals happen to be in your statistics class who do not behave at all like typical students in the population—they are too bright, too lazy, or whatever. If so, we should not believe what such a sample indicates about our law of nature because the evidence it provides will be misleading and our conclusions will be wrong! Therefore, as you'll see, researchers always deal with the possibility that their conclusions about the population might be incorrect because their sample is unrepresentative.

Nonetheless, after identifying the population and sample, the next step is to define the specific situation and behaviors to observe and measure. We do this by selecting our *variables.*

Obtaining Data by Measuring Variables

In our example research, we asked: Does studying statistics improve your learning of them? Now we must decide what we mean by "studying" and how to measure it, and what we mean by "learning" and how to measure it. In research the factors we measure that influence behaviors—as well as the behaviors themselves—are called variables. A **variable** is anything that, when measured, can produce two or more different scores. A few of the variables found in behavioral research include your age, race, gender, and intelligence; your personality type or political affiliation; how anxious, angry, or aggressive you are; how attractive you find someone; how hard you will work at a task; or how accurately you recall a situation.

Variables fall into two general categories. If a score indicates the *amount* of a variable that is present, the variable is a *quantitative* variable. A person's height, for example, is a quantitative variable. Some variables, however, cannot be measured in amounts, but instead a score *classifies* an individual on the basis of some characteristic. Such variables are called *qualitative,* or classification, variables. A person's gender, for example, is a qualitative variable, because the "score" of male or female indicates a quality, or category.

For our study, we might measure "studying" using such variables as how much effort is put into studying or the number of times a chapter is read, but say we select the variable of the number of hours spent studying for a particular statistics test. We might measure "learning" by measuring how well statistical results can be interpreted or how quickly a specific procedure can be performed, but say we select the variable of grades on the statistics test.

As in any research, we then study the law of nature by studying the *relationship* between our variables.

Understanding Relationships

If nature relates those mental activities that we call studying to those mental activities that we call learning, then different amounts of learning should occur with different amounts of studying. In other words, there should be a relationship between studying and learning. A **relationship** is a pattern in which, as the scores on one variable change, the corresponding scores on the other variable change in a consistent manner. In our example, we predict the relationship in which the longer you study, the higher your test grade will be.

Say that we asked some students how long they studied for a test and their subsequent grades on the test. We might obtain the data in Table 2.1.[1] To see the relationship, first look at those people who studied for 1 hour and see their grades. Then look at those whose score is 2 hours and see their grades. And so on. These data form a relationship because, as study-time scores change (increase), test grades also change in a consistent fashion (also increase). Further, when study-time scores do not change (for example, Gary and Bob both studied for 1 hour), grades do not change either (they both received Fs). In statistics, we use the term *association* when talking about relationships. Here, low study times are associated with low test grades and high study times are associated with high test grades.

TABLE 2.1

Scores Showing a Relationship between the Variables of Study Time and Test Grades

Student	Study Time in Hours	Test Grades
Gary	1	F
Bob	1	F
Sue	2	D
Jane	2	D
Tony	3	C
Sidney	3	C
Ann	4	B
Rose	4	B
Lou	5	A

REMEMBER In a *relationship,* as the scores on one variable change, the scores on the other variable change in a consistent manner.

Because this relationship occurs in the sample data, we have evidence that the amount that people study *does* make a difference in their test grades. Therefore, assuming that the sample is representative, we can generalize this finding to the broader population so that we can talk about how people learn in general. In the same way, most research investigates relationships because a relationship is the telltale sign of a law of nature at work: When nature ties behaviors or events together, we see a relationship between the variables that measure those behaviors and events.

Thus, an important step in any research is to determine if there is a relationship in the sample data that matches the relationship that we predict. A major use of statistical procedures is to help us understand the relationship, examining the scores and the pattern they form. The simplest relationships fit one of two patterns. Sometimes the pattern fits "The more you *X*, the more you *Y*," with higher *X* scores paired with higher *Y* scores. Thus, the old saying "The bigger they are, the harder they fall" describes such a relationship, as does "The more often you speed, the more traffic tickets you accumulate." At other times, the pattern fits "The more you *X,* the *less* you *Y,* with higher

---

[1]The data presented in this book are a work of fiction. Any resemblance to real data is purely a coincidence.

X scores paired with lower *Y* scores." Thus, that other old saying "The more you practice statistics, the less difficult they are" describes a relationship, as does "The more alcohol you consume, the less coordinated you are."

Relationships may also form more complicated patterns where, for example, more *X* at first leads to more *Y,* but beyond a certain point more *X* leads to *less Y*. For example, the more you exercise, the better you feel, but beyond a certain point more exercise leads to feeling less well, as pain and exhaustion set in.

Although the above examples involve quantitative variables, relationships can also involve qualitative variables. For example, men typically are taller than women. If you think of male and female as "scores" on the variable of gender, then this is a relationship, because as gender scores change (going from male to female), height scores tend to decrease. We can study any combination of qualitative and quantitative variables in a relationship.

The Strength of a Relationship The data back in Table 2.1 show a perfectly consistent association between study time and test grades: In a perfectly consistent relationship, each score on one variable is paired with only one score on the other variable. In Table 2.1, all those who studied the same amount received the same grade. In the real world, however, not everyone who studies the same amount will receive the same grade. (Life is not fair.) However, a relationship can be present even if there is only some *degree* of consistency so that, as the scores on one variable change, the scores on the other variable *tend* to change in a consistent fashion. The degree of consistency in a relationship is called its *strength,* and a less consistent relationship is called a *weaker* relationship. For example, Table 2.2 shows two relationships between the number of hours spent studying and the number of *errors* made on a test.

First look at Part A on the left side of Table 2.2. Again note the error scores paired with each study-time score. Two aspects of the data produce a less consistent relationship: (1) Not everyone who studies the same amount receives the same error score (1 hour of study produced 13, 12, or 11 errors), and (2) sometimes the same error score is paired with *different* studying scores (11 errors occur with 1 and 2 hours). Nonetheless, a reasonably clear pattern is still here in which one batch of similar error scores

TABLE 2.2

Data Showing a Stronger (A) and Weaker (B) Relationship between Study Time and Number of Errors on a Test

A

Student	X Study Time in Hours	Y Errors on Test
1	1	13
2	1	12
3	1	11
4	2	11
5	2	10
6	2	10
7	3	10
8	3	9
9	4	9
10	4	8
11	5	7
12	5	6

B

Student	X Study Time in Hours	Y Errors on Test
1	1	13
2	1	11
3	1	9
4	2	12
5	2	10
6	2	9
7	3	9
8	3	7
9	4	9
10	4	7
11	5	8
12	5	6

tends to occur at one study time, but a *different,* lower batch of similar error scores tends to occur at the next study time. Therefore, this is a reasonably strong relationship.

The data in Part B of Table 2.2 show a weaker relationship: (1) Each study time is paired with a wider range of error scores (here 1 hour of study produced anywhere between 13 and 9 errors), and (2) the same error scores occur with a greater variety of study times (here 9 errors occur with 1, 2, 3 or 4 hours). These aspects produce greater overlap between the error scores at one study time and those at the next, so there is closer to the *same* batch of error scores at each study time. This produces a pattern of decreasing errors that is harder to see.

Thus, the **strength of a relationship** is the extent to which one or close to one value of *Y* tends to be consistently associated with only one value of *X*. Conversely, in a weaker relationship, a greater variety of *Y* scores is associated with each *X* score and/or the same *Y* score is paired with different *X* scores.

> *REMEMBER*   A stronger relationship occurs the more that one group of similar *Y* scores is associated with one *X* score and a different group of similar *Y* scores is associated with the next *X* score.

Two factors produce a relationship that is not perfectly consistent. First, extraneous influences are operating. For example, say that distracting noises occurred while someone studied for 1 hour, but not when someone else studied for 1 hour. Because of this, their studying might not be equally effective, resulting in different error scores paired with the same study-time score. Second, individual differences are operating. **Individual differences** refer to the fact that no two individuals are identical because of differences in genetics, experience, intelligence, personality, and many other variables. Thus, test performance will be influenced by a person's intelligence, aptitude, and motivation. Because students exhibit individual differences in these characteristics, they will each be influenced differently by the same amount of studying and so will produce different error scores at the same study-time score.

Theoretically, a relationship can have any degree of strength. However, perfectly consistent relationships do not occur in real research because individual differences and extraneous variables are always operating. Despite this, the less consistent relationships back in Table 2.2 still support our original hypothesis about how nature operates: They show that, *at least to some degree*, nature does relate studying and test errors as we predicted. Therefore, our next step would be to measure the degree to which nature does this. Likewise, in any research, it is never enough to say that you have observed a relationship; you must also determine the strength of the relationship. (Later chapters discuss statistical procedures for describing the strength of a relationship.)

> *REMEMBER*   Research is concerned not only with the existence of a relationship but also with the *strength* of the relationship.

When No Relationship Is Present   At the other extreme, when there is no consistent pattern between two variables, there is no relationship. For example, there is not (I think) a relationship between the number of chocolate bars people consume each day and the number of times they blink each minute. If we measure individuals on these two variables, we might have the data shown in Table 2.3. Mentally draw in horizontal lines so that you look at the batch of eye-blink scores paired with one chocolate score at a time. Here there is no consistent change in the blink scores as the scores on the chocolate variable change. Instead, very similar—but not identical—groups of blinking scores are paired with each chocolate score. Because there is no relationship in this sample, we do not have evidence that these variables are linked in nature.

TABLE 2.3

Scores Showing No
Relationship between
Number of Chocolate
Bars Consumed per Day
and Number of Eye
Blinks per Minute

Participant	X Chocolate Bars per Day	Y Eye Blinks per Minute
1	1	12
2	1	10
3	1	8
4	2	11
5	2	10
6	2	8
7	3	12
8	3	10
9	3	9
10	4	11
11	4	10
12	4	8

REMEMBER   A relationship is not present when virtually the same batch of scores from one variable is paired with every score on the other variable.

A QUICK REVIEW

- A relationship is present when, as the scores on one variable change, the scores on another variable tend to change in a consistent fashion.

MORE EXAMPLES

Below, Sample A shows a perfect relationship: One Y score occurs at only one X. Sample B shows a less consistent relationship: Sometimes different Ys occur at a particular X, and the same Y occurs with different Xs. Sample C shows no relationship: The same Ys tend to show up at every X.

	A		B		C
X	Y	X	Y	X	Y
1	20	1	12	1	12
1	20	1	15	1	15
1	20	1	20	1	20
2	25	2	20	2	20
2	25	2	30	2	12
2	25	2	40	2	15
3	30	3	40	3	20
3	30	3	40	3	15
3	30	3	50	3	12

For Practice

Which samples show a perfect, inconsistent, or no relationship?

A		B		C		D	
X	Y	X	Y	X	Y	X	Y
2	4	80	80	33	28	40	60
2	4	80	79	33	20	40	60
3	6	85	76	43	27	45	60
3	6	85	75	43	20	45	60
4	8	90	71	53	20	50	60
4	8	90	70	53	28	50	60

Answers

A: Perfect Relationship; B: Inconsistent Relationship; C and D: No Relationship

Graphing Relationships   It is important that you be able to recognize a relationship and its strength when looking at a graph. In a graph we have the X and Y axes and the X and Y scores, but how do we decide which variable to call X or Y? In any study we implicitly ask this question: For a *given* score on one variable, I wonder what scores occur on the other variable? The variable you identify as your "given" is then called the *X* variable (plotted on the *X* axis). Your other, "I wonder" variable is your *Y* variable (plotted on the *Y* axis). Thus, if we ask, "For a given amount of study time, what error scores occur?" then study time is the *X* variable, and errors is the *Y* variable. But if we ask, "For a given error score, what study time occurs?" then errors is the *X* variable, and study time is the *Y* variable.

Once you've identified your *X* and *Y* variables, describe the relationship using this general format: "Scores on the *Y* variable change **as a function of** changes in the *X* variable." So far we have discussed relationships involving "test scores as a function of study time" and "number of eye blinks as a function of amount of chocolate consumed." Likewise, if you hear of a study titled "Differences in Career Choices as a Function of Personality Type," you would know that we had wondered what career choices (the *Y* scores) were associated with each of several particular, given personality types (the *X* scores).

> **REMEMBER** The "given" variable in a study is designated the *X* variable, and we describe a relationship using the format "changes in *Y as a function of changes in X*."

Recall from Chapter 1 that a "dot" on a graph is called a *data point.* Then, to read a graph, read from left to right along the *X* axis and ask, "As the scores on the *X* axis increase, what happens to the scores on the *Y* axis?" Figure 2.1 shows the graphs from four sets of data.

FIGURE 2.1

Plots of data points from four sets of data

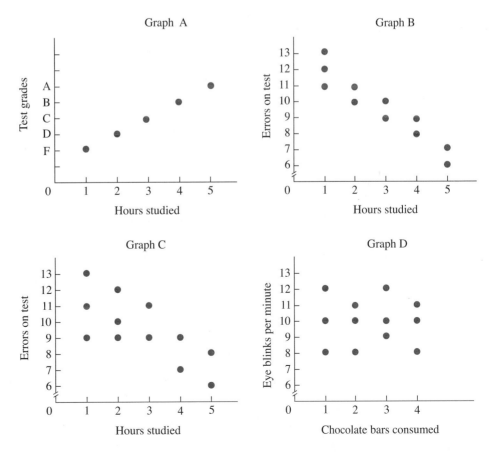

Graph A shows the original test-grade and study-time data from Table 2.1. Here, as the *X* scores increase, the data points move upwards, indicating higher *Y* scores, so this shows that as the *X* scores increase, the *Y* scores also increase. Further, because everyone who obtained a particular *X* obtained the same *Y,* the graph shows perfectly consistent association because there is one data point at each *X.*

Graph B shows test errors as a function of the number of hours studied from Table 2.2A. Here increasing *X* scores are associated with decreasing values of *Y.* Further, because several different error scores occurred with each study-time score, we see a vertical spread of different data points above each *X.* This shows that the relationship is not perfectly consistent.

Graph C shows the data from Table 2.2B. Again, decreasing *Y* scores occur with increasing *X* scores, but here there is greater vertical spread among the data points above each *X.* This indicates that there are greater differences among the error scores at each study time, indicating a weaker relationship. For any graph, whenever the data points above each *X* are more vertically spread out, it means that the *Y* scores differ more, and so a weaker relationship is present.

Graph D shows the eye-blink and chocolate data from Table 2.3, in which there was no relationship. The graph shows this because the data points in each group are at about the same height, indicating that about the same eye-blink scores were paired with each chocolate score. Whenever a graph shows an essentially flat pattern, it reflects data that do not form a relationship.

APPLYING DESCRIPTIVE AND INFERENTIAL STATISTICS

Statistics help us make sense out of data, and now you can see that "making sense" means understanding the scores and the relationship that they form. However, because we are always talking about samples and populations, we distinguish between *descriptive statistics,* which deal with samples, and *inferential statistics,* which deal with populations.

Descriptive Statistics

Because relationships are never perfectly consistent, researchers are usually confronted by many different scores that may have a relationship hidden in them. The purpose of descriptive statistics is to bring order to this chaos. **Descriptive statistics** are procedures for organizing and summarizing sample data so that we can communicate and describe their important characteristics. (When you see *descriptive,* think *describe.*)

As you'll see, these "characteristics" that we describe are simply the answers to questions that we would logically ask about the results of any study. Thus, for our study-time research, we would use descriptive statistics to answer: What scores occurred? What is the average or typical score? Are the scores very similar to each other or very different? For the relationship, we would ask: Is a relationship present? Do error scores tend to increase or decrease with more study time? How consistently do errors change? And so on.

On the one hand, descriptive procedures are useful because they allow us to quickly and easily get a general understanding of the data without having to look at every single score. For example, hearing that the average error score for 1 hour of study is 12 simplifies a bunch of different scores. Likewise, you can summarize the overall relationship by mentally envisioning a graph that shows data points that follow a downward slanting pattern.

On the other hand, however, there is a cost to such summaries, because they will not precisely describe *every* score in the sample. (Above, not everyone who studied 1 hour scored 12.) Less accuracy is the price we pay for a summary, so descriptive statistics always imply "generally," "around," or "more or less."

Descriptive statistics also have a second important use. A major goal of behavioral science is to be able to predict when a particular behavior will occur. This translates into predicting individuals' scores on a variable that measures the behavior. To do this we use a relationship, because it tells us the high or low *Y* scores that tend to naturally occur with a particular *X* score. Then, by knowing someone's *X* score and using the relationship, we can predict his or her *Y* score. Thus, from our previous data, if I know the number of hours you have studied, I can predict the errors you'll make on the test, and I'll be reasonably accurate. (The common descriptive statistics are discussed in the next few chapters.)

> **REMEMBER**   *Descriptive statistics* are used to summarize and describe the important characteristics of sample data and to predict an individual's *Y* score based on his or her *X* score.

Inferential Statistics

After answering the above questions for our sample, we want to answer the same questions for the population being represented by the sample. Thus, although technically descriptive statistics are used to describe samples, their logic is also applied to populations. Because we usually cannot measure the scores in the population, however, we must *estimate* the description of the population, based on the sample data.

But remember, we cannot automatically assume that a sample is representative of the population. Therefore, before we draw any conclusions about the relationship in the population, we must first perform inferential statistics. **Inferential statistics** are procedures for deciding whether sample data accurately represent a particular relationship in the population. Essentially, inferential procedures are for deciding whether to *believe* what the sample data seem to indicate about the scores and relationship that would be found in the population. Thus, as the name implies, inferential procedures are for making *inferences* about the scores and relationship found in the population.

If the sample is deemed representative, then we use the descriptive statistics computed from the sample as the basis for estimating the scores that would be found in the population. Thus, if our study-time data pass the inferential "test," we will infer that a relationship similar to that in our sample would be found if we tested everyone after they had studied 1 hour, then tested everyone after studying 2 hours, and so on. Likewise, we would predict that when people study for 1 hour, they will make around 12 errors and so on. (We discuss inferential procedures in the second half of this book.)

After performing the appropriate descriptive and inferential procedures, we stop being a "statistician" and return to being a behavioral scientist: We interpret the results in terms of the underlying behaviors, psychological principles, sociological influences, and so on, that they reflect. This completes the circle because then we *are* describing how nature operates.

> **REMEMBER**   *Inferential statistics* are for deciding whether to believe what the sample data indicate about the scores that would be found in the population.

Statistics versus Parameters

Researchers use the following system so that we know when we are describing a sample and when we are describing a population. A number that is the answer from a descriptive procedure (describing a sample of scores) is called a **statistic.** Different statistics describe different characteristics of sample data, and the symbols for them are letters from the English alphabet. On the other hand, a number that describes a characteristic of a population of scores is called a **parameter.** The symbols for different parameters are letters from the Greek alphabet.

Thus, for example, the average in your statistics class is a sample average, a descriptive *statistic* that is symbolized by a letter from the English alphabet. If we then estimate the average in the population, we are estimating a *parameter,* and the symbol for a population average is a letter from the Greek alphabet.

> **REMEMBER** *Descriptive procedures* result in *statistics,* which describe sample data and are symbolized using the English alphabet. *Inferential procedures* are for estimating *parameters,* which describe a population of scores and are symbolized using the Greek alphabet.

UNDERSTANDING EXPERIMENTS AND CORRELATIONAL STUDIES

All research generally focuses on demonstrating a relationship. Although we discuss a number of descriptive and inferential procedures, only a few of them are appropriate for a particular study. Which ones you should use depends on several issues. First, your choice depends on what it is you want to know—what question about the scores do you want to answer?

Second, your choice depends on the specific research design being used. A study's **design** is the way the study is laid out: how many samples there are, how the participants are tested, and the other specifics of how a researcher goes about demonstrating a relationship. Different designs require different statistical procedures. Therefore, part of learning *when* to use different statistical procedures is to learn with what type of design a procedure is applied. To begin, research can be broken into two major types of designs because, essentially, there are two ways of demonstrating a relationship: *experiments* and *correlational studies.*

Experiments

In an **experiment** the researcher actively changes or manipulates one variable and then measures participants' scores on another variable to see if a relationship is *produced.* For example, say that we examine the amount of study time and test errors in an experiment. We decide to compare 1, 2, 3, and 4 hours of study time, so we randomly select four samples of students. We ask one sample to study for 1 hour, administer the test, and count the number of errors that each participant makes. We have another sample study for 2 hours, administer the test, and count their errors, and so on. Then we look to see if we have produced the relationship where, as we increase study time, error scores tend to decrease.

To select the statistical procedures you'll use in a particular experiment, you must understand the components of an experiment.

The Independent Variable An **independent variable** is the variable that is changed or manipulated by the experimenter. Implicitly, it is the variable that we think *causes* a change in the other variable. In our studying experiment, we manipulate study time because we think that longer studying causes fewer errors. Thus, amount of study time is our independent variable. Or, in an experiment to determine whether eating more chocolate causes people to blink more, the experimenter would manipulate the independent variable of the amount of chocolate a person eats. You can remember the independent variable as the variable that occurs *independently* of the participants' wishes (we'll have some participants study for 4 hours whether they want to or not).

Technically, a *true* independent variable is manipulated by doing something *to* participants. However, there are many variables that an experimenter cannot manipulate in this way. For example, we might hypothesize that growing older causes a change in some behavior. But we can't *make* some people 20 years old and make others 40 years old. Instead, we would manipulate the variable by selecting one sample of 20-year-olds and one sample of 40-year-olds. Similarly, if we want to examine whether gender is related to some behavior, we would select a sample of females and a sample of males. In our discussions, we will *call* such variables independent variables because the experimenter controls them by controlling a characteristic of the samples. Statistically, all independent variables are treated the same. (Technically, though, such variables are called *quasi-independent variables.*)

Thus, the experimenter is always in control of the independent variable, either by determining what is done to each sample or by determining a characteristic of the individuals in each sample. In essence, a participant's "score" on the independent variable is assigned by the experimenter. In our examples, we, the researchers, decided that one group of students will have a score of 1 hour on the variable of study time or that one group of people will have a score of 20 on the variable of age.

Conditions of the Independent Variable   An independent variable is the *overall* variable that a researcher examines; it is potentially composed of many different amounts or categories. From these the researcher selects the conditions of the independent variable. A **condition** is a specific amount or category of the independent variable that creates the specific situation under which participants are examined. Thus, although our independent variable is amount of study time—which could be any amount—our conditions involve only 1, 2, 3, or 4 hours. Likewise, 20 and 40 are two conditions of the independent variable of age, and male and female are each a condition of the independent variable of gender. A condition is also known as a **level** or a **treatment:** By having participants study for 1 hour, we determine the specific "level" of studying that is present, and this is one way we "treat" the participants.

The Dependent Variable   The **dependent variable** is used to measure a participant's behavior under each condition. A participant's high or low score is supposedly caused or influenced by—*depends on*—the condition that is present. Thus, in our studying experiment, the number of test errors is the dependent variable because we believe that errors depend on the amount of study. If we manipulate the amount of chocolate people consume and measure their eye blinking, eye blinking is our dependent variable. Or, if we studied whether 20- or 40-year-olds are more physically active, then activity level is our dependent variable. (*Note:* The dependent variable is also called the *dependent measure,* and we obtain *dependent scores.*)

A major component of your statistics course will be for you to read descriptions of various experiments and, for each, to identify its components. Use Table 2.4 for help. (It is also reproduced inside the front cover.) As shown, from the description, find the variable that the researcher manipulates in order to influence a behavior—it is the *independent variable*, and the amounts of the variable that are present are the *conditions*. The behavior that is to be influenced is measured by the *dependent variable*, and the amounts of the variable that are present are indicated by the *scores*. All statistical analyses are applied to only the scores from this variable.

TABLE 2.4

Summary of Identifying an Experiment's Components

Researcher's Activity		Role of Variable		Name of Variable		Amounts of Variable Present		Compute Statistics?
Researcher Manipulates variable	→	Variable influences a behavior	→	Independent variable	→	Conditions (Levels)	→	No
Researcher measures variable	→	Variable measures behavior that is influenced	→	Dependent Variable	→	Scores (Data)	→	Yes

REMEMBER In an experiment, the researcher manipulates the *conditions* of the *independent variable* and, under each, measures participants' behavior by measuring their scores on the *dependent variable.*

Drawing Conclusions from Experiments The purpose of an experiment is to produce a relationship in which, as we change the conditions of the independent variable, participants' scores on the dependent variable tend to change in a consistent fashion. To see the relationship and organize your data, always diagram your study as shown in Table 2.5. Each column in the table is a condition of the independent variable (here, amount of study time) under which we tested some participants. Each number in a column is a participant's score on the dependent variable (here, number of test errors).

To see the relationship, remember that a condition is a participant's "score" on the independent variable, so participants in the 1-hour condition all had a score of 1 hour paired with their dependent (error) score of 13, 12, or 11. Likewise, participants in the 2-hour condition scored "2" on the independent variable, while scoring 9, 8, or 7 errors. Now, look for the relationship as we did previously, first looking at the error scores paired with 1 hour, then looking at the error scores paired with 2 hours, and so on. Essentially, as amount of study time increased, participants produced a different, lower batch of error scores. Thus, a relationship is present because, as study time increases, error scores tend to decrease.

For help envisioning this relationship, we would graph the data points as we did previously. Notice that in any experiment we are asking, "For a given condition of the independent variable, I wonder what dependent scores occur?" Therefore, the independent variable is always our X variable, and the dependent variable is our Y variable. Likewise, we always ask, "Are there consistent changes in the dependent variable *as a function of* changes in the independent variable?" (Chapter 4 discusses special techniques for graphing the results of experiments.)

TABLE 2.5

Diagram of an Experiment Involving the Independent Variable of Number of Hours Spent Studying and the Dependent Variable of Number of Errors Made on a Statistics Test

Each column contains participants' dependent scores measured under one condition of the independent variable.

Independent Variable: Number of Hours Spent Studying

	Condition 1: 1 Hour	Condition 2: 2 Hours	Condition 3: 3 Hours	Condition 4: 4 Hours
Dependent Variable: Number of Errors Made on a Statistics Test	13 12 11	9 8 7	7 6 5	5 3 2

For help summarizing such an experiment, we have specific *descriptive* procedures for summarizing the scores in each condition and for describing the relationship. For example, it is simpler if we know the average error score for each hour of study. Notice, however, that we apply descriptive statistics only to the *dependent* scores. Above, we do not know what error score will be produced in each condition so errors is our "I Wonder" variable that we need help making sense of. We do not compute anything about the conditions of the independent variable because we created and controlled them. (Above, we have no reason to average together 1, 2, 3, and 4 hours.) Rather, the conditions simply create separate groups of dependent scores that we examine.

> **REMEMBER**   We apply descriptive statistics only to the scores from the dependent variable.

Then the goal is to infer that we'd see a similar relationship if we tested the entire population in the experiment, and so we have specific *inferential* procedures for experiments to help us make this claim. If the data pass the inferential test, then we use the sample *statistics* to estimate the corresponding population *parameters* we would expect to find. Thus, Table 2.5 shows that participants who studied for 1 hour produced around 12 errors. Therefore, we would infer that if the population of students studied for 1 hour, their scores would be close to 12 also. But our sample produced around 8 errors after studying for 2 hours, so we would infer the population would also make around 8 errors when in this condition. And so on. As this illustrates, the goal of any experiment is to demonstrate a relationship in the population, describing the different group of dependent scores associated with each condition of the independent variable. Then, because we are describing how everyone scores, we can return to our original hypothesis and add to our understanding of how these behaviors operate in nature.

Correlational Studies

Not all research is an experiment. Sometimes we conduct a correlational study. In a **correlational study** we simply measure participants' scores on two variables and then determine whether a relationship is present. Unlike in an experiment in which the researcher actively attempts to *make* a relationship happen, in a correlational design the researcher is a passive observer who looks to see if a relationship *exists* between the two variables. For example, we used a correlational approach back in Table 2.1 when we simply asked some students how long they studied for a test and what their test grade was. Or, we would have a correlational design if we asked people their career choices and measured their personality, asking "Is career choice related to personality type?" (As we'll see, correlational studies examine the "correlation" between variables, which is another way of saying they examine the relationship.)

> **REMEMBER**   In a *correlational study,* the researcher simply measures participants' scores on two variables to determine if a relationship exists.

As usual, we want to first describe and understand the relationship that we've observed in the sample, and correlational designs have their own descriptive statistical procedures for doing this. Then, to describe the relationship that would be found in the population, we have specific correlational inferential procedures. Finally, as with an experiment, we would translate the relationship back to the original hypothesis about studying and learning that we began with, so that we can add to our understanding of nature.

A QUICK REVIEW

- In an experiment, the researcher changes the conditions of the independent variable and then measures participants' behavior using the dependent variable.

- In a correlational design, the researcher measures participants on two variables.

MORE EXAMPLES

In a study, participants' relaxation scores are measured after they've been in a darkened room for either 10, 20, or 30 minutes. This is an experiment because the researcher controls the length of time in the room. The independent variable is length of time, the conditions are 10, 20, or 30 minutes, and the dependent variable is relaxation.

A survey measures participants' patriotism and also asks how often they've voted. This is a correlational design because the researcher passively measures both variables.

For Practice

1. In an experiment, the _____ is changed by the researcher to see if it produces a change in participants' scores on the _____

2. To see if drinking influences one's ability to drive, participants' level of coordination is measured after drinking 1, 2, or 3 ounces of alcohol. The independent variable is _____, the conditions are _____, and the dependent variable is _____.

3. In an experiment, the _____ variable reflects participants' behavior or attributes.

4. We measure the age and income of 50 people to see if older people tend to make more money. What type of design is this?

Answers

1. independent variable; dependent variable
2. amount of alcohol; 1, 2, or 3 ounces; level of coordination
3. dependent
4. correlational

A Word about Causality

When people hear of a relationship between *X* and *Y,* they tend to automatically conclude that it is a *causal relationship,* with changes in *X* causing the changes in *Y.* This is not necessarily true (people who weigh more tend to be taller, but being heavier does not *make* you taller!). The problem is that, *coincidentally,* some additional variable may be present that we are not aware of, and it may actually be doing the causing. For example, we've seen that less study time *appears* to cause participants to produce higher error scores. But perhaps those participants who studied for 1 hour coincidentally had headaches and the actual cause of their higher error scores was not lack of study time but headaches. Or, perhaps those who studied for 4 hours happened to be more motivated than those in the other groups, and this produced their lower error scores. Or, perhaps some participants cheated, or the moon was full, or who knows! Researchers try to eliminate these other variables, but we can never be certain that we have done so.

Our greatest confidence in our conclusions about the causes of behavior come from experiments because they provide the greatest opportunity to control or eliminate those other, potentially causal variables. Therefore, we discuss the relationship in an experiment as if changing the independent variable "causes" the scores on the dependent variable to change. The quotation marks are there, however, because we can never definitively *prove* that this is true; it is always possible that some hidden variable was present that was actually the cause.

Correlational studies provide little confidence in the causes of a behavior because this design involves little control of other variables that might be the actual cause. Therefore, we never conclude that changes in one variable cause the other variable to change based on a correlational study. Instead, it is enough that we simply describe how nature relates the variables. Changes in *X* might cause changes in *Y,* but we have no convincing evidence of this.

Recognize that statistics do not solve the problem of causality. That old saying that "You can prove anything with statistics" is totally incorrect! When people think logically, statistics do not *prove* anything. No statistical procedure can prove that one variable causes another variable to change. Think about it: How could some formula written on a piece of paper "know" what causes particular scores to occur in nature?

Thus, instead of proof, any research merely provides *evidence* that supports a particular conclusion. How well the study controls other variables is part of the evidence, as are the statistical results. This evidence helps us to *argue* for a certain conclusion, but it is not "proof" because there is always the possibility that we are wrong. (We discuss this issue further in Chapter 7.)

THE CHARACTERISTICS OF SCORES

We have one more important issue to consider when deciding on the particular descriptive or inferential procedure to use in an experiment or correlational study. Although participants are always measured, different variables can produce scores that have different underlying mathematical characteristics. The particular mathematical characteristics of the scores also determine which descriptive or inferential procedure to use. Therefore, always pay attention to two important characteristics of the variables: the type of *measurement scale* involved and whether the scale is *continuous* or *discrete*.

The Four Types of Measurement Scales

Numbers mean different things in different contexts. The meaning of the number 1 on a license plate is different from the meaning of the number 1 in a race, which is different still from the meaning of the number 1 in a hockey score. The kind of information that a score conveys depends on the *scale of measurement* that is used in measuring it. There are four types of measurement scales: *nominal, ordinal, interval,* and *ratio.*

With a **nominal scale,** each score does not actually indicate an amount; rather, it is used for identification. (When you see *nominal,* think *name.*) License plate numbers and the numbers on football uniforms reflect a nominal scale. The key here is that nominal scores indicate only that one individual is qualitatively different from another, so in research, nominal scores classify or categorize individuals. For example, in a correlational study, we might measure the political affiliation of participants by asking if they are Democrat, Republican, or "Other." To simplify these names we might replace them with nominal scores, assigning a 5 to Democrats, a 10 to Republicans, and so on (or we could use any other numbers). Then we might also measure participants' income, to determine whether as party affiliation "scores" change, income scores also change. Or, if an experiment compares the conditions of male and female, then the independent variable is a nominal, categorical variable, where we might assign a "1" to identify each male, and a "2" to identify each female. Because we assign the numbers arbitrarily, they do not have the mathematical properties that numbers normally have. For example, here the number 1 does not indicate more than 0 but less than 2 as it usually does.

A different approach is to use an **ordinal scale.** Here the scores indicate rank order— anything that is akin to 1st, 2nd, 3rd . . . is ordinal. (*Ordinal* sounds like *ordered.*)

In our studying example, we'd have an ordinal scale if we assigned a 1 to students who scored best on the test, a 2 to those in second place, and so on. Then we'd ask, "As study times change, do students' ranks also tend to change?" Or, if an experiment compares the conditions of first graders to second graders, then this independent variable involves an ordinal scale. The key here is that ordinal scores indicate only a relative amount—identifying who scored relatively high or low. Also, there is no zero in ranks, and the same amount does not separate every pair of adjacent scores: 1st may be only slightly ahead of 2nd, but 2nd may be miles ahead of 3rd.

A third approach is to use an **interval scale.** Here each score indicates an actual quantity, and an equal amount separates any adjacent scores. (For interval scores, remember *equal* intervals between them.) However, although interval scales do include the number 0, it is not a *true zero*—it does not mean *none* of the variable is present. Therefore, the key here is that you can have less than zero, so an interval scale allows negative numbers. For example, temperature (in Celsius or Fahrenheit) involves an interval scale: Because 0° does not mean that zero heat is present, you can have even less heat at −1°. In research, interval scales are common with intelligence or personality tests: A score of zero does not mean zero intelligence or zero personality. Or, in our studying research we might determine the average test score and then assign students a zero if they are average, a +1, +2, and so on, for the amount they are above average, and a −1, −2, and so on, for the amount they are below average. Then we'd see if more positive scores tend to occur with higher study times. Or, if we create conditions based on whether participants are in a positive, negative, or neutral mood, then this independent variable reflects an interval scale.

Notice that with an interval scale, it is incorrect to make "ratio" statements that compare one score to another score. For example, at first glance it seems that 4°C is twice as warm as 2°C. However, if we measure the *same* physical temperatures using the Fahrenheit scale, we would have about 35° and 39°, respectively. Now one temperature is not twice that of the other. Essentially, if we don't know the true amount of a variable that is present at 0, then we don't know the true amount that is present at any other score.

Only with our final scale of measurement, a **ratio scale,** do the scores reflect the true amount of the variable that is present. Here the scores measure an actual amount, there is an equal unit of measurement, and 0 truly means that none of the variable is present. The key here is that you cannot have negative numbers because you cannot have less than nothing. Also, only with ratio scores can we make "ratio" statements, such as "4 is twice as much as 2." (So for *ratio,* think *ratio*!) We used ratio scales in our previous examples when measuring the number of errors and the number of hours studied. Likewise, in an experiment, if we compare the conditions of having people on diets consisting of either 1000, 1500, or 2000 calories a day, then this independent variable involves a ratio scale.

We can study relationships that involve any combination of the above scales.

> *REMEMBER* The *scale of measurement* reflected by scores may be *nominal, ordinal, interval,* or *ratio.*

Continuous versus Discrete Scales

A measurement scale may be either continuous or discrete. A **continuous scale** allows for fractional amounts; it "continues" between the whole-number amounts, so decimals make sense. The variable of age is continuous because someone can be 19.6879 years old. On the other hand, some variables involve a **discrete scale,** which are measured only in whole amounts. Here, decimals do not make sense. For example, whether you are male or female or in first or second grade are discrete variables because you can be in one group or the other, but not in-between. (Some variables may seem to involve fractions—such as shoe size—but they are still discrete variables, because smaller divisions are not possible and, again, there is no in between.)

Note: when a discrete variable has only two possible categories or scores, it is called a **dichotomous variable.** Male/female or living/dead are dichotomous variables.

Usually researchers assume that nominal or ordinal variables are discrete and that interval or ratio variables are at least *theoretically* continuous. For example, intelligence tests are designed to produce whole-number scores. But, theoretically, an IQ of 95.6 makes sense, so intelligence is a theoretically continuous (interval) variable. Likewise, it sounds strange if the government reports that the average family has 2.4 children because no one has .4 of a child. However, it makes sense to treat this ratio variable as if it is continuous, because we can interpret what it means if this year the average is 2.4 children, but last year the average was 2.8. (I've heard that a recent survey showed the average American home contains 2.5 people and 2.7 televisions!)

> *REMEMBER*   Whether a variable is *continuous* or *discrete* and whether it is measured using a *nominal, ordinal, interval,* or *ratio* scale are factors that determine which statistical procedure to apply.

To help you remember the four scales of measurement, Table 2.6 summarizes their characteristics.

TABLE 2.6

Summary of Types of Measurement Scales

Each column describes the characteristics of the scale.

| | Type of Measurement Scale | | | |
	Nominal	*Ordinal*	*Interval*	*Ratio*
What Does the Scale Indicate?	Quality	Relative quantity	Quantity	Quantity
Is There an Equal Unit of Measurement?	No	No	Yes	Yes
Is There a True Zero?	No	No	No	Yes
How Might the Scale be Used in Reasearch?	To identify males and females as 1 and 2	To judge who is 1st, 2nd, etc., in aggressiveness	To convey the results of intelligence and personality tests	To count the number of correct answers on a test
Additional Examples	Telephone numbers Social Security numbers	Letter grades Elementary school grade	Checkbook balance Individual's standing relative to class average	Weight Distance traveled

A QUICK REVIEW

- *Nominal* scales identify categories and *ordinal* scales reflect rank order. Both *interval* and *ratio* scales measure actual quantities, but negative numbers can occur with interval scales and not with ratio scales.

- *Interval* and *ratio* scales are assumed to be continuous scales, which allow fractional amounts; *nominal* and *ordinal* scales are assumed to be discrete scales, which do not allow fractional amounts.

continued

MORE EXAMPLES

If your grade on an essay exam is based on the number of correct statements you include, then a ratio scale is involved; if it is based on how much your essay is better or worse than what the professor expected, an interval scale is involved; if it indicates that yours was relatively one of the best or worst essays in the class, this is an ordinal scale (as is pass/fail, which is dichotomous); if it is based on the last digit of your ID number, then a nominal scale is involved. If you can receive one grade or another, but nothing in between, it involves a discrete scale; if fractions are possible, it involves a continuous scale.

For Practice

1. Whether you are ahead or behind when gambling involves a _____ scale.
2. The number of hours you slept last night involves a _____ scale.
3. Your blood type involves a _____ scale.
4. Whether you are a lieutenant or major in the army involves a _____ scale.
5. A _____ scale allows fractions; a _____ scale allows only whole amounts.

Answers

1. interval   4. ordinal
2. ratio   5. continuous; discrete
3. nominal

STATISTICS IN PUBLISHED RESEARCH: USING STATISTICAL TERMS

You have already begun to learn the secret language found in published research. You'll frequently encounter such terms as *relationship, independent* and *dependent variable, condition,* or *statistic.* Also, that phrase "as a function of" is common. Often it is in the title of a graph, so seeing "Agility as a Function of Age" indicates that the graph shows the relationship between X scores that measure participants' ages and Y scores that measure agility. The phrase is also used in the title of reports. For example, "Anxiety When Dating as a Function of Introversion Level" indicates that the researcher wondered if people are more or less anxious about going on a date, depending on the particular (given) amount of introversion that they exhibit.

The reason that published research seems to involve a secret language is because many details are left out. Implicitly it is assumed that the reader of a report (you) has taken a statistics course and so understands the terminology of statistics and research. This means that most of the terms that we'll discuss are seldom defined in published reports. Therefore, for you to understand research and apply statistical procedures (let alone understand this book), these terms need to become part of your everyday vocabulary.

PUTTING IT ALL TOGETHER

As you proceed through this course, however, don't let the terminology and details obscure your ultimate purpose. Keep things in perspective by remembering the overall logic of research, which can be summarized by the following five steps:

1. Based on a hypothesized law of nature, we design either an experiment or a correlational study to measure variables and to observe the predicted relationship in the sample.

2. We use descriptive statistics to understand the scores and the relationship in the sample.

3. We use inferential procedures to decide whether the sample accurately represents the scores and relationship that we would find if we could study everyone in the population.

4. By describing the scores and relationship that would be found in the population, we are actually describing the behavior of everyone in a particular group in a particular situation.

5. By describing the behavior of everyone in a particular situation, we *are* describing how a law of nature operates.

CHAPTER SUMMARY

1. The group of all individuals to which research applies is the *population.* The subset of the population that is actually measured is the *sample*.

2. Usually, participants are selected using random sampling so that all scores in the population have the same chance of being selected. The sample should be representative. By chance, however, a sample may be unrepresentative.

3. A *variable* is anything that, when measured, can produce two or more different scores. Variables may be *quantitative,* measuring a quantity or amount, or *qualitative,* measuring a quality or category.

4. A *relationship* occurs when a change in scores on one variable is associated with a consistent change in scores on another variable.

5. The term *individual differences* refers to the fact that no two individuals are identical.

6. Because of individual differences and external influences, relationships can have varying *strengths*.

7. The "given" variable in any relationship is designated the *X* variable, and we describe a relationship using the format "changes in *Y as a function of* changes in *X.*"

8. *Descriptive statistics* are used to organize, summarize, and describe sample data, and to predict an individual's *Y* score using the relationship with *X. Inferential statistics* are for deciding whether the sample data actually represent the relationship that occurs in the population.

9. A *statistic* is a number that describes a characteristic of a sample of scores, symbolized using a letter from the English alphabet. A statistic is used to infer or estimate the corresponding *parameter.* A parameter is a number that describes a characteristic of a population of scores, symbolized using a letter from the Greek alphabet.

10. A study's *design* is the particular way in which the study is laid out.

11. In an *experiment,* we manipulate the *independent variable* and then measure participants' scores on the *dependent variable.* A specific amount or category of the independent variable is called a *condition, treatment,* or *level*.

12. In a *correlational study,* neither variable is actively manipulated. Scores on both variables are simply measured and then the relationship is described.

13. In any type of research, if a relationship is observed, it may or may not mean that changes in one variable *cause* the other variable to change.

14. The four *scales of measurement* are (a) a *nominal scale,* in which numbers name or identify a quality or characteristic; (b) an *ordinal scale,* in which numbers indicate rank order; (c) an *interval scale,* in which numbers measure a specific amount, but with no true zero; or (d) a *ratio scale,* in which numbers measure a specific amount and 0 indicates truly zero amount.

15. A *continuous variable* can be measured in fractional amounts. A *discrete variable* is measured only in whole amounts. A *dichotomous variable* is a discrete variable that has only two amounts or categories.

KEY TERMS

as a function of *19*   level *23*
condition *23*   nominal scale *27*
continuous scale *28*   ordinal scale *27*
correlational study *25*   parameter *22*
dependent variable *23*   participant *14*
descriptive statistics *20*   population *13*
design *22*   ratio scale *28*
dichotomous variable *29*   relationship *15*
discrete scale *29*   sample *13*
experiment *22*   statistic *21*
independent variable *22*   strength of a relationship *17*
individual differences *17*   treatment *23*
inferential statistics *21*   variable *14*
interval scale *28*

REVIEW QUESTIONS

(Answers for odd-numbered problems are in Appendix D.)

1. (a) What is a *population*? (b) What is a *sample*? (c) How are samples used to make conclusions about populations? (d) What are researchers really referring to when they talk about the population?

2. What do you see when (a) a relationship exists between two variables? (b) No relationship is present?

3. What does the *strength* of a relationship refer to?

4. What pattern in the Y scores will produce a weaker relationship?

5. What are the two aspects of a study to consider when deciding on the particular descriptive or inferential statistics that you should employ?

6. What is the difference between an experiment and a correlational study?

7. What is the difference between the independent variable and the conditions of the independent variable?

8. In an experiment, what is the dependent variable?

9. What is the general purpose of all research, whether experiments or correlational studies?

10. (a) What are descriptive statistics used for? (b) What are inferential statistics used for?

11. (a) What is the difference between a statistic and a parameter? (b) What types of symbols are used for statistics and parameters?

12. (a) Define the four scales of measurement. (b) What are *continuous* and *discrete variables*? (c) Which scales of measurement are usually assumed to be discrete and which are assumed to be continuous?

APPLICATION QUESTIONS

13. A student named Foofy (who you'll be taking statistics with) conducted a survey. In her sample, 83% of mothers employed outside the home would rather be home raising children. She concluded that "the statistical analyses prove that most working women would rather be at home." What is the problem with this conclusion?

14. In study A, a researcher gives participants various amounts of alcohol and then observes any decrease in their ability to walk. In study B, a researcher notes the various amounts of alcohol that participants drink at a party and then observes any decrease in their ability to walk. (a) Which study is an experiment, and which is a correlational study? Why? (b) Which study will be best for showing that drinking alcohol causes an impairment in walking? Why?

15. Another student in your class, Poindexter, conducted a survey of college students about their favorite beverage. Based on what the sample said, he concluded that most college students prefer carrot juice to other beverages! What statistical argument can you give for not accepting this conclusion?

16. In each of the following experiments, identify the independent variable, the conditions of the independent variable, and the dependent variable: (a) Studying whether scores on a final exam are influenced by background music that is soft, loud, or absent. (b) Comparing freshmen, sophomores, juniors, and seniors with respect to how much fun they have while attending college. (c) Studying whether being first born, second born, or third born is related to intelligence. (d) Examining whether length of daily exposure to a sun lamp (15 minutes versus 60 minutes) alters self-reported depression. (e) Studying whether being in a room with blue walls, green walls, red walls, or beige walls influences the number of aggressive acts produced by adolescents.

17. List the scales of measurement, starting with the scale that provides the most precise information about the amount of a variable present and ending with the scale that provides the least precise information.

18. Using the terms *sample, population, variable, statistics,* and *parameter,* summarize the steps a researcher follows, starting with a hypothesis and ending with a conclusion about a nature.

19. For the following data sets, which show a relationship?

Sample A		Sample B		Sample C		Sample D	
X	Y	X	Y	X	Y	X	Y
1	10	20	40	13	20	92	71
1	10	20	42	13	19	93	77
1	10	22	40	13	18	93	77
2	20	22	41	13	17	95	75
2	20	23	40	13	15	96	74
3	30	24	40	13	14	97	73
3	30	24	42	13	13	98	71

20. Which sample in question 19 shows the strongest relationship? How do you know?

21. Which of the graphs below depict a relationship? How do you know?

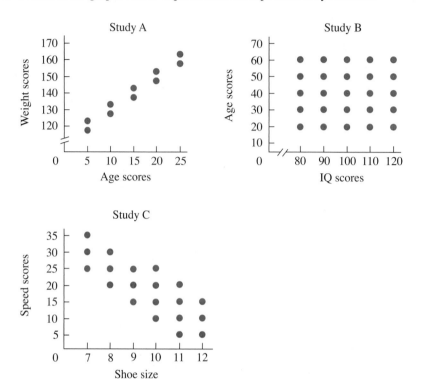

22. Which study in question 21 demonstrates the strongest relationship? How do you know?

23. In question 21, why is each relationship a telltale sign that a law of nature is at work?

24. (a) Poindexter says that Study A in problem 21 examines age scores as a function of weight scores. Is he correct? (b) He also claims that in Study C the researcher is asking, "For a given shoe size, what speed scores occur?" Is he correct? (c) If the studies in question 21 were conducted as experiments, in each, which variable is the independent variable and which is the dependent variable?

25. Complete the chart below to identify the characteristics of each variable.

Variable	Qualitative or Quantitative	Continuous, Discrete, or Dichotomous	Type of Measurement Scale
Gender	_____	_____	_____
Academic major	_____	_____	_____
Number of minutes before and after an event	_____	_____	_____
Restaurant ratings (best, next best, etc.)	_____	_____	_____
Speed (miles per hr)	_____	_____	_____
Dollars in your pocket	_____	_____	_____
Change in weight (in lb)	_____	_____	_____
Checking account balance	_____	_____	_____
Reaction time	_____	_____	_____
Letter grades	_____	_____	_____
Clothing size	_____	_____	_____
Registered voter	_____	_____	_____
Therapeutic approach	_____	_____	_____
Schizophrenia type	_____	_____	_____
Work absences	_____	_____	_____
Words recalled	_____	_____	_____

3 Frequency Distributions and Percentiles

So, we're off into the wonderful world of descriptive statistics. Recall that descriptive statistics tell us the obvious things we would ask about the relationship and scores in a sample. Before we examine the relationship between two variables, however, we first summarize the scores from *each* variable alone. Then, two important things that we always wish to know are: Which scores occurred, and how often did each occur? The way to answer this is to organize the scores into tables and graphs based on what are called *frequency distributions.* In this chapter, you'll see how to create various kinds of frequency distributions and how to use a frequency distribution to derive additional information about the scores.

Before we get to that, however, here are some terms and symbols you'll encounter in this chapter.

NEW STATISTICAL NOTATION

The scores we initially measure in a study are called the raw scores. Descriptive statistics help us to "boil down" the raw scores into an interpretable, "digestible" form. There are several ways to do this, but the starting point is to count the number of times each score occurred. The number of times a score occurs is the score's **frequency**, symbolized by the lowercase *f*. (Always pay attention to whether a symbol is upper- or lowercase.) If we count the frequency of every score in the data, we create a frequency distribution. A **distribution** is the general name that researchers have for any organized set of data. As you'll see, there are several ways to create a frequency distribution, so we will combine the term *frequency* (and *f*) with other terms and symbols.

In most statistical procedures, we also count the total number of scores. The symbol for the total number of scores in a set of data is the uppercase *N*. An *N* of 10 means we have 10 scores, or *N* = 43 means we have 43 scores. Note that *N* is not the number of different scores, so even if all 43 scores in a sample are the same score, *N* still equals 43.

> **REMEMBER** The *frequency* of a score is symbolized by *f*. The total *number* of scores in the data is symbolized by *N*.

WHY IS IT IMPORTANT TO KNOW ABOUT FREQUENCY DISTRIBUTIONS?

Presenting data in a graph or table is important for two reasons. First, it answers our question about the different scores that occurred in our data and it does this in an organized manner. You'll also see that we have names for some commonly occurring distributions so that we can easily communicate and *envision* a picture of even very large sets of data. Therefore, always create a table or graph of your data. As the saying goes, "A picture is worth a thousand words," and nowhere is this more appropriate than when trying to make sense out of data. Second, the procedures discussed here are important because they are the building blocks for other descriptive and inferential statistics. (You will be using what you learn here throughout the remainder of this book.)

As you'll see, we can organize data in one of four ways: using each score's *simple frequency, relative frequency, cumulative frequency,* or *percentile.*

SIMPLE FREQUENCY DISTRIBUTIONS

The most common way to organize scores is to create a simple frequency distribution. A **simple frequency distribution** shows the number of times each score occurs in a set of data. The symbol for a score's **simple frequency** is simply *f*. To find *f* for a score, count how many times the score occurs. If three participants scored 6, then the frequency of 6 (its *f*) is 3. Creating a simple frequency distribution involves counting the frequency of every score in the data. One way to see a distribution is in a table.

TABLE 3.1

Simple Frequency Distribution Table

The left-hand column identifies each score, and the right-hand column contains the frequency with which the score occurred.

Score	f
17	1
16	0
15	4
14	6
13	4
12	1
11	1
10	1

Presenting Simple Frequency in a Table

Let's begin with the following raw scores. (They might measure a variable from a correlational study, or they may be dependent scores from an experiment.)

14	14	13	15	11	15	13	10	12
13	14	13	14	15	17	14	14	15

In this disorganized arrangement, it is difficult to make sense out of these scores. See what happens, though, when we arrange them into the simple frequency table shown in Table 3.1.

We have several rules for making a frequency table. Start with a score column and an *f* column. The score column has the highest score in the data at the top of the column. Below that are all *possible* whole-number scores in decreasing order, down to the lowest score that occurred. Thus, the highest score is 17, the lowest score is 10, and although no one obtained a score of 16, we still include it. Opposite each score in the *f* column is the score's frequency: In this sample there is one 17, zero 16s, four 15s, and so on.

Not only can we easily see the frequency of each score, but we can also determine the combined frequency of several scores by adding together their individual fs. For example, the score of 13 has an f of 4, and the score of 14 has an f of 6, so their combined frequency is 10.

Notice that, although there are 8 scores in the score column, N is not 8. There are 18 scores in the original sample, so N is 18. You can see this by adding together the frequencies in the f column: The 1 person scoring 17 plus the 4 people scoring 15 and so on adds up to the 18 people in the sample. In a frequency distribution, the sum of the frequencies always equals N.

REMEMBER The sum of all frequencies in a sample equals N.

That's how to create a simple frequency distribution. Such a distribution is also called a *regular frequency distribution* or a plain old *frequency distribution*.

A QUICK REVIEW

▪ A frequency distribution shows the number of times participants obtained each score.

MORE EXAMPLES

The scores 15, 16, 13, 16, 15, 17, 16, 15, 17, and 15, contain one 13, no 14s, four 15s, and so on, producing the following frequency table:

Scores	f
17	2
16	3
15	4
14	0
13	1

For Practice

1. What is the difference between f and N?
2. Create a frequency table for these scores: 7, 9, 6, 6, 9, 7, 7, 6, and 6.

3. What is the N here?
4. What is the frequency of 6 and 7 together?

Answers

1. f is the number of times a score occurs; N is the total number of scores in the data.

2.

Scores	f
9	2
8	0
7	3
6	4

3. $N = 9$
4. $f = 3 + 4 = 7$

Graphing a Simple Frequency Distribution

When researchers talk of a frequency distribution, they often imply a graph. Essentially, it shows the relationship between each score and the frequency with which it occurs. We ask, "For a given score, what is its corresponding frequency?", so we place the scores on the X axis and frequency on the Y axis.

REMEMBER A graph of a frequency distribution shows the scores on the X axis and their frequency on the Y axis.

Recall that a variable will involve one of four types of measurement scales—nominal, ordinal, interval, or ratio. The type of scale involved determines whether we graph a frequency distribution as a *bar graph,* a *histogram*, or a *polygon*.

Bar Graphs   Recall that in nominal data each score identifies a category, and in ordinal data each score indicates rank order. A frequency distribution of nominal or ordinal scores is graphed by creating a **bar graph**. In a bar graph, a vertical bar is centered over each score on the X axis, and *adjacent bars do not touch*.

Figure 3.1 shows two bar graphs of simple frequency distributions. Say that the upper graph is from a survey in which we counted the number of participants in each category of the nominal variable of political party affiliation. The X axis is labeled using the "scores" of political party, and because this is a nominal variable, they can be arranged in any order. In the frequency table, we see that six people were Republicans, so we draw a bar at a height (frequency) of 6 and so on.

Say that the lower graph is from a survey in which we counted the number of participants having different military ranks (an ordinal variable). Here the X axis is labeled from left to right, which corresponds to low to high. Again, the height of each bar is the score's frequency.

> *REMEMBER*   Create a *bar graph* to show the frequency distribution of nominal or ordinal scores.

FIGURE 3.1

Simple frequency bar graphs for nominal and ordinal data

The height of each bar indicates the frequency of the corresponding score on the x axis.

Nominal Variable of Political Affiliations	
Party	*f*
Communist	1
Socialist	3
Democrat	8
Republican	6

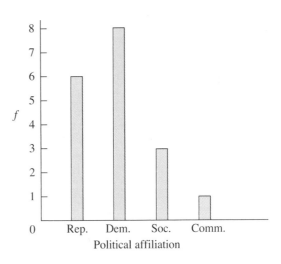

Ordinal Variable of Military Rank	
Party	*f*
General	3
Colonel	8
Lieutenant	4
Sergeant	5

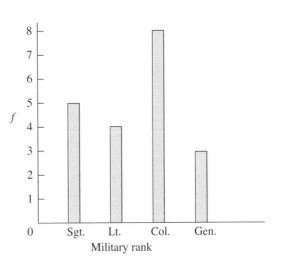

The reason we create bar graphs with nominal and ordinal scales is because researchers assume that both are *discrete* scales: You can be in one group or the next, but not in-between. The space between the bars in a bar graph indicates this. Later we will see bar graphs in other contexts and this same rule always applies:

Create a bar graph whenever the *X* variable is discrete.

On other hand, recall that interval and ratio scales are assumed to be *continuous*: They allow fractional amounts that continue between the whole numbers. To communicate this, these scales are graphed using continuous figures.

Histograms   Create a histogram when plotting a frequency distribution containing a *small number* of different interval or ratio scores. A **histogram** is similar to a bar graph except that in a *histogram adjacent bars touch*. For example, say that we measured the number of parking tickets some people received, obtaining the data in Figure 3.2. Again, the height of each bar indicates the corresponding score's frequency. Although you cannot have a fraction of a ticket, this ratio variable is theoretically continuous (e.g., you can talk about an average of 3.14 tickets per person). By having no gap between the bars in our graph, we communicate that there are no gaps when measuring this *X* variable.

Polygons   Usually, we don't create a histogram when we have a large *number* of different interval or ratio scores, such as if our participants had from 1 to 50 parking tickets. The 50 bars would need to be very skinny, so the graph would be difficult to read. We have no rule for what number of scores is too large, but when a histogram is unworkable, we create a frequency polygon. Construct a **frequency polygon** by placing a data point over each score on the *X* axis at a height corresponding to the appropriate frequency. Then connect the data points using straight lines. To illustrate this, Figure 3.3 shows the previous parking ticket data plotted as a frequency polygon. Because each line *continues* between two adjacent data points, we communicate that our measurements continue between the two scores on the *X* axis and therefore that this is a continuous variable. Later we will create graphs in other contexts that also involve connecting data points with straight lines. This same rule always applies:

Connect adjacent data points with straight lines whenever the *X* variable is continuous.

FIGURE 3.2

Histogram showing the simple frequency of parking tickets in a sample

Score	f
7	1
6	4
5	5
4	4
3	6
2	7
1	9

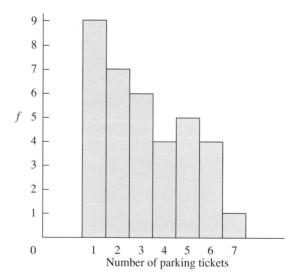

FIGURE 3.3

Simple frequency
polygon showing the
frequency of parking
tickets in a sample

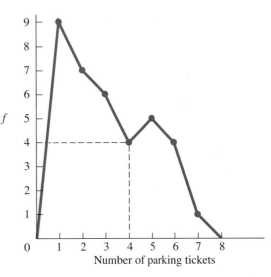

Score	f
7	1
6	4
5	5
4	4
3	6
2	7
1	9

Notice that a polygon includes on the X axis the next score above the highest score in the data and the next score below the lowest score (in Figure 3.3, scores of 0 and 8 are included). These added scores have a frequency of 0, so the polygon touches the X axis. In this way, we create a complete geometric figure—a polygon—with the X axis as its base.

> **REMEMBER** Create a *histogram* or *polygon* to plot the frequency distribution for an interval or ratio variable.

Often in statistics you must a read a polygon to determine a score's frequency, so be sure you can do this: Locate the score on the X axis and then move upward until you reach the line forming the polygon. Then, moving horizontally, locate the frequency of the score. For example, as shown by the dashed line in Figure 3.3, the score of 4 has an f equal to 4.

> **REMEMBER** The height of the polygon above any score corresponds to that score's frequency.

Table 3.2 reviews the rules for constructing bar graphs, histograms, and polygons.

TABLE 3.2

When to create a bar
graph, histogram, or
polygon

*Consider the scale of
measurement of scores on
the X axis.*

Graph	When Used?	How Produced?
Bar graph	With nominal or ordinal scores	Adjacent bars do not touch
Histogram	With small range of interval/ratio scores	Adjacent bars do touch
Polygon	With large range of interval/ratio scores	Straight lines; add points above and below actual scores

A QUICK REVIEW

▪ Create a *bar graph* with nominal or ordinal scores, a *histogram* with a few interval/ratio scores, and a *polygon* with many different interval/ratio scores.

MORE EXAMPLES

After a survey, to graph (1) the frequency of males versus females (a nominal variable), create a bar graph; (2) the number of people who are first born, second born, etc. (an ordinal variable), create a bar graph; (3) the frequency of participants falling into each of five salary ranges (a few ratio scores), create a histogram; (4) the frequency for each individual salary reported (many ratio scores), create a polygon.

For Practice

1. A _____ has a separate, discrete bar above each score, a _____ contains bars that touch, and a _____ has dots connected with straight lines.

2. To show the number of freshmen, sophomores, and juniors who are members of a fraternity, plot a _____.

3. To show the frequency of people who are above average weight by either 0, 5, 10, or 15 pounds, plot a _____ .

4. To show the number of people preferring chocolate or vanilla ice cream in a sample, plot a _____ .

5. To show the number of people who are above average weight by each amount between 0 and 100 pounds, plot a _____ .

Answers

1. bar graph; histogram; polygon
2. bar graph
3. histogram
4. bar graph
5. polygon

TYPES OF SIMPLE FREQUENCY DISTRIBUTIONS

Research often produces scores that form frequency polygons having one of several common shapes, and so we have names to identify them. Each shape comes from an idealized distribution of a population. By far the most important frequency distribution is the *normal distribution*. (This is the big one, folks.)

The Normal Distribution

Figure 3.4 shows the polygon of the ideal normal distribution. (Let's say these are test scores from a population of college students.) Although specific mathematical properties define this polygon, in general it is a bell-shaped curve. But don't call it a bell curve (that's so pedestrian!). Call it a **normal curve** or a **normal distribution** or say that the scores are *normally distributed*.

Because it represents an ideal population, a normal curve is different from the choppy polygon we saw previously. First, the curve is smooth because a population produces so many different scores that the individual data points are too close together for straight lines to connect them. Instead, the data points themselves form a smooth line. Second, because the curve reflects an infinite number of scores, we cannot label the *Y* axis with specific frequencies. Simply remember that the higher the curve is above a score, the higher is the score's frequency. Finally, regardless of how high or low an *X* score might be, theoretically it might sometimes occur. Therefore, as we read to the left or to the right on the *X* axis, the frequencies approach—but never reach—a frequency of zero, so the curve approaches but never actually touches the *X* axis.

FIGURE 3.4

The ideal normal curve

Scores farther above and below the middle scores occur with progressively lower frequencies.

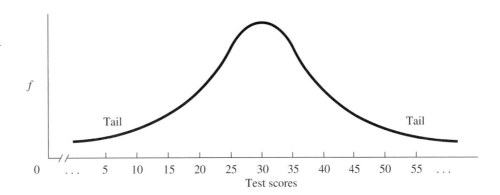

As you can see from Figure 3.4, the normal distribution has the following characteristics. The score with the highest frequency is the middle score between the highest and lowest scores. (Here it is the score of 30.) The normal curve is *symmetrical*, meaning that the left half below the middle score is a mirror image of the right half above the middle score. As we proceed away from the middle score either toward the higher or lower scores, the frequencies at first decrease slightly. Farther from the middle score, however, the frequencies decrease more drastically, with the highest and lowest scores having relatively low frequency.

In statistics the scores that are relatively far above and below the middle score of the distribution are called the "extreme" scores. Then, the far left and right portions of a normal curve containing the low-frequency, extreme scores are called the **tails** of the distribution. In Figure 3.4, the tails are roughly below the score of 15 and above the score of 45.

The reason the normal curve is important is because it is a very common distribution in psychology and other behavioral sciences: For most of the variables that we study, the scores naturally form a curve similar to this, with most of the scores around the middle score, and with progressively fewer higher or lower scores. Because of this, the normal curve is also very common in our upcoming statistical procedures. Therefore, before you proceed, be sure that you can read the normal curve. Can you see in Figure 3.4 that the most frequent scores are between 25 and 35? Do you see that a score of 15 has a relatively low frequency and a score of 45 has the same low frequency? Do you see that there are relatively few scores in the tail above 50 or in the tail below 10? Above all, you must be able to see this in your sleep.

> **On a normal distribution, the farther a score is from the central score of the distribution, the less frequently the score occurs.**

A distribution may not match the previous curve exactly, but it can still meet the mathematical definition of a normal distribution. Consider the three curves in Figure 3.5. Curve *B* is generally what we think of as the ideal normal distribution. Curve *A* is skinny relative to the ideal because only a few scores around the middle score have a relatively high frequency. On the other hand, Curve *C* is fat relative to the ideal because more scores farther below and above the middle have a high frequency. Because these curves generally have that bell shape, however, for statistical purposes their differences are not critical.

Other Common Frequency Polygons

Not all data form a normal distribution and then the distribution is called *nonnormal*. One common type of nonnormal distribution is a "skewed distribution." A *skewed distribution* is similar to a normal distribution except that it *has only one pronounced tail*.

FIGURE 3.5

Variations of bell-shaped curves

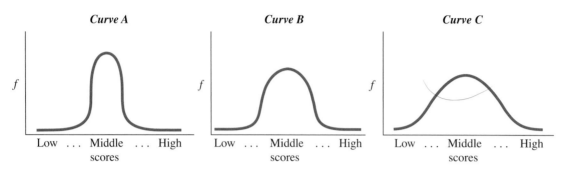

As shown in Figure 3.6, a distribution may be either negatively skewed or positively skewed.

A **negatively skewed distribution** contains extreme low scores that have a low frequency but does not contain low-frequency, extreme high scores. The polygon on the left in Figure 3.6 shows a negatively skewed distribution. This pattern might be found, for example, by measuring the running speed of professional football players. Most would tend to run at higher speeds, but a relatively few linemen lumber in at the slower speeds. (To remember negatively skewed, remember that the pronounced tail is over the lower scores, sloping toward zero, toward where the negative scores would be.)

On the other hand, a **positively skewed distribution** contains extreme high scores that have low frequency but does not contain low-frequency, extreme low scores. The right-hand polygon in Figure 3.6 shows a positively skewed distribution. This pattern might be found, for example, if we measured participants' "reaction time" for recognizing words. Usually, scores will tend to be rather low, but every once in a while a person will "fall asleep at the switch," requiring a large amount of time and thus producing a high score. (To remember positively skewed, remember that the tail slopes away from zero, toward where the higher, *positive* scores are located.)

> **REMEMBER** Whether a *skewed* distribution is *negative* or *positive* corresponds to whether the distinct tail slopes toward or away from zero.

Another type of nonnormal distribution is a *bimodal distribution*, shown in the left-hand side of Figure 3.7. A **bimodal distribution** is a symmetrical distribution containing two distinct humps, each reflecting relatively high-frequency scores. At the center

FIGURE 3.6

Idealized skewed distributions

The direction in which the distinctive tail slopes indicates whether the skew is positive or negative.

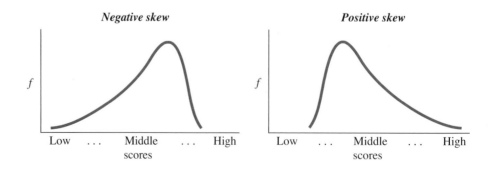

FIGURE 3.7

Idealized bimodal and
rectangular distributions

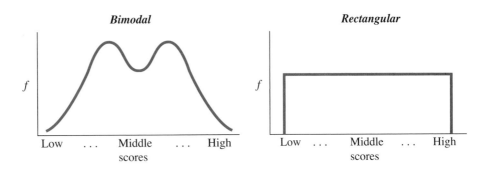

of each hump is one score that occurs more frequently than the surrounding scores, and technically the center scores have the same frequency. Such a distribution would occur with test scores, for example, if most students scored at 60 or 80, with fewer students failing or scoring in the 70s or 90s.

Finally, a third type of distribution is a *rectangular distribution*, as shown in the right-hand side of Figure 3.7. A **rectangular distribution** is a symmetrical distribution shaped like . . . are you ready? . . . like a rectangle! There are no discernible tails because the frequencies of all scores are the same.

Labeling Frequency Distributions

You need to know the names of the previous distributions because descriptive statistics describe the characteristics of data, and one very important characteristic is the shape of the distribution that the data form. Thus, although I might have data containing many different scores, if, for example, I tell you they form a normal distribution, you can mentally *envision* the distribution and quickly and easily understand what the scores are like: Few scores are very low or very high, with the most common, frequent scores in the middle. Therefore, the first step when examining any data is to identify the shape of the simple frequency distribution that they form.

> *REMEMBER*   The shape of the frequency distribution that scores form is an important characteristic of the data.

Recognize, however, that data in the real world will never form the perfect shapes that we've discussed. Instead, the scores will form a bumpy, rough approximation to the ideal distribution. For example, data never form a perfect normal curve and, at best, only come close to that shape. However, rather than drawing a different, approximately normal curve every time, we simplify the task by envisioning the ideal normal curve as our one "model" of any distribution that generally has this shape. Likewise, we envision the ideal shape when discussing the other common curves that we've seen.

Thus, we apply the names of our ideal distributions to actual data as a way of summarizing and communicating their general shape. For example, Figure 3.8 contains some frequency distributions that might be produced in research, and the corresponding labels we might use. (Notice that we even apply these names to histograms or bar graphs.) We assume that the sample represents a population that more closely fits the corresponding ideal polygon: If we measure the population, the additional scores and their corresponding frequencies should "fill in" the sample curve, smoothing it out to be closer to the ideal curve.

FIGURE 3.8

Simple frequency distributions of sample data with appropriate labels

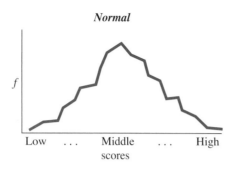

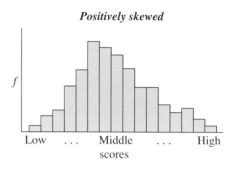

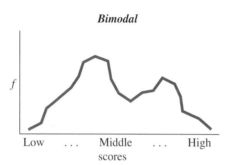

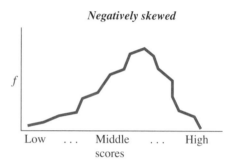

A QUICK REVIEW

- The most common frequency distributions are *normal distributions, negatively or positively skewed distributions, bimodal distributions*, and *rectangular distributions*.

MORE EXAMPLES

The variable of intelligence (IQ) usually forms a normal distribution: The most common scores are in the middle, with higher or lower IQs occurring progressively less often. If IQ was positively skewed, there would be only one distinct tail, located at the higher scores. If IQ was negatively skewed, there would be only a distinct tail at the lower scores. If IQ formed a bimodal distribution, there would be two distinct parts of the curve containing the highest-frequency scores. If IQ formed a rectangular distribution, each score would have the same frequency.

For Practice

1. Arrange the scores below from most frequent to least frequent.

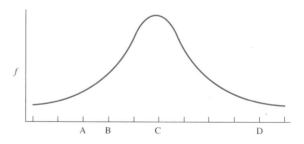

continued

2. What label should be given to each of the following?

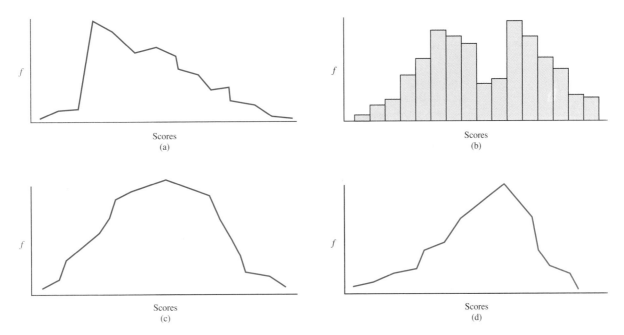

Scores
(a)

Scores
(b)

Scores
(c)

Scores
(d)

Answers

1. C, B, A, D
2. a. positively skewed; b. bimodal; c. normal;
   d. negatively skewed

RELATIVE FREQUENCY AND THE NORMAL CURVE

We will return to simple frequency distributions—especially the normal curve—throughout the remainder of this course. However, counting the frequency of scores is not the only thing we do. Another important procedure is to describe scores using *relative frequency*. **Relative frequency** is the proportion of *N* that is made up by a score's simple frequency. Recall that a proportion indicates a fraction of the total, so relative frequency indicates the fraction of the entire sample that is made up by the times that a score occurs. Thus, whereas simple frequency is the *number* of times a score occurs, relative frequency is the *proportion* of time the score occurs. The symbol for relative frequency is *rel. f*.

We'll first calculate relative frequency using a formula so that you understand its math, although later we'll compute it using a different approach. Here is your first statistical formula.

The formula for computing a score's relative frequency is

$$rel. f = \frac{f}{N}$$

This says that to compute relative frequency, divide the frequency (f) by the total number of scores (N). For example, if a score occurred four times (f) in a sample of 10 scores (N), then filling in the formula gives

$$rel.\, f = \frac{f}{N} = \frac{4}{10} = .40$$

The score has a relative frequency of .40, meaning that the score occurred .40 of the time in the sample.

As you can see here, one reason that we compute relative frequency is simply because it can be easier to interpret than simple frequency. Interpreting that a score has a frequency of 4 is difficult because we have no frame of reference—is this often or not? However, we can easily interpret the relative frequency of .40 because it means that the score occurred .40 of the time.

> **REMEMBER** *Relative frequency* indicates the proportion of the time (out of N) that a score occurred.

We can also begin with *rel. f* and compute the corresponding simple frequency. *To transform relative frequency into simple frequency, multiply the relative frequency times N.* Thus, if *rel. f* is .4 and N is 10, multiply .4 times 10 and the answer is 4; the score occurs four times in this sample.

Finally, sometimes we transform relative frequency to percent. Converting relative frequency to percent gives the percent of the time that a score occurred. *To transform relative frequency to percent, multiply the rel. f times* 100. Above, the *rel. f* was .40, so $(.40)(100) = 40\%$. Thus, 40% of the sample had this score. Conversely, *to transform percent into relative frequency, divide the percent by* 100. Thus 40% / 100 = .40.

Presenting Relative Frequency in a Table or Graph

A distribution showing the relative frequency of all scores is called a **relative frequency distribution**. To create a relative frequency table, first create a simple frequency table, as we did previously. Then add a third column labeled "*rel. f.*"

For example, look at Table 3.3. To compute *rel. f*, we need N, which here is 20. Then the score of 1, for example, has $f = 4$, so its relative frequency is 4/20, or .20. And so on.

We can also determine the combined relative frequency of several scores by adding their frequencies together: In Table 3.3, a score of 1 has a relative frequency of .20, and a score of 2 has a relative frequency of .50, so together, their relative frequency is .20 1 .50, or .70; participants having a 1 or 2 compose .70 of our sample. (To check your table, remember that, except for rounding error, the sum of all relative frequencies in a distribution should equal 1: All scores together should constitute 100% of the sample.)

We graph a relative frequency distribution using the same rules as with simple frequency: Create a bar graph if the scores involve a nominal or ordinal scale, and create a

TABLE 3.3

Relative Frequency Distribution

The left-hand column identifies the scores, the middle column shows each score's frequency, and the right-hand column shows each score's relative frequency.

Score	f	rel. f
6	1	.05
5	0	.00
4	2	.10
3	3	.15
2	10	.50
1	4	.20
Total: 20		1.00 = 100%

FIGURE 3.9

Examples of relative frequency distributions using the data in Table 3.3

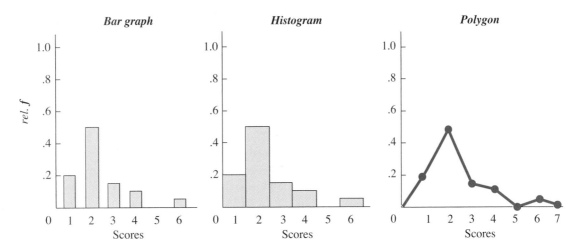

histogram or polygon if the scores involve an interval or ratio scale. Figure 3.9 presents examples using the data from Table 3.3. The only novelty here is that the Y axis reflects relative frequency, so it is labeled in increments between 0 and 1.0.

Finding Relative Frequency Using the Normal Curve

Although relative frequency is an important component of statistics, we will not emphasize the previous formula. (You're welcome.) Instead, most of the time our data will be normally distributed, and then we will use the normal curve to determine relative frequency.

To understand this approach, think about a normal curve in a different way. Imagine that you are flying in a helicopter over a parking lot. The X and Y axes are laid out on the ground, and the people who received a particular score are standing in line in front of the marker for their score. The lines of people are packed so tightly together that, from the air, you only see the tops of many heads in a "sea of humanity." If you painted a line that went behind the last person in line at each score, you would have the normal curve shown in Figure 3.10.

From this perspective, the height of the curve above any score reflects the number of people standing in line at that score. Thus, in Figure 3.10, the score of 30 has the highest frequency because the longest line of people is standing at this score in the parking lot.

FIGURE 3.10

Parking lot view of the ideal normal curve.

The height of the curve above any score reflects the number of people standing in line at that score.

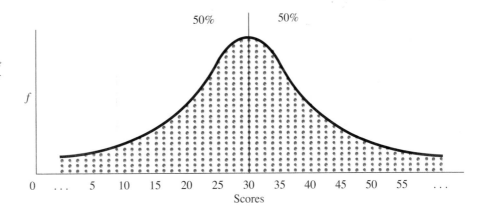

The reason for using this "parking lot view" is so that you think of the normal curve as a picture of something solid: The space under the curve between the curve and the *X* axis has *area* that represents individuals and their scores. The entire parking lot contains everyone in the sample and 100% of the scores. Therefore, any portion of the parking lot—any portion of the space under the curve—corresponds to that portion of the sample.

For example, notice that in Figure 3.10 a vertical line is drawn through the middle score of 30, and so .50 of the parking lot is to the left of the line. Because the complete parking lot contains all participants, a part that is .50 of it contains 50% of the participants. (We can ignore those relatively few people who are straddling the line.) Participants are standing to the left of the line because they received scores of 29, 28, and so on. So, in total, 50% of the participants received scores below 30. Now turn this around: If 50% of the participants obtained scores below 30, then the scores below 30 occurred 50% of the time. Thus, the scores below 30 have a combined relative frequency of .50.

This logic is so simple it almost sounds tricky: if you have one-half of the parking lot, then you have one-half of the participants and thus one-half of the scores, so those scores occur .50 of the time. Or, if you have 25% of the parking lot, then you have 25% of the participants and 25% of the scores, so those scores occur .25 of the time.

This is how we describe what we have done using statistical terminology: The total space occupied by the everyone in the parking lot is called the *total area under the normal curve*. We identify some particular scores and determine the area of the corresponding portion of the polygon above those scores. We then compare the area of this portion to the total area to determine the **proportion of the total area under the curve** that we have selected. Then, as we've seen,

> **The proportion of the total area under the normal curve that is occupied by a group of scores corresponds to the combined relative frequency of those scores.**

Of course, statisticians don't fly around in helicopters, eyeballing parking lots, so here's a different example: Say that by using a ruler and protractor, we determine that in Figure 3.11 the entire polygon occupies an area of 6 square inches on the page. This total area corresponds to all scores, which is *N*. Say that the area under the curve between the scores of 30 and 35 covers 2 square inches. This area is due to the number of times these scores occur. Therefore, the scores between 30 and 35 occupy 2 out of the 6 square inches created by all scores, so these scores constitute 2/6, or .33, of the entire distribution. Thus, the scores between 30 and 35 constitute .33 of our *N*, so they have a relative frequency of .33.

We could obtain this answer by using the formula for relative frequency if, using *N* and each score's *f*, we computed the *rel. f* for each score between 30 and 35 and then added them together. However, the advantage of using the area under the curve is that we can get the answer without knowing *N* or the simple frequencies of these scores.

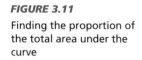

FIGURE 3.11

Finding the proportion of the total area under the curve

The complete curve occupies 6 square inches, with scores between 30 and 35 occupying 2 square inches.

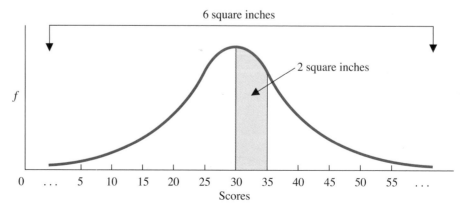

In fact, whatever the variable might be, whatever the *N* might be, and whatever the actual frequency of each score is, we know that the area these scores occupy is 33% of the total area, and that's all we need to know to determine their relative frequency. This is especially useful because, as you'll see in Chapter 6, statisticians have created a system for easily finding the area under any part of the normal curve. Therefore, we can easily determine the relative frequency for scores in any part of a normal distribution. (No, you won't need a ruler and a protractor.) Until then, simply remember this:

> *REMEMBER* The total area under the normal curve corresponds to the times that all scores occur, so a *proportion of the total area under the curve* corresponds to the proportion of time some of the scores occur, which is their relative frequency.

A QUICK REVIEW

- Relative frequency is the proportion of the time that a score occurs.
- The area under the normal curve corresponds to 100% of a sample, so a proportion of the curve will contain that proportion of the scores, which is their relative frequency.

MORE EXAMPLES

Below, the shaded area is .15 of the total curve (so 15% of people in the parking lot are standing at these scores). Thus, scores between 55 and 60 occur .15 of the time, so their combined relative frequency is .15. Above the score of 70 is .50 of the curve, so scores above 70 have a combined relative frequency of .50.

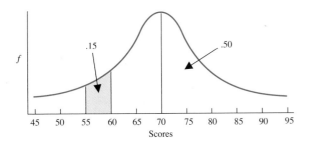

For Practice

1. If a score occurs 23% of the time, its relative frequency is _____ .
2. If a score's relative frequency is .34, it occurs _____ percent of the time.
3. If scores occupy .20 of the area under the curve, they have a relative frequency of _____ .
4. Say that the scores between 15 and 20 have a relative frequency of .40. They make up _____ of the area under the normal curve.

Answers

1. 23%/100 5 .23
2. (.34)(100) 5 34%
3. .20
4. .40

COMPUTING CUMULATIVE FREQUENCY AND PERCENTILE

Researchers have one other approach for organizing the scores in addition to computing simple and relative frequency. Sometimes we want to know a score's standing *relative* to the other scores. For example, it may be most informative to know that 30 people scored above 80 or that 60 people scored below 80. When we seek such information, the convention in statistics is to count the number of scores *below* the score, computing either *cumulative frequency* or *percentile*.

TABLE 3.4

Cumulative Frequency
Distribution

*The left-hand column identi-
fies the scores, the center column contains the simple
frequency of each score, and
the right-hand column contains the cumulative
frequency of each score.*

Score	f	cf
17	1	20
16	2	19
15	4	17
14	6	13
13	4	7
12	0	3
11	2	3
10	1	1

Computing Cumulative Frequency

Cumulative frequency is the frequency of all scores at or below a particular score. The symbol for cumulative frequency is *cf*. To compute a score's cumulative frequency, we add the simple frequencies for all scores below the score to the frequency for the score, to get the frequency of scores at or below the score.

For example, Table 3.4 shows a "cumulative frequency distribution" created from a simple frequency table to which we add a *cf* column. Begin with the lowest score. Here, no one scored below 10, and one person scored 10, so we have a *cf* of 1 (one person scored 10 or below 10). Next, there were two scores of 11. We add this *f* to the previous *cf* for 10, so the *cf* for 11 is 3 (three people scored at 11 or below 11). Next, no one scored at 12, but three people scored below 12, so the *cf* for 12 is also 3. And so on, each time adding the frequency for a score to the cumulative frequency for the score immediately below it.

As a check, verify that the *cf* for the highest score equals *N*: Here all 20 participants obtained either the highest score or a score below it.

Nowadays, researchers seldom create graphs showing cumulative frequency, so we won't. In fact, cumulative frequency is not the most common way to summarize scores. But cumulative frequency is the first step in computing *percentiles*, which are very common.

Computing Percentiles

We've seen that the *proportion* of time a score occurs provides a frame of reference that is easier to interpret than the *number* of times a score occurs. Therefore, our final procedure is to transform cumulative frequency into a percent of the total. A score's **percentile** is the percent of all scores in the data that are at or below the score. Thus, for example, if the score of 80 is at the 75th percentile, this means that 75% of the sample scored at or below 80.

Usually, we will already know a score's *cf*. Then

> **The formula for finding the percentile for a score with a known *cf* is**
>
> $$\text{Score's Percentile} = \left(\frac{cf}{N}\right)(100)$$

TABLE 3.5

Percentiles

*The right-hand column
contains the percentile of
each score.*

Score	f	cf	Percentile
17	1	20	100
16	2	19	95
15	4	17	85
14	6	13	65
13	4	7	35
12	0	3	15
11	2	3	15
10	1	1	5

This says to first divide the score's *cf* by *N*, which transforms the *cf* into a proportion of the total sample. Then we multiply this times 100, which converts it into a percent of the total. Thus, if a score has a *cf* of 5 and *N* is 10, then $(5/10)(100) = 50$, so the score is at the 50th percentile.

Table 3.5 shows the previous cumulative frequency table (where $N = 20$) transformed to percentiles. With one person scoring 10 or below, $(1/20)(100)$ equals 5, so 10 is at the 5th percentile. The three people scoring 11 or below are at the 15th percentile and so on. The highest score is, within rounding error, the 100th percentile, because 100% of the sample has the highest score or below.

For finding the score at a percentile not shown in our table, the precise way is with a computer program (such as SPSS). However, a quick way to find an approximate percentile is to use the area under the normal curve.

Finding Percentile Using the Area Under the Normal Curve

Percentile describes the scores that are *lower* than a particular score, and on the normal curve, lower scores are to the *left* of a particular score. Therefore, the percentile for a given score corresponds to the percent of the total area under the curve that is to the left of the score. For example, on the distribution in Figure 3.12, 50% of the curve is to the left of the middle score of 30. Because scores to the left of 30 are below it, 50% of the distribution is below 30 (in the parking lot, 50% of the people are standing to the left of the line and all of their scores are less than 30). Thus, the score of 30 is at the 50th percentile. Likewise, to find the percentile for the score of 20 in Figure 3.12, we would find the percent of the total area that is to the left of 20. Say that we find that 15% of the curve is to the left of 20; then 20 is at the 15th percentile.

We can also work the other way to find the score at a given percentile. Say that we seek the score at the 85th percentile. We would measure over until 85% of the area under the curve is to the left of a certain point. If, as shown in Figure 3.12, the score of 45 is at that point, then 45 is at the 85th percentile.

Notice that we make a slight change in our definition of percentile when we use the normal curve. *Technically*, a percentile is the percent of scores *at* or below a score. However, in everyday use, the *at* may be dropped. Then percentile becomes the percent of scores *below* a particular score. This is acceptable if we are describing a large sample or a population because those participants *at* the score are a negligible portion of the total (remember that we ignored those relatively few people who were straddling the line). Thus, in Figure 3.12, the score of 30 is at the 50th percentile, so we say that 50% of the scores are below 30 and 50% are above it.

However, if we are describing a *small* sample, we should not ignore those *at* the score, because those participants may actually constitute a sizable portion of the sample. Say that 10% have this score. If we conclude that 50% are above and 50% are below, with 10% at the score, we have the impossible total of 110%! Therefore, with small samples, percentile is calculated and defined as the percent of scores *at or below* a particular score. Because of this distinction, you should use the area under the normal curve to compute percentile when you have a large sample or a population that also fits the normal curve.

FIGURE 3.12

Normal distribution showing the area under the curve to the left of selected scores

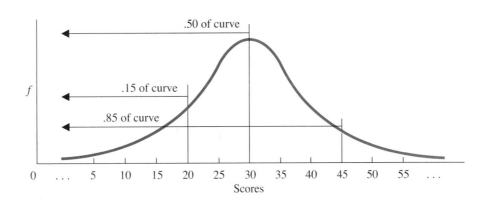

A QUICK REVIEW

- Cumulative frequency (*cf*) indicates the number of participants in a sample that scored at or below a particular score.
- Percentile indicates the percent of a sample that scored at or below a particular score.
- The proportion of the normal curve to the left of a score is the proportion of participants scoring below that score, which translates into the score's percentile.

MORE EXAMPLES

If the score of 80 has a *cf* of 35, it means that 35 participants had a score of 80 or below 80. If 80 is at the 90th percentile, it means that 90% of the sample had a score of 80 or below 80 and, on a normal distribution, 80 is located over toward the right-hand tail, with 90% of the curve located to the left of it.

For Practice

On an exam, 10 students scored 19, 15 students scored 20, no students scored 21, and everyone else scored 21.

1. The *cf* for the score of 20 is ____ .
2. The *cf* for the score of 21 is ____ .
3. If 60 students took the exam, the percentile for the score of 20 is ____ .
4. What does this percentile indicate?
5. In a normal curve showing these grades, how much of the curve is to the right of (above) 20 and how much is to the left of (below) 20?

Answers
1. 15
2. 15
3. (15/60) = .25. Then (.25)(100) = 25th percentile.
4. 25% of the class scored 20 or below 20.
5. 75% of the curve is above 20, and 25% of the curve is below 20.

STATISTICS IN PUBLISHED RESEARCH: APA PUBLICATION RULES

The rules that we've used for creating tables and graphs are part of the procedures for creating research publications established by the American Psychological Association (APA). We will also follow "APA format" when we discuss how to report various statistics. (*Note*: Computer programs, such as SPSS, sometimes *do not* operate according to these rules.)

You won't always see frequency tables and graphs in published reports, because they are very expensive to print. Instead, often researchers simply state that "the scores were normally distributed" or whatever, and you are expected to mentally envision the distribution.

One way in which researchers shrink the size of published tables and graphs is to create a *grouped distribution*.

A WORD ABOUT GROUPED FREQUENCY DISTRIBUTIONS

In the previous examples, we examined each score individually, creating **ungrouped distributions**. When we have too many scores to produce a manageable ungrouped distribution, we create a grouped distribution. In a **grouped distribution**, scores are combined to form small groups, and then we report the total *f*, *rel. f*, *cf*, or percentile of all scores in each group.

For example, look at the grouped distribution shown in Table 3.6. In the score column, "0-4" contains the scores 0, 1, 2, 3, 4, and "5–9" contains scores 5 through 9, and so on. The *f* for each group is the sum of the frequencies for the scores in that group. Thus, the scores between 0 and 4 have a total *f* of 7, while, for the highest scores between 40 and 44, the total *f* is 2. Likewise, the combined relative frequency of scores between 0 and 4 is .28, while for scores between 40 and 44 it is .08. Each cumulative

TABLE 3.6

Grouped Distribution Showing *f, rel.f, cf,* and Percentile

The left-hand column identifies the lowest and highest score in each group.

Score	f	rel. f	cf	Percentile
40–44	2	.08	25	100
35–39	2	.08	23	92
30–34	0	.00	21	84
25–29	3	.12	21	84
20–24	2	.08	18	72
15–19	4	.16	16	64
10–14	1	.04	12	48
5–9	4	.16	11	44
0–4	7	.28	7	28

frequency is the number of scores that are at 44 or below the *highest* score in the group. Thus, 7 scores are at 4 or below while 25 scores are at 44 or below. (Because 44 is the highest score, we know that N is 25.) Finally, each percentile indicates the percent of scores that are below the highest score in the group, so the score of 4 is at the 28th percentile, and the score of 44 is at the 100th percentile.

Appendix A.1 shows examples of graphs of grouped distributions (Figure A.1) with the details of how to create grouped distributions.

PUTTING IT ALL TOGETHER

All of the procedures in this chapter indicate how often certain scores occur, but each provides a slightly different perspective that allows you to interpret the data in a slightly different way. Which particular procedure you should use is determined by which provides the most useful information. However, you may not automatically know which is the best technique for a given situation. So, understand that researchers often *explore* their data, and you should too. Try different techniques and then choose the approach that allows you to make the most sense out of your data.

Using the SPSS Appendix: Appendix B shows you how to use the SPSS computer program to produce the distributions that we've discussed. As described in Appendix B.1 you'll first need to label your variables and input your raw scores. Then, as in Appendix B.2, you can create frequency tables and plot bar graphs, histograms and polygons. You can also compute percentiles.

CHAPTER SUMMARY

1. The number of scores in the data is symbolized by N.

2. A *simple frequency distribution* shows the frequency of each score. The symbol for simple frequency is f.

3. When graphing a simple frequency distribution, if the variable involves a nominal or an ordinal scale, create a *bar graph*. If the variable involves a few different interval or ratio scores, create a *histogram*. With many different interval or ratio scores, create a *polygon*.

4. In a *normal distribution* forming a *normal curve*, extreme high and low scores are relatively infrequent, scores closer to the middle score are more frequent, and the middle score occurs most frequently. The low-frequency, extreme low and extreme high scores are in the *tails* of a normal distribution.

5. A *negatively skewed distribution* contains low-frequency, extreme low scores, but not low-frequency, extreme high scores. A *positively skewed distribution* contains low-frequency, extreme high scores, but not low-frequency, extreme low scores.

6. A *bimodal distribution* is symmetrical, with two areas showing relatively high-frequency scores. In a *rectangular distribution*, all scores have the same frequency.

7. The *relative frequency* of a score, symbolized by *rel. f*, is the proportion of time that the score occurred. A *relative frequency distribution* is graphed in the same way as a simple frequency distribution except that the *Y* axis is labeled in increments between 0 and 1.0.

8. The *proportion of the total area under the normal curve* occupied by particular scores equals the combined relative frequency of those scores.

9. The *cumulative frequency* of a score, symbolized by *cf*, is the frequency of all scores at or below the score.

10. *Percentile* is the percent of all scores at or below a given score. On the normal curve the percentile of a score is the percent of the area under the curve to the left of the score.

11. In an *ungrouped distribution*, the *f*, *rel. f*, *cf*, or percentile of each individual score is reported.

12. In a *grouped distribution*, different scores are grouped together, and the *f*, *rel. f*, *cf*, or percentile for each group is reported.

KEY TERMS

f   *N*   *rel. f*   *cf*
bar graph *39*
bimodal distribution *44*
cumulative frequency *52*
distribution *36*
frequency *36*
frequency polygon *40*
grouped distribution *54*
histogram *40*
negatively skewed distribution *44*
normal curve *42*
normal distribution *42*

percentile *52*
positively skewed distribution *44*
proportion of the total area under the curve *50*
rectangular distribution *45*
relative frequency *47*
relative frequency distribution *48*
simple frequency *37*
simple frequency distribution *37*
tail *43*
ungrouped distribution *54*

REVIEW QUESTIONS

(Answers for odd-numbered problems are in Appendix D.)

1. What do each of the following symbols mean? (a) *N*; (b) *f*; (c) *rel. f*; (d) *cf*.
2. (a) What is the difference between a bar graph and a histogram? (b) With what kind of data is each used?

3. (a) What is the difference between a histogram and a polygon? (b) With what kind of data is each used?

4. (a) What is the difference between a score's simple frequency and its relative frequency? (b) What is the difference between a score's cumulative frequency and its percentile?

5. (a) What is the advantage of computing relative frequency instead of simple frequency? (b) What is the advantage of computing percentile instead of cumulative frequency?

6. (a) What is the difference between a skewed distribution and a normal distribution? (b) What is the difference between a bimodal distribution and a normal distribution? (c) What does a rectangular distribution indicate about the frequencies of the scores?

7. What is the difference between a positively skewed distribution and a negatively skewed distribution?

8. (a) Why must the cf for the highest score in a sample equal N? (b) Why must the sum of all fs in a sample equal N?

9. What is the difference between graphing a relationship as we did in Chapter 2 and graphing a frequency distribution?

10. What is the difference between how we use the proportion of the total area under the normal curve to determine relative frequency and how we use it to determine percentile.

11. What does it mean when a score is in a tail of a normal distribution?

12. (a) How is percentile defined in a small sample? (b) How is percentile defined for a large sample or population when calculated using the normal curve?

APPLICATION QUESTIONS

13. In reading psychological research, you encounter the following statements. Interpret each one. (a) "The IQ scores were approximately normally distributed." (b) "A bimodal distribution of physical agility scores was observed." (c) "The distribution of the patients' memory scores was severely negatively skewed."

14. From the data 1, 4, 5, 3, 2, 5, 7, 3, 4, and 5, Poindexter created the following frequency table. What five things did he do wrong?

Score	f	cf
1	1	0
2	1	1
3	2	3
4	2	5
5	3	8
7	1	9
	N = 6	

15. The distribution of scores on your statistics test is positively skewed. What does this indicate about the difficulty of the test?

16. (a) On a normal distribution of exam scores, Poindexter scored at the 10th percentile, so he claims that he outperformed 90% of his class. Why is he correct or incorrect? (b) Because Foofy's score is in a tail of the distribution, she claims she had one of the highest scores on the exam. Why is she correct or incorrect?

17. Interpret each of the following. (a) In a small sample, you scored at the 35th percentile. (b) Your score has a *rel. f* of .40. (c) Your score is in the upper tail of the normal curve. (d) Your score is in the left-hand tail of the normal curve. (e) Your score has a *cf* of 50. (f) From the normal curve, your score is at the 60th percentile.

18. Draw a normal curve and identify the approximate location of the following scores. (a) You have the most frequent score. (b) You have a low-frequency score, but the score is higher than most. (c) You have one of the lower scores, but it has a relatively high frequency. (d) Your score seldom occurred.

19. The following shows the distribution of final exam scores in a large introductory psychology class. The proportion of the total area under the curve is given for two segments.

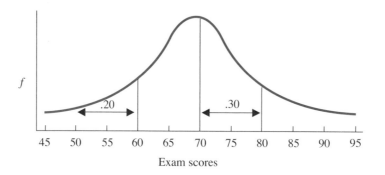

(a) Order the scores 45, 60, 70, 72, and 85 from most frequent to least frequent. (b) What is the percentile of a score of 60? (c) What proportion of the sample scored below 70? (d) What proportion scored between 60 and 70? (e) What proportion scored above 80? (f) What is the percentile of a score of 80?

20. What is the advantage and disadvantage of using grouped frequency distributions?

21. Organize the ratio scores below in a table showing simple frequency, relative frequency, and cumulative frequency.

    49 52 47 52 52 47 49 47 50
    51 50 49 50 50 50 53 51 49

22. (a) Draw a simple frequency polygon using the data in question 23. (b) Draw a relative frequency histogram of these data.

23. Organize the interval scores below in a table showing simple frequency, cumulative frequency, and relative frequency.

    16 11 13 12 11 16 12 16 15
    16 11 13 16 12 11

24. Using the data in question 25, draw the appropriate graph to show (a) simple frequency and (b) relative frequency.

INTEGRATION QUESTIONS

25. Describe each scale of measurement. (Ch. 2)

26. (a) Which scales of measurement are assumed to be discrete; what does this mean? (b) Which scales are assumed to be continuous; what does this mean? (Ch. 2)

27. What type of graph should you create when counting the frequency of: (a) The brands of cell phones owned by students? Why? (b) The different body weights reported in a statewide survey? Why? (c) The people falling into one of eight salary ranges? Why? (d) The number of students who were absent from a class either at the beginning, middle, or end of the semester. (Chs. 2, 3)

28. An experimenter studies vision in low light by having participants sit in a darkened room for either 5, 15, or 25 minutes and then tests their ability to correctly identify 20 objects. (a) What is the independent variable here? (b) What are the conditions? (c) What is the dependent variable? (d) You would use the scores from which variable to create a frequency distribution? (Chs. 2, 3)

29. Our N is 50, and for some scores we have selected the proportion of the area under the curve is .60. (a) What percent of the time do we expect these scores to occur? (b) How many of our participants do we expect to have these scores? (Chs. 1, 3)

■ ■ SUMMARY OF FORMULAS

1. The formula for computing a score's relative frequency is

$$rel.\ f = \frac{f}{N}$$

2. The formula for finding the percentile for a score with a known cf is:

$$\text{Score's Percentile} = \left(\frac{cf}{N}\right)(100)$$

4 Measures of Central Tendency: The Mean, Median, and Mode

GETTING STARTED

To understand this chapter, recall the following:

- From Chapter 2, the logic of statistics and parameters and the difference between an independent and a dependent variable.
- From Chapter 3, when to create bar graphs or polygons, how to interpret polygons, and how to use the area under the normal curve.

Your goals in this chapter are to learn

- What *measures of central tendency* tell us about data.
- What the *mean, median*, or *mode* is and when each is appropriate.
- How a sample mean is used.
- What *deviations around the mean* are.
- How to interpret and graph the results of an experiment.

The frequency distributions discussed in the previous chapter are important because the shape of a distribution is an important characteristic of data for us to know. Therefore, the first step in any statistical analysis is to determine the distribution's shape. Then, however, we compute individual numbers—*statistics*—that each describe an important characteristic of the data. This chapter discusses statistics that describe the important characteristic called *central tendency*. The following sections present (1) the concept of central tendency, (2) the three ways to compute central tendency, and (3) how we use them to summarize and interpret data.

But first . . .

NEW STATISTICAL NOTATION

A new important symbol is Σ, the Greek letter called sigma. Sigma is the symbol for *summation*. It is used in conjunction with a symbol for scores, so you will see such notations as ΣX. In words, ΣX is pronounced **sum of X** and literally means to find the sum of the X scores. Thus, ΣX for the scores 5, 6, and 9 is 20, and in code we would say, $\Sigma X = 20$. Notice that we do not care whether each X is a different score. If the scores are 4, 4, and 4, then $\Sigma X = 12$.

REMEMBER The symbol ΣX indicates to *sum* the X scores.

WHY IS IT IMPORTANT TO KNOW ABOUT CENTRAL TENDENCY?

Recall that descriptive statistics tell us the obvious things we would ask about a sample of scores. So think about what questions you ask your professor about the grades after you've taken an exam. Your first question is how did you do, but your second question is how did everyone else do? Did everyone score high, low, or what? Central tendency is important because it answers this most basic question about data: Are the scores generally high scores or low scores? You need this information to understand both how the class performed and how you performed relative to everyone else. But it is difficult to do this by looking at the individual scores or at the frequency distribution. Instead, it is much better if you know something like the class average; an average on the exam of 80 versus 30 is very understandable. Therefore, in virtually all research, we first compute a statistic that shrinks the data down into one number that summarizes everyone's score. This statistic is called a *measure of central tendency.* This is the only way to easily describe and interpret the sample as a whole.

WHAT IS CENTRAL TENDENCY?

To understand central tendency, first change your perspective of what a score indicates. For example, if I am 70 inches tall, don't think of this as indicating that I have 70 inches of height. Instead, think of any variable as an infinite continuum—a straight line—and think of a score as indicating a participant's *location* on that line. Thus, my score locates me at the address labeled 70 inches. If my brother is 60 inches tall, then he is located at the point marked 60 on the height variable. The idea is not so much that he is 10 inches shorter than I am, but rather that we are separated by a *distance* of 10 units—in this case, 10 "inch" units. In statistics, scores are *locations,* and the difference between any two scores is the *distance* between them.

From this perspective, a frequency distribution shows the *location* of the scores. For example, Figure 4.1 shows height scores from two samples, one containing low scores and one containing higher scores. In our parking lot view of the normal curve, participants' scores determine *where* they stand. A high score puts them on the right side of the lot, a low score puts them on the left side, and a middle score puts them in a crowd in the middle. Further, if we have two distributions containing different scores, then the *distributions* have different locations on the variable.

Thus, when we ask, "Are the scores generally high scores or low scores?" we are actually asking, "*Where* on the variable is the distribution located?" A **measure of central tendency** is a score that summarizes the *location* of a distribution on a variable. Listen

FIGURE 4.1

Two sample polygons on the variable of height

Each polygon indicates the locations of the scores and their frequencies.

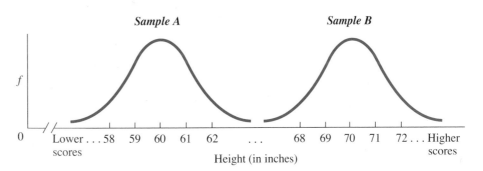

to its name: It is a score that indicates where the *center* of the distribution *tends* to be located. Thus, a measure of central tendency is a number that is a *summary* that you can think of as indicating *where* on the variable most scores are located; or the score that everyone scored *around;* or the *typical* score; or the score that serves as the *address* for the distribution as a whole.

So, in Sample A back in Figure 4.1, most of the scores are in the neighborhood of 59, 60, and 61 inches, so a measure of central tendency will indicate that the distribution is centered at 60 inches. In Sample B, the scores are distributed around 70 inches.

Notice that the above example again illustrates how to use descriptive statistics to *envision* the important aspects of the distribution *without* looking at every individual score. If a researcher told you only that one normal distribution is centered at 60 and the other is centered at 70, you could envision Figure 4.1 and have a general idea of what's in the data. Thus, although you'll see other statistics that add to this mental picture, measures of central tendency are at the core of summarizing data.

> **REMEMBER** The first step in summarizing any set of data is to compute the appropriate *measure of central tendency.*

We will discuss three common measures of central tendency. The trick is to compute the correct one so that you *accurately* envision where most scores in the data are located. Which measure of central tendency you should calculate depends on two factors:

1. The scale of measurement used so that the summary makes sense given the nature of the scores.
2. The shape of the frequency distribution the scores produce so that the measure accurately summarizes the distribution.

In the following sections, we first discuss the *mode,* then the *median,* and finally the *mean.*

THE MODE

One way to describe where most of the scores in a distribution are located is to find the one score that occurs most frequently. The most frequently occurring score is called the **mode.** (We have no accepted symbol for the mode.) For example, say we've collected some test scores and arranged them from lowest to highest: 2, 3, 3, 4, 4, 4, 4, 5, 5, and 6. The score of 4 is the mode because it occurs more frequently than any other score. The left-hand distribution in Figure 4.2 shows that the mode does summarize these data because most of the scores are "around" 4. Also, notice that the scores form a roughly normal curve, with the highest point at the mode. When a polygon has one hump, such as on the normal curve, the distribution is called **unimodal,** indicating that one score qualifies as the mode.

Data may not always produce a single mode. For example, consider the scores 2, 3, 4, 5, 5, 5, 6, 7, 8, 9, 9, 9, 10, 11, and 12. Here two scores, 5 and 9, are tied for the most frequently occurring score. This sample is plotted on the right in Figure 4.2. In Chapter 3, such a distribution was called **bimodal** because it has two modes. Describing this distribution as bimodal and identifying the two modes does summarize where most of the scores tend to be located—most are either around 5 or around 9.

FIGURE 4.2

(a) A unimodal distribution and (b) a bimodal distribution

Each vertical line marks a highest point on the distribution. This indicates a most frequent score, which is a mode.

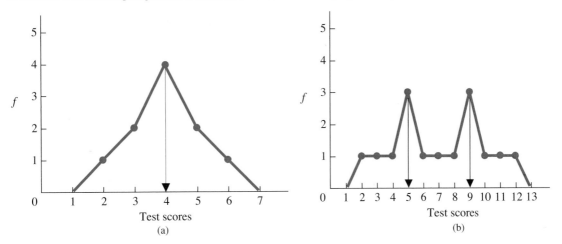

Uses of the Mode

The mode is typically used to describe central tendency when the scores reflect a nominal scale of measurement (when participants are categorized using a qualitative variable). For example, say that we asked some people their favorite flavor of ice cream. The way to summarize such data would be to indicate the most frequently occurring category: Reporting that the mode was a preference for "Goopy Chocolate" is very informative. Also, you have the option of reporting the mode along with other measures of central tendency when describing other scales of measurement because it's always informative to know the "modal score."

> **REMEMBER** The *mode* is the most frequently occurring score in the data and is usually used to summarize nominal scores.

There are, however, two potential limitations with the mode. First, the distribution may contain many scores that are all tied at the same highest frequency. With more than two modes, we fail to summarize the data. In the most extreme case, we might obtain a rectangular distribution such as 4, 4, 5, 5, 6, 6, 7, and 7. Here there is no mode. A second problem is that the mode does not take into account any scores other than the most frequent score(s), so it may not accurately summarize where *most* scores in the distribution are located. For example, say that we obtain the skewed distribution containing 7, 7, 7, 20, 20, 21, 22, 22, 23, and 24. The mode is 7, but this is misleading. Most of the scores are not *around* 7 and instead are up in the low 20s.

Because of these limitations, we usually rely on one of the other measures of central tendency when we have ordinal, interval, or ratio scores.

THE MEDIAN

Often a better measure of central tendency is the median. The **median** is simply another name for the score at the 50th percentile. Recall that 50% of a distribution is at or below the score at the 50th percentile. Thus, if the median is 10, then 50% of the scores are either at or below 10. (Note that the median will not always be one of the scores that occurred.) The median is typically a better measure of central tendency than the mode because (1) only one score can be the median and (2) the median will usually be around where most of the scores in the distribution tend to be located. The symbol for the median is usually its abbreviation, Mdn.

As we discussed in the previous chapter, when researchers are dealing with a large distribution they may ignore the relatively few scores *at* a percentile, so they may say that 50% of the scores are below the median and 50% are above it. To visualize this, recall that a score's percentile equals the *proportion of the area under the curve* that is to the left of—below—the score. Therefore, the 50th percentile is the score that separates the lower 50% of the distribution from the upper 50%. For example, look at Graph A in Figure 4.3. Because 50% of the area under the curve is to the left of the line, the score at the line is the 50th percentile, so that score is the median.

In fact, the median is the score below which 50% of the area of *any* polygon is located. For example, in the skewed distribution in Graph B of Figure 4.3, if .50 of the area under the curve is to the left of the vertical line, then the score at the line is the median.

There are several ways to calculate the median. When scores form a perfect normal distribution, the median is also the most frequent score, so it is the same score as the mode. When scores are approximately normally distributed, the median will be close to the mode.

When data are not at all normally distributed, however, there is no easy way to determine the point below which .50 of the area under the curve is located. Also, recall that using the area under the curve is not accurate with a small sample. In these situations, we can *estimate* the median using the following system. Arrange the scores from lowest to highest. With an odd number of scores, the score in the middle position is the approximate median. For example, for the nine scores 1, 2, 3, 3, 4, 7, 9, 10, and 11, the score in the middle position is the fifth score, so the median is the score of 4. On the other hand, if N is an even number, the average of the two scores in the middle is the approximate median. For example, for the ten scores 3, 8, 11, 11, 12, 13, 24, 35, 46, and 48, the middle scores are at position 5 (the score of 12) and position 6 (the score of 13). The average of 12 and 13 is 12.5, so the median is approximately 12.5.

To precisely calculate the median, consult an advanced textbook for the formula, or as in Appendix B.3, use the SPSS computer program, which is the easiest solution.

FIGURE 4.3

Location of the median in a normal distribution (A) and in a positively skewed distribution (B)

The vertical line indicates the location of the median, with one-half of the distribution on each side of it.

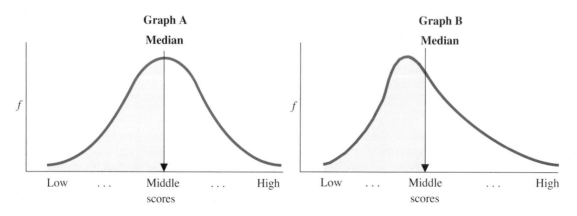

Uses of the Median

The median is not used to describe nominal data: To say, for example, that 50% of our participants preferred "Goopy Chocolate" or *below* is more confusing than informative. On the other hand, the median is the preferred measure of central tendency when the data are ordinal scores. For example, say that a group of students ranked how well a college professor teaches. Reporting that the professor's median ranking was 3 communicates that 50% of the students rated the professor as number 1, 2, or 3. Also, as you'll see later, the median is preferred when interval or ratio scores form a very skewed distribution. (Again you have the option of reporting the median with other types of data when appropriate, simply because it is informative.)

> **REMEMBER** The *median* (Mdn) is the score at the 50th percentile and is used to summarize ordinal or highly skewed interval or ratio scores.

Computing the median still ignores some information in the data because it reflects only the frequency of scores in the lower 50% of the distribution, without considering their mathematical values or considering the scores in the upper 50%. Therefore, the median is not our first choice for describing the central tendency of normal distributions of interval or ratio scores.

A QUICK REVIEW

- The mode is the most frequent score in the data.
- The median is the score at the 50th percentile. If N is an odd number, the median is the middle score. If N is even, the median is the average of the middle scores.

MORE EXAMPLES

For the scores 1, 4, 2, 3, 5, and 3, the mode is 3. The median is found in two steps. First, order the scores: 1, 2, 3, 3, 4, and 5. Then, N is even, so the median is the average of the scores in the third and fourth positions: The average of 3 and 3 is 3. However, for the scores 3, 4, 7, 9, and 10, $N = 5$, so the median is at the middle position, which is 7.

For Practice

1. What is the mode in 4, 6, 8, 6, 3, 6, 8, 7, 9, and 8?
2. What is the median in the above scores?
3. With what types of scores is the mode most appropriate?
4. With what types of scores is the median most appropriate?

Answers

1. In this bimodal data, both 6 and 8 are modes.
2. $N = 10$; the median is the average of 6 and 7, which is 6.50.
3. With nominal scores
4. With ordinal or skewed interval/ratio scores

THE MEAN

By far the all-time most common measure of central tendency in behavioral research is the mean. The **mean** is the score located at the mathematical center of a distribution. Although technically we call this statistic the arithmetic mean, it is what most people call the average. Compute a mean in the same way that you compute an average: Add up all the scores and then divide by the number of scores you added. Unlike the mode or the median, the mean includes every score, so it does not ignore any information in the data.

Let's first compute the mean in a sample. Usually, we use X to stand for the raw scores in a sample and then the symbol for a *sample* mean is $\overline{X}$. It is pronounced "the sample mean" (not "bar X": bar X sounds like the name of a ranch!). Get in the habit of thinking of $\overline{X}$ as a quantity itself so that you understand statements such as "the size of $\overline{X}$" or "this $\overline{X}$ is larger than that $\overline{X}$."

To compute $\overline{X}$, recall that the symbol for "add up all the scores" is ΣX and that the symbol for the number of scores is N. Then

> **The formula for computing a sample mean is**
>
> $$\overline{X} = \frac{\Sigma X}{N}$$

As an example, consider the scores 3, 4, 6, and 7. Adding the scores together produces $\Sigma X = 20$, and N is 4. Thus, $\overline{X} = 20/4 = 5$. Saying that the mean of these scores is 5 indicates that the mathematical center of this distribution is located at the score of 5. (As here, the mean may be a score that does not actually occur in the data.)

What is the mathematical center of a distribution? Think of the center as the balance point. Thus, visualize a teeter-totter on a playground. The left-hand side of Figure 4.4 shows 3, 4, 6, and 7 sitting on the teeter-totter, and the mean of 5 balances the distribution. The right-hand side of Figure 4.4 shows how the mean is the balance point even when the distribution is not symmetrical (the score of 1 has an f of 2). Here the mean is 4 (because $\Sigma X/N = 20/5 = 4$), and it balances the distribution.

Uses of the Mean

Computing the mean is appropriate whenever getting the "average" of the scores makes sense. Therefore, do not use the mean when describing nominal data. (For example, if we count the number of males versus females, an average of the genders would be meaningless.) Likewise, do not compute the mean with ordinal scores (it is strange to say that, on average, runners came in 5.7th in a race). This leaves the mean to describe interval or ratio data.

In addition, however, also consider the shape of the distribution. The mathematical center of the distribution must also be the point where most of the scores are located. This will be the case when we have a *symmetrical* and *unimodal* distribution. For example, say that a simple creativity test produced the scores of 5, 6, 2, 1, 3, 4, 5, 4, 3, 7, and 4, which are shown in Figure 4.5. Here, $\Sigma X = 44$ and $N = 11$, so the mean

FIGURE 4.4

The mean as the balance point of a distribution

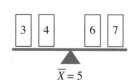

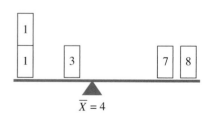

FIGURE 4.5

Location of the mean on a symmetrical distribution.

The vertical line indicates the location of the mean score.

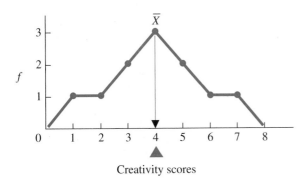

Creativity scores

score is 4. Computing the mean is appropriate here because it *is* the point *around* which *most* of the scores are located: Most often the scores are at or near 4.

Notice that Figure 4.5 shows an approximately normal distribution. *Always compute the mean to summarize a normal or approximately normal distribution:* The mean is the mathematical center of any distribution, and in a normal distribution, most of the scores *are* located around this central point. Therefore, the mean is an accurate summary and provides an accurate address for the distribution.

REMEMBER   Describe the central tendency of a normal distribution of interval or ratio scores by computing the *mean*.

A QUICK REVIEW

- The mean is the average score, located at the mathematical center of the distribution.
- The mean is appropriate for a symmetrical distribution of interval or ratio scores.

MORE EXAMPLES

To find the mean of the scores 3, 4, 6, 8, 7, 3, and 5:
$\Sigma X = 3 + 4 + 6 + 8 + 7 + 3 + 5 = 36$, and $N = 7$.
Then $\overline{X} = 36/7 = 5.1428$; this rounds to 5.14.

For Practice

1. What is the symbol for the sample mean?
2. What is the mean of 7, 6, 1, 4, 5, and 2?

3. With what data is the $\overline{X}$ appropriate?
4. How is a mean interpreted?

Answers

1. $\overline{X}$
2. $\Sigma X = 25$, $N = 6$, $\overline{X} = 4.1666$, rounding to 4.17.
3. With normally distributed or symmetrical distributions of interval or ratio scores
4. It is the center or balance point of the distribution.

Comparing the Mean, Median, and Mode

In a perfect normal distribution, all three measures of central tendency are located at the same score. For example, above in Figure 4.5 the mean of 4 also splits the curve in half, so 4 is the median. Also, the mean of 4 has the highest frequency, so 4 is the mode. If a distribution is only roughly normal, then the mean, median, and mode will be close to the same score. In this case, you might think that any measure of central tendency would be good enough. Not true! The mean uses all information in the scores, and most of the *inferential* procedures we'll see involve the mean. Therefore, the rule is that the mean is the preferred statistic to use with interval or ratio data unless it clearly provides an inaccurate summary of the distribution.

The mean will inaccurately describe a *skewed* (nonsymmetrical) distribution. You have seen this happen if you've ever obtained one low grade in a class after receiving many high grades—your average drops like a rock. The low score produces a negatively skewed distribution, and the mean gets pulled away from where most of your grades are, toward that low grade. What hurts is then telling someone your average because it's misleading: It sounds as if all of your grades are relatively low, even though you have only that one zinger.

The mean is always pulled toward the tail of any skewed distribution because it must balance the entire distribution. You can see this starting with the symmetrical distribution containing the scores 1, 2, 2, 2, and 3. The mean is 2, and this accurately describes most scores. However, adding the score of 20 skews the sample. Now the mean is pulled up to 5. But! Most of these scores are not at or near 5. As this illustrates, although the mean is always at the mathematical center, in a skewed distribution, that center is not where most of the scores are located.

The solution is to use the *median* to summarize a skewed distribution. Figure 4.6 shows the relative positions of the mean, median, and mode in skewed distributions. In both graphs, the mean is pulled toward the extreme tail and is not where most scores are located. The mode is toward the side away from the extreme tail and so the distribution is not centered here either. Thus, of the three measures, the median most accurately reflects the central tendency—the overall address—of a skewed distribution.

> **REMEMBER** Use the mean to summarize normal distributions of interval or ratio scores; use the median to summarize skewed distributions.

It is for the above reasons that the government uses the median to summarize such skewed distributions as yearly income or the price of houses. For example, recent U.S. census data indicated that the median income is slightly above $50,000. But a relatively small number of corporate executives, movie stars, professional athletes, and the like make millions! If we include these salaries and compute the average income, it is about $68,000. However, this is misleading because most people do not earn at or near this higher figure. Instead, the median is a better summary of this distribution.

Believe it or not, we've now covered the basic measures of central tendency. In sum, the first step in summarizing data is to compute a measure of central tendency to describe the score *around* which the distribution tends to be located.

- Compute the mode with nominal data or with a distinctly bimodal distribution of any type of scores.

- Compute the median with ordinal scores or with a skewed distribution of interval or ratio scores.

- Compute the mean with a normal distribution of interval or ratio scores.

FIGURE 4.6

Measures of central tendency for skewed distributions

The vertical lines show the relative positions of the mean, median, and mode.

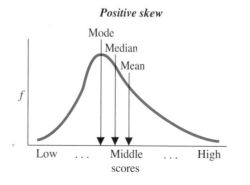

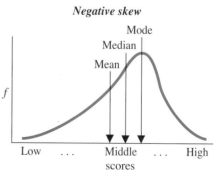

Most often the data in behavioral research are summarized using the mean. This is because most often we measure variables using interval or ratio scores and most often with such scores, "mother nature" produces a reasonably normal distribution. Because the mean is used so extensively, we will delve further into its characteristics and uses in the following sections.

TRANSFORMATIONS AND THE MEAN

Recall that we perform transformations to make scores easier to work with or to make scores from different variables comparable. The simplest transformation is to add, subtract, multiply, or divide each score by a constant. This brings up the burning question: How do transformations affect the mean?

If we add a constant (K) to each raw score in a sample, the new mean of the transformed scores will equal the old mean of the raw scores plus K. For example, the scores 7, 8, and 9 have a mean of 8. Adding 5 to each score produces 12, 13, and 14. The new mean is 13. The old mean of 8 plus the constant 5 also equals 13. Thus, the rule is that $new\ \overline{X} = old\ \overline{X} + K$. The same logic applies for other mathematical operations. When subtracting K from each score, $new\ \overline{X} = old\ \overline{X} - K$. When multiplying each score by K, $new\ \overline{X} = (old\ \overline{X})K$. When dividing each score by K, $new\ \overline{X} = (old\ \overline{X})/K$.

The above rules also apply to the median and to the mode. In essence, using a constant merely changes the location of each score on the variable by K points, so we also move the "address" of the distribution by the same amount.

DEVIATIONS AROUND THE MEAN

To understand why the mean is used so extensively in statistics and research, you must first understand why the mean is the center of a distribution. The mean is the center score because it is just as far from the scores above it as it is from the scores below it. That is, the *total* distance that some scores lie above the mean equals the *total* distance that other scores lie below the mean.

The distance separating a score from the mean is called the score's **deviation**, indicating the amount the score "deviates" from the mean. A score's deviation is equal to the score *minus* the mean, or in symbols:

> **The formula for computing a score's deviation is**
>
> $X - \overline{X}$

Thus, if the sample mean is 47, a score of 50 deviates by $+3$, because $50 - 47$ is $+3$. A score of 40 deviates from the mean of 47 by -7, because $40 - 47 = -7$.

> *REMEMBER*   Always subtract the mean *from* the raw score when computing a score's *deviation*.

Do not, however, think of deviations as simply positive or negative numbers. Think of a deviation as having two components: The number, which indicates *distance from* the mean (which is always positive), and the sign, which indicates *direction from the mean*. For example, the table in Figure 4.7 shows that we computed the deviation scores for our previous creativity scores by subtracting the mean of 4 from each score. In the polygon in Figure 4.7 the X axis is labeled twice, once using each creativity score and, underneath, using its corresponding deviation score. Thus, you can see that a positive deviation indicates that the raw score is larger than the mean and graphed to the right of the mean. A negative deviation indicates that the score is less than the mean and graphed to the left of the mean. The size of the deviation (regardless of its sign) indicates the distance the raw score lies from the mean: the *larger* the deviation, the *farther* into the tail the score is from the mean. A deviation of 0 indicates that an individual's raw score equals the mean.

Now you can see why the mean is the mathematical center of any distribution. If we add together all of the positive deviations we have the total distance that some scores are above the mean. Adding all of the negative deviations, we have the total distance that other scores are below the mean. If we add all of the positive and negative deviations together, we have what is called the **sum of the deviations around the mean**. As in the table in Figure 4.7, the total of the positive deviations will always equal the total of the negative deviations. Therefore:

The sum of the deviations around the mean always equals zero.

Thus, the mean is the center of a distribution because, in total, it is an equal distance form the scores above and below it. Therefore, the half of the distribution that is below the mean balances with the half of the distribution that is above the mean.

The sum of the deviations around the mean always equals zero, *regardless* of the shape of the distribution. For example, in the skewed sample of 4, 5, 7 and 20, the mean is 9, which produces deviations of −5, −4, −2, and +11, respectively. Their sum is again zero.

Some of the formulas you will see in later chapters involve something similar to computing the sum of the deviations around the mean. The statistical code for finding the sum of the deviations around the mean is $\Sigma(X - \overline{X})$. Always work inside parentheses first, so this says to first find the deviation for each score, $(X - \overline{X})$. The Σ indicates to then find the sum of the deviations. Thus, say we have the scores 3, 4, 6, and 7. With $\overline{X} = 5$, then $\Sigma(X - \overline{X}) = -2 + -1 + 1 + 2$, which equals zero.

REMEMBER $\Sigma(X - \overline{X})$ is the symbol for the sum of the deviations around the mean.

FIGURE 4.7

Deviations Around the Mean

The mean is subtracted from each score, resulting in the score's deviation.

Score	Minus	Mean Score	Equals	Deviation
1	−	4	=	−3
2	−	4	=	−2
3	−	4	=	−1
3	−	4	=	−1
4	−	4	=	0
4	−	4	=	0
4	−	4	=	0
5	−	4	=	+1
5	−	4	=	+1
6	−	4	=	+2
7	−	4	=	+3
			Sum =	0

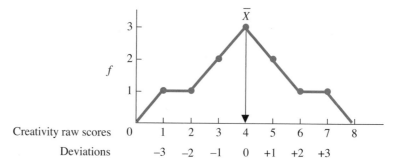

The importance of the sum of the deviations equaling zero is that this makes the mean the score that, *literally,* everyone in the sample scored *around,* with scores above or below it to the same extent. Therefore, we think of the mean as the typical score because it more or less describes everyone's score, with the same amounts of more and less. This is why the mean is such a useful tool for summarizing a distribution. It is also why we use the mean when we are predicting any individual scores.

Using the Mean to Predict Scores

Recall that a goal of behavioral science is to predict a behavior in a particular situation. This translates into predicting the scores found in that situation. When we don't know anything else, the mean is our best prediction about the score that any individual obtains. Because it is the central, typical score, we act as if all the scores were the mean score, and so we predict that score every time. This is why, if the class average on an exam is 80, you would predict that each student's grade is 80. Further, for any students who were absent, you'd predict that they will score an 80 as well. Likewise, if your friend has a B average in college, you would predict that he or she received a B in every course. For any future course, you'd also predict a B.

However, not every score in a sample will equal the mean, so our predictions will sometimes be wrong. To measure the amount of our error when predicting unknown scores, we measure how well we can predict the known scores in our data. The amount of error in any single prediction is the difference between what someone actually gets (X) and what we predict he or she gets $(\overline{X})$. In symbols, this difference is $X - \overline{X}$. We've seen that this is called a *deviation,* but alter your perspective here: In this context, a deviation is the amount of error we have when we predict the mean as someone's score.

> *REMEMBER* When we use the mean to predict scores, a deviation $(X - \overline{X})$ indicates our error: the difference between the $\overline{X}$ we predict for someone and the X that he or she actually gets.

If we determine the amount of error in every prediction, our total error is equal to the sum of the deviations. As we've seen, in any data the sum of the deviations is always zero. Thus, by predicting the mean score every time, the errors in our predictions will, over the long run, cancel out to equal zero. For example, the test scores 70, 75, 85, and 90 have a $\overline{X}$ of 80. One student scored the 70, but we would predict he scored 80, so we would be wrong by -10. But, another student scored the 90; by predicting an 80 for her, we would be off by $+10$. In the same way, our errors for the sample will cancel out so that the total error is zero. Likewise, we assume that other participants will behave similarly to those in our sample, so that using the mean to predict any unknown scores should also result in a total error of zero.

If we predict any score other than the mean, the total error will be *greater* than zero. A total error of zero means that, over the long run, we *overestimate* by the same amount that we *underestimate.* A basic rule of statistics is that if we can't perfectly describe every score, then the next best thing is to have our errors balance out. There is an old joke about two statisticians shooting at a target. One hits 1 foot to the left of the target, and the other hits 1 foot to the right. "Congratulations," one says. "We got it!" Likewise, if we cannot perfectly describe every score, then we want our errors—our over- and underestimates—to balance out to zero. Only the mean provides this capability.

Of course, although our *total* error will equal zero, any *individual* prediction may be very wrong. Later chapters will discuss how to reduce these errors. For now, however, remember that unless we have additional information about the scores, the mean is our best prediction of any score.

> *REMEMBER* We use the mean as everyone's predicted score because, over the long run, the over- and underestimates will cancel out so that the total error in our predictions equals zero.

A QUICK REVIEW

- A deviation is the difference between a score and the mean (or $(X - \overline{X})$) and indicates a score's location relative to the mean.

- When using the mean to predict a score, the score's deviation indicates the amount of error in the prediction. The total error over all such predictions is $\Sigma(X - \overline{X})$, the sum of the deviations around the mean, which equals zero.

More Examples

In some data, $\overline{X} = 30$. Therefore, scores above 30 will produce positive deviations which will cancel out with the negative deviations from the scores below the mean, so that the sum of the deviations equals zero. For every participant, we'd predict a score of 30. Over the long run, our over- and under-estimates will cancel out. In symbols, each error is the difference between the actual X and the $\overline{X}$ that we predict, so the total error is the sum of the deviations, $\Sigma(X - \overline{X})$, which is zero.

For Practice

1. By performing $X - \overline{X}$, you are computing a _____.
2. By performing $\Sigma(X - \overline{X})$, you are computing _____.
3. By saying that $\Sigma(X - \overline{X}) = 0$, you are saying that the mean is located _____ relative to the scores in a sample.
4. By saying that $\Sigma(X - \overline{X}) = 0$, you are saying that when predicting someone's score is the mean, our errors _____.

Answers

1. deviation
2. the sum of the deviations around the mean
3. in the center of the distribution
4. balance out to equal a total of zero

DESCRIBING THE POPULATION MEAN

Recall that ultimately we seek to describe the population of scores we would find in a given situation. Populations are unwieldy, so we must also summarize them. Usually we have interval or ratio scores that form at least an approximately normal distribution, so we usually describe the population using the mean.

The symbol for the mean of a population is the Greek letter μ (pronounced "mew"). Thus, to indicate that the population mean is 143, we'd say that $\mu = 143$. The symbol μ simply shows that we're talking about a population instead of a sample, but a mean is a mean, so a population mean has the same characteristics as a sample mean: μ is the average score in the population, it is the center of the distribution, and the sum of the deviations around μ equals zero. Thus, μ is the score around which everyone in the population scored, it is the typical score, and it is the score that we predict for any individual in the population.

> *REMEMBER* The symbol for a population mean is μ.

How do we determine the value of μ? If all the scores in the population are known to us, we compute μ using the same formula that we used to compute $\overline{X}$:

$$\mu = \frac{\Sigma X}{N}$$

Usually, however, a population is infinitely large, so we cannot directly compute μ. Instead, we estimate μ based on the mean of a random sample. If, for example, a sample's mean in a particular situation is 99, then, assuming the sample accurately represents the population, we would estimate that μ in this situation is also 99. We make such an inference because it is a population with a mean of 99 that is most likely to produce a sample with a mean of 99. That is, a $\overline{X}$ of 99 indicates a sample with mostly scores around 99 in it. What population would be most likely to provide these scores? The population containing mostly scores around 99—where the population mean is 99. Thus, we assume that most scores in a sample are located where most scores in the population are located, so a sample mean is a good estimate of μ.

SUMMARIZING RESEARCH

Now you can understand how measures of central tendency are used in research. Usually we compute the mean because we have normally distributed interval or ratio scores. Thus, we might compute the mean number of times our participants exhibit a particular behavior, or compute their mean response to a survey question. In a *correlational study* we compute their mean score on the X variable and the mean score on the Y variable. Using such means, we can describe the typical score and predict the scores of other individuals, including those of the entire population.

Instead we might compute the median or the mode if we have other types of scores or distributions. When considering the shape of the distribution, we are usually concerned with the shape of distribution for the *population*, because ultimately that is what we want to describe. How do we know its shape? The first step in conducting a study is to read relevant published research reports. From these you will learn many things, including what other researchers say about the population and how they compute central tendency.

Summarizing an Experiment

We perform similar steps when summarizing the results of an experiment. Remember though that the results that need summarizing are the scores from the *dependent variable*. Therefore, it is the characteristics of the dependent scores that determine whether we compute the mean, median, or mode.

> *REMEMBER* The measure of central tendency to compute in an experiment is determined by the type of scale used to measure the *dependent* variable.

Usually it is appropriate to compute the mean, and we do so for *each condition* of the independent variable. For example, here is a very simple study. Say that we think people will make more mistakes when recalling a long list of words than when recalling a short list. We create three conditions of the independent variable of list length. In one condition, participants read a list containing 5 words and then recall it. In another condition, participants read a 10-item list and recall it, and in a third condition, they read a 15-item list and recall it. For each participant, we measure the dependent variable of number of errors made in recalling the list.

First, look at the individual scores shown in Table 4.1. A relationship is present here because a different and higher set of error scores occurs in each condition. Most experiments involve a much larger *N*, however, so to see the relationship buried in the raw scores, we compute a measure of central tendency.

In our memory experiment, the variable of recall errors is a ratio variable that is assumed to form an approximately normal distribution. Therefore, we compute the mean score in each condition by computing the mean of the scores in each *column*. These are shown under each column in Table 4.1.

Normally in a research report you will not see the individual raw scores. Therefore, to interpret the mean in any study, simply envision the scores that would typically produce such a mean. For example, when $\overline{X} = 3$, envision a normal distribution of scores above and below 3, with most scores close to 3. Likewise, for each mean, essentially envision the kinds of raw scores shown in our columns. Thus, the means show that recalling a 5-item list resulted in one distribution located around three errors, but recalling a 10-item list produced a different distribution at around six errors, and recalling a 15-item list produced still another distribution at around nine errors. Further, we use the mean score to describe the individual scores in each condition. In Condition 1, for example, we'd predict that any participant would make about three errors.

Most important is the fact that, by looking at the means alone, we see that a relationship is present here: as the conditions change (from 5 to 10 to 15 items in a list), the scores on the dependent variable also change (from around 3, to around 6, to around 9 errors, respectively).

Note, however, that not *all* means must change for a relationship to be present. For example, we might find that only the mean in the 5-item condition is different from the mean in the 15-item condition. We still have a relationship if, at least *sometimes*, as the conditions of the independent variable change, the dependent scores also change.

> **REMEMBER** We summarize an experiment usually by computing the mean of the dependent scores in each condition. A relationship is present if the means from two or more conditions are different.

The above logic also applies to the median or mode. For example, say that we study political party affiliation as a function of a person's year in college. Our dependent variable is political party, a nominal variable, so the mode is the appropriate measure of central tendency. We might see that freshmen most often claim to be Republican, but the mode for sophomores is Democrat; for juniors, Socialist; and for seniors, Communist. These data reflect a relationship because they indicate that as college level changes, political affiliation tends to change. Likewise, say that the median income for freshmen is lower than the median income for sophomores, which is lower than for upperclassmen. This tells us that the location of the distribution of incomes is different for each class, so we know that the income "scores" of individuals are changing as their year in college changes.

TABLE 4.1

Errors Made by Participants Recalling a 5-, 10-, or 15-Item List

The mean of each condition is under each column.

Condition 1: 5-Item List	Condition 2: 10-Item List	Condition 3: 15-Item List
2	5	7
3	6	9
4	7	11
$\overline{X} = 3$	$\overline{X} = 6$	$\overline{X} = 9$

Graphing the Results of an Experiment

Recall that the independent variable involves the conditions "given" to participants so it is plotted on the X axis. The dependent variable is plotted on the Y axis. However, because we want to summarize the data, usually we do *not* plot the individual scores. Rather, we plot either the mean, median, or mode of the dependent scores from each condition.

> *REMEMBER*   Always plot the conditions of the independent variable on the X axis and the mean, median, or mode of the dependent scores on the Y axis.

Note: Do not be confused by the fact that we use X to represent the scores when computing the means. We still plot them on the Y axis.

We complete the graph by creating either a *line graph* or a *bar graph*. The type of graph to select is determined by the characteristics of the *independent variable*.

Line Graphs   Create a line graph when the independent variable is an interval or a ratio variable. In a **line graph** adjacent data points are connected with straight lines.

For example, our independent variable of list length involves a ratio scale. Therefore, we create the line graph shown on the left in Figure 4.8. Note that the label on the Y axis is "mean recall errors." Then we place a data point above the 5-item condition opposite the mean of 3 errors, another data point above the 10-item condition at the mean of 6 errors, and so on. We use straight lines to connect the data points here for the same reason we did when producing polygons: Anytime the variable on the X axis involves an interval or ratio scale, we assume that it is a *continuous* variable and therefore we draw lines. The lines show that the relationship *continues* between the points shown on the X axis. For example, we assume that if there had been a 6-item list, the mean error score would fall on the line connecting the means for the 5- and 10-item lists. (The APA publication guidelines do not recommend using histograms to summarize experiments.)

FIGURE 4.8

Line graphs showing (A) the relationship for mean errors in recall as a function of list length and (B) the data points we envision around each mean

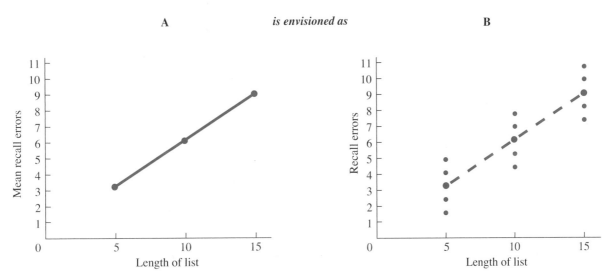

The previous Figure 4.8 conveys the same information as the means did back in Table 4.3, where a different mean indicates a different distribution of scores. We envision these distributions as shown in the right-hand graph. Each mean implies a sample of scores and their corresponding data points are *around*—above and below—the mean's data point. Because the vertical positions of the means change as the conditions change, we know that the raw scores also change, so a relationship is present.

Notice that you can easily spot such a relationship because the different means produce a line graph that is not horizontal. *On any graph, if the summary data points form a line that is not horizontal, it indicates that the individual* Y *scores are changing as the* X *scores change, so a relationship is present.* On the other hand, say that each condition had produced a mean of 5 errors. As shown below on the left in Figure 4.9, this results in a horizontal (flat) line, indicating that as list length changes, the mean error score stays the same. This implies that (as in the figure on the right) the individual scores stay the same regardless of the condition, so no relationship is present. Thus, *on any graph, if the summary data points form a horizontal line, it indicates that the individual* Y *scores do not change as the* X *scores change, and so a relationship is not present.*

Bar Graphs Create a **bar graph** when the independent variable is a nominal or an ordinal variable. Notice that the rule here is the same as it was in Chapter 3: Create a bar graph whenever the X scores are nominal or ordinal scores. With experiments, we place a bar above each condition on the X axis to the height on the Y axis that corresponds to the mean for that condition. As usual, adjacent bars do not touch: Recall that nominal or ordinal scales are *discrete*, meaning that you can have one score or the other, but nothing in-between. The spaces between the bars communicate this.

For example, say that we compared the recall errors made by psychology majors, English majors, and physics majors. The independent variable of college major is a nominal variable, so we have the bar graph shown in Figure 4.10. Because the tops of the bars do not form a horizontal line, we know that different means and thus different scores occur in each condition. We can again envision that we would see individual error scores at around 8 for physics majors, around 4 for psychology majors, and around 12 for English majors. Thus, the scores change as a function of college major, so a relationship is present.

FIGURE 4.9

Line graphs showing (A) no relationship for mean errors in recall as a function of list length and (B) the data points we envision around each mean

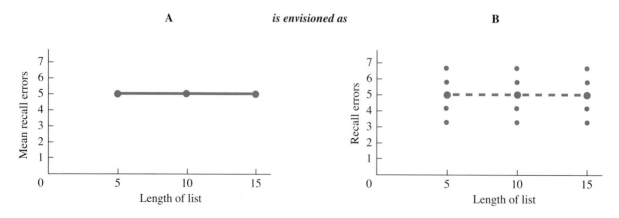

FIGURE 4.10

Bar graph showing mean errors in recall as a function of college major

The height of each bar corresponds to the mean score for the condition.

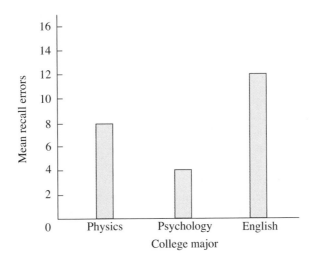

Note: In a different experiment we might have measured a nominal or an ordinal *dependent* variable. In that case we would plot the mode or median on the *Y* axis for each condition. Then, again depending on the characteristics of the independent variable, we would create either a line or bar graph.

REMEMBER   The scale of measurement of the dependent variable determines the measure of central tendency to calculate. The scale of the independent variable determines the type of graph to create.

A QUICK REVIEW

- Graph the independent variable on the *X* axis and the mean, median, or mode of the dependent scores on the *Y* axis.

- Create a line graph when the independent variable is interval or ratio; produce a bar graph when it is nominal or ordinal.

MORE EXAMPLES

We ask males and females to rate their satisfaction with an instructor and the mean scores are 20 and 30, respectively. To graph this, gender is a nominal independent variable, so plot a bar graph, with the labels "male" and "female" on *X* and the means for each on *Y*. Instead, say we measure the satisfaction scores of students tested with either a 10-, 40-, or 80-question final exam and, because the scores form skewed distributions, compute the median scores. Test length is a ratio independent variable, so plot a line graph, with the labels 10, 40, and 80 on *X* and the median of each condition on *Y*.

For Practice

1. The independent variable is plotted on the ____ axis, and the dependent variable is plotted on the ____ axis.

2. A ____ shows a data point above each *X*, with adjacent data points connected with straight lines.

3. A ____ shows a discrete bar above each *X*.

4. The characteristics of the ____ variable determine whether to plot a line or bar graph.

5. Create a bar graph with ____ or ____ variables.

6. Create a line graph with ____ or ____ variables.

Answers

1. *X*; *Y*
2. line graph
3. bar graph
4. independent
5. nominal; ordinal
6. interval; ratio

Inferring the Relationship in the Population

Previously we summarized the results of our list-learning experiment in terms of the sample data. But this is only part of the story. The big question remains: Do these data reflect a law of nature? Do longer lists produce more recall errors for everyone in the population?

Recall that to make inferences about the population, we must first compute the appropriate inferential statistics. Assuming that our data passed the inferential test, we can conclude that each sample mean represents the population mean that would be found for that condition. The mean for the 5-item condition was 3, so we infer that if everyone in the population recalled a 5-item list, the mean error score (μ) would be 3. Similarly, if the population recalled a 10-item list, μ would equal the condition's mean of 6, and if the population recalled a 15-item list, μ would be 9. Because we are describing how the scores change in the population, we are describing the relationship in the population.

In research publications, we essentially assume that a graph for the population would be similar to our previous line graph of the sample data. However, so that you can understand some statistical concepts we will discuss later, you should envision the relationship in the population in the following way. Each sample mean provides a good estimate of the corresponding μ, indicating approximately *where* on the dependent variable each population would be located. Further, we've assumed that recall errors are normally distributed. Thus, we can envision the relationship in the population as the normal distributions shown in Figure 4.11. (*Note:* These are frequency distributions, so the *dependent* scores of recall errors are on the X axis.) As the conditions of the independent variable change, scores on the dependent variable increase so that there is a different, higher population of scores for each condition. Essentially, the distributions change from one centered around 3 errors, to one around 6 errors, to one around 9 errors. (The overlap among the distributions simply shows that some people in one condition make the same number of errors as other people in adjacent conditions.) On the other hand, say that the sample means from our conditions were the same, such as all equal to 5. Then we would envision the one, same population of scores, located at the μ of 5, that we'd expect regardless of our conditions.

Remember that the population of scores reflects the behavior of everyone. If, as the independent variable changes, *everyone's* behavior changes, then we have learned about a law of nature involving that behavior. Above, we believe that everyone's recall behavior tends to change as list length changes, so we have evidence of how human memory generally works in this situation. That's about all there is to it; we have basically achieved the goal of our research.

The process of arriving at the above conclusion sounds easy because it is easy. In essence, statistical analysis of most experiments involves three steps: (1) Compute each sample mean (and other descriptive statistics) to summarize the scores and the relationship found in the experiment, (2) Perform the appropriate inferential procedure to determine whether the data are representative, and (3) Determine the location of the

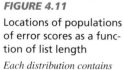

FIGURE 4.11

Locations of populations of error scores as a function of list length

Each distribution contains the scores we would expect to find if the population were tested under a condition.

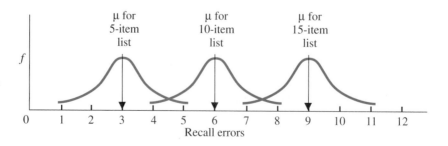

population of scores that you expect would be found for each condition by estimating each μ. Once you've described the expected population for each condition, you are basically finished with the statistical analysis.

>*REMEMBER* Envision a normal distribution located at the μ for each condition when envisioning the relationship in the population from an experiment.

STATISTICS IN PUBLISHED RESEARCH: USING THE MEAN

When reading a research article, you may be surprised by how seldom the word *relationship* occurs. The phrase you will see instead is "a difference between the means." As we saw, when the means from two conditions are different numbers, then each condition is producing a different batch of scores, so a relationship is present. Thus, researchers often communicate that they have found a relationship by saying that they have found a difference between two or more means. If they have not found a relationship, then they say no difference was found.

One final note: Research journals that follow APA's publication guidelines do not use the statistical symbol for the sample mean $(\overline{X})$. Instead, the symbol is *M*. When describing the population mean, however, μ is used. (For the median, *Mdn* is used.)

PUTTING IT ALL TOGETHER

The mean is *the* measure of central tendency in behavioral research. You should understand the mode and the median, but the most important topics in this chapter involve the mean, because they form the basis for virtually everything else that we will discuss. In particular, remember that the mean is the basis for interpreting most studies. You will always want to say something like "the participants scored around 3" in a particular condition because then you are describing their typical *behavior* in that situation. Thus, regardless of what other fancy procedures we discuss, remember that to make sense out of your data you must ultimately return to identifying *around* where the scores in each condition are located.

Using the SPSS Appendix: See Appendix B.3 to compute the mean, median, and mode for one sample of data, along with other descriptive statistics we'll discuss. For experiments, we will obtain the mean for each condition as part of performing the experiment's inferential procedure.

CHAPTER SUMMARY

1. *Measures of central tendency* summarize the location of a distribution on a variable, indicating where the center of the distribution tends to be.

2. The *mode* is the most frequently occurring score or scores in a distribution. It is used primarily to summarize nominal data.

3. The *median,* symbolized by Mdn, is the score at the 50th percentile. It is used primarily with ordinal data and with skewed interval or ratio data.

4. The *mean* is the average score located at the mathematical center of a distribution. It is used with interval or ratio data that form a symmetrical, unimodal distribution, such as the normal distribution. The symbol for a sample mean is $\overline{X}$, and the symbol for a population mean is μ.

5. Transforming raw scores by using a *constant* results in a new value of the mean, median, or mode that is equal to the one that would be obtained if the transformation were performed directly on the old value.

6. The amount a score *deviates* from the mean is computed as $(X - \overline{X})$.

7. The *sum of the deviations around the mean*, $\Sigma(X - \overline{X})$, equals zero. This makes the mean the best score to use when predicting any individual score, because the *total error* across all such estimates will equal zero.

8. In graphing the results of an experiment, the independent variable is plotted on the X axis and the dependent variable on the Y axis. A *line graph* is created when the independent variable is measured using a ratio or an interval scale. A *bar graph* is created when the independent variable is measured using a nominal or an ordinal scale.

9. On a graph, if the summary data points form a line that is not horizontal, then the individual Y scores change as a function of changes in the X scores, and a relationship is present. If the data points form a horizontal line, then the Y scores do not change as a function of changes in the X scores, and a relationship is not present.

10. A random sample mean $(\overline{X})$ is the best estimate of the corresponding population's mean (μ). The $\overline{X}$ in each condition of an experiment is the best estimate of the μ that would be found if the population was tested under that condition.

11. We conclude that a relationship in the population is present when we infer different values of μ, implying different distributions of dependent scores, for two or more conditions of the independent variable.

KEY TERMS

ΣX Mdn $\overline{X}$ $\Sigma(X - \overline{X})$ μ

bar graph *76*
bimodal distribution *62*
deviation *69*
line graph *75*
mean *66*

measure of central tendency *61*
median *63*
mode *62*
sum of the deviations around the mean *70*
sum of *X* *60*
unimodal distribution *62*

REVIEW QUESTIONS

(Answers for odd-numbered problems are in Appendix D.)

1. What does a measure of central tendency indicate?
2. What two aspects of the data determine which measure of central tendency to use?
3. What is the mode, and with what type of data is it most appropriate?
4. What is the median, and with what type of data is it most appropriate?
5. What is the mean, and with what type of data is it most appropriate?
6. Why is it best to use the mean with a normal distribution?
7. Why is it inappropriate to use the mean with a skewed distribution?
8. Which measure of central tendency is used most often in behavioral research? Why?
9. What two pieces of information about the location of a score does a deviation score convey?

10. Why do we use the mean of a sample to predict any score that might be found in that sample?

11. (a) What is the symbol for a score's deviation from the mean? (b) What is the symbol for the sum of the deviations? (c) What does it mean to say "the sum of the deviations around the mean equals zero"?

12. What is μ and how do we usually determine its value?

APPLICATION QUESTIONS

13. For the following data, compute (a) the mean and (b) the mode.

     55   57   59   58   60   57   56   58   61   58   59

14. (a) In question 13, what is your best estimate of the median (without computing it)? (b) Explain why you think your answer is correct. (c) Calculate the approximate median using the method described in this chapter.

15. A researcher collected the following sets of data. For each, indicate the measure of central tendency she should compute: (a) the following IQ scores: 60, 72, 63, 83, 68, 74, 90, 86, 74, and 80; (b) the following error scores: 10, 15, 18, 15, 14, 13, 42, 15, 12, 14, and 42; (c) the following blood types: A−, A−, O, A+, AB−, A+, O, O, O, and AB+; (d) the following grades: B, D, C, A, B, F, C, B, C, D, and D.

16. You misplaced two of the scores in a sample, but you have the data indicated below. What should you guess the missing scores to be? Why?

     7   12   13   14   11   14   13   13   12   11

17. On a normal distribution of scores, four participants obtained the following deviation scores: −5, 0, +3, and +1. (a) Which person obtained the lowest raw score? How do you know? (b) Which person's raw score had the lowest frequency? How do you know? (c) Which person's raw score had the highest frequency? How do you know? (d) Which person obtained the highest raw score? How do you know?

18. In a normal distribution of scores, five participants obtained the following deviation scores: +1, −2, +5, and −10. (a) Which score reflects the highest raw score? (b) Which score reflects the lowest raw score? (c) Rank-order the deviation scores in terms of their frequency, starting with the score with the lowest frequency.

19. For the following experimental results, interpret specifically the relationship between the independent and dependent variables:

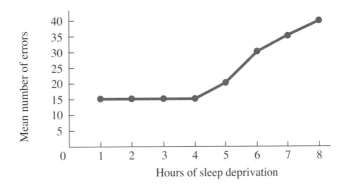

20. (a) In question 19, give a title to the graph, using "as a function of." (b) If you participated in the study in question 19 and had been deprived of 5 hours of sleep, how many errors do you think you would make? (c) If we tested all people in the world after 5 hours of sleep deprivation, how many errors do you think each would make? (d) What symbol stands for your prediction in part c?

21. Foofy says a deviation of $+5$ is always better than a deviation of -5. Why is she correct or incorrect?

22. You hear that a line graph of data from the Grumpy Emotionality Test slants downward as a function of increases in the amount of sunlight present on the day participants were tested. (a) What does this tell you about the mean scores for the conditions? (b) What does this tell you about the raw scores for each condition? (c) Assuming the samples are representative, what does this tell you about the μ's? (d) What do you conclude about whether there is a relationship between emotionality and sunlight in nature?

23. You conduct a study to determine the impact that varying the amount of noise in an office has on worker productivity. You obtain the following productivity scores.

Condition 1: Low Noise	Condition 2: Medium Noise	Condition 3: Loud Noise
15	13	12
19	11	9
13	14	7
13	10	8

(a) Assuming that productivity scores are normally distributed ratio scores, compute the summaries of this experiment. (b) Draw the appropriate graph for these data. (c) Assuming that the data are representative, draw how we would envision the populations produced by this experiment. (d) What conclusions should you draw from this experiment?

24. Assume that the data in question 25 reflect a highly skewed interval variable. (a) Compute the summaries of these scores. (b) What conclusion would you draw from the sample data? (c) What conclusion would you draw about the populations produced by this experiment?

INTEGRATION QUESTIONS

25. (a) How do you recognize the independent variable of an experiment? (b) How do you recognize the dependent variable? (Ch. 2)

26. (a) What is the rule for when to make a bar graph in *any* type of study? (b) Variables using what scales meet this rule? (c) How do you recognize such scales? (d) What is the rule for when to connect data points with lines? (e) Variables using what scales meet this rule? (f) How do you recognize such scales? (Chs. 2, 3)

27. When graphing the results of an experiment: (a) Which variable is plotted on the X axis? (b) Which variable is plotted on the Y axis. (c) When do you produce a bar graph or a line graph? (Chs. 3, 4)

28. Foofy conducts an experiment in which participants are given 1, 2, 3, 4, 5, or 6 hours of training on a new computer statistics program. They are then tested on the speed with which they can complete several analyses. She summarizes her results by computing that the mean number of training hours per participant is 3.5, and so she expects μ would also be 3.5. Is she correct? If not, what should she do? (Chs. 2, 4)

29. For each of the experiments below, determine (1) which variable should be plotted on the Y axis and which on the X axis, (2) whether the researcher should use a line graph or a bar graph to present the data, and (3) how she should summarize scores on the dependent variable: (a) a study of income as a function of age; (b) a study of politicians' positive votes on environmental issues as a function of the presence or absence of a wildlife refuge in their political district; (c) a study of running speed as a function of carbohydrates consumed; (d) a study of rates of alcohol abuse as a function of ethnic group. (Chs. 2, 4)

30. Using *independent* and *dependent*: In an experiment, the characteristics of the _____ variable determine the measure of central tendency to compute, and the characteristics of the _____ variable determine the type of graph to produce. (Chs. 2, 4)

■ ■ ■ SUMMARY OF FORMULAS

1. The formula for the sample mean is

$$\overline{X} = \frac{\Sigma X}{N}$$

2. The formula for a score's deviation is $X - \overline{X}$

3. To estimate the median, arrange the scores in rank order. If N is an odd number, the score in the middle position is roughly the median. If N is an even number, the average of the two scores in the middle positions is roughly the median.

Measures of Variability: Range, Variance, and Standard Deviation

GETTING STARTED

To understand this chapter, recall the following:

- From Chapter 4, what the mean is, what $\overline{X}$ and μ stand for, and what deviations are.

Your goals in this chapter are to learn

- What is meant by *variability*.

- What the *range* indicates.

- When the *standard deviation* and *variance* are used and how to interpret them.

- How to compute the standard deviation and variance when describing a sample, when describing the population, and when estimating the population.

- How variance is used to measure errors in prediction and what is meant by *accounting for variance*.

So far you've learned that applying descriptive statistics involves considering the shape of the frequency distribution formed by the scores and then computing the appropriate measure of central tendency. This information simplifies the distribution and allows you to envision its general properties.

But not everyone will behave in the same way, and so there may be many, very different scores. Therefore, to have a complete description of any set of data, you must also answer the question "Are there large differences or small differences among the scores?" This chapter discusses the statistics for describing the differences among scores, which are called *measures of variability*. The following sections discuss (1) the concept of variability, (2) how to compute statistics that describe variability, and (3) how to use these statistics in research.

First, though, here are a few new symbols and terms.

NEW STATISTICAL NOTATION

A new symbol you'll see is ΣX^2, which indicates to find the **sum of the squared Xs:** First square each X and then find the sum of the squared Xs. Thus, to find ΣX^2 for the scores 2, 2, and 3, we have $2^2 + 2^2 + 3^2$, which becomes $4 + 4 + 9$, which equals 17.

We have a similar looking operation called the **squared sum of X** that is symbolized by (ΣX^2). Work inside the parentheses first, so first find the sum of the X scores and then square that sum. Thus, to find $(\Sigma X)^2$ for the scores 2, 2, and 3, we have $(2 + 2 + 3)^2$, which is $(7)^2$, which is 49. Notice that for the same scores of 2, 2, and 3, ΣX^2 produced 17, while $(\Sigma X)^2$ produced the different answer of 49. Be careful when dealing with these terms.

REMEMBER ΣX^2 indicates the *sum of squared Xs,* and $(\Sigma X)^2$ indicates the *squared sum of* X.

With this chapter we begin using *subscripts.* Pay attention to subscripts because they are part of the symbols for certain statistics.

Finally, some statistics will have two different formulas, a *definitional formula* and a *computational formula.* A definitional formula defines a statistic and helps you to understand it. Computational formulas are the formulas to use when actually computing a statistic. Trust me, computational formulas give exactly the same answers as definitional formulas, but they are much easier and faster to use.

WHY IS IT IMPORTANT TO KNOW ABOUT MEASURES OF VARIABILITY?

Computing a measure of variability is important because without it a measure of central tendency provides an incomplete description of a distribution. The mean, for example, only indicates the central score and where the most frequent scores are. You can see what's missing by looking at the three samples in Table 5.1. Each has a mean of 6, so if you didn't look at the distributions, you might think that they are identical. However, sample A contains scores that differ greatly from each other and from the mean. Sample B contains scores that differ less from each other and from the mean. In sample C no differences occur among the scores.

Thus, to completely describe a set of data, we need to know not only the central tendency but also how much the individual scores differ from each other and from the center. We obtain this information by calculating statistics called measures of variability. **Measures of variability** describe the extent to which scores in a distribution *differ* from each other. With many, large differences among the scores, our statistic will be a larger number, and we say the data are more *variable* or show greater *variability.*

Measures of variability communicate three related aspects of the data. First, the opposite of variability is consistency. Small variability indicates few and/or small differences among the scores, so the scores must be consistently close to each other (and reflect that similar *behaviors* are occurring). Conversely, larger variability indicates that scores (and behaviors) were inconsistent. Second, recall that a score indicates a *location* on a variable and that the difference between two scores is the *distance* that separates them. From this perspective, by measuring differences, measures of variability indicate how *spread out* the scores and the distribution are. Third, a measure of variability tells us how accurately the measure of central tendency describes the distribution. Our focus will be on the mean, so the greater the variability, the more the scores are spread out, and the less accurately they are summarized by the one, mean score. Conversely, the smaller the variability, the closer the scores are to each other and to the mean.

Thus, by knowing the variability in the samples in Table 5.1, we'll know that sample C contains consistent scores (and behaviors) that are close to each other, so 6 accurately represents them. Sample B contains scores that differ more—are more spread out—so they are less consistent, and so 6 is not so accurate a summary. Sample A contains very inconsistent scores that are spread out far from each other, so 6 does not describe most of them.

You can see the same aspects of variability with larger distributions. For example, consider the three distributions in Figure 5.1

TABLE 5.1

Three Different Distributions
Having the Same Mean Score

Sample A	Sample B	Sample C
0	8	6
2	7	6
6	6	6
10	5	6
12	4	6
$\overline{X} = 6$	$\overline{X} = 6$	$\overline{X} = 6$

FIGURE 5.1

Three variations of the normal curve

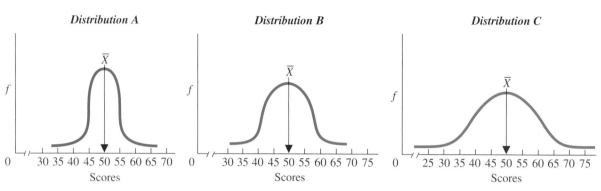

using our a "parking lot approach." If our statistics indicate very small variability, we should envision a polygon similar to Distribution A: It is narrow or "skinny" because most people in the parking lot are standing in line at their scores close to the mean (with few at, say, 45 and 55). We envision such a distribution, because it produces small differences among the scores, and thus will produce a measure of variability that is small. However, a relatively larger measure of variability indicates a polygon similar to Distribution B: It is more spread out and wider, because longer lines of people are at scores farther above and below the mean (more people scored near 45 and 55). This will produce more frequent and/or larger differences among the scores and this produces a measure of variability that is larger. Finally, very large variability suggests a distribution similar to Distribution C: It is very wide because people are standing in long lines at scores that are farther into the tails (scores near 45 and 55 occur very often). Therefore, frequently scores are anywhere between very low and very high, producing many large differences, which produce a large measure of variability.

> *REMEMBER*   *Measures of variability* communicate the differences among the scores, how consistently close to the mean the scores are, and how spread out the distribution is.

A QUICK REVIEW

- Measures of *variability* describe the amount that scores differ from each other.
- When scores are variable, we see frequent and/or large differences among the scores, indicating that participants are behaving inconsistently.

MORE EXAMPLES

If a survey produced high variability, then each person had a rather different answer—and score—than the

next. This produces a wide normal curve. If the variability on an exam is small, then many students obtained either the same or close to the same score. This produces a narrow normal curve.

For Practice

1. When researchers measure the differences among scores, they measure _____.
2. The opposite of variability is _____.

continued

3. When the variability in a sample is large, are the scores close together or very different from each other?

4. If a distribution is wide or spread out, then the variability is _____.

Answers
1. variability
2. consistency
3. different
4. large

THE RANGE

One way to describe variability is to determine how far the lowest score is from the highest score. The descriptive statistic that indicates the distance between the two most extreme scores in a distribution is called the **range.**

> **The formula for computing the range is**
>
> Range = highest score − lowest score

For example, the scores of 0, 2, 6, 10, 12 have a range of 12 − 0 = 12. The less variable scores of 4, 5, 6, 7, 8 have a range of 8 − 4 = 4. The perfectly consistent sample of 6, 6, 6, 6, 6 has a range of 6 − 6 = 0.

Thus, the range does communicate the spread in the data. However, the range is a rather crude measure. It involves only the two most extreme scores it is based on the least typical and often least frequent scores. Therefore, we usually use the range as our sole measure of variability only with nominal or ordinal data.

With nominal data we compute the range by counting the number of categories we have. For example, say we ask participants their political party affiliation: We have greater consistency if only 4 parties are mentioned than if 14 parties are reported. With ordinal data the range is the distance between the lowest and highest rank: If 100 runners finish a race spanning only the positions from first through fifth, this is a close race with many ties; if they span 75 positions, the runners are more spread out.

It is also informative to report the range along with the following statistics that are used with interval and ratio scores.

UNDERSTANDING THE VARIANCE AND STANDARD DEVIATION

Most behavioral research involves interval or ratio scores that form a normal distribution. In such situations (when the mean is appropriate), we use two similar measures of variability, called the *variance* and the *standard deviation*.

Understand that we *use* the variance and the standard deviation to describe how different the scores are from each other. We *calculate* them, however, by measuring how much the scores differ from the mean. Because the mean is the center of a distribution, when scores are spread out from each other, they are also spread out from the mean. When scores are close to each other, they are also close to the mean.

This brings us to an important point. The mean is the score *around* which a distribution is located. *The variance and standard deviation allow us to quantify "around."* For example, if the grades in a statistics class form a normal distribution with a mean of 80, then you know that most of the scores are around 80. But are most scores between 79 and 81 or between 60 and 100? By showing how spread out scores are from the mean, the variance and standard deviation define "around."

> *REMEMBER* The *variance* and *standard deviation* are two measures of variability that indicate how much the scores are spread out around the mean.

Mathematically, the *distance* between a score and the mean is the *difference* between them. Recall from Chapter 4 that this difference is symbolized by $X - \overline{X}$, which is the amount that a score *deviates* from the mean. Thus, a score's deviation indicates how far it is spread out from the mean. Of course, some scores will deviate by more than others, so it makes sense to compute something like the average amount the scores deviate from the mean. Let's call this the "average of the deviations." The larger the average of the deviations, the greater the variability.

To compute an average, we sum the scores and divide by N. We *might* find the average of the deviations by first computing $X - \overline{X}$ for each participant and then summing these deviations to find $\Sigma(X - \overline{X})$. Finally, we'd divide by N, the number of deviations. Altogether, the formula for the average of the deviations would be[1]

$$\text{Average of the deviations} = \frac{\Sigma(X - \overline{X})}{N}$$

We *might* compute the average of the deviations using this formula, except for a *big* problem. Recall that the sum of the deviations around the mean, $\Sigma(X - \overline{X})$, always equals zero because the positive deviations cancel out the negative deviations. This means that the numerator in the above formula will always be zero, so the average of the deviations will always be zero. So much for the average of the deviations!

But remember our purpose here: We want a statistic *like* the average of the deviations so that we know the average amount the scores are spread out around the mean. But, because the average of the deviations is always zero, we calculate slightly more complicated statistics called the variance and standard deviation. *Think* of them, however, as each producing a number that indicates something like the average or typical amount that the scores differ from the mean.

> *REMEMBER* Interpret the variance and standard deviation as roughly indicating the average amount the raw scores deviate from the mean.

The Sample Variance

If the problem with the average of the deviations is that the positive and negative deviations cancel out, then a solution is to *square* each deviation. This removes all negative signs, so the sum of the squared deviations is not necessarily zero and neither is the average squared deviation.

By finding the average squared deviation, we compute the variance. The **sample variance** is the average of the squared deviations of scores around the sample mean. The symbol for the sample variance is S_X^2. Always include the squared sign because it is part of the symbol. The capital S indicates that we are describing a sample, and the subscript X indicates that it is computed for a sample of X scores.

---

[1]In advanced statistics there is a very real statistic called the "average deviation." This isn't it.

REMEMBER   The symbol S_X^2 stands for the variance in a sample of scores.

The formula for the variance is similar to the previous formula for the average deviation except that we add the squared sign. The definitional formula for the sample variance is

$$S_X^2 = \frac{\Sigma(X - \overline{X})^2}{N}$$

Although we will see a better, computational formula later, we will use this one now so that you understand the variance. Say that we measure the ages of some children, as shown in Table 5.2. The mean age is 5 so we first compute each deviation by subtracting this mean from each score. Next, we square each deviation. Then adding the squared deviations gives $\Sigma(X - \overline{X})^2$, which here is 28. The N is 7, so

$$S_X^2 = \frac{\Sigma(X - \overline{X})^2}{N} = \frac{28}{7} = 4$$

Thus, in this sample, the variance equals 4. In other words, the average squared deviation of the age scores around the mean is 4.

The good news is that the variance is a legitimate measure of variability. The bad news, however, is that the variance does not make much sense as the "average deviation." There are two problems. First, squaring the deviations makes them very large, so the variance is unrealistically large. To say that our age scores differ from their mean by an *average* of 4 is silly because not one score actually deviates from the mean by this much. The second problem is that variance is rather bizarre because it measures in squared units. We measured ages, so the scores deviate from the mean by 4 *squared* years (whatever that means!).

Thus, it is difficult to interpret the variance as the "average of the deviations." The variance is not a waste of time, however, because it is used extensively in statistics. Also, variance does communicate the *relative* variability of scores. If one sample has $S_X^2 = 1$ and another has $S_X^2 = 3$, you know that the second sample is more variable because it has a larger average squared deviation. Thus, think of variance as a number that generally communicates how variable the scores are: The larger the variance, the more the scores are spread out.

The measure of variability that more directly communicates the "average of the deviations" is the *standard deviation*.

The Sample Standard Deviation

The sample variance is always an unrealistically large number because we square each deviation. To solve this problem, we take the square root of the variance. The answer is called the standard deviation. The **sample standard deviation** is the square root of the

TABLE 5.2

Calculation of Variance Using the Definitional Formula

Participant	*Age Score*	−	$\overline{X}$	=	$(X - \overline{X})$	$(X - \overline{X})^2$
1	2	−	5	=	−3	9
2	3	−	5	=	−2	4
3	4	−	5	=	−1	1
4	5	−	5	=	0	0
5	6	−	5	=	1	1
6	7	−	5	=	2	4
7	8	−	5	=	3	9
	$N = 7$					$\Sigma(X - \overline{X})^2 = 28$

sample variance. (Technically, the standard deviation equals the square root of the average squared deviation! But that's why we'll think of it as somewhat like the "average deviation".) Conversely, squaring the standard deviation produces the variance.

The symbol for the sample standard deviation is S_X (which is the square root of the symbol for the sample variance: $\sqrt{S_X^2} = S_X$).

> **REMEMBER** The symbol S_X stands for the sample standard deviation.

To create the definitional formula here, we simply add the square root sign to the previous defining formula for variance. The definitional formula for the sample standard deviation is

$$S_X = \sqrt{\frac{\Sigma(X - \overline{X})^2}{N}}$$

The formula shows that to compute S_X we first compute everything inside the square root sign to get the variance. In our previous age scores the variance was 4. Then we find the square root of the variance to get the standard deviation. In this case,

$$S_X = \sqrt{4}$$

so

$$S_X = 2$$

The standard deviation of the age scores is 2.

The standard deviation is as close as we come to the "average of the deviations," and there are three related ways to interpret it. First, our S_X of 2 indicates that the age scores differ from the mean by an "average" of 2. Some scores deviate by more and some by less, but overall the scores deviate from the mean by close to an average of 2. Further, the standard deviation measures in the same units as the raw scores, so the scores differ from the mean age by an "average" of 2 *years.*

Second, the standard deviation allows us to gauge how consistently close together the scores are and, correspondingly, how accurately they are summarized by the mean. If S_X is relatively large, then we know that a large proportion of scores are relatively far from the mean. If S_X is small, then more of the scores are close to the mean, and relatively few are far from it.

And third, the standard deviation indicates how much the scores below the mean deviate from it and how much the scores above the mean deviate from it, so the standard deviation indicates how much the scores are spread out *around* the mean. To see this, we find the scores at "plus 1 standard deviation from the mean" ($+1S_X$) and "minus 1 standard deviation from the mean" ($-1S_X$). For example, our age scores of 2, 3, 4, 5, 6, 7, and 8 produced $\overline{X} = 5$ and $S_X = 2$, The score that is $+1S_X$ from the mean is the score at 5 + 2, or 7. The score that is $-1S_X$ from the mean is the score at 5 – 2, or 3. Looking at the individual scores, you can see that it is accurate to say that the majority of the scores are between 3 and 7.

> **REMEMBER** The *standard deviation* indicates the "average deviation" from the mean, the consistency in the scores, and how far scores are spread out around the mean.

In fact, the standard deviation is mathematically related to the normal curve so that computing the scores at $-1S_X$ and $+1S_X$ is especially useful. For example, say that in the statistics class with a mean of 80, the S_X is 5. The score at $80 - 5$ (at $-1S_X$) is 75, and the score at $80 - 5$ (at $+1S_X$) is 85. Figure 5.2 shows about where these scores are located on a normal distribution.

(Here is how to visually locate about where the scores at $-1S_X$ and $+1S_X$ are on any normal curve. Above the scores close to the mean, the curve forms a downward convex shape (∩). As you travel toward each tail, the curve changes its pattern to an upward convex shape (∪) The points at which the curve changes its shape are called *inflection points*. The scores under the inflection points are the scores that are 1 standard deviation away from the mean.)

Now here's the important part: These inflection points produce the characteristic bell shape of the normal curve so that about 34% of the *area under the curve* is always between the mean and the score that is at an inflection point. Recall that area under the curve translates into the relative frequency of scores. Therefore, *about 34% of the scores in a normal distribution are between the mean and the score that is 1 standard deviation from the mean.* Thus, in Figure 5.2, approximately 34% of the scores are between 75 and 80, and 34% of the scores are between 80 and 85. Altogether, about 68% of the scores are between the scores at $+1S_X$ and $-1S_X$ from the mean. Thus, 68% of the statistics class has scores between 75 and 85. Conversely, about 16% of the scores are in the tail below 75, and 16% are above 85. Thus, saying that most scores are between 75 and 85 is an accurate summary because the majority of scores (68%) are here.

> **REMEMBER** Approximately 34% of the scores in a normal distribution are between the mean and the score that is 1 standard deviation from the mean.

In summary, here is how the standard deviation (and variance) add to our description of a distribution. If we know that data form a normal distribution and that, for example, the mean is 50, then we know where the center of the distribution is and what the typical score is. With the variability, we can *envision* the distribution. For example, Figure 5.3 shows the three curves you saw at the beginning of this chapter. Say that I tell you that S_X is 4. This indicates that participants who did not score 50 missed it by an "*average*" of 4 and that most (68%) of the scores fall in the relatively narrow range between 46 $(50 - 4)$ and 54 $(50 + 4)$. Therefore, you should envision something like Distribution A: The high-frequency raw scores are bunched close to the mean, and the middle 68% of the curve is the narrow slice between 46 and 54. If, however, S_X is 7, you know that the scores are more inconsistent because participants missed 50 by an "average" of 7, and 68% of the scores fall in the wider range between 43 to 57. Therefore, you'd envision Distribution B: A larger average deviation is produced when scores further above

FIGURE 5.2

Normal distribution showing scores at plus or minus 1 standard deviation

With $S_X = 5$, the score of 75 is at $-1S_X$, and the score of 85 is at $+1S_X$. The percentages are the approximate percentages of the scores falling into each portion of the distribution.

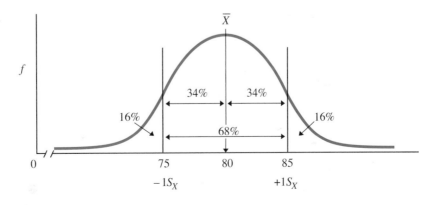

FIGURE 5.3

Three variations of the normal curve

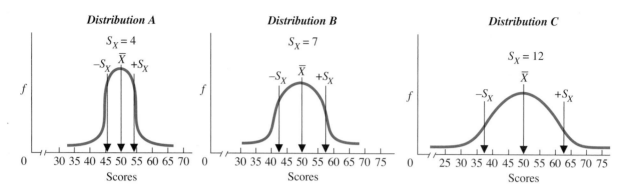

or below the mean have higher frequency, so the high-frequency part of the distribution is now spread out between 43 and 57. Lastly, say that S_X is 12; you know that large/-frequent differences occur, with participants missing 50 by an "average" of 12 points, so 68% of the scores are between 38 and 62. Therefore, envision Distribution C: Scores frequently occur that are way above or below 50, so the middle 68% of the distribution is relatively wide and spread out between 38 and 62.

A QUICK REVIEW

- The sample variance (S_X^2) and the sample standard deviation (S_X) are the two statistics to use with the mean to describe variability.

- The standard deviation is *interpreted* as the average amount that scores deviate from the mean.

MORE EXAMPLES

For the the scores 5, 6, 7, 8, 9, the $\overline{X} = 7$. The variance (S_X^2) is the average squared deviation of the scores around the mean (here, $S_X = 2$). The standard deviation is the square root of the variance. S_X is interpreted as the average deviation of the scores: Here, $S_X = 1.41$, so when participants missed the mean, they were above or below 7 by an "average" of 1.41. Further, in a normal distribution, about 34% of the scores would be between the $\overline{X}$ and 8.41 $(7 + 1.41)$. About 34% of the scores would be between the $\overline{X}$ and 5.59 $(7 - 1.41)$.

For Practice

1. The symbol for the sample variance is _____.
2. The symbol for the sample standard deviation is _____.

3. What is the difference between computing the standard deviation and the variance?

4. In sample A, $S_X = 6.82$; in sample B, $S_X = 11.41$. Sample A is _____ (more/less) variable and most scores tend to be _____ (closer to/farther from) the mean.

5. If $\overline{X} = 10$ and $S_X = 2$, then 68% of the scores fall between _____ and _____.

Answers
1. S_X^2
2. S_X
3. The standard deviation is the square root of the variance.
4. less; closer
5. 8; 12

COMPUTING THE SAMPLE VARIANCE AND SAMPLE STANDARD DEVIATION

The previous definitional formulas for the variance and standard deviation are important because they show that the core computation is to measure how far the raw scores are from the mean. However, by reworking them, we have less obvious but faster computational formulas.

Computing the Sample Variance

The computational formula for the sample variance is derived from its previous definitional formula: we've replaced the symbol for the mean with its formula and then reduced the components.

> **The computational formula for the sample variance is**
>
> $$S_X^2 = \frac{\Sigma X^2 - \dfrac{(\Sigma X)^2}{N}}{N}$$

Use this formula only when describing a *sample*. It says to first find ΣX, then square that sum, and then divide by N. Subtract that result from ΣX^2. Finally, divide by N again.

For example, for our age scores in Table 5.3, the ΣX is 35, ΣX^2 is 203, and N is 7. Putting these quantities in the formula, we have

$$S_X^2 = \frac{\Sigma X^2 - \dfrac{(\Sigma X)^2}{N}}{N} = \frac{203 - \dfrac{(35)^2}{7}}{7}$$

The squared sum of X is 35^2, which is 1225, so

$$S_X^2 = \frac{203 - \dfrac{1225}{7}}{7}$$

TABLE 5.3

Calculation of Variance Using the Computational Formula

Now, 1225 divided by 7 equals 175, so

$$S_X^2 = \frac{203 - 175}{7}$$

Because 203 minus 175 equals 28, we have

$$S_X^2 = \frac{28}{7}$$

Finally, after dividing, we have

$$S_X^2 = 4$$

Thus, again, the sample variance for these age scores is 4.

Do not read any further until you can work this formula!

X *Score*	**X²**
2	4
3	9
4	16
5	25
6	36
7	49
8	64
$\Sigma X = 35$	$\Sigma X^2 = 203$

Computing the Sample Standard Deviation

The computational formula for the standard deviation merely adds the square root symbol to the previous formula for the variance.

The computational formula for the sample standard deviation is

$$S_X = \sqrt{\dfrac{\Sigma X^2 - \dfrac{(\Sigma X)^2}{N}}{N}}$$

Use this formula only when computing the sample standard deviation. For example, for the age scores in Table 5.3, we saw that ΣX is 35, ΣX^2 is 203, and N is 7. Thus,

$$S_X = \sqrt{\dfrac{203 - \dfrac{(35)^2}{7}}{7}}$$

As we saw, the computations inside the square root symbol produce the variance, which is 4, so we have

$$S_X = \sqrt{4}$$

After finding the square root, the standard deviation of the age scores is

$$S_X = 2$$

Finally, be sure that your answer makes sense when computing S_X (and S_X^2). First, variability cannot be a negative number because you are measuring the *distance* scores are from the mean, and the formulas involve *squaring* each deviation. Second, watch for answers that don't fit the data. For example, if the scores range from 0 to 50, the mean should be around 25. Then the largest deviation is about 25, so the "average" deviation will be much less than 25. However, it is also unlikely that S_X is something like .80: If there are only two deviations of 25, imagine how many tiny deviations it would take for the average to be only .80.

Strange answers may be correct for strange distributions, but always check whether they seem sensible. A rule of thumb is

> **For any roughly normal distribution, the standard deviation should equal about one-sixth of the range.**

A QUICK REVIEW

- ΣX^2 indicates to find the *sum of the squared Xs.*
- $(\Sigma X)^2$ indicates to find the *squared sum of* X.

MORE EXAMPLES

For the scores 5, 6, 7, 8, 9:

1. To find the variance,

$$\Sigma X = 5 + 6 + 7 + 8 + 9 = 35$$
$$\Sigma X^2 = 5^2 + 6^2 + 7^2 + 8^2 + 9^2 = 255$$
$$N = 5$$

continued

so

$$S_X^2 = \frac{\Sigma X^2 - \frac{(\Sigma X)^2}{N}}{N} = \frac{255 - \frac{(35)^2}{5}}{5}$$

$$S_X^2 = \frac{255 - \frac{1225}{5}}{5} = \frac{225 - 245}{5}$$

$$S_X^2 = \frac{10}{5} = 2.00$$

2. To find the standard deviation, perform the above steps and then find the square root, so

$$S_X = \sqrt{\frac{\Sigma X^2 - \frac{(\Sigma X)^2}{N}}{N}} = \sqrt{\frac{255 - \frac{1225}{5}}{5}}$$

$$= \sqrt{2.00} = 1.41$$

For Practice

For the scores 2, 4, 5, 6, 6, 7:

1. What is $(\Sigma X)^2$?
2. What is ΣX^2?
3. What is the variance?
4. What is the standard deviation?

Answers

1. $(30)^2 = 900$
2. $2^2 + 4^2 + 5^2 + 6^2 + 7^2 = 166$
3. $S_X^2 = \dfrac{166 - \frac{900}{6}}{6} = 2.667$
4. $S_X = \sqrt{2.667} = 1.63$

Mathematical Constants and the Standard Deviation

As discussed in Chapter 4, sometimes we transform scores by either adding, subtracting, multiplying, or dividing by a constant. What effects do such transformations have on the standard deviation and variance? The answer depends on whether we add (subtracting is adding a negative number) or multiply (dividing is multiplying by a fraction).

Adding a constant to all scores merely shifts the entire distribution to higher or lower scores. We do not alter the relative position of any score, so we do not alter the spread in the data. For example, take the scores 4, 5, 6, 7, and 8. The mean is 6. Now add the constant 10. The resulting scores of 14, 15, 16, 17, and 18 have a mean of 16. Before the transformation, the score of 4 was 2 points away from the mean of 6. In the transformed data, the score is now 14, but it is still 2 points away from the new mean of 16. In the same way, each score's distance from the mean is unchanged, so the standard deviation is unchanged. If the standard deviation is unchanged, the variance is also unchanged.

Multiplying by a constant, however, *does* alter the relative positions of scores and therefore changes the variability. If we multiply the scores 4, 5, 6, 7, and 8 by 10, they become 40, 50, 60, 70, and 80. The original scores that were 1 and 2 points from the mean of 6 are now 10 and 20 points from the new mean of 60. Each transformed score produces a deviation that is 10 times the original deviation, so the new standard deviation is also 10 times greater. (Note that this rule does not apply to the variance. The new variance will equal the square of the new standard deviation.)

> *REMEMBER*   Adding or subtracting a constant does not alter the variability of scores, but multiplying or dividing by a constant does alter the variability.

THE POPULATION VARIANCE AND THE POPULATION STANDARD DEVIATION

Recall that our ultimate goal is to describe the population of scores. Sometimes researchers will have access to a population of scores, and then they directly calculate the actual population variance and standard deviation. The symbol for the known and *true* **population standard deviation** is σ_X (the σ is the lowercase Greek letter *s*, called sigma). Because the squared standard deviation is the variance, the symbol for the true **population variance** is σ_X^2. (In each case, the subscript X indicates a population of X scores.)

The definitional formulas for σ_X and σ_X^2 are similar to those we saw for a sample:

<table>
<tr><td style="text-align:center">*Population Standard Deviation*</td><td style="text-align:center">*Population Variance*</td></tr>
<tr><td style="text-align:center">

$$\sigma_X = \sqrt{\frac{\Sigma(X - \mu)^2}{N}}$$

</td><td style="text-align:center">

$$\sigma_X^2 = \frac{\Sigma(X - \mu)^2}{N}$$

</td></tr>
</table>

The only novelty here is that we determine how far each score deviates from the population mean, μ. Otherwise the population standard deviation and variance tell us exactly the same things about the population that we saw previously for a sample: Both are ways of measuring how much, "on average," the scores differ from μ, indicating how much the scores are spread out in the population. And again, 34% of the population will have scores between μ and the score that is $+1\sigma_X$ above μ, and another 34% will have scores between μ and the score that is $-1\sigma_X$ below μ, for a total of 68% falling between these two scores.

> **REMEMBER** The symbols σ_X^2 and σ_X are used when describing the *known true population variability.*

The previous defining formulas are important for showing you what σ_X and σ_X^2 are. We won't bother with their computing formulas, because these symbols will appear for you only as a *given*, when much previous research allows us to *know* their values.

However, at other times, we will not *know* the variability in the population. Then we will estimate it using a sample.

Estimating the Population Variance and Population Standard Deviation

We use the variability in a sample to estimate the variability that we would find if we could measure the population. However, we do *not* use the previous formulas for the sample variance and standard deviation as the basis for this estimate. These statistics (and the symbols S_X and S_X^2) are used *only* to describe the variability in a sample. They are *not* for estimating the corresponding population parameters.

To understand why this is true, say that we measure an entire population of scores and compute its true variance. We then draw many samples from the population and compute the sample variance of each. Sometimes a sample will not perfectly represent the population so that the sample variance will be either smaller or larger than the population variance. The problem is that, over many samples, more often than not the sample variance will *underestimate* the population variance. The same thing occurs when using the standard deviation.

In statistical terminology, the formulas for S_X^2 and S_X are called the **biased estimators:** They are biased toward underestimating the true population parameters. This is a problem because, as we saw in the previous chapter, if we cannot be accurate, we at least want our under- and overestimates to cancel out over the long run. (Remember the statisticians shooting targets?) With the biased estimators, the under- and overestimates do not cancel out. Instead, they are *too often too small* to use as estimates of the population.

The sample variance (S_X^2) and the sample standard deviation (S_X) are perfectly accurate for describing a *sample*, but their formulas are not designed for estimating the population. To accurately estimate a population, we should have a sample of random scores, so here we need a sample of random deviations. Yet, when we measure the variability of a sample, we use the mean as our reference point, so we encounter the restriction that the sum of the deviations must equal zero. Because of this, not all deviations in the sample are "free" to be random and to reflect the variability found in the population. For example, say that the mean of five scores is 6 and that four of the scores are 1, 5, 7, and 9. Their deviations are -5, -1, $+1$, and $+3$, so the sum of their deviations is -2. Therefore, the final score must be 8, because it must have a deviation of $+2$ so that the sum of all deviations is zero. Thus, the deviation for this score is determined by the other scores and is not a random deviation that reflects the variability found in the population. Instead, only the deviations produced by the four scores of 1, 5, 7, and 9 reflect the variability found in the population. The same would be true for any four of the five scores. Thus, in general, out of the N scores in a sample, only $N - 1$ of them (the N of the sample minus 1) actually reflect the variability in the population.

The problem with the biased estimators $(S_X$ and $S_X^2)$ is that these formulas divide by N. Because we divide by too large a number, the answer tends to be too small. Instead, we should divide by $N - 1$. By doing so, we compute the **unbiased estimators** of the population variance and standard deviation. The definitional formulas for the unbiased estimators of the population variance and standard deviation are

<table>
<tr><td align="center">***Estimated Population Variance***</td><td align="center">***Estimated Population Standard Deviation***</td></tr>
<tr><td align="center">$$s_X^2 = \frac{\Sigma(X - \overline{X})^2}{N - 1}$$</td><td align="center">$$s_X = \sqrt{\frac{\Sigma(X - \overline{X})^2}{N - 1}}$$</td></tr>
</table>

Notice we can call them the **estimated population standard deviation** and the **estimated population variance.** These formulas are almost the same as the previous defining formulas that we used with samples: The standard deviation is again the square root of the variance, and in both the core computation is to determine the amount each score deviates from the mean and then compute something like an "average" deviation. The only novelty here is that, when calculating the estimated population standard deviation or variance, the final division involves $N - 1$.

The symbol for the unbiased estimator of the standard deviation is the lowercase s_X, and the symbol for the unbiased estimator of the variance is the lowercase s_X^2. To keep all of your symbols straight, remember that the symbols for the sample involve the capital or big S, and in those formulas you divide by the "big" value of N. The symbols for estimates of the population involve the lowercase or small s, and here you divide by the smaller quantity, $N - 1$. Further, the *small s* is used to estimate the *small* Greek s called σ. Finally, think of s_X^2 and s_X as the inferential variance and the inferential standard deviation, because the *only* time you use them is to infer the variance or standard deviation of the population based on a sample. Think of S_X^2 and S_X as the descriptive variance and standard deviation because they are used to describe the sample.

> *REMEMBER* S_X^2 and S_X describe the variability in a *sample;* s_X^2 and s_X estimate the variability in the *population.*

For future reference, the quantity $N - 1$ is called the degrees of freedom. The **degrees of freedom** is the number of scores in a sample that are free to reflect the variability in the population. The symbol for degrees of freedom is *df,* so here $df = N - 1$.

In the final analysis, you can think of $N - 1$ as simply a correction factor. Because $N - 1$ is a smaller number than N, dividing by $N - 1$ produces a slightly larger answer. Over the long run, this larger answer will prove to be a more accurate estimate of the population variability.

Computing the Estimated Population Variance and Standard Deviation

The only difference between the computational formula for the estimated population variance and the previous computational formula for the sample variance is that here the final division is by $N - 1$.

> **The computational formula for the estimated population variance is**
>
> $$S_X^2 = \frac{\Sigma X^2 - \dfrac{(\Sigma X)^2}{N}}{N - 1}$$

Notice that in the numerator we still divide by N.

In previous examples, our age scores of 3, 5, 2, 4, 6, 7, and 8 produced $\Sigma X^2 = 203$, and $\Sigma X = 35$. The N was 7 so $N - 1$ equals 6. Putting these quantities into the above formula gives

$$s_X^2 = \frac{\Sigma X^2 - \dfrac{(\Sigma X)^2}{N}}{N - 1} = \frac{203 - \dfrac{(35)^2}{7}}{6}$$

Work through this formula the same way you did for the sample variance: 35^2 is 1225, and 1225 divided by 7 equals 175, so

$$s_X^2 = \frac{203 - 175}{6}$$

Now 203 minus 175 equals 28, so

$$s_X^2 = \frac{28}{6}$$

and the final answer is

$$S_X^2 = 4.67$$

This answer is slightly larger than the sample variance for these age scores, which was $S_X^2 = 4$. Although 4 accurately describes the sample, we estimate the variance of the population is 4.67. In other words, if we could compute the true population variance, we would expect σ_X^2 to be 4.67.

A standard deviation is always the square root of the corresponding variance. Therefore, the formula for the estimated population standard deviation involves merely adding the square root sign to the previous formula for the variance.

> **The computational formula for the estimated population standard deviation is**
>
> $$s_X = \sqrt{\dfrac{\Sigma X^2 - \dfrac{(\Sigma X)^2}{N}}{N-1}}$$

For our age scores, the estimated population variance was $s_X^2 = 4.67$. Then, s_X is $\sqrt{4.67}$ which is 2.16. Thus, if we could compute the standard deviation using the entire population of scores, we would expect σ_X to be 2.16.

Interpreting the Estimated Population Variance and Standard Deviation

Interpret the estimated population variance and standard deviation in the same way as S_X^2 and S_X, except that now they describe how much we *expect* the scores to be spread out in the population, how consistent or inconsistent we *expect* the scores to be, and how accurately we *expect* the population to be summarized by μ.

Notice that, assuming a sample is representative, we have pretty much reached our ultimate goal of describing the population of scores. If we can assume that the distribution is normal, we have described its overall shape. The sample mean $(\overline{X})$ provides a good estimate of the population mean (μ). So, for example, based on a statistics class with a mean of 80, we'd infer that the population would score at a μ of 80. The size of s_X (or s_X^2) estimates how spread out the population is, so if s_X turned out to be 6, we'd expect that the "average amount" the individual scores deviate from the μ of 80 is about 6. Further, we'd expect about 34% of the scores to fall between 74 and 80 (between μ and the score at $-1s_X$) and about 34% of the scores to fall between 80 and 86 (between μ and the score at $+1s_X$) for a total of 68% of the scores between 74 and 86. With this picture in mind, and because scores reflect behaviors, we have a good idea of how most individuals in the population behave in this situation (which is why we conduct research the first place).

A QUICK REVIEW

- The symbols s_X and s_X^2 refer to the *estimated population standard deviation* and *variance,* respectively. When computing them, the final division involves $N-1$.

MORE EXAMPLES

For the scores 5, 6, 7, 8, 9, to estimate the population variability:

$\Sigma X = 35$; $\Sigma X^2 = 255$. $N = 5$, so $N - 1 = 4$.

$$s_X^2 = \frac{\Sigma X^2 - \dfrac{(\Sigma X)^2}{N}}{N-1} = \frac{255 - \dfrac{(35)^2}{5}}{4}$$

$$s_X^2 = \frac{255 - \dfrac{(1225)}{5}}{4} = \frac{255 - 245}{4}$$

$$s_X^2 = \frac{10}{4} = 2.50$$

continued

The standard deviation is the square root of the variance, so

$$s_X = \sqrt{\dfrac{255 - \dfrac{(35)^2}{5}}{4}} = \sqrt{2.50} = 1.58$$

For Practice

1. The symbols for the biased population estimators are ____ and ____.

2. The symbols for the unbiased population estimators are ____ and ____.

3. When do you compute the unbiased estimators?

4. When do you compute the biased estimators?

5. Compute the estimated population variance and standard deviation for the scores 1, 2, 2, 3, 4, 4, and 5.

Answers

1. S_X^2; S_X

2. s_X^2; s_X

3. To estimate the population standard deviation and variance

4. To describe the sample standard deviation and variance

5. $s_X^2 = \dfrac{75 - \dfrac{(21)^2}{7}}{6} = 2.00$; $S_X = \sqrt{2.00} = 1.41$

A SUMMARY OF THE VARIANCE AND STANDARD DEVIATION

To keep track of all of the symbols, names, and formulas for the different statistics you've seen, remember that *variability* refers to the differences between scores and that the *variance* and *standard deviation* are two methods for describing variability. In every case, we are finding the difference between each score and the mean and then calculating an answer that is somewhat like the "average deviation."

Organize your thinking about measures of variability using Figure 5.4. Any standard deviation is merely the square root of the corresponding variance. We compute the descriptive versions when the scores are available: When describing the sample, we calculate S_X^2 and S_X. When describing the population we calculate σ_X^2 and σ_X. When the population of scores is unavailable, we *infer* the variability of the population based on a sample by computing the *unbiased estimators*, s_X^2 and s_X. These "inferential" formulas require a final division by $N - 1$ instead of by N.

With these basics in hand, you are now ready to apply the variance and standard deviation to research.

APPLYING THE VARIANCE AND STANDARD DEVIATION TO RESEARCH

As we've seen, we usually summarize data by computing the mean to describe the typical score and the standard deviation to describe how consistently close the other scores were to it. Thus, the mean from a study might describe the number of times that participants exhibited a particular behavior, but a small standard deviation indicates that they consistently did so. Or, in a survey, the mean might describe the typical opinion held by participants, but a large standard deviation indicates substantial disagreement among them.

We also compute the mean and standard deviation in each *condition* of an experiment. For example, in Chapter 4 we tested the influence of recalling a 5- 10- or 15- item list. Say that we obtained the recall scores in Table 5.4. In published research you would not see the individual scores, so you would know only that, for example, a

FIGURE 5.4

Organizational chart of descriptive and inferential measures of variability

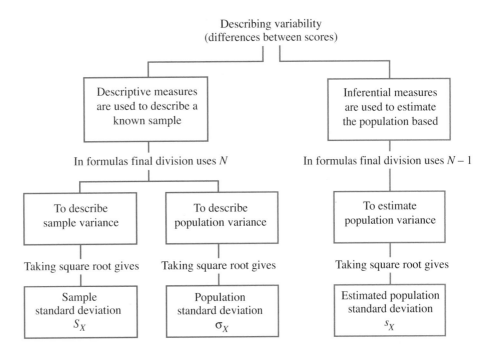

5-item list produced scores around the mean of 3. By also considering the variability, you would also know that these scores differed from this mean by an "average" of only .82. In the 15-item condition, however, scores were spread out by almost twice as much, differing from the mean by 1.63. Therefore, we know that scores were closer to the mean in the 5-item condition, so 3 is a more accurate summary here than 9 is for the 15-item condition. Also, because these recall scores reflect a behavior, we know that memory behavior is more consistent when people recall a 5-item list, with relatively large differences in their behavior when recalling a 15-item list.

Variability and the Strength of a Relationship

Measures of variability also tell us about the *strength* of the overall relationship that an experiment demonstrates. Recall from Chapter 2 that a stronger relationship is more consistent. In an experiment, this translates into everyone in a condition having the same score or close to the same score. In other words, using the terminology of this chapter we would say that a strong relationship occurs when there is little *variability* among the scores within each condition. For example, all three standard deviations in Table 5.4 are relatively small. This indicates that, as shown, the raw scores within each condition are relatively close to each other. Therefore, the overall relationship between list length and recall scores is rather strong.

Conversely, larger standard deviations per condition (say in the neighborhood of 3 or 4) would indicate that a particular list length produced a wide variety of recall scores that are much more spread out around their mean. Therefore, we would describe this as a less consistent, *weaker* relationship.

TABLE 5.4

Mean and Standard Deviation in Each Condition of Recalling 5-, 10-, or 15-Item Lists

5-Item List	10-Item List	15-Item List
3	5	9
4	5	11
2	8	7
$\overline{X} = 3$	$\overline{X} = 6$	$\overline{X} = 9$
$S_X = .82$	$S_X = 1.41$	$S_X = 1.63$

Later we'll see more objective techniques for describing the strength of a relationship in experiments (and in correlational studies). For now:

> *REMEMBER* The strength of a relationship is determined by the variability of the dependent scores (the Y scores) that are paired with each condition (each X score).

A third use of variability is that it communicates the amount of *error* we have when predicting participants' scores.

Variability and Errors in Prediction

You know that the mean is the best score to predict as any participant's score, so, for example, we'd predict a recall score of 3 for anyone in the 5-item condition. However, sometimes our predictions will be wrong. To determine our errors when predicting unknown scores, we determine how well we can predict the known scores in the data. As in Chapter 4, the amount of error in *one* prediction is the difference between what someone actually gets (X) and what we predict he or she gets (the $\overline{X}$). This difference is $X - \overline{X}$, a *deviation*. Because some predictions will contain more error than others, we want to find the average error, so we need the "average deviation." As you've seen, the closest we get to the average deviation is to compute the variance and standard deviation.

Thus, we have a novel way to view S_X and S_X^2: Because they measure the difference between each score and the mean, they also measure the *"average"* error in our predictions when we predict the mean for all participants. For example, back in Table 5.4, the mean in the 15-item condition is 9 and the standard deviation is 1.63. This indicates that the scores differ from the mean by an "average" of 1.63, so if we predict that all participants in this condition score 9, on average we'll be "off" by about 1.63. If S_X was larger, at say 4, then we'd know that participants' scores are farther from the mean, so we'd have greater error when predicting that they scored at 9.

Similarly, the sample variance is somewhat like the average deviation, although less directly. This is too bad because, technically, *variance is the proper way to measure the errors in our prediction*. In fact, variance is sometimes called *error* or *error variance*. Thus, when $S_X = 1.63$, the variance is 1.63^2, which is 2.66. This indicates that when we predict that participants in the 15-item condition scored 9, our "average error"—as measured by the variance—is about 2.66. Although this number may seem strange, simply remember that the larger the variance, the larger the error, and the smaller the variance, the smaller the error.

> *REMEMBER* When we predict that participants obtained the mean score, our "average error" is measured by the variance.

The same logic applies to the population. If the population is known, then we'll predict anyone's score is μ, and our errors in prediction equal σ_X^2. Or, if we must estimate the population using the sample, then we'll use the sample mean to estimate the μ we predict for everyone, and we estimate that our errors in prediction will equal s_X^2.

> *REMEMBER* Summarizing data using the standard deviation and variance indicates the consistency of the scores and behavior, the strength of the relationship, and the "average error" when using the mean to predict scores.

Accounting for Variance

Finally, we have one other use of the term *variance*. In research reports you will encounter such phrases as *accounting for variance* or the **variance accounted for**. They are used when researchers describe the usefulness of a relationship when we use it to predict scores. Because a relationship shows the particular Y scores that are naturally paired with an X, if we know participants' X, we know the Y around which they tend to score. Thus, to some extent we can predict when individuals have one Y score and when other individuals have a *different* Y score. One way to measure the differences in scores is to compute the variance. If we compute the variance of all Y scores in a study, this reflects all of the differences in scores that we want to predict, so this is the variance that we want to account for.

How well a relationship helps us to predict the different Y scores is the extent that it "explains" or "accounts" for the variance in Y scores.

For example, back in Table 5.4 we have a total of nine scores, so their differences produce the variance in recall scores that we want to account for. However, the relationship with list length tends to group similar scores together. Therefore, we know when participants score around 3 (when they recall a 5-item list) and when they produce a different score of, say, 9 (when they recall a 15-item list). By considering list length, our predictions seem very close to each person's actual score, so we seem to be close to predicting many of the differences among the nine scores. Therefore, in our lingo, we would say that the variable of list length seems to "account for" a sizable portion of the variance in recall scores. However, we still have some *error* in our predictions because not everyone scored exactly the score we'd predict. Therefore, some differences among scores are not predicted, so we say that some of the variance in recall scores is *not* accounted for.

On the other hand, consider when a relationship is weaker, such as the relationship between someone's gender and his or her height. We would predict the average man's height for any man and the average woman's height for any woman. However, there is a wide range of women's and men's heights, so our predictions each time may not be very close to someone's actual height. Therefore, this relationship is not all that much help in predicting someone's exact height, and so it would be described as accounting for little of the variance in height.

As these examples illustrate, more consistent relationships account for a greater amount of the variance. Chapters 8 and 12 discuss ways to precisely measure the amount of variance accounted for. Until then,

> *REMEMBER* How well a relationship helps us to predict the different Y scores in the data is the extent that it *accounts for the variance* in Y scores.

STATISTICS IN PUBLISHED RESEARCH: REPORTING VARIABILITY

The standard deviation is most often reported in published research because it most directly indicates how consistently close the scores are to the mean and because it allows us to easily determine the middle 68% of the distribution. However, as if you haven't seen enough symbols already, journals that follow APA guidelines do not use the statistical symbols we have used for the sample standard deviation. Instead, the symbol for the sample standard deviation is *SD*.

We do not, however, use symbols in sentences, so we would not say "The large SD was . . . ." Instead, symbols usually appear in parentheses. For example, recall that the symbol for the sample mean is *M,* so in a report of our list-length study, you might see this: "The fewest errors were produced when recalling 5-item lists ($M = 3.00$, $SD = .83$) with the 10-item condition producing more . . . "

PUTTING IT ALL TOGETHER

At this point the three steps in analyzing any set of data should be like a reflex for you. (1) Consider the scale of measurement used and the shape of the distribution formed by the scores. (2) Describe around where most participants scored, usually by computing the $\overline{X}$ for each group or for each condition of an experiment. (3) Describe the variability—how spread out the scores are—around each mean, usually by computing the sample standard deviation. With this information, you are largely finished with descriptive statistics because you know the important characteristics of the sample data and you'll be ready to draw inferences about the corresponding population.

Using the SPSS Appendix Section B.3 in Appendix B shows how to compute the mean, median, and mode, as well as the variance, standard deviation, and range for a sample of scores. Later we will compute the mean and standard deviation in each condition of an experiment as part of performing inferential statistics.

CHAPTER SUMMARY

1. *Measures of variability* describe how much the scores differ from each other, or how much the distribution is spread out.

2. The *range* is the difference between the highest and the lowest score.

3. The *variance* is used with the mean to describe a normal distribution of interval or ratio scores. It is the average of the squared deviations of scores around the mean.

4. The *standard deviation* is also used with the mean to describe a normal distribution of interval/ratio scores. It is the square root of the variance. It can be thought of as somewhat like the "average" amount that scores deviate from the mean.

5. Transforming scores by adding or subtracting a constant does not alter the standard deviation. Transforming scores by multiplying or dividing by a constant alters the standard deviation by the same amount as if we had multiplied or divided the original standard deviation by the constant.

6. There are three versions of the formula for variance: $S_{\overline{X}}^2$ describes how far the sample scores are spread out around $\overline{X}$, $\sigma_{\overline{X}}^2$ describes how far the population of scores is spread out around μ, and $s_{\overline{X}}^2$ is computed using sample data but is the inferential, unbiased estimate of how far the scores in the population are spread out around μ.

7. There are three versions of the formula for the standard deviation: S_X describes how far the sample scores are spread out around $\overline{X}$, σ_X describes how far the population is spread out around μ, and S_X is computed using sample data but is the inferential, unbiased estimate of how far the scores in the population are spread out around μ.

8. The formulas for the descriptive measures of variability (for S_X^2 and S_X) use N as the final denominator. The inferential formulas (for s_X^2 and s_X) use $N - 1$. The quantity $N - 1$ is the *degrees of freedom* in the sample.

9. On a normal distribution, approximately 34% of the scores are between the mean and the score that is a distance of one standard deviation from the mean. Therefore, approximately 68% of the distribution lies between the two scores that are plus and minus one standard deviation from the mean.

10. We summarize an experiment usually by computing the mean and standard deviation in each condition. When the standard deviations are relatively small, the scores in the conditions are similar, and so a more consistent—*stronger*—relationship is present.

11. When we predict that participants obtained the mean score, our error in predictions is determined by the variability in the scores. In this context the variance and standard deviation measure the differences between the participants' actual scores (X) and the score we predict for them $(\overline{X})$, so we are computing an answer that is somewhat like the "average" error in our predictions.

12. The amount that a relationship with X helps us to predict the different Y scores in the data is the extent that X *accounts for the variance* in Y scores.

KEY TERMS

ΣX^2 $(\Sigma X)^2$ S_X^2 S_X σ_X^2 σ_X s_X^2 s_X df

biased estimator 96
degrees of freedom 97
estimated population standard
   deviation 97
estimated population variance 97
measures of variability 85
population standard deviation 96

population variance 96
variance accounted for 103
range 87
sample standard deviation 89
sample variance 88
sum of the squared Xs 84
squared sum of X 84
unbiased estimator 97

REVIEW QUESTIONS

(Answers for odd-numbered problems are in Appendix D.)

1. In any research, why is describing the variability important?
2. What do measures of variability communicate about (a) the size of differences among the scores in a distribution? (b) how consistently the participants behaved? (c) the size of our "average error" when we predict that participants obtained the mean?
3. (a) What is the range? (b) Why is it not the most accurate measure of variability? (c) When is it used as the sole measure of variability?
4. (a) What do both the variance and the standard deviation tell you about a distribution? (b) Which measure will you usually want to compute? Why?
5. (a) What is the mathematical definition of the variance? (b) Mathematically, how is a sample's variance related to its standard deviation and vice versa?

6. (a) What do S_X, s_X, and σ_X have in common in terms of what they communicate? (b) How do they differ in terms of their use?

7. Why are your estimates of the population variance and standard deviation always larger than the corresponding values that describe a sample from that population?

8. In an experiment, how does the size of S_X in each condition suggest the strength of the relationship?

9. (a) How do we determine the scores that mark the middle 68% of a sample? (b) How do we determine the scores that mark the middle 68% of a *known* population? (c) How do we *estimate* the scores that mark the middle 68% of an unknown population?

10. (a) What is the phrase used to convey that by knowing participants' X score in a relationship it helps us to more accurately predict their Y? (b) How do we describe it when our predictions are closer to participants' actual scores?

APPLICATION QUESTIONS

11. In a condition of an experiment, a researcher obtains the following creativity scores:

    3 2 1 0 7 4 8 6 6 4

    In terms of creativity, interpret the variability of these data using the following: (a) the range, (b) the variance, and (c) the standard deviation.

12. If you could test the entire population in question 11, what would you expect each of the following to be? (a) The typical, most common creativity score; (b) the variance; (c) the standard deviation; (d) the two scores between which about 68% of all creativity scores occur in this situation.

13. In Question 11: (a) What are the scores at $-1S_X$ and $+1S_X$ from the mean? (b) If N is 1000, how many people do you expect will score between 1.59 and 6.61? (c) How many people do you expect will score below 1.59?

14. As part of studying the relationship between mental and physical health, you obtain the following heart rates:

    73 72 67 74 78 84 79 71 76 76
    79 81 75 80 78 76 78

    In terms of differences in heart rates, interpret these data using the following: (a) the range, (b) the variance, and (c) the standard deviation.

15. If you could test the population in question 14, what would you expect each of the following to be? (a) The shape of the distribution; (b) the typical, most common rate; (c) the variance; (d) the standard deviation; (e) the two scores between which about 68% of all heart rates fall.

16. Foofy has a normal distribution of scores ranging from 2 to 9. (a) She computed the variance to be $-.06$. What should you conclude about this answer, and why? (b) She recomputes the standard deviation to be 18. What should you conclude, and why? (c) She recomputes the variance to be 1.36. What should you conclude, and why? (d) If she computed that $S_X = 0$ and $S_X^2 = 2$, what would you conclude?

17. From his statistics grades, Guchi has a $\overline{X}$ of 60 and $S_X = 20$. Pluto has a $\overline{X}$ of 60 and $S_X = 5$. (a) Who is the more inconsistent student, and why? (b) Who is more accurately described as a 60 student, and why? (c) For which student can you more accurately predict the next test score, and why? (d) Who is more likely to do either extremely well or extremely poorly on the next exam?

18. Indicate whether by knowing someone's score on the first variable, the relationship accounts for a large or small amount of the variance in the second variable. (a) For children ages 1 to 6, using age to predict height ; (b) for ages 30 to 40, using age to predict the driving ability of adults; (c) using a students' hours of studying to predict final exam grades; (d) using students' hair color to predict final exam grades.

19. Consider the results of this experiment:

Condition A	Condition B	Condition C
12	33	47
11	33	48
11	34	49
10	31	48

(a) What "measures" should you compute to summarize the experiment? (b) These are ratio scores. Compute the appropriate descriptive statistics and summarize the relationship in the sample data. (c) How consistent does it appear the participants were in each condition? (d) Does this relationship account for much of the variance in the scores?

20. Say that you conducted the experiment in question 19 on the entire population. (a) Summarize the relationship that you'd expect to observe. (b) How consistently do you expect participants to behave in each condition?

21. In two studies, the mean is 40 but in Study A, $S_X = 5$, and in Study B, $S_X = 10$. (a) What is the difference in the appearance of the distributions from these studies? (b) Where do you expect the majority of scores to fall in each study?

22. Consider these ratio scores from an experiment:

Condition 1	Condition 2	Condition 3
18	8	3
13	11	9
9	6	5

(a) What should you do to summarize the experiment? (b) Summarize the relationship in the sample data. (c) How consistent were participants in each condition?

23. Say that you conducted the experiment in question 22 on the entire population. (a) Summarize the relationship that you'd expect to observe. (b) How consistently do you expect participants to behave in each condition?

24. Comparing the results in questions 19 and 22, which experiment produced the stronger relationship? How do you know?

INTEGRATION QUESTIONS

25. What are the three major pieces of information we need in order to summarize the scores in any data? (Ch. 3, 4, 5)

26. What is the difference between what a measure of central tendency tells us and what a measure of variability tells us? (Chs. 4, 5)

27. What is a researcher communicating with each of the following statements? (a) "The line graph of the means was relatively flat, although the variability in each condition was quite large." (b) "For the sample of men ($M = 14$ and $SD = 3$) we conclude . . . " (c) "We expect that in the population the average score is 14 and the standard deviation is 3.5 . . . " (Chs. 4, 5)

28. For each of the following, identify the conditions of the independent variable, the dependent variable, their scales of measurement, which measure of central tendency and variability to compute and which scores you would use in the computations. (a) We test whether participants laugh longer (in seconds) to jokes told on a sunny or rainy day. (b) We test babies whose mothers were or were not recently divorced, measuring whether the babies lost weight, gained weight, or remained the same. (c) We compare a group of adult children of alcoholics to a group whose parents were not alcoholics. In each, we measure participants' income. (d) We count the number of creative ideas produced by participants who are paid either 5, 10, or 50 cents per idea. (e) We measure the number of words in the vocabulary of 2-year-olds as a function of whether they have 0, 1, 2, or 3 older siblings. (f) We compare people 5 years after they have graduated from either high school, a community college, or a four-year college. Considering all participants at once, we rank order their income. (Chs. 2, 4, 5)

29. For each experiment in question 28, indicate the type of graph you would create, and how you would label the X and Y axes. (Chs. 2, 4)

■ ■ ■ SUMMARY OF FORMULAS

1. The formula for the range is

   Range = highest score − lowest score

2. The computational formula for the sample variance is

$$S_X^2 = \frac{\Sigma X^2 - \dfrac{(\Sigma X)^2}{N}}{N}$$

3. The computational formula for the sample standard deviation is

$$S_X = \sqrt{\frac{\Sigma X^2 - \dfrac{(\Sigma X)^2}{N}}{N}}$$

4. The computational formula for estimating the population variance is

$$s_X^2 = \frac{\Sigma X^2 - \dfrac{(\Sigma X)^2}{N}}{N - 1}$$

5. The computational formula for estimating the population standard deviation is

$$s_X = \sqrt{\frac{\Sigma X^2 - \dfrac{(\Sigma X)^2}{N}}{N - 1}}$$

6

z-Scores and the Normal Curve Model

GETTING STARTED

To understand this chapter, recall the following:

- From Chapter 3, that relative frequency is the proportion of time that scores occur, and that it corresponds to the proportion of the area under the normal curve; that a percentile equals the percent of the area under the curve to the left of a score.

- From Chapter 4, that the larger a score's deviation, the farther into the tail of the distribution it lies and the lower is its simple frequency and relative frequency.

- From Chapter 5, that S_X and σ_X indicate the "average" deviation of scores around $\overline{X}$ and μ, respectively.

Your goals in this chapter are to learn

- What a *z-score* is and what it tells you about a raw score's relative standing.

- How the standard normal curve is used with *z*-scores to determine relative frequency, simple frequency, and percentile.

- The characteristics of the *sampling distribution of means* and what the *standard error of the mean* is.

- How computing *z*-scores for sample means is used to determine their relative frequency.

The techniques discussed in the preceding chapters for graphing, measuring central tendency, and measuring variability comprise the descriptive procedures used in most behavioral research. In this chapter, we'll combine these procedures to answer another question about data: How does any one particular score compare to the other scores in a sample or population? We answer this question by transforming raw scores into *z-scores*.

In the following sections, we discuss (1) the logic of *z*-scores and their simple computation, (2) how *z*-scores are used to describe individual scores, and (3) how *z*-scores are used to describe sample means.

NEW STATISTICAL NOTATION

Statistics often involve negative and positive numbers. Sometimes, however, we ignore a number's sign. The size of a number, regardless of its sign, is the *absolute value* of the number. When we do not ignore the sign, you'll encounter the symbol $\pm$, which means "plus or minus." Saying "± 1," means $+1$ or -1. Saying "the scores between ± 1," means all possible scores from -1, through 0, up to and including $+1$.

WHY IS IT IMPORTANT TO KNOW ABOUT z-SCORES?

Recall that we transform raw scores to make different variables comparable and to make scores within the same distribution easier to interpret. The "z-transformation" is the Rolls-Royce of transformations because with it we can compare and interpret scores from virtually *any* normal distribution of interval or ratio scores.

Why do we need to do this? Because researchers usually *don't* know how to interpret someone's raw score: Usually, we won't know whether, in nature, a score should be considered high or low, good, bad, or what. Instead, the best we can do is compare a score to the other scores in the distribution, describing the score's relative standing. **Relative standing** reflects the systematic evaluation of a score relative to the sample or population in which the score occurs. The way to calculate the relative standing of a score is to transform it into a z-score. As you'll see, with z-scores we can easily determine the underlying raw score's location in a distribution, its relative and simple frequency, and its percentile. All of this helps us to know whether the individual's raw score was *relatively* good, bad, or in-between.

UNDERSTANDING z-SCORES

To see how z-scores reflect relative standing, let's say that we conduct a study at Prunepit University in which we measure the "attractiveness" of a sample of men. The scores form the normal curve in Figure 6.1, with a $\overline{X} = 60$. Of these scores, we especially want to interpret those of three men: Slug, who scored 35; Binky, who scored 65; and Biff, who scored 90. Using the statistics you've learned, you already know how to do this. Let's review.

What would we say to Slug? "Bad news, Slug. Your score places you far below average in attractiveness. What's worse, down in the tail, the height of the curve above your score indicates a low *frequency,* so not many men received this low score. Also, the proportion of the area under the curve at your score is small, so the *relative frequency*— the proportion of all men receiving your score—is low. Finally, Slug, your *percentile* is low, so a small percentage scored below you while a large percentage scored above you. So Slug, scores such as yours are relatively infrequent, and few scores are lower than yours."

FIGURE 6.1

Frequency distribution of attractiveness scores at Prunepit U

Scores for three individuals are identified on the X axis.

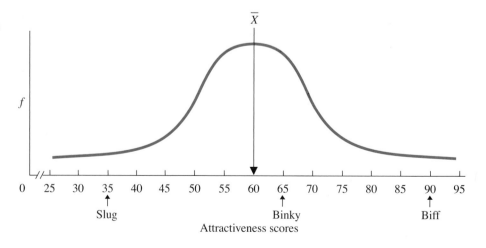

For Binky, there's good news and bad news. "The good news, Binky, is that your score of 65 is above the mean, which is also the median; you are better-looking than more than 50% of these men. The bad news is that you are not *far* above the mean. Also, the area under the curve at your score is relatively large, and thus the relative frequency of equally attractive men is large. What's worse, a relatively large part of the distribution has higher scores."

And then there's Biff. "Yes, Biff, your score of 90 places you well above average in attractiveness. In fact, as you have repeatedly told everyone, you are one of the most attractive men around. Also, the area under the curve at your score is quite small, so only a small proportion of men are equally attractive. Finally, the area under the curve to the left of your score is relatively large, so if we cared to figure it out, we'd find that you are at a very high percentile, with only a small percentage above you."

These descriptions are based on each man's relative standing because, considering our "parking lot" approach to the normal curve, we literally determined where each stands in the parking lot compared to everyone else. However, there are two problems with these descriptions. First, they were somewhat subjective and imprecise. Second, to get them we had to look at all scores in the distribution. However, recall that the point of statistics is to accurately *summarize* our data so that we don't need to look at every score. The way to obtain the above information, but more precisely and without looking at every score, is to compute each man's z-score.

Our description of each man above was based on how far above or below the mean his raw score appeared to be. To precisely determine this distance, our first calculation is to determine a score's deviation, which equals $X - \overline{X}$. For example, Biff's score of 90 deviates by +30 (because $90 - 60 = +30$.) Likewise, Slug's score of 35 deviates by $35 - 60 = -25$. Such deviations *sound* impressive, but are they? We have the same problem with deviations that we had with raw scores; we don't necessarily know whether a particular deviation should be considered large or small. However, looking at the distribution, we see that only a few scores deviate by such large amounts and *that* is what makes them impressive. Thus, a score is impressive if it is far from the mean, and "far" is determined by how often other scores deviate from the mean by that amount.

Therefore, to interpret a score's location, we need to compare its deviation to all deviations; we need a *standard* to compare to each *deviation;* we need the *standard deviation*! As you know, we think of the standard deviation as our way of computing the "average deviation." By comparing a score's deviation to the standard deviation, we can describe the location of the score in terms of this average deviation. Thus, say that, the sample standard deviation for the attractiveness scores is 10. Biff's deviation of +30 is equivalent to 3 standard deviations, so Biff's raw score is located 3 standard deviations above the mean. Thus, his raw score is impressive because it is three times as far above the mean as the "average" amount that scores were about the mean.

By transforming Biff's deviation into standard deviation units, we have computed his z-score. A **z-score** is the distance a raw score is from the mean when measured in standard deviations. The symbol for a z-score in a sample or population is z.

A z-score always has two components: (1) either a positive or negative sign which indicates whether the raw score is above or below the mean, and (2) the absolute value of the z-score which indicates how far the score lies from the mean when measured in standard deviations. So, Biff is *above* the mean by 3 standard deviations, so his z-score is +3. If he had been below the mean by this amount, he would have $z = -3$.

Thus, like any raw score, a z-score is a *location* on the distribution. However, the important part is that a z-score also simultaneously communicates its *distance* from the mean. By knowing where a score is relative to the mean, we know the score's relative standing within the distribution.

> **REMEMBER** A *z-score* describes a raw score's location in terms of how far above or below the mean it is when measured in standard deviations.

Computing z-Scores

Above, we computed Biff's z-score in two steps. First, we found the score's deviation by subtracting the mean from the raw score. Then we divided the score's deviation by the standard deviation. So,

> **The formula for transforming a raw score in a sample into a z-score is**
>
> $$z = \frac{X - \overline{X}}{S_X}$$

(This is both the definitional and the computational formula.) We are computing a z-score from a *sample* of scores, so we use the descriptive sample standard deviation, S_X (the formula with the final division by N). When starting from scratch with a sample of raw scores, first compute $\overline{X}$ and S_X and then substitute their values into the formula.

To find Biff's z-score, we substitute his raw score of 90, the $\overline{X}$ of 60, and the S_X of 10 into the formula:

$$z = \frac{X - \overline{X}}{S_X} = \frac{90 - 60}{10}$$

Find the deviation in the numerator first and always subtract $\overline{X}$ from X. This gives

$$z = \frac{+30}{10}$$

After dividing,

$$z = +3.00$$

Likewise, Binky's raw score was 65, so

$$z = \frac{X - \overline{X}}{S_X} = \frac{65 - 60}{10} = \frac{+5}{10} = +.50$$

Binky's raw score is literally one-half of one standard deviation above the mean.

And finally, Slug's raw score is 35, so

$$z = \frac{X - \overline{X}}{S_X} = \frac{35 - 60}{10} = \frac{-25}{10} = -2.50$$

Here, 35 minus 60 results in a deviation of *minus* 25, so his z-score is -2.50. Slug's raw score is 2.5 standard deviations *below* the mean.

Of course, a raw score that equals the mean produces a z-score of 0, because it is zero distance from itself. For example, an attractiveness score of 60 will produce an X and $\overline{X}$ that are the same number, so their difference is 0.

We can also compute a *z*-score for a score in a population, if we know the population mean (μ) and the true standard deviation of the population (σ_X). (We never compute *z*-scores using the estimated population standard deviation, s_X) The logic here is the same as in the previous formula, but using these symbols gives

---

The formula for transforming a raw score in a population into a *z*-score is

$$z = \frac{X - \mu}{\sigma_X}$$

Now the answer indicates how far the raw score lies from the population mean when measured using the population standard deviation. For example, say that in the population of attractiveness scores, $\mu = 60$ and $\sigma_X = 10$. Biff's raw score of 90 is again a $z = (90 - 60)/10 = +3.00$, but now this is his location in the *population* of scores.

Notice that the size of a *z*-score will depend on both the size of the raw score's deviation *and* the size of the standard deviation. Biff's deviation of $+30$ was impressive because the standard deviation was only 10. If the standard deviation had been 30, then Biff would have had $z = (90 - 60)/30 = +1.00$. Now he is not so impressive because his deviation equals the "average" deviation, indicating that his raw score is among the more common scores.

Computing a Raw Score When *z* Is Known

Sometimes we know a *z*-score and want to find the corresponding raw score. For example, say that another man, Bucky, scored $z = +1$. What is his raw score? With $\overline{X} = 60$ and $S_X = 10$, his *z*-score indicates that he is 1 standard deviation above the mean. In other words, he is 10 points above 60, so his raw score is 70. What did we just do? We multiplied his *z*-score times S_X and then added the mean.

The formula for transforming a *z*-score in a sample into a raw score is

$$X = (z)(S_X) + \overline{X}$$

For Bucky's *z*-score of $+1$

$$X = (+1)(10) + 60$$

so

$$X = +10 + 60$$

so

$$X = 70$$

To check this answer, compute the *z*-score for the raw score of 70. You should end up with the *z*-score you started with: $+1$.

Finally, say that Fuzzy has a $z = -1.30$ (with $\overline{X} = 60$ and $S_X = 10$). Then his raw score is

$$X = (-1.30)(10) + 60$$

so

$$X = -13 + 60$$

Adding a negative number is the same as subtracting its positive value, so

$$X = 47$$

Fuzzy has a raw score of 47.

The above logic is also used to transform a z-score into its corresponding raw score in the population. Using the symbols for the population gives

> **The formula for transforming a z-score in a population into a raw score is**
>
> $$X = (z)(\sigma_X) + \mu$$

Here, we multiply the z-score times the population standard deviation and then add μ. So, say that Fuzzy is from the population where $\mu = 60$ and $\sigma_X = 10$. For his z of -1.30 we have: $X = (-1.30)(10) + 60 = 47$. He has a raw score of 47 in the population.

After transforming a raw score or z-score, always check whether your answer makes sense. At the very least, raw scores smaller than the mean must produce negative z-scores, and raw scores larger than the mean must produce positive z-scores. When working with z-score, always pay close attention to the positive or negative sign! Further, as you'll see, we seldom obtain z-scores greater than $+3$ or less than -3. Although they are possible, be very skeptical if you compute such a z-score, and double-check your work.

A QUICK REVIEW

- A $+z$ indicates that the raw score is above the mean, a $-z$ that it is below the mean.

- The absolute value of z indicates the score's distance from the mean, measured in standard deviations.

MORE EXAMPLES

In a sample, $\overline{X} = 25$ and $S_X = 5$. To find z for $X = 32$:

$$z = \frac{X - \overline{X}}{S_X} = \frac{32 - 25}{5} = \frac{+7}{5} = +1.40$$

To find the raw score for $z = -.43$:

$$X = (z)(S_X) + \overline{X} = (-.43)(5) + 25$$
$$= -2.15 + 25 = 22.85$$

For Practice

With $\overline{X} = 50$ and $S_X = 10$,

1. What is z for $X = 44$?
2. What X produces $z = -1.30$?

With $\mu = 100$ and $\sigma_X = 16$,

3. What is the z for a score of 132?
4. What X produces $z = +1.4$?

Answers

1. $z = (44 - 50)/10 = -.60$
2. $X = (-1.30)(10) + 50 = 37$
3. $z = (132 - 100)/16 = +2.00$
4. $X = (+1.4)(16) + 100 = 122.4$

INTERPRETING z-SCORES USING THE z-DISTRIBUTION

The reason that z-scores are so useful is that they directly communicate the relative standing of a raw score. The way to see this is to first envision any sample or population as a z-distribution. A **z-distribution** is the distribution produced by transforming all raw scores in the data into z-scores. For example, say that our attractiveness scores produce the z-distribution shown in Figure 6.2. The X axis is also labeled using the original raw scores to show that by creating a z-distribution, we only change the way that we identify each score. Saying that Biff has a z of +3 is merely another way to say that he has a raw score of 90. He is still at the same point on the distribution, so Biff's z of +3 has the same frequency, relative frequency, and percentile as his raw score of 90.

By envisioning such a z-distribution, you can see how z-scores form a standard way to communicate relative standing. The z-score of 0 indicates that the raw score equals the mean. A "+" indicates that the z-score (and raw score) is above and graphed to the right of the mean. Positive z-scores become increasingly larger as we proceed farther to the right. Larger positive z-scores (and their corresponding raw scores) occur less frequently. Conversely, a "−" indicates that the z-score (and raw score) is below and graphed to the left of the mean. Negative z-scores become increasingly larger as we proceed farther to the left. Larger negative z-scores (and their corresponding raw scores) occur less frequently. However, as shown, most of the z-scores are between +3 and −3.

Do not be misled by negative z-scores: A raw score that is farther below the mean is a *smaller* raw score, but it produces a negative z-score whose absolute value is *larger.* Thus, for example, a z-score of −2 corresponds to a *lower* raw score than a z-score of −1. Also, a negative z-score is not automatically a bad score. For some variables, the goal is to have as low a raw score as possible (for example, errors on a test). With these variables, larger negative z-scores are best.

FIGURE 6.2

z-distribution of attractiveness scores at Prunepit U

The labels on the X axis show first the raw scores and then the z-scores.

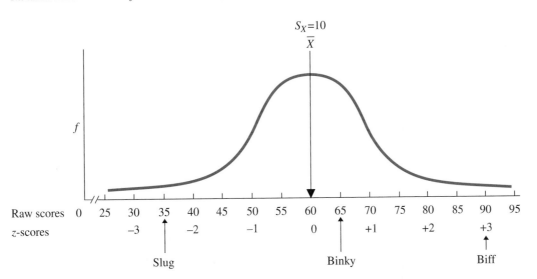

> **REMEMBER** On a normal distribution, the larger the z-score, whether positive or negative, the farther the raw score is from the mean and the less frequently the raw score and the z-score occur.

Figure 6.2 illustrates three important characteristics of any z-distribution.

1. *A z-distribution has the same shape as the raw score distribution.* Only when the underlying raw score distribution is normal will its z-distribution be normal.

2. *The mean of any z-distribution is 0.* Whatever the mean of the raw scores is, it transforms into a z-score of 0.

3. *The standard deviation of any z-distribution is 1.* Whether the standard deviation in the raw scores is 10 or 100, it is still one standard deviation, which transforms into an amount in z-scores of 1.

Now you can see why z-scores are so useful: *All* normal z-distributions are similar, so a particular z-score will convey the *same* information in every distribution. Therefore, the way to interpret the individual scores from *any* normally distributed variable is to envision a z-distribution similar to Figure 6.2. Then, for example, if we know that z is 0, we know that the corresponding raw score is at the mean (and at the median and mode). Also, recall from Chapter 5 that the two raw scores that are $\pm 1S_X$ from the mean delineate the middle 68% of the curve. Now you know that being $\pm 1S_X$ from the mean produces z-scores of ± 1. Therefore, approximately 68% of the scores on *any* normal distribution will be between the z-scores at ± 1. Likewise, any other z-score will always be in the *same* relative location, so if z is +.50, then, like Binky's, the raw score is slightly above the mean and slightly above the 50th percentile, where scores still have a high simple and relative frequency. But, if z is +3, then, like Biff's, the raw score is one of the highest possible scores in the upper tail of the distribution, having a low frequency, a low relative frequency and a very high percentile. And so on.

> **REMEMBER** The way to interpret the raw scores in any sample or population is to determine their relative standing by envisioning them as a z-distribution.

As you'll see in the following sections, in addition to describing relative standing as above, z distributions have two additional uses: (1) comparing scores from different distributions and (2) computing the relative frequency of scores.

USING z-SCORES TO COMPARE DIFFERENT VARIABLES

An important use of z-scores is when we compare scores from different variables. Here's a new example. Say that Cleo received a grade of 38 on her statistics quiz and a grade of 45 on her English paper. These scores reflect different kinds of tasks, so it's like comparing apples to oranges. The solution is to transform the raw scores from each class into z-scores. Then we can compare Cleo's relative standing in English to her relative standing in statistics, so we are no longer comparing apples and oranges.

Note: The z-transformation equates or standardizes different distributions, so z-scores are often referred to as **standard scores**.

Say that for the statistics quiz the $\overline{X}$ was 30 and the S_X was 5. Cleo's grade of 38 becomes $z = +1.6$. For the English paper, the $\overline{X}$ was 40 and the S_X was 10, so her 45 becomes $z = +.5$. A z-score of $+1.6$ is farther above any mean than a z-score of

$z = +.5$. Thus, Cleo did relatively better in statistics because she is farther above the statistics mean than she is above the English mean.

Another student, Attila, obtained raw scores that produced $z = -2$ in statistics and $z = -1$ in English. In which class did he do better? His *z*-score of -1 in English is relatively better, because it is less distance below the mean.

We can also see these results in Figure 6.3, in which the *z*-distributions from both classes are plotted on one set of axes. (The greater height of the English distribution reflects a larger *N*, with a higher *f* for each score.) Notice that the raw scores from each class are spaced differently along the *X* axis, because the classes have different standard deviations. However, *z*-scores always increment by *one* standard deviation, whether it equals 5 points in statistics or 10 points in English. Therefore, the spacing of the *z*-scores is the same for the two classes and so they are comparable. Now we can see that Cleo scored better in statistics than in English but that Attila scored better in English than in statistics.

> **REMEMBER** To compare raw scores from two different variables, transform the scores into *z*-scores.

USING *z*-SCORES TO DETERMINE THE RELATIVE FREQUENCY OF RAW SCORES

A third important use of *z*-scores is for computing the relative frequency of raw scores. Recall that relative frequency is the proportion of time that a score occurs, and that relative frequency can be computed using *the proportion of the total area under the curve*. We can use the *z*-distribution to determine relative frequency because, as we've seen, when raw scores produce the same *z*-score they are at the same location on their distributions. By being at the same location, a *z*-score delineates the same proportion of the curve, cutting off the same "slice" of the distribution every time. Thus, *the relative frequency at particular z-scores will be the same on all normal z-distributions*.

For example, 50% of the raw scores on a normal curve are to the left of the mean, and scores to the left of the mean produce negative *z*-scores. In other words, the

FIGURE 6.3

Comparison of distributions for statistics and English grades, plotted on the same set of axes

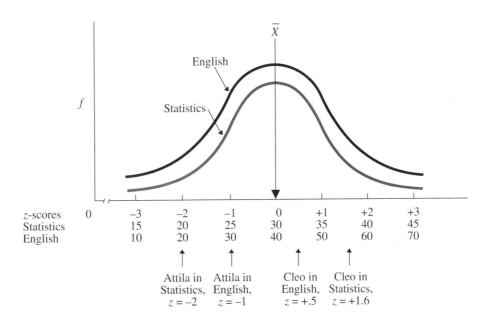

negative z-scores make up 50% of any distribution. Thus, the negative z-scores back in Figure 6.3 constitute 50% of their respective distributions, which corresponds to a relative frequency of .50. On *any* normal z-distribution, the relative frequency of the negative z-scores is .50.

Having determined the relative frequency of the z-scores, we work backwards to identify the corresponding raw scores. In the statistics distribution in Figure 6.3, students having negative z-scores have raw scores ranging between 15 and 30, so the relative frequency of scores between 15 and 30 is .50. In the English distribution, students having negative z-scores have raw scores between 10 and 40, so the relative frequency of these scores is .50.

Similarly, approximately 68% of the scores always fall between the z-scores of $+1$ and -1. Thus, in Figure 6.3, students having z-scores between ± 1 constitute approximately 68% of each distribution. Working backwards to the raw scores we see that statistics grades between 25 and 35 constitute approximately 68% of the statistics distribution, and English grades between 30 and 50 constitute approximately 68% of the English distribution.

In the same way, we can determine the relative frequencies for any set of scores. Thus, in a normal distribution of IQ scores (whatever the $\overline{X}$ and S_X may be), we know that those IQs producing negative z-scores have a relative frequency of .50, and that about 68% of all IQ scores will fall between the scores at z-scores of ± 1. The same will be true for a distribution of running speeds, a distribution of personality test scores, or for *any* normal distribution.

We can also use z-scores to determine the relative frequency of scores in any other portion of a distribution. To do so, we employ the *standard normal curve*.

The Standard Normal Curve

Because all normal z-distributions are similar, we don't need to draw a different z-distribution for every set of raw scores. Instead, we envision one standard curve that, in fact, is called the standard normal curve. The **standard normal curve** is a perfect normal z-distribution that serves as our model of any approximately normal z-distribution. It is used in this way: Most data produce only an approximately normal distribution, producing a roughly normal z-distribution. However, to simplify things, we operate as if the z-distribution always fits one, perfect normal curve, which is the standard normal curve. We use this curve to first determine the relative frequency of particular z-scores. Then, as we did above, we work backwards to determine the relative frequency of the corresponding raw scores. This is the relative frequency we would *expect,* if our data formed a perfect normal distribution. Usually, this provides a reasonably accurate description of our data, although how accurate we are depends on how closely the data conform to the true normal curve. Therefore, the standard normal curve is most accurate when (1) we have a large sample (or population) of (2) interval or ratio scores that (3) come close to forming a normal distribution.

The first step is to find the relative frequency of the z-scores and for that we look at the area under the standard normal curve. Statisticians have already determined the proportion of the area under various parts of the normal curve, as shown in Figure 6.4. The numbers above the X axis indicate the proportion of the total area between the z-scores. The numbers below the X axis indicate the proportion of the total area between the mean and the z-score. (Don't worry, you won't need to memorize them.)

Each proportion is also the relative frequency of the z-scores—and raw scores—located in that part of the curve. For example, between a z of 0 and a z of $+1$ is .3413

FIGURE 6.4

Proportions of total area under the standard normal curve

The curve is symmetrical: 50% of the scores fall below the mean, and 50% fall above the mean.

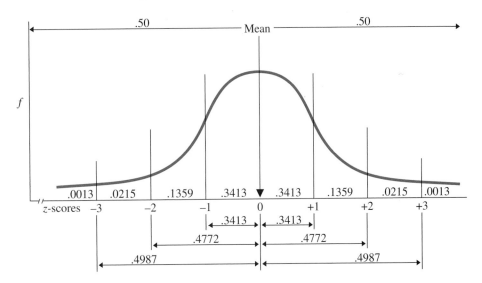

(or 34.13%) of the area, so, as we've seen, about 34% of the scores are here. Or, scores between a *z* of +1 and a *z* of +2 occur .1359 of the time, and this added to .3413, gives a total of 0.4772 of the scores between the mean and *z* = +2. Finally, with .3413 of the scores between the mean and *z* = −1 and .3413 of the scores between the mean and *z* = +1, a total of .6826 of the area under the curve is between ±1. (See, approximately 68% of the distribution really *is* between ±1S_X from the mean!)

We can also add together nonadjacent portions of the curve. For example, out in the lower tail beyond *z* = −2 is .0228 of the area under the curve (because .0215 + .0013 = .0228). Likewise, in the upper tail beyond *z* = +2 is also .0228 of the area under the curve. Thus, a total of .0456 (or 4.56%) of all scores fall in the tails beyond *z* = ±2.

Finally, notice that *z*-scores beyond +3 or beyond −3 occur only .0013 of the time, for a total of .0026 (.26 of 1%!), which is why the range of *z* is essentially between ±3. This also explains why Chapter 5 said that the S_X should be about one-sixth of the range of the raw scores. The range is roughly between *z*s of ±3, a distance of six times the standard deviation. If the range is six times the standard deviation, then the standard deviation is one-sixth of the range.

> **REMEMBER** For any approximately normal distribution, transform the raw scores to *z*-scores and use the *standard normal curve* to find the relative frequency of the scores.

We usually apply the standard normal curve in one of four ways. *First, we can find relative frequency.* Most often we begin with a particular raw score in mind and then compute its *z*-score (using our original *z*-score formula). For example, in our original attractiveness scores, say that another man, Cubby, has a raw score of 80, which, with $\overline{X}$ = 60 and S_X = 10, is a *z* of +2. We can envision Cubby's location as in Figure 6.5. We might first ask what proportion of scores are expected to fall between the mean and Cubby's score. We see that .4772 of the total area falls between the mean and *z* = +2. Therefore, we expect .4772 or 47.72%, of the attractiveness scores to fall between the mean and Cubby's score of 80. (Conversely, .0228 of the area under the curve—and scores—are above his score.)

FIGURE 6.5

Location of Cubby's score on the z-distribution of attractiveness scores

Cubby's raw score of 80 is a z-score of +2.

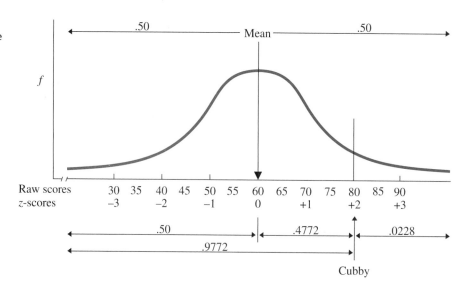

Second, we can find simple frequency. We might ask how *many* people scored between the mean and Cubby's score. Then we would convert the above relative frequency to simple frequency by multiplying the N of the sample times the relative frequency. Say that our N was 1000. If we expect .4772 of the scores to fall here, then $(.4772)(1000) = 477.2$, so we expect about 477 people to have scores between the mean and Cubby's score.

Third, we can find a raw score's percentile. Recall that a percentile is the percent of all scores below—graphed to the left of—a score. After computing the z-score for a raw score, first see if it is above or below the mean. As in Figure 6.5, the mean is also the median (the 50th percentile). A positive z-score is above the mean, so Cubby's z-score of +2 is above the 50th percentile. In fact, his score is above the .4772 that fall between the mean and his score. Thus, adding the .50 of the scores below the mean to the .4772 of the scores between the mean and his score gives a total of .9772 of all scores below Cubby's. We usually round off percentile to a whole number, so Cubby's raw score of 80 is at the 98th percentile. Conversely, anyone scoring above the raw score of 80 would be in about the top 2% of scores.

On the other hand, say that Elvis obtained an attractiveness score of 40, producing a z-score of -2. As in Figure 6.6, a total of .0228 (2.28%) of the distribution is *below* (to the left of) Elvis's score. With rounding, Elvis ranks at the 2nd percentile.

Fourth, we can find the raw score at a specific percentile. We can also work in the opposite direction to find the raw score located at a particular percentile (or relative frequency). Say that we had started by asking what attractiveness score is at the 2nd percentile (or we had asked below what raw score is .0228 of the distribution?). As in Figure 6.6, a z-score of -2 is at the 2nd percentile, or below $z = -2$ is .0228 of the distribution. Then to find the raw score at this z, we use a formula for transforming a z-score into a raw score. For example, we saw that in a sample, $X = (z)(S_X) + \overline{X}$. We'd find that the corresponding attractiveness score is 40.

Using the z-Table

So far our examples have involved whole-number z-scores, although with real data a z-score may contain decimals. However, fractions of z-scores do *not* result in proportional divisions of the previous areas. Instead, find the proportion of the total area

FIGURE 6.6

Location of Elvis's score on the z-distribution of attractiveness scores

Elvis is at approximately the 2nd percentile.

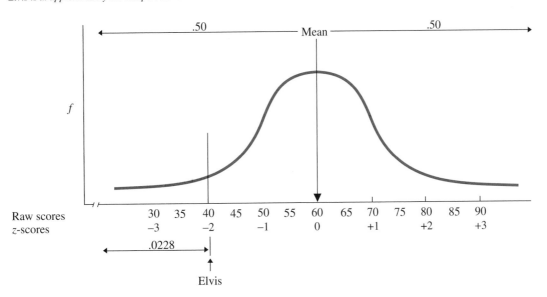

under the standard normal curve for any two-decimal z-score by looking in Table 1 of Appendix C. A portion of this "z-table" is reproduced in Table 6.1.

Say that you seek the area under the curve that is above or below a $z = +1.63$. First, locate the z in column A, labeled "z." Then column B, labeled "Area Between the Mean and z," contains the proportion of the area under the curve between the mean and the z identified in column A. Thus, .4484 of the curve is between the mean and the z of $+1.63$. Because this z is positive, we place this area between the mean and the z on the *right-hand side* of the distribution, as shown in Figure 6.7. Column C, labeled "Area Beyond z in the Tail," contains the proportion of the area under the curve that is in the tail beyond the z-score. Thus, .0516 of the curve is in the right-hand tail of the distribution beyond the z of $+1.63$ (also shown in Figure 6.7).

Notice that the z-table contains no positive or negative signs. *You* must decide whether z is positive or negative, based on the problem you're working. Thus, if our z had been -1.63, columns B and C would provide the respective areas on the left-hand side of Figure 6.7. (As shown here, always sketch the normal curve, locate the mean, and identify the portions of the curve you're working with. This greatly simplifies the problem.)

If you get confused when using the z-table, look at the normal distribution at the top of the table, like in Table 6.1. The different shaded portions indicate the area under the curve described in each column.

To work in the opposite direction to find the z-score that corresponds to a particular proportion, read the columns in the reverse order. First, find the proportion in column B or C, that you seek, and then identify the corresponding z-score in

TABLE 6.1

Sample Portion of the z-Table

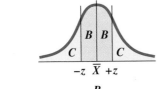

A	B	C
	Area Between	**Area**
z	**Mean and z**	**Beyond z in Tail**
1.60	.4452	.0548
1.61	.4463	.0537
1.62	.4474	.0526
1.63	.4484	.0516
1.64	.4495	.0505
1.65	.4505	.0495

FIGURE 6.7

Distribution showing the area under the curve for $z = -1.63$ and $z = +1.63$

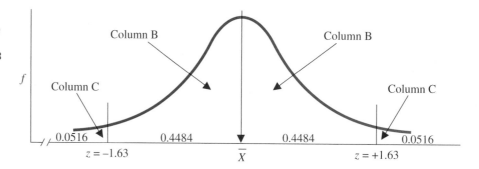

column A. For example, say that you seek the z-score corresponding to 44.84% of the curve between the mean and z. Find .4484 in column B and then, in column A, the z-score is 1.63.

Sometimes, you will need a proportion that is not given in the table, or you'll need the proportion corresponding to a three-decimal z-score. In such cases, round to the nearest value in the z-table or, to compute the precise value, perform "linear interpolation" (described in Appendix A.2).

Use the information from the z-tables as we have done previously. For example, say that we want to examine Bucky's raw score, which transforms into the *positive* z-score of $+1.63$, located at the right-hand side back in Figure 6.7. If we seek the proportion of scores above his score, then from column C we expect that .0516 of the scores are above this score. If we seek the relative frequency of scores between his score and the mean, from column B we expect that .4484 of the scores are between the mean up to his raw score. Then we can also compute simple frequency or percentile as discussed previously. Or, if we began by asking what raw score demarcates .4484 or .0516 of the curve, we would first find these proportions in column B or C, respectively, then find the z-score of $+1.63$ in column A, and then use our formula to transform the z-score to the corresponding raw score.

Table 6.2 summarizes these procedures.

TABLE 6.2

Summary of Steps When Using the z-Tables

If You Seek	First, You Should	Then You
Relative frequency of scores between $\overline{X}$ and X	transform X to z	find its area in column B*
Relative frequency of scores beyond X in tail	transform X to z	find its area in column C*
X that marks a given *rel. f* between X and $\overline{X}$	Find *rel. f* in column B	transform its z to X
X that marks a given *rel. f* beyond X in tail	find *rel. f* in column C	transform its z to X
Percentile of an X above $\overline{X}$	transform X to z	find its area in column B and add .50
Percentile of an X below $\overline{X}$	transform X to z	find its area in column C

*To find the simple frequency of the scores, multiply *rel. f* times N.

A QUICK REVIEW

- To find the relative frequency of scores above or below a raw score, transform it into a z-score. From the z-tables, find the proportion of the area under the curve above or below that z.
- To find the raw score at a specified relative frequency, find the proportion in the z-tables and transform the corresponding z into its raw score.

MORE EXAMPLES

With $\overline{X} = 40$ and $S_X = 4$,

To find the relative frequency of scores above 45: $z = (X - \overline{X})/S_X = (45 - 40)/4 = +1.25$. Saying "above" indicates in the upper tail, so from column C the relative frequency is .1056.

To find the percentile of the score of 41.5: $z = (41.5 - 40)/4 = +0.38$. Between this positive z and $\overline{X}$ in column B is .1480. This score is at the 65th percentile because $.1480 + .50 = .6480 = .65$.

To find the proportion below $z = -0.38$: "Below" indicates the lower tail, so from column C is .3520.

To find the score above the mean with 15% of the scores between it and the mean (or the score at the 65th percentile): From column B, the proportion closest to .15 is .1517 so $z = +0.39$. Then $X = (+0.39)(4.0) + 40 = 41.56$.

For Practice

For a sample: $\overline{X} = 65$, $S_X = 12$, and $N = 1000$.

1. What is the relative frequency of scores below 59?
2. What is the percentile of 75?
3. How many scores are between the mean and 70?
4. What raw score delineates the top 3%?

Answers

1. $z = (59 - 65)/12 = -.50$; "below" is the lower tail, so from column C is .3085.
2. $z = (75 - 65)/12 = +.83$; between z and the mean, from column B, is .2967. Then $.2967 + .50 = .7967 = $ 80th percentile.
3. $z = (70 - 65)/12 = +.42$; from column B is .1628; $(.1628)(1000)$ gives about 163 scores.
4. The "top" is the upper tail, so from column C the closest to .03 is .0301, with $z = +1.88$; so $X = (+1.88)(12) + 65 = 87.56$.

STATISTICS IN PUBLISHED RESEARCH: USING z-SCORES

A common use of z-scores is with diagnostic psychological tests such as intelligence or personality tests. Often, however, test results are also shared with people who do not understand z-scores; imagine someone learning that he or she has a *negative* personality score! Therefore, z-scores are often transformed to more user-friendly scores. A famous example of this is the Scholastic Aptitude Test (SAT). To eliminate negative scores and decimals, sub-test scores are transformed so that the mean is about 500 and the standard deviation is about 100. We've seen that z-scores are usually between ±3, which is why SAT scores are limited to between 200 and 800 on each part. (You may hear of higher scores but this occurs by adding together the subtest scores.)

The normal curve and z-scores are also used when researchers create a "statistical definition" of a psychological or sociological attribute. When debating such issues as what a genius is or how to define "abnormal," researchers often rely on relative standing. For example, the term "genius" might be defined as scoring above a z of +2 on an intelligence test. We've seen that only about 2% of any distribution is above this score, so we have defined genius as being in the top 2% on the intelligence test. Or, "abnormal" might be defined as having a z-score below −2 on a personality inventory. Such scores are statistically abnormal because they are very infrequent, extremely low scores.

Finally, when instructors "curve grades" it means they are assigning grades based on relative standing, and the curve they usually use is the normal curve and *z*-scores. If the instructor defines A students as the top 2%, then students with *z*-scores greater than +2 receive As. If B students are the next 13%, then students having *z*-scores between +1 and +2 receive Bs, and so on.

USING *z*-SCORES TO DESCRIBE SAMPLE MEANS

Now we will discuss how to compute a *z*-score for a sample mean. This procedure is very important because all inferential statistics involve computing something like a *z*-score for our sample data. We'll elaborate on this procedure in later chapters but, for now, simply understand how to compute a *z*-score for a sample mean and then apply the standard normal curve model.

To see how the procedure works, say that we give a part of the SAT to a sample of 25 students at Prunepit U. Their mean score is 520. Nationally, the mean of *individual* SAT scores is 500 (and σ_X is 100), so it appears that at least some Prunepit students scored relatively high, pulling the mean to 520. But how do we interpret the performance of the sample as a whole? The problem is the same as when we examined individual raw scores: Without a frame of reference, we don't know whether a particular sample mean is high, low, or in-between.

The solution is to evaluate a sample mean by computing its *z*-score. Previously, a *z*-score compared a particular raw score to the other scores that occur in this situation. Now we'll compare our sample mean to the other sample means that occur in this situation. Therefore, the first step is to take a small detour and create a distribution showing these other means. This distribution is called the *sampling distribution of means.*

The Sampling Distribution of Means

If the national average SAT score is 500, then, in other words, the μ of the population of SAT scores is 500. Because we randomly selected a sample of 25 students and obtained their SAT scores, we essentially drew a sample of 25 scores from this population. To evaluate our sample mean, we first create a distribution showing all other possible means we might have obtained.

One way to do this would be to record all SAT scores from the population on slips of paper and deposit them into a very large hat. We could then hire a statistician to sample this population. So that we can see *all* possible sample means that might occur the statistician would sample the population an infinite number of times: She would randomly select a sample with the same size *N* as ours (25), compute the sample mean, replace the scores in the hat, draw another 25 scores, compute the mean, and so on. (She'd get very bored, so the pay would have to be good.)

Even though the μ of individual SAT scores is 500, our "bored statistician" would not obtain a sample mean equal to 500 every time. There are a variety of SAT scores in the population and sometimes the luck of the draw would produce an unrepresentative mix of them: sometimes a sample would contain too many high scores and not enough low scores compared to the population, so the sample mean would be above 500 to some degree. At other times, a sample would contain too many low scores and not enough high scores, so the mean would be below 500 to some degree. Therefore, over the long run, the statistician would obtain many different sample means. To see them all, she would create a frequency polygon, which is called the sampling distribution of means. The **sampling distribution of means** is the frequency distribution of all possible sample means that occur when an infinite number of samples of the same size N are randomly selected from one raw score population.

Our SAT sampling distribution of means is shown in Figure 6.8. This is similar to a distribution of raw scores, except that here each score on the X axis is a sample mean. (Still think of the distribution as a parking lot full of people, except that now each person is the captain of a sample, having the sample's mean score.)

The sampling distribution is the *population* of sample means so its mean is symbolized by μ and it stands for the average sample mean. (Yes, that's right, it's the mean of the means!) Here the μ of the sampling distribution equals 500, which was also the μ of the raw scores. To the right of μ are the sample means the statistician obtained that are greater than 500, and to the left of μ are the sample means that were less than 500. However, the sampling distribution forms a normal distribution. This is because most scores in the population are close to 500, so most of the time the statistician will get a sample containing scores that are close to 500, so the sample mean will be close to 500. Less frequently, the statistician will obtain a strange sample containing mainly scores that are farther below or above 500, producing means that are farther below or above 500. Once in a great while, some very unusual samples will be drawn, resulting in sample means that deviate greatly from 500. However, because the individual SAT scores are balanced around 500, over the long run, the sample means created from those scores will also be balanced around 500, so the average mean (μ) will equal 500.

The story about the bored statistician is useful because it helps you to understand what a sampling distribution is. Of course, in reality, we cannot "infinitely" sample a population. However, we know that the sampling distribution would look like Figure 6.8 because of the central limit theorem. The **central limit theorem** is a statistical principle that defines the mean, the standard deviation, and the shape of a sampling distribution. From the central limit theorem, we know that the sampling distribution of means always (1) forms an approximately normal distribution, (2) has a μ equal to the μ of the underlying raw score population from which the sampling distribution was created, and (3), as you'll see shortly, has a standard deviation that is mathematically related to the standard deviation of the raw score population.

The importance of the central limit theorem is that with it we can describe the sampling distribution from any variable *without* actually having to *infinitely* sample the population of raw scores. All we need to know is (1) that the raw score population

FIGURE 6.8

Sampling distribution of random sample means of SAT scores

The X axis is labeled to show the different values of $\overline{X}$ we obtain when we sample a population where the mean in 500.

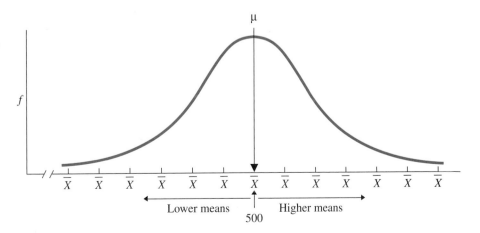

forms a normal distribution of ratio or interval scores (so that computing the mean is appropriate) and (2) what the μ and σ_X of the raw score population is. Then we'll know the important characteristics of the sampling distribution of means.

> **REMEMBER** The central limit theorem allows us to envision the *sampling distribution of means,* which shows all means that occur through exhaustive sampling of a raw score population.

Why do we want to see the sampling distribution? Remember that we took a small detour, but the original problem was to evaluate our Prunepit mean of 520. Once we envision the distribution back in Figure 6.8, we have a model of the frequency distribution of all sample means that occur when measuring SAT scores. Then we can use it to determine the relative standing of our sample mean. To do so, we simply determine *where* a mean of 520 falls on the X axis of the sampling distribution in Figure 6.8 and then interpret the curve accordingly. If 520 lies close to 500, then it is a frequent, common mean when sampling SAT scores (the bored statistician frequently obtained this result). But if 520 lies toward the tail of the distribution, far from 500, then it is a more infrequent and unusual sample mean (the statistician seldom found such a mean).

The sampling distribution is a normal distribution, and you already know how to determine the location of any "score" on a normal distribution: We use—you guessed it—z-scores. That is, we determine how far the sample mean is from the mean of the sampling distribution when measured using the standard deviation of the distribution. This will tell us the sample mean's relative standing among all possible means that occur in this situation.

To calculate the z-score for a sample mean, we need one more piece of information: the standard deviation of the sampling distribution.

The Standard Error of the Mean

The standard deviation of the sampling distribution of means is called the **standard error of the mean**. (The term *standard deviation* was already taken.) Like a standard deviation, the standard error of the mean can be thought of as the "average" amount that the sample means deviate from the μ of the sampling distribution. That is, in some sampling distributions, the sample means may be very different from one another and, "on average," deviate greatly from the average sample mean. In other distributions, the $\overline{X}$s may be very similar and deviate little from μ.

For the moment, we'll discuss the *true* standard error of the mean, as if we had actually computed it using the entire sampling distribution. Its symbol is $\sigma_{\overline{X}}$. The σ indicates that we are describing a population, but the subscript $\overline{X}$ indicates that we are describing a population of sample means—what we call the sampling distribution of means. The central limit theorem tells us that $\sigma_{\overline{X}}$ can be found using the following formula:

The formula for the true standard error of the mean is

$$\sigma_{\overline{X}} = \frac{\sigma_X}{\sqrt{N}}$$

Notice that the formula involves σ_X, the true standard deviation of the underlying raw score population, and N, our sample size.

The size of $\sigma_{\overline{X}}$ depends first on the size of σ_X. This is because with more variable raw scores the statistician often gets a very different set of scores from one sample to the next, so the sample means will be very different (and $\sigma_{\overline{X}}$ will be larger). But, if the raw scores are not so variable, then different samples will tend to contain the same scores, and so the means will be similar (and $\sigma_{\overline{X}}$ will be smaller). Second, the size of $\sigma_{\overline{X}}$ depends on the size of N. With a very small N (say 2), it is easy for each sample to be different from the next, so the sample means will differ (and $\sigma_{\overline{X}}$ will be larger). However, with a large N, each sample will be more like the population, so all sample means will be closer to the population mean (and $\sigma_{\overline{X}}$ will be smaller).

To compute $\sigma_{\overline{X}}$ for the sampling distribution of SAT means, we know that σ_X is 100 and our N is 25. Thus, using the above formula we have

$$\sigma_{\overline{X}} = \frac{\sigma_X}{\sqrt{N}} = \frac{100}{\sqrt{25}}$$

The square root of 25 is 5, so

$$\sigma_{\overline{X}} = \frac{100}{5}$$

and thus

$$\sigma_{\overline{X}} = 20$$

A $\sigma_{\overline{X}}$ of 20 indicates that in our SAT sampling distribution, our individual sample means differ from the μ of 500 by something like an "average" of 20 points.

Notice that although the individual SAT scores differ by an "average" (σ_X) of 100, their sample means differ by only an "average" ($\sigma_{\overline{X}}$) of 20. This is because the bored statistician will often encounter a variety of high and low scores in each sample, but they will usually balance out to produce means at or close to 500. Therefore, the sample means will not be as spread out around 500 as the individual scores are. Likewise, every sampling distribution is less spread out than the underlying raw score population used to create it.

Now, at last, we can calculate a *z*-score for our sample mean.

Computing a *z*-Score for a Sample Mean

We use this formula to compute a *z*-score for a sample mean:

> **The formula for the transforming a sample mean into a *z*-score is**
>
> $$z = \frac{\overline{X} - \mu}{\sigma_{\overline{X}}}$$

In the formula, $\overline{X}$ stands for our sample mean, μ stands for the mean of the sampling distribution (which equals the mean of the underlying raw score population) and $\sigma_{\overline{X}}$ stands for the standard error of the mean. As we did with individual scores, this formula measures how far a score is from the mean of a distribution, measured using the standard deviation. Here, however, we are measuring how far the sample mean score is from the mean of the sampling distribution, measured using the "standard deviation" called the standard error.

For the sample from Prunepit U, $\overline{X} = 520$, $\mu = 500$, and $\sigma_{\overline{X}} = 20$, so

$$z = \frac{\overline{X} - \mu}{\sigma_{\overline{X}}} = \frac{520 - 500}{20} = \frac{+20}{20} = +1.00$$

Thus, a sample mean of 520 has a z-score of $+1$ on the SAT sampling distribution of means that occurs when N is 25.

Another Example Combining All of the Above   Say that over at Podunk U, a sample of 25 SAT scores produced a mean of 460. With $\mu = 500$ and $\sigma_X = 100$, we'd again envision the SAT sampling distribution we saw back in Figure 6.8. To find the z-score, first, compute the standard error of the mean ($\sigma_{\overline{X}}$):

$$\sigma_{\overline{X}} = \frac{\sigma_X}{\sqrt{N}} = \frac{100}{\sqrt{25}} = \frac{100}{5} = 20$$

Then find z:

$$z = \frac{\overline{X} - \mu}{\sigma_{\overline{X}}} = \frac{460 - 500}{20} = \frac{-40}{20} = -2$$

The Podunk sample has a z-score of -2 on the sampling distribution of SAT means.

Describing the Relative Frequency of Sample Means

Everything we said previously about a z-score for an individual score applies to a z-score for a sample mean. Thus, because our original Prunepit mean has a z-score of $+1$, we know that it is above the μ of the sampling distribution by an amount equal to the "average" amount that sample means deviate above μ. Therefore, we know that, although they were not stellar, our Prunepit students did outperform a substantial proportion of comparable samples. Our sample from Podunk U, however, has a z-score of -2, so its mean is very low compared to other means that occur in this situation.

And here's the nifty part: Because the sampling distribution of means always forms at least an approximately normal distribution, if we transformed *all* of the sample means into z-scores, we would have a roughly normal z-distribution. Recall that the *standard normal curve* is our model of *any* roughly normal z-distribution. Therefore, we can apply the standard normal curve to a sampling distribution.

> **REMEMBER**   The standard normal curve model and the z-table can be used with any sampling distribution, as well as with any raw score distribution.

Figure 6.9 shows the standard normal curve applied to our SAT sampling distribution of means. Once again, larger positive or negative z-scores indicate that we are farther into the tails of the distribution, and the corresponding proportions are the same proportions we used to describe raw scores. Therefore, as we did then, we can use the standard normal curve (and the z-table) to determine the proportion of the area under any part of the curve. This proportion is also the expected relative frequency of the corresponding sample means in that part of the sampling distribution.

FIGURE 6.9

Proportions of the standard normal curve applied to the sampling distribution of SAT means

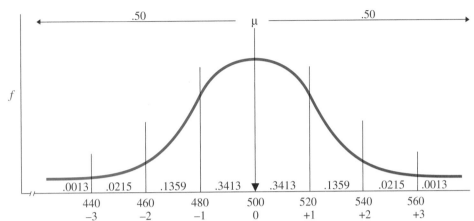

For example, the Prunepit sample has a z of $+1$. As in Figure 6.9 (and from column B of the z-table), .3413 of all scores fall between the mean and z of $+1$ on any normal distribution. Therefore, .3413 of all SAT sample means are expected to fall between the μ and the sample mean at a z of $+1$. Because here the μ is 500 and a z of $+1$ is at the sample mean of 520, we can also say that .3413 of all SAT sample means are expected to be between 500 and 520 (when N is 25).

However, say that we asked about sample means above our sample mean. As in column C of the z-table, above a z of $+1$ is .1587 of the distribution. Therefore, we expect that .1587 of SAT sample means will be above 520. Similarly, the Podunk U sample mean of 460 has a z of -2. As in column B of the z-table, a total of .4772 of a distribution falls between the mean and this z-score. Therefore, we expect .4772 of SAT means to be between 500 and 460, with (as in column C) only .0228 of the means below 460.

We can use this same procedure to describe sample means from any normally distributed variable.

Summary of Describing a Sample Mean with a *z*-Score

To describe a sample mean from any raw score population, follow these steps:

1. Envision the sampling distribution of means (or better yet, draw it) as a normal distribution with a μ equal to the μ of the underlying raw score population.

2. Locate the sample mean on the sampling distribution, by computing its z-score.

   a. Using the σ_X of the raw score population and your sample N, compute the standard error of the mean: $\sigma_{\bar{X}} = \sigma_X/\sqrt{N}$

   b. Compute z by finding how far your $\bar{X}$ is from the μ of the sampling distribution, measured in standard error units: $z = (\bar{X} - \mu)/\sigma_{\bar{X}}$

3. Use the z-table to determine the relative frequency of scores above or below this z-score, which is the relative frequency of sample means above or below your mean.

A QUICK REVIEW

- To describe a sample mean, compute its z-score and use the z-table to determine the relative frequency of sample means above or below it.

MORE EXAMPLES

On a test, $\mu = 100$, $\sigma_X = 16$, and our $N = 64$. What proportion of sample means will be above $\overline{X} = 103$?

First, compute the standard error of the mean ($\sigma_{\overline{X}}$):

$$\sigma_{\overline{X}} = \frac{\sigma_X}{\sqrt{N}} = \frac{16}{\sqrt{64}} = \frac{16}{8} = 2$$

Next, compute z:

$$z = \frac{\overline{X} - \mu}{\sigma_{\overline{X}}} = \frac{103 - 100}{2} = \frac{+3}{2} = +1.5$$

Finally, examine the z-table: The area *above* this z is the upper tail of the distribution, so from column C is 0.0668. This is the proportion of sample means expected to be above a mean of 103.

For Practice

A population of raw scores has $\mu = 75$ and $\sigma_X = 22$; our $N = 100$ and $\overline{X} = 80$

1. The μ of the sampling distribution here equals _____.
2. The symbol for the standard error of the mean is _____, and here it equals _____.
3. The z-score for a sample mean of 80 is _____.
4. How often will sample means between 75 and 80 occur in this situation?

Answers

1. 75
2. $\sigma_{\overline{X}}$; $22/\sqrt{100} = 2.20$
3. $z = (80 - 75)/2.20 = +2.27$
4. From Column B: .4884 of the Time

PUTTING IT ALL TOGETHER

The most important concept for you to understand is that any normal distribution of scores can be described using the standard normal curve model and z-scores. To paraphrase a famous saying, a normal distribution is a normal distribution is a normal distribution. Any normal distribution contains the same proportions of the total area under the curve between z-scores. Therefore, whenever you are discussing individual scores or sample means, *think z-scores* and use the previous procedures.

You will find it *very* beneficial to sketch the normal curve when working on z-score problems. For raw scores, label where the mean is and about where the specified raw score or z-score is, and identify the area that you seek. At the least, this will instantly tell you whether you seek information from column B or column C in the z-table. For sample means, first draw and identify the raw score population that the bored statistician would sample, and then draw and label the above parts of the sampling distribution.

By the way, what was Biff's percentile?

Using the SPSS Appendix As described in Appendix B.3, SPSS will simultaneously transform an entire sample of scores into z-scores. We enter in the raw scores and SPSS produces a set of z-scores in a column next to where we typed our raw scores.

CHAPTER SUMMARY

1. The *relative standing* of a score reflects a systematic evaluation of the score relative to a sample or population. A *z-score* indicates a score's relative standing by indicating the distance the score is from the mean when measured in standard deviations.

2. A positive z-score indicates that the raw score is above the mean; a negative z-score indicates that the raw score is below the mean. The larger the absolute value of z, the farther the raw score is from the mean, so the less frequently the z-score and raw score occur.

3. z-scores are used to describe the relative standing of raw scores, to compare raw scores from different variables, and to determine the relative frequency of raw scores.

4. A *z-distribution* is produced by transforming all raw scores in a distribution into z-scores.

5. The *standard normal curve* is a perfect normal z-distribution that is our model of the z-distribution that results from data that are approximately normally distributed, interval or ratio scores.

6. The *sampling distribution of means* is the frequency distribution of all possible sample means that occur when an infinite number of samples of the same size N are randomly selected from one raw score population.

7. The *central limit theorem* shows that in a sampling distribution of means (a) the distribution will be approximately normal, (b) the mean of the sampling distribution will equal the mean of the underlying raw score population, and (c) the variability of the sample means is related to the variability of the raw scores.

8. The true *standard error of the mean* ($\sigma_{\bar{X}}$) is the standard deviation of the sampling distribution of means.

9. The location of a sample mean on the sampling distribution of means can be described by calculating a z-score. Then the standard normal curve model can be applied to determine the expected relative frequency of the sample means that are above or below the z-score.

10. Biff's percentile was 99.87.

KEY TERMS

$\pm$ z $\sigma_{\bar{X}}$
central limit theorem *125*
relative standing *110*
sampling distribution of means *125*
standard error of the mean *126*

standard normal curve *118*
standard score *116*
z-distribution *115*
z-score *111*

REVIEW QUESTIONS

(Answers for odd-numbered questions are in Appendix D.)

1. (a) What does a z-score indicate? (b) Why are z-scores important?
2. On what two factors does the size of a z-score depend?
3. What is a z-distribution?
4. What are the three general uses of z-scores with individual raw scores?

5. (a) What is the standard normal curve? (b) How is it applied to a set of data? (c) What three criteria should be met for it to give an accurate description of the scores in a sample?
6. (a) What is a sampling distribution of means? (b) When is it used? (c) Why is it useful?
7. What three things does the central limit theorem tell us about the sampling distribution of means? (b) Why is the central limit theorem so useful?
8. What does the standard error of the mean indicate?
9. (a) What are the steps for using the standard normal curve to find a raw score's relative frequency or percentile? (b) What are the steps for finding the raw score that cuts off a specified relative frequency or percentile? (c) What are the steps for finding a sample mean's relative frequency?

APPLICATION QUESTIONS

10. In an English class last semester, Foofy earned a 76 ($\overline{X} = 85$, $S_X = 10$). Her friend, Bubbles, in a different class, earned a 60 ($\overline{X} = 50$, $S_X = 4$). Should Foofy be bragging about how much better she did? Why?
11. Poindexter received a 55 on a biology test ($\overline{X} = 50$) and a 45 on a philosophy test ($\overline{X} = 50$). He is considering whether to ask his two professors to curve the grades using z-scores. (a) Does he want the S_X to be large or small in biology? Why? (b) Does he want the S_X to be large or small in philosophy? Why?
12. Foofy computes z-scores for a set of normally distributed exam scores. She obtains a z-score of -3.96 for 8 out of 20 of the students. What do you conclude?
13. For the data,

    9   5   10   7   9   10   11   8   12   7   6   9

    (a) Compute the z-score for the raw score of 10. (b) Compute the z-score for the raw score of 6.
14. For the data in question 13, find the raw scores that correspond to the following: (a) $z = +1.22$; (b) $z = -0.48$.
15. Which z-score in each of the following pairs corresponds to the lower raw score? (a) $z = +1.0$ or $z = +2.3$; (b) $z = -2.8$ or $z = -1.7$; (c) $z = -.70$ or $z = -.20$; (d) $z = 0.0$ or $z = -2.0$.
16. For each pair in question 15, which z-score has the higher frequency?
17. In a normal distribution, what proportion of all scores would fall into each of the following areas? (a) Between the mean and $z = +1.89$; (b) below $z = -2.30$; (c) between $z = -1.25$ and $z = +2.75$; (d) above $z = +1.96$ and below -1.96.
18. For a distribution in which $\overline{X} = 100$, $S_X = 16$, and $N = 500$: (a) What is the relative frequency of scores between 76 and the mean? (b) How many participants are expected to score between 76 and the mean? (c) What is the percentile of someone scoring 76? (d) How many subjects are expected to score above 76?
19. Poindexter may be classified as having a math dysfunction—and not have to take statistics—if he scores below the 25th percentile on a diagnostic test. The μ of the test is 75 ($\sigma_X = 10$). Approximately what raw score is the cutoff score for him to avoid taking statistics?
20. For an IQ test, we know the population $\mu = 100$ and the $\sigma_X = 16$. We are interested in creating the sampling distribution when $N = 64$. (a) What does that

sampling distribution of means show? (b) What is the shape of the distribution of IQ means and the mean of the distribution? (c) Calculate $\sigma_{\bar{X}}$ for this distribution. (d) What is your answer in part (c) called, and what does it indicate? (e) What is the relative frequency of sample means above 101.5?

21. Someone has two job offers and must decide which to accept. The job in City A pays $47,000 and the average cost of living there is $65,000, with a standard deviation of $15,000. The job in City B pays $70,000, but the average cost of living there is $85,000, with a standard deviation of $20,000. Assuming salaries are normally distributed, which is the better job offer? Why?

22. Suppose you own shares of a company's stock, the price of which has risen so that, over the past ten trading days, its mean selling price is $14.89. Over the years, the mean price of the stock has been $10.43 ($\sigma_X = \$5.60$.) You wonder if the mean selling price over the next ten days can be expected to go higher. Should you wait to sell, or should you sell now?

23. A researcher develops a test for selecting intellectually gifted children, with a μ of 56 and a σ_X of 8. (a) What percentage of children are expected to score below 60? (b) What percentage of scores will be above 54? (c) A gifted child is defined as being in the top 20%. What is the minimum test score needed to qualify as gifted?

24. Using the test in question 23, you measure 64 children, obtaining a $\bar{X}$ of 57.28. Slug says that because this $\bar{X}$ is so close to the μ of 56, this sample could hardly be considered gifted. (a) Perform the appropriate statistical procedure to determine whether he is correct. (b) In what percentage of the top scores is this sample mean?

25. A researcher reports that a sample mean produced a relatively large positive or negative z score. (a) What does this indicate about that mean's relative frequency? (b) What graph did the researcher examine to make this conclusion? (c) To what was the researcher comparing his mean?

INTEGRATION QUESTIONS

26. What does a relatively small standard deviation indicate about the scores in a sample? (b) What does this indicate about how accurately the mean summarizes the scores. (c) What will this do to the z-score for someone who is relatively far from the mean? Why? (Chs. 5, 6)

27. (a) With what type of data is it appropriate to compute the mean and standard deviation? (b) With what type of data is it appropriate to compute z-scores? (Chs. 4, 5, 6)

28. (a) What is the difference between a proportion and a percent? (b) What are the mathematical steps for finding a specified percent of N? (Ch. 1)

29. (a) We find that .40 of a sample of 500 people score above 60. How many people scored above 60? (b) In statistical terms, what are we asking about a score when we ask how many people obtained the score? (c) We find that 35 people out of 50 failed an exam. What proportion of the class failed? (d) What percentage of the class failed? (Chs. 1, 3)

30. What is the difference between the normal distributions we've seen in previous chapters and (a) a z-distribution and (b) a sampling distribution of means? (Chs. 3, 6)

■ ■ SUMMARY OF FORMULAS

1. The formula for transforming a raw score in a sample into a z-score is

$$z = \frac{X - \overline{X}}{S_X}$$

2. The formula for transforming a z-score in a sample into a raw score is

$$X = (z)(S_X) + \overline{X}$$

3. The formula for transforming a raw score in a population into a z-score is

$$z = \frac{X - \mu}{\sigma_X}$$

4. The formula for transforming a z-score in a population into a raw score is

$$X = (z)(\sigma_X) + \mu$$

5. The formula for transforming a sample mean into a z-score on the sampling distribution of means is

$$z = \frac{\overline{X} - \mu}{\sigma_{\overline{X}}}$$

6. The formula for the true standard error of the mean is

$$\sigma_{\overline{X}} = \frac{\sigma_X}{\sqrt{N}}$$

7

The Correlation Coefficient

GETTING STARTED

To understand this chapter, recall the following:

- From Chapter 2, that in a relationship, particular Y scores tend to occur with a particular X, a more consistent relationship is "stronger," and we can use someone's X to predict what his/her Y will be.

- From Chapter 5, that greater variability indicates a greater variety of scores is present and so greater variability produces a weaker relationship. Also that the phrase "accounting for variance" refers to accurately predicting Y scores.

Your goals in this chapter are to learn

- The logic of correlational research and how it is interpreted.

- How to read and interpret a *scatterplot* and a *regression line*.

- How to identify the *type* and *strength* of a relationship.

- How to interpret a *correlation coefficient*.

- When to use the *Pearson r* and the *Spearman r_S*.

- The logic of inferring a population correlation based on a sample correlation.

Recall that in research we want to not only demonstrate a relationship but also describe and summarize the relationship. The one remaining type of descriptive statistic for us to discuss is used to summarize relationships, and it is called the *correlation coefficient*. In the following sections, we'll consider when these statistics are used and what they tell us. Then we'll see how to compute the two most common versions of the correlation coefficient. First, though, a few more symbols.

NEW STATISTICAL NOTATION

Correlational analysis requires scores from two variables. Then, X stands for the scores on one variable, and Y stands for the scores on the other variable. Usually each pair of X–Y scores is from the same participant. If not, there must be a rational system for pairing the scores (for example, pairing the scores of roommates). Obviously we must have the same number of X and Y scores.

We use the same conventions for Y that we've previously used for X. Thus, ΣY is the sum of the Y scores, ΣY^2 is the sum of the squared Y scores, and $(\Sigma Y)^2$ is the squared sum of the Y scores.

You will also encounter three other notations. First, $(\Sigma X)(\Sigma Y)$ indicates to first find the sum of the Xs and the sum of the Ys and then multiply the two sums together. Second, ΣXY, called the sum of the cross products, says to first multiply each X score in a pair times its corresponding Y score and then sum all of the resulting products.

REMEMBER $(\Sigma X)(\Sigma Y)$ says to multiply the sum of X times the sum of Y. ΣXY says to multiply each X times its paired Y and then sum the products.

Finally, D stands for the numerical *difference* between the X and Y scores in a pair, which you find by subtracting one from the other.

Now, on to the correlation coefficient.

WHY IS IT IMPORTANT TO KNOW ABOUT CORRELATION COEFFICIENTS?

Recall that a relationship is present when, as the X scores increase, the corresponding Y scores change in a consistent fashion. Whenever we find a relationship, we then want to know its characteristics: What pattern is formed, how consistently do the scores change together, and what direction do the scores change? The best—and easiest—way to answer these questions is to compute a correlation coefficient. The **correlation coefficient** is the descriptive statistic that, in a single number, summarizes and describes the important characteristics of a relationship. The correlation coefficient *quantifies* the pattern in a relationship, examining *all X–Y* pairs at once. No other statistic does this. Thus, the correlation coefficient is important because it simplifies a complex relationship involving many scores into one, easily interpreted statistic. Therefore, in any research where a relationship is found, always calculate the appropriate correlation coefficient.

As a starting point, the correlation coefficients discussed in this chapter are most commonly associated with correlational research.

UNDERSTANDING CORRELATIONAL RESEARCH

Recall that a common research design is the correlational study. The term *correlation* is synonymous with *relationship,* so in a correlational design we examine the relationship between variables. (Think of *correlation* as meaning the shared, or "co," relationship between the variables.) The relationship can involve scores from virtually any variable, regardless of how we obtain them. Often we use a questionnaire or observe participants, but we may also measure scores using any of the methods used in experiments.

Recall that correlational studies differ from experiments in terms of *how* we demonstrate the relationship. For example, say that we hypothesize that as people drink more coffee they become more nervous. To demonstrate this in an experiment, we might assign some people to a condition in which they drink 1 cup of coffee, assign others to a 2-cup condition and assign still others to a 3-cup condition. Then we would measure participants' nervousness and see if more nervousness is related to more coffee. Notice that, by creating the conditions, we (the researchers) determine each participant's X score because we decide whether their "score" will be 1, 2, or 3 cups on the coffee variable.

In a correlational design, however, we do *not* manipulate any variables, so we do not determine participants' X scores. Rather, the scores on both variables reflect an amount or category of a variable that a participant has *already* experienced. Therefore, we simply measure the two variables and describe the relationship that is present. Thus, we might ask participants the amount of coffee they have consumed today and measure how nervous they are.

Recognize that computing a correlation coefficient does not create a correlational *design:* It is the absence of manipulation that creates the design. In fact, in later chapters we will compute correlation coefficients in experiments. However, correlation coefficients are most often used as the primary descriptive statistic in correlational research, and you must be careful when interpreting the results of such a design.

Drawing Conclusions from Correlational Research

People often mistakenly think that a correlation automatically indicates causality. However, recall from Chapter 2 that the existence of a relationship does not necessarily indicate that changes in *X cause* the changes in *Y*. A relationship—a *correlation*—can exist, even though one variable does not cause or influence the other. Two requirements must be met to confidently conclude that *X* causes *Y*.

First, *X must occur before Y*. However, in correlational research, we do not always know which factor occurred first. For example, if we simply measure the coffee drinking and nervousness of some people after the fact, it may be that participants who were already more nervous *then* tended to drink more coffee. Therefore, maybe greater nervousness actually caused greater coffee consumption. In any correlational study, it is possible that *Y* causes *X*.

Second, *X must be the only variable that can influence Y*. But, in correlational research, we do little to control or eliminate other potentially causal variables. For example, in the coffee study, some participants may have had less sleep than others the night before testing. Perhaps the lack of sleep caused those people to be more nervous *and* to drink more coffee. In any correlational study, some other variable may cause both *X* and *Y* to change. (Researchers often refer to this as "the third variable problem.")

Thus, a correlation by itself does not indicate causality. You must also consider the research method used to demonstrate the relationship. In experiments we apply the independent variable *first,* and we control other potential causal variables, so experiments provide better evidence for identifying the causes of a behavior.

Unfortunately, this issue is often lost in the popular media, so be skeptical the next time some one uses *correlation* and *cause* together. The problem is that people often ignore that a relationship may be a meaningless coincidence. For example, here's a relationship: As the number of toilets in a neighborhood increases, the number of crimes committed in that neighborhood also increases. Should we conclude that indoor plumbing causes crime? Of course not! Crime tends to occur more frequently in the crowded neighborhoods of large cities. Coincidentally, there are more indoor toilets in such neighborhoods.

The problem is that it is easy to be trapped by more mysterious relationships. Here's a serious example: A particular neurological disease occurs more often in the colder, northern areas of the United States than in the warmer, southern areas. Do colder temperatures cause this disease? Maybe. But, for all the reasons given above, the mere existence of this relationship is not evidence of causality. The north also has fewer sunny days, burns more heating oil, and differs from the south in many other ways. One of these variables might be the cause, while coincidentally, colder temperatures are also present.

Thus, a correlational study is not used to infer a causal relationship. It is possible that changes in *X* might cause changes in *Y*, but we will have no convincing evidence of this. Instead, correlational research is used to simply describe how nature relates the variables, without identifying the cause.

> *REMEMBER* We should not infer causality from correlational designs, because X may cause Y, Y may cause X, or a third variable may cause both X and Y.

Distinguishing Characteristics of Correlational Analysis

There are four major differences between how we handle data in a correlational analysis versus in an experiment. First, back in our coffee experiment, we would examine the *mean* nervousness score (Y) for each condition of the amount of coffee consumed (X). With correlational data, however, we typically have a large range of different X scores: People would probably report many amounts of coffee beyond only 1, 2, or 3 cups. Comparing the mean nervousness scores for many groups would be very difficult. Therefore, in correlational procedures, we do not compute a mean Y score at each X. Instead, the correlation coefficient summarizes the *entire* relationship at once.

A second difference is that, because we examine all pairs of X–Y scores, correlational procedures involve *one* sample: *In correlational designs, N always stands for the number of pairs of scores in the data.*

Third, we will not use the terms *independent* and *dependent variable* with a correlational study (although some researchers argue that these terms are acceptable here). Part of our reason is that either variable may be called X or Y. How do we decide? Recall that in a relationship the X scores are the "given" scores. Thus, if we ask, "For a given amount of coffee, what are the nervousness scores?" then amount of coffee is X, and nervousness is Y. Conversely, if we ask, "For a given nervousness score, what is the amount of coffee consumed?" then nervousness is X, and amount of coffee is Y. Further, recall that, in a relationship, particular Y scores naturally occur at a particular X. Therefore, if we know someone's X, we can predict his or her corresponding Y. The procedures for doing this are described in the next chapter, where the X variable is called the *predictor variable,* and the Y variable is called the *criterion variable.* As you'll see, researchers used correlational techniques to identify X variables that are "good predictors" of Y scores.

Finally, as in the next section, we graph correlational data by creating a *scatterplot*.

Plotting Correlational Data: The Scatterplot

A **scatterplot** is a graph that shows the location of each data point formed by a pair of X–Y scores. Figure 7.1 contains the scores and resulting scatterplot showing the relationship between coffee consumption and nervousness. It shows that people drinking 1 cup have nervousness scores around 1 or 2, but those drinking 2 cups have higher nervousness scores around 2 or 3, and so on. Thus, we see that one batch of data points (and Y scores) tend to occur with one X, and a different batch of data points (and thus different Y scores) are at a different X.

Real research typically involves a larger N and the data points will not form such a clear pattern. In fact, notice the strange data point produced by $X = 3$ and $Y = 9$. A data point that is relatively far from the majority of data points in the scatterplot is referred to as an **outlier**—it *lies out* of the general pattern. Why an outlier occurs is usually a mystery to the researcher.

Notice that the scatterplot does summarize the data somewhat. In the table, two people had scores of 1 on coffee consumption and nervousness, but the scatterplot shows

FIGURE 7.1

Scatterplot showing nervousness as a function of coffee consumption

Each data point is created using a participant's coffee consumption as the X score and nervousness as the Y score.

Cups of Coffee: X	Nervousness Scores: Y
1	1
1	1
1	2
2	2
2	3
3	4
3	5
3	9
4	5
4	6
5	8
5	9
6	9
6	10

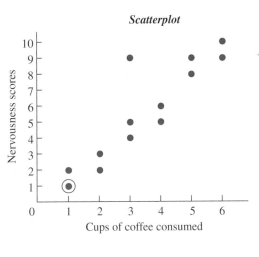

Scatterplot

one data point for them. (As shown, some researchers circle such a data point to indicate that points are on top of each other.) In a larger study, many participants with a particular *X* score may obtain the same *Y,* so the number of data points may be considerably smaller than the number of pairs of raw scores.

When you conduct a correlational study, always begin the analysis by creating a scatterplot. The scatterplot allows you to see the relationship that is present and to map out the best way to summarize it. (Also, you can see whether the extreme scores from any outliers may be biasing your computations.) Published reports of correlational studies, however, often do *not* show the scatterplot. Instead, from the description provided, you should *envision* the scatterplot, and then you will understand the relationship formed by the data. You get the description of the scatterplot from the correlation coefficient. A correlation coefficient communicates two important characteristics of a relationship: the *type* of relationship that is present and the *strength* of the relationship.

> *REMEMBER*   A correlation coefficient is a statistic that communicates the *type* and *strength* of relationship.

TYPES OF RELATIONSHIPS

The **type of relationship** that is present in a set of data is the overall direction in which the *Y* scores change as the *X* scores change. There are two general types of relationships: *linear* and *nonlinear* relationships.

Linear Relationships

The term *linear* means "straight line," and a linear relationship forms a pattern that follows *one* straight line. This is because in a **linear relationship**, as the *X* scores increase, the *Y* scores tend to change in only *one* direction. To understand this, first look at the data points in the scatterplot on the left in Figure 7.2. This shows the relationship between the hours that students study and their test performance. A scatterplot that slants in only one direction like this indicates a linear relationship: it indicates that as students study longer, their grades tend only to increase. The scatterplot on the right in Figure 7.2 shows the relationship between the hours that students watch television and their test scores. It too is a linear relationship, showing that, as students watch more television, their test scores tend only to decrease.

For our discussions, we will summarize a scatterplot by drawing a line around its outer edges. (Published research will *not* show this.) As in Figure 7.2, a scatterplot that forms an *ellipse* that slants in one direction indicates a linear relationship: by slanting, it indicates that the *Y* scores are changing as the *X* scores increase; slanting in one direction indicates it is linear relationship.

Further, as shown, we can also summarize a relationship by drawing a line through the scatterplot. (Published research *will* show this.) The line is called the *regression line*. While the correlation coefficient is the *statistic* that summarizes a relationship, the regression line is the *line* on a graph that summarizes the relationship. We will discuss the procedures for drawing the line in the next chapter, but for now, the **regression line** summarizes a relationship by passing through the center of the scatterplot. That is, although all data points are not *on* the line, the distance that some are above the line equals the distance that others are below it, so the regression line passes through the center of the scatterplot. Therefore, think of the regression line as showing the linear—straight line—relationship hidden in the data: It is how we visually summarize the general pattern in the relationship.

> **REMEMBER** The *regression line* summarizes a relationship by passing through the center of the scatterplot.

The difference between the scatterplots in Figure 7.2 illustrates the two subtypes of linear relationships that occur, depending on the *direction* in which the *Y* scores change. The study–test relationship is a positive relationship. In a **positive linear relationship,** as the *X* scores increase, the *Y* scores also tend to increase. Thus, low *X* scores are paired with low *Y* scores, and high *X* scores are paired with high *Y* scores. Any relationship that fits the pattern "the more *X*, the more *Y*" is a positive linear relationship.

FIGURE 7.2

Scatterplots showing positive and negative linear relationships

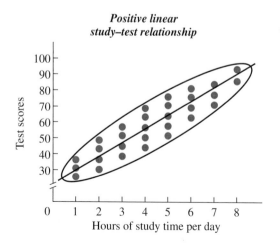
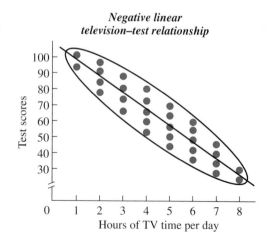

(Remember *positive* by remembering that as the *X* scores increase, the *Y* scores change in the direction away from zero, toward higher *positive* scores.)

On the other hand, the television–test relationship is a negative relationship. In a **negative linear relationship,** as the *X* scores increase, the *Y* scores tend to decrease. Low *X* scores are paired with high *Y* scores, and high *X* scores are paired with low *Y* scores. Any relationship that fits the pattern "the more *X,* the less *Y*" is a negative linear relationship. (Remember *negative* by remembering that as the *X* scores increase, the *Y* scores change toward zero, heading toward *negative* scores.)

Note: The term *negative* does not mean that there is something wrong with a relationship. It merely indicates the direction in which the *Y* scores change as the *X* scores increase.

Nonlinear Relationships

If a relationship is not linear, then it is nonlinear. *Nonlinear* means that the data cannot be summarized by *one straight* line. Another name for a nonlinear relationship is a curvilinear relationship. In a **nonlinear, or curvilinear, relationship,** as the *X* scores change, the *Y* scores do not tend to *only* increase or *only* decrease: At some point, the *Y* scores change their direction of change.

Nonlinear relationships come in many different shapes, but Figure 7.3 shows two common ones. The scatterplot on the left shows the relationship between a person's age and the amount of time required to move from one place to another. Very young children move slowly, but as age increases, movement time decreases. Beyond a certain age, however, the time scores change direction and begin to increase. (Such a relationship is called *U-shaped.*) The scatterplot on the right shows the relationship between the number of alcoholic drinks consumed and feeling well. At first, people tend to feel better as they drink, but beyond a certain point, drinking more makes them feel progressively worse. (Such a scatterplot reflects an *inverted U-shaped relationship.*) Curvilinear relationships may be more complex than those above, producing a wavy pattern that repeatedly changes direction. To be nonlinear, however, a scatterplot does not need to be *curved.* A scatterplot might be best summarized by straight regression

FIGURE 7.3
Scatterplots showing nonlinear relationships

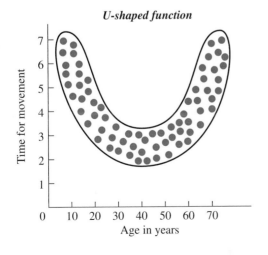
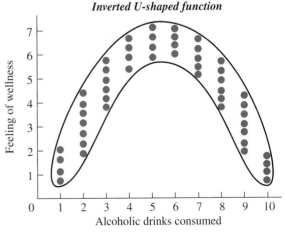

lines that form a V, an inverted V, or any other shape. It would still be nonlinear as long as it does not fit *one* straight line.

Notice that the terms *linear* and *nonlinear* are also used to describe relationships found in experiments. If, as the amount of the independent variable (X) increases, the dependent scores (Y) also increase, then it is a positive linear relationship. If the dependent scores decrease as the independent variable increases, it is a negative relationship. And if, as the independent variable increases, the dependent scores change their direction of change, it is a nonlinear relationship.

How the Correlation Coefficient Describes the Type of Relationship

Remember that the correlation coefficient is a number that we compute using our data. We communicate that the data form a linear relationship first because we compute a *linear* correlation coefficient—a coefficient whose formula is designed to summarize a linear relationship. (Behavioral research focuses primarily on linear relationships, so we'll discuss only them.) How do you know whether data form a linear relationship? If the scatterplot generally follows a straight line, then linear correlation is appropriate. Also, sometimes, researchers describe the extent to which a nonlinear relationship has a linear component and somewhat fits a straight line. Here, too, linear correlation is appropriate. However, do not try to summarize a nonlinear relationship by computing a linear correlation coefficient. This is like putting a round peg into a square hole: The data won't fit a straight line very well, and the correlation coefficient won't accurately describe the relationship.

The correlation coefficient communicates not only that we have a linear relationship but also whether it is positive or negative. Sometimes our computations will produce a negative number (with a minus sign), indicating that we have a negative relationship. Other data will produce a positive number (and we place a plus sign with it), indicating that we have a positive relationship. Then, with a positive correlation coefficient we envision a scatterplot that slants upward as the X scores increase. With a negative coefficient we envision a scatterplot that slants downward as the X scores increase.

The other characteristic of a relationship communicated by the correlation coefficient is the *strength* of the relationship.

STRENGTH OF THE RELATIONSHIP

Recall that the **strength of a relationship** is the extent to which one value of Y is consistently paired with one and only one value of X. The size of the coefficient that we compute (ignoring its sign) indicates the strength of the relationship. The largest value you can obtain is 1, indicating a perfectly consistent relationship. (*You cannot beat perfection so you can never have a coefficient greater than 1!*) The smallest possible value is 0, indicating that no relationship is present. Thus, when we include the positive or negative sign, the correlation coefficient may be any value between -1 and $+1$. The *larger* the absolute value of the coefficient, the *stronger* the relationship. In other words, the closer the coefficient is to ± 1, the more consistently one value of Y is paired with one and only one value of X.

> **REMEMBER** A *correlation coefficient* has two components: The sign indicates either a positive or a negative linear relationship; the absolute value indicates the strength of the relationship.

Correlation coefficients do not, however, measure in units of "consistency." Thus, if one correlation coefficient is $+.40$ and another is $+.80$, we *cannot* conclude that one relationship is twice as consistent as the other. Instead, we evaluate any correlation coefficient by comparing it to the extreme values of 0 and ± 1. The starting point is a perfect relationship.

Perfect Association

A correlation coefficient of $+1$ or -1 describes a perfectly consistent linear relationship. Figure 7.4 shows an example of each. (In this and the following figures, first look at the scores to see how they pair up. Then look at the scatterplot. Other data having the same correlation coefficient produce similar patterns, so we envision similar scatterplots.)

Here are four interrelated ways to think about what a correlation coefficient tells you about the relationship. First, it indicates *the relative degree of consistency*. A coefficient of ± 1 indicates that *everyone* who obtains a particular X score obtains one and only one value of Y. Every time X changes, the Y scores all change to one new value.

Second, and conversely, *the coefficient communicates the variability in the Y scores paired with an X*. When the coefficient is ± 1, only one Y is paired with an X, so there is no variability—no differences—among the Y scores paired with each X.

Third, *the coefficient communicates how closely the scatterplot fits the regression line*. Because a coefficient equal to ± 1 indicates zero variability or *spread* in the Y

FIGURE 7.4

Data and scatterplots reflecting perfect positive and negative correlations

Perfect positive coefficient = +1.0

X	Y
1	2
1	2
1	2
3	5
3	5
3	5
5	8
5	8
5	8

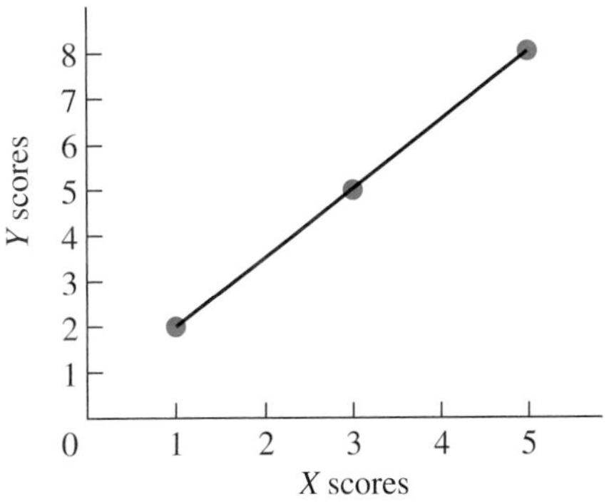

Perfect negative coefficient = −1.0

X	Y
1	8
1	8
1	8
3	5
3	5
3	5
5	2
5	2
5	2

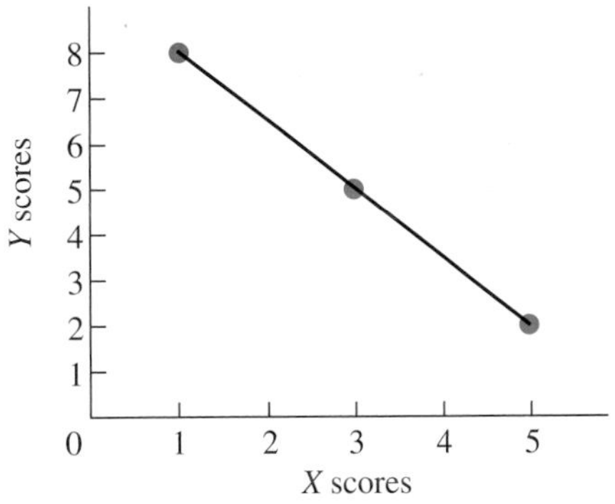

scores at each *X,* we know that their data points are on top of one another. And, because it is a perfect straight-line relationship, all data points will lie *on* the regression line.

Fourth, *the coefficient communicates the relative accuracy of our predictions* when we predict participants' *Y* scores by using their *X* scores. A coefficient of ±1 indicates perfect accuracy in predictions: because only one *Y* score occurs with each *X* we will *know* every participants' *Y* score every time. Look at the positive relationship back in Figure 7.4: We will always know when people have a *Y* score of 2 (when they have an *X* of 1), and we will know when they have a *different Y* of 5 or 8 (when they have an *X* of 3 or 5, respectively). The same accuracy is produced in the negative relationship.

Note: In statistical lingo, because we can perfectly predict the *Y* scores here, we would say that these *X* variables are perfect "predictors" of *Y*. Further, recall from Chapter 5 that the variance is a way to measure differences among scores. When we can accurately predict when *different Y* scores will occur, we say we are "accounting for the variance in *Y*." A better predictor (*X*) will account for more of the variance in *Y*. To communicate the perfect accuracy in predictions with correlations of ±1, we would say that "100% of the variance is accounted for."

> **REMEMBER** The correlation coefficient communicates the consistency of the relationship, the variability of the *Y* scores at each *X*, the shape of the scatterplot, and our accuracy when using *X* to predict *Y* scores.

Intermediate Association

A correlation coefficient that does not equal ±1 indicates that the data form a linear relationship to only some degree. The closer the coefficient is to ±1, however, the closer the data are to forming a perfect relationship, and the closer the scatterplot is to forming a straight line. Therefore, the way to interpret any other value of the correlation coefficient is to compare it to ±1.

For example, Figure 7.5 shows data that produce a correlation coefficient of $+.98$. Again interpret the coefficient in four ways. First, *consistency:* A coefficient less than ±1 indicates that not every participant at a particular *X* had the same *Y*. However, a coefficient of $+.98$ is close to $+1$, so there is close to perfect consistency. That is, even though different values of *Y* occur with the same *X*, the *Y* scores are relatively close to each other.

Second, *variability:* By indicating reduced consistency, this coefficient indicates that there is now variability (differences) among the *Y* scores at each *X*. However, because

FIGURE 7.5

Data and scatterplot reflecting a correlation coefficient of $+.98$

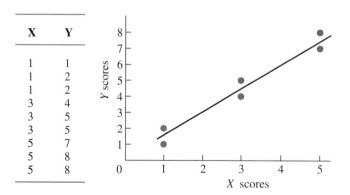

+.98 is close to +1 (close to the situation where there is zero variability) we know that the variability in our *Y* scores is close to zero and relatively small.

Third, *the scatterplot*: Because there is variability in the *Y*s at each *X,* not all data points fall *on* the regression line. Back in Figure 7.5, variability in *Y* scores results in a group of data points at each *X* that are vertically spread out above and below the regression line. However, a coefficient of +.98 is close to +1, so we know that the data points are close to, or hug, the regression line, resulting in a scatterplot that is a narrow, or skinny, ellipse.

Fourth, *predictions:* When the correlation coefficient is not ±1, knowing participants' *X* scores allows us to predict only *around* what their *Y* score will be. For example, in Figure 7.5, for an *X* of 1 we'd predict that a person has a *Y* around 1 or 2, but we won't know which. In other words, we will have some error in our predictions. However, a coefficient of +.98 is close to +1 (close to the situation where there is zero error). This indicates that our predicted *Y* scores will be close to the actual *Y* scores that participants obtained, and so our error will be small. With predictions that are close to participants' *Y* scores, we would describe this *X* variable as "a good predictor of *Y*." Further, because we will still know when *Y* scores around 1 or 2 occur and when *different Y*s around, say, 4 or 5 occur, this *X* variable still "accounts for" a sizable portion of the variance among all *Y* scores.

The key to understanding the strength of any relationship is this:

> **As the variability—differences—in the *Y* scores paired with an *X* becomes larger, the relationship becomes weaker.**

The correlation coefficient communicates this because, as the variability in the *Y*s at each *X* becomes larger, the value of the correlation coefficient approaches 0. Figure 7.6 shows data that produce a correlation coefficient of only −.28. (The fact that this is a negative relationship has nothing to do with its strength.) Here the spread in the *Y* scores (the variability) at each *X* is relatively large. This does two things that are contrary to a relationship. First, instead of seeing a different *Y* scores at *different X*s, we see very different *Y*s for individuals who have the *same X*. Second, instead of seeing one value of *Y* at only one *X*, the *Y* scores at different *X*s overlap, so we see one value of *Y* paired with *different* values of *X*. Thus, the weaker the relationship, the more the *Y* scores tend to change when *X* does not, and the more the *Y* scores tend to stay the same when *X* does change.

Thus, it is the variability in *Y* at each *X* that determines the consistency of a relationship, which in turn determines the characteristics we've examined. Thus, a coefficient

FIGURE 7.6

Data and scatterplot reflecting a correlation coefficient of −.28.

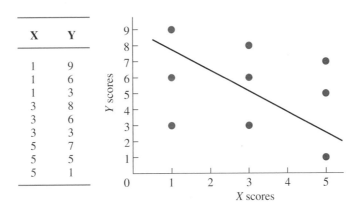

X	Y
1	9
1	6
1	3
3	8
3	6
3	3
5	7
5	5
5	1

of −.28 is not very close to ±1, so, as in Figure 7.6, we know that (1) only barely does one value or close to one value of *Y* tend to be associated with one value of *X*; (2) conversely, the variability among the *Y* scores at every *X* is relatively large; (3) the large differences among *Y* scores at each *X* produce data points on the scatterplot at each *X* that are vertically spread out, producing a "fat" scatterplot that does not hug the regression line; and (4) because each *X* is paired with a wide variety of *Y* scores, knowing participants' *X* will not allow us to accurately predict their *Y*. Instead, our prediction errors will be large because we have only a very general idea of when higher *Y* scores tend to occur and when lower *Y* scores occur. Thus, this *X* is a rather poor "predictor" because it "accounts" for little of the variance among *Y* scores.

> **REMEMBER** Greater *variability* in the *Y* scores at each *X* reduces the strength of a relationship and the size of the correlation coefficient.

Zero Association

The lowest possible value of the correlation coefficient is 0, indicating that no relationship is present. Figure 7.7 shows data that produce such a coefficient. When no relationship is present, the scatterplot is circular or forms an ellipse that is parallel to the *X* axis. Likewise, the regression line is a horizontal line.

A scatterplot like this is as far from forming a slanted straight line as possible, and a correlation coefficient of 0 is as far from ±1 as possible. Therefore, this coefficient tells us that no *Y* score tends to be consistently associated with only one value of *X*. Instead, the *Y*s found at one *X* are virtually the same as those found at any other *X*. This also means that knowing someone's *X* score will not in any way help us to predict the corresponding *Y*. (We can account for none of the variance in *Y*.) Finally, this coefficient indicates that the spread in *Y* at any *X* equals the overall spread of *Y* in the data, producing a scatterplot that is a circle or horizontal ellipse that in no way hugs the regression line.

> **REMEMBER** The larger a correlation coefficient (whether positive or negative), the stronger the linear relationship, because the less the *Y*s are spread out at each *X*, and so the closer the data come to forming a straight line.

FIGURE 7.7

Data and scatterplot reflecting a correlation coefficient of 0.

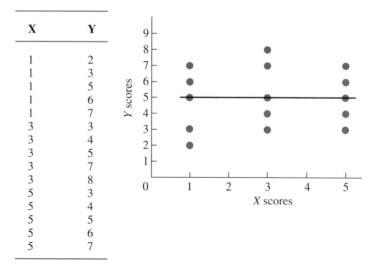

X	Y
1	2
1	3
1	5
1	6
1	7
3	3
3	4
3	5
3	7
3	8
5	3
5	4
5	5
5	6
5	7

A QUICK REVIEW

- As *X* scores increase, in a positive linear relationship, the *Y* scores tend to increase, and in a negative linear relationship, the *Y* scores tend to decrease.
- The larger the correlation coefficient, the more consistently one *Y* occurs with one *X,* the smaller the variability in *Y*s at an *X,* the more accurate our predictions, and the narrower the scatterplot.

MORE EXAMPLES

A coefficient of +.84 indicates (1) as *X* increases, *Y* consistently increases; (2) everyone at a particular *X* shows little variability in *Y* scores; (3) by knowing an individual's *X,* we can closely predict his/her *Y* score; and (4) the scatterplot is a narrow ellipse, with the data points lying near the upward slanting regression line. However, a coefficient of +.38 indicates (1) as *X* increases, *Y* somewhat consistently increases; (2) a wide variety of *Y* scores paired with a particular *X;* (3) knowing an *X* score does not produce accurate predictions of the paired *Y* score; and (4) the scatterplot is a wide ellipse around the upward slanting regression line.

For Practice

1. In a _____ relationship, as the *X* scores increase, the *Y* scores increase or decrease only. This is not true in a _____ relationship.
2. The more that you smoke cigarettes, the lower is your healthiness. This is a _____ linear relationship, producing a scatterplot that slants _____ as *X* increases.
3. The more that you exercise, the better is your muscle tone. This is a _____ linear relationship, producing a scatterplot that slants _____ as *X* increases.
4. In a stronger relationship the variability among the *Y* scores at each *X* is _____, producing a scatterplot that forms a _____ ellipse.
5. The _____ line summarizes the scatterplot.

Answers
1. linear; nonlinear
2. negative; down
3. positive; up
4. smaller; narrower
5. regression

THE PEARSON CORRELATION COEFFICIENT

Now that you understand the correlation coefficient, we can discuss its computation. However, statisticians have developed a number of correlation coefficients having different names and formulas. Which one is used in a particular study depends on the nature of the variables and the scale of measurement used to measure them. By far the most common correlation coefficient in behavioral research is the Pearson correlation coefficient. The **Pearson correlation coefficient** describes the linear relationship between two interval variables, two ratio variables, or one interval and one ratio variable. (Technically, its name is the *Pearson Product Moment Correlation Coefficient.*) The symbol for the Pearson correlation coefficient is the lowercase *r.* (All of the example coefficients in the previous section were *r*s.)

Mathematically *r* compares how consistently each value of *Y* is paired with each value of *X.* In Chapter 6, you saw that we compare scores from different variables by transforming them into *z*-scores. Computing *r* involves transforming each *Y* score into a *z*-score (call it z_Y), transforming each *X* score into a *z*-score (call it z_X), and then determining the "average" amount of correspondence between all pairs of *z*-scores. The Pearson correlation coefficient is defined as

$$r = \frac{\sum (z_X z_Y)}{N}$$

Multiplying each z_X times the paired z_Y, summing the products, and then dividing by N produces the average correspondence between the scores.

Luckily, the computational formula for r does all of that at once. It is derived from the above formula by replacing the symbols z_X and z_Y with their formulas and then, in each, replacing the symbols for the mean and standard deviation with their formulas. This produces a monster of a formula. After reducing it, we have the smaller monster below.

> ### The computational formula for the Pearson correlation coefficient is
>
> $$r = \frac{N(\Sigma XY) - (\Sigma X)(\Sigma Y)}{\sqrt{[N(\Sigma X^2) - (\Sigma X)^2][N(\Sigma Y^2) - (\Sigma Y)^2]}}$$

In the numerator, N (the number of pairs) is multiplied times ΣXY. From this, subtract the quantity obtained by multiplying (ΣX) times (ΣX). In the denominator, in the left brackets, multiply N times ΣX^2 and from that subtract $(\Sigma X)^2$. In the right bracket, multiply N times ΣY^2 and from that subtract $(\Sigma Y)^2$. Multiply the answers in the two brackets together and find the square root. Then divide the denominator into the numerator and, voilà, the answer is the Pearson r.

As an example, say that we ask ten people the number of times they visited a doctor in the last year and the number of glasses of orange juice they drink daily. We obtain the data in Figure 7.8. To describe the linear relationship between juice drinking and doctor visits, (two ratio variables,) we compute r.

Table 7.1 shows a good way to organize your computations. In addition to the columns for X and Y, create columns containing X^2, Y^2, and XY. Sum all columns. Then square ΣX and ΣY.

FIGURE 7.8

The relationship between number of glasses of orange juice consumed daily and number of yearly doctor visits.

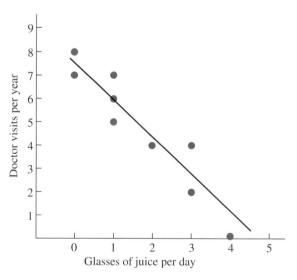

Participant	Juice Scores: X	Doctor Visits: Y
1	0	8
2	0	7
3	1	7
4	1	6
5	1	5
6	2	4
7	2	4
8	3	4
9	3	2
10	4	0

TABLE 7.1

Sample Data for Computing the *r* Between Orange Juice Consumed (the *X* variable) and Doctor Visits (the *Y* variable)

| Participant | Glasses of Juice per Day | | Doctor Visits per Year | | |
	X	X²	Y	Y²	XY
1	0	0	8	64	0
2	0	0	7	49	0
3	1	1	7	49	7
4	1	1	6	36	6
5	1	1	5	25	5
6	2	4	4	16	8
7	2	4	4	16	8
8	3	9	4	16	12
9	3	9	2	4	6
10	4	16	0	0	0
$N = 10$	$\Sigma X = 17$	$\Sigma X^2 = 45$	$\Sigma Y = 47$	$\Sigma Y^2 = 275$	$\Sigma XY = 52$
	$(\Sigma X)^2 = 289$		$(\Sigma Y)^2 = 2209$		

Putting these quantities in the formula for *r* we get

$$r = \frac{N(\Sigma XY) - (\Sigma X)(\Sigma Y)}{\sqrt{[N(\Sigma X^2) - (\Sigma X)^2][N(\Sigma Y^2) - (\Sigma Y)^2]}} = \frac{10(52) - (17)(47)}{\sqrt{[10(45) - 289][10(275) - 2209]}}$$

In the numerator, multiplying 10 times 52 is 520. Also, 17 times 47 is 799. Now we have

$$r = \frac{520 - 799}{\sqrt{[10(45) - 289][10(275) - 2209]}}$$

Complete the numerator: 799 *from* 520 is −279. (Note the negative sign.)

In the denominator, first perform the operations within each bracket. In the left bracket, 10 times 45 is 450. From that subtract 289, obtaining 161. In the right bracket, 10 times 275 is 2750. From that subtract 2209, obtaining 541. We have

$$r = \frac{-279}{\sqrt{[161][541]}}$$

Now multiply the quantities in the brackets together: 161 times 541 equals 87,101. After taking the square root we have

$$r = \frac{-279}{295.129}$$

Divide and there you have it: $r = -.95$.

Thus, the correlation coefficient between orange juice drinks and doctor visits is −.95. (*Note:* We usually round the coefficient to two decimals.) Had this been a positive relationship, *r* would not be negative and we would include the + sign. Instead, on a scale of 0 to ±1, a −.95 indicates that this is an extremely strong, negative linear relationship. Therefore, we envision a very narrow, downward slanting scatterplot like that back in Figure 7.8. We know that each amount of orange juice is associated with a very small range of doctor visits, and as juice scores increase, doctor visits consistently decrease. Further, based on participants' juice scores, we can very accurately predict their doctor visits. (Orange juice is an extremely good "predictor" of doctor visits, accounting for a substantial portion of the variance in these Y scores.) Of course if the correlation were this large in real life, we'd all be drinking a lot more orange juice, incorrectly thinking that this would *cause* fewer doctor visits.

> REMEMBER Compute the *Pearson correlation coefficient* to describe the linear relationship between interval and/or ratio variables.

Recognize that this correlation coefficient describes the relationship in our *sample*. Ultimately we will want to describe the laws of nature, inferring the correlation coefficient we would expect to find if we could measure everyone in the population. However, before we can do this, we must perform the appropriate *inferential* procedure (discussed in Chapter 11). Only if our sample correlation coefficient passes the inferential test will we then talk about how this relationship occurs in nature.

A QUICK REVIEW

- The Pearson correlation coefficient (r) describes the linear relationship between two interval and/or ratio variables.

MORE EXAMPLES

X	Y
1	3
1	2
2	4
2	5
3	5
3	6

To compute r for the above scores:

$\Sigma X = 12, (\Sigma X)^2 = 144, \Sigma X^2 - 28, \Sigma Y = 25,$
$(\Sigma Y)^2 = 625, \Sigma Y^2 = 155, \Sigma XY = 56$ and $N = 6$

$$r = \frac{N(\Sigma XY) - (\Sigma X)(\Sigma Y)}{\sqrt{[N(\Sigma X^2) - (\Sigma X)^2][N(\Sigma Y^2) - (\Sigma Y)^2]}}$$

$$r = \frac{6(56) - (12)(25)}{\sqrt{[6(28) - 144][6(115) - 625]}}$$

In the numerator, 6 times 56 is 336, and 12 times 25 is 300, so

$$r = \frac{336 - 300}{\sqrt{[(6(28) - 144][6(115) - 625]}}$$

$$r = \frac{+36}{\sqrt{[6(28) - 144][6(115) - 625]}}$$

In the denominator, 6 times 28 is 168; 6 times 115 is 690, so

$$r = \frac{+36}{\sqrt{[168 - 144][690 - 625]}} = \frac{+36}{\sqrt{[24][65]}}$$

$$r = \frac{+36}{\sqrt{1560}} = \frac{+36}{39.497} = +.91$$

For Practice

Compute r for the following:

X	Y
1	1
1	3
2	2
2	4
3	4

Answer

$$r = \frac{5(28) - (9)(14)}{\sqrt{[5(19) - 81][5(46) - 196]}}$$

$$= \frac{+14}{\sqrt{[14][34]}} = +.64$$

THE SPEARMAN RANK-ORDER CORRELATION COEFFICIENT

Another very common correlation coefficient is used when we have *ordinal* scores (when we have the equivalent of 1st, 2nd, 3rd, etc., on each variable). The **Spearman rank-order correlation coefficient** describes the linear relationship between two variables when measured by ranked scores. The symbol for the Spearman correlation coefficient is r_S. (The subscript s stands for Spearman.)

Sometimes we compute r_S because we have initially assigned each participant a rank on each of two variables. Or, if we want to correlate one ranked variable with one interval or ratio variable, we transform the interval or ratio scores into ranked scores (we might give the participant with the highest score a 1, the next highest score is ranked 2, and so on). Either way that we obtain the ranks, r_S tells us the extent to which the ranks on one variable consistently match the ranks on the other variable to form a linear relationship. If every participant has the same rank on both variables, r_S will equal $+1$. If everyone's rank on one variable is the opposite of his or her rank on the other variable, r_S will equal -1. With only some degree of consistent pairing of the ranks, r_S will be between 0 and ±1. If there is no consistent pairing, r_S will equal 0.

Ranked scores often occur in behavioral research because a variable is difficult to measure quantitatively. Instead we must evaluate participants by asking observers to make subjective judgments that are then used to rank order the participants. For example, say that we ask two observers to judge how aggressively children behave while playing. Each observer assigns the rank of 1 to the most aggressive child, 2 to the second-most aggressive child, and so on. Because r_S describes the consistency with which rankings match, one use of r_S is to determine the extent to which the two observers' rankings agree.

Figure 7.9 shows the ranked scores and the resulting scatterplot that the two observers might produce for nine children. Notice that we treat each observer as a variable. Judging from the scatterplot, it appears that they form a positive relationship. To describe this relationship, we compute r_S.

Note: If you have any "tied ranks" (when two or more participants receive the same score on the *same* variable) you must first adjust them as described in the section "Resolving Tied Ranks" in Chapter 15.

FIGURE 7.9

Sample data for computing r_S between rankings assigned to children by observer A and observer B

Participant	Observer A: X	Observer B: Y
1	4	3
2	1	2
3	9	8
4	8	6
5	3	5
6	5	4
7	6	7
8	2	1
9	7	9

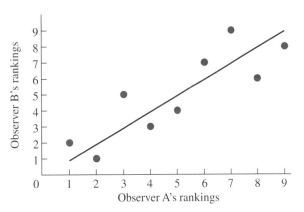

Here is the formula for r_S.

> **The computational formula for the Spearman rank-order correlation coefficient is**
>
> $$r_S = 1 - \frac{6(\Sigma D^2)}{N(N^2 - 1)}$$

The logic of the formula here is similar to that in the previous Pearson formula, except that r_S accommodates the peculiarities of ranks (e.g., zero cannot occur). This is why the formula always contains the 6 in the numerator. The D in the numerator stands for the difference between the two ranks in each X–Y pair, and N is the number of pairs of ranks. Note that after dealing with the fraction, the final step is to subtract from 1.

A good way to organize your computations is shown in Table 7.2. For the column labeled D, either subtract every X from its paired Y or, as shown, every Y from its X. Then compute D^2 by squaring the difference in each pair. Finally, determine the sum of the squared differences, ΣD^2 (here ΣD^2 is 18). You will also need N, the number of X–Y pairs (here 9), and N^2 (here 81). Filling in the formula gives

$$r_S = 1 - \frac{6(\Sigma D^2)}{N(N^2 - 1)} = 1 - \frac{6(18)}{9(81 - 1)}$$

In the numerator, 6 times 18 is 108. In the denominator, $81 - 1$ is 80, and 9 times 80 is 720. Now

$$r_S = 1 - \frac{108}{720}$$

After dividing

$$r_S = 1 - .15$$

Subtracting yields

$$r_S = +.85$$

TABLE 7.2

Data Arrangement for Computing r_S

Participant	Observer A: X	Observer B: Y	D	D²
1	4	3	1	1
2	1	2	−1	1
3	9	8	1	1
4	8	6	2	4
5	3	5	−2	4
6	5	4	1	1
7	6	7	−1	1
8	2	1	1	1
9	7	9	−2	4
				$\Sigma D^2 = 18$

Thus, on a scale of 0 to ± 1, these rankings form a consistent linear relationship to the extent that $r_S = +.85$. This tells us that a child receiving a particular rank from one observer tended to receive very close to the same rank from the other observer. Therefore, the data form a rather narrow scatterplot that tends to hug the regression line. (The r_S must also pass the inferential test in Chapter 11 before we can draw any inferences about it.)

> *REMEMBER* Compute the *Spearman correlation coefficient* to describe the linear relationship between two ordinal variables.

A QUICK REVIEW

- The Spearman correlation coefficient (r_S) describes the type and strength of the linear relationship between two sets of ranks.

MORE EXAMPLES

To determine r_S for the following ranks, find the D of each X–Y pair, and then D^2 and N.

X	Y		D	D²
1	1	=	0	0
2	2	=	0	0
4	3	=	1	1
3	6	=	−3	9
6	5	=	1	1
5	4	=	1	1
			$\Sigma D^2 =$	12
			$N =$	6

$$r_s = 1 - \frac{6(\Sigma D^2)}{N(N^2 - 1)} = 1 - \frac{6(12)}{6(36 - 1)}$$

$$= 1 - \frac{72}{210} = 1 - .343 = +.66$$

For Practice

1. When do we compute r_S ?
2. The first step in computing r_S is to compute each ____?

For the ranks:

X	Y
1	2
2	1
3	3
4	5
5	4

3. The $\Sigma D^2 = $ ____ and $N = $ ____?
4. The $r_S = $ ____?

Answers

1. When we have ordinal scores.
2. D
3. $\Sigma D^2 = 4$; $N = 5$
4. $r_S = +.80$

THE RESTRICTION OF RANGE PROBLEM

As you learn about conducting research, you'll learn of potential mistakes to avoid that otherwise can lead to problems with your statistical conclusions. One important mistake to avoid with all correlation coefficients is called the **restriction of range problem**. It occurs when we have data in which the range between the lowest and highest scores on one or both variables is limited. This will produce a correlation coefficient that is *smaller* than it would be if the range were not restricted. Here's why.

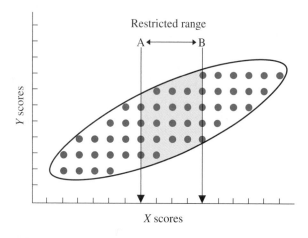

FIGURE 7.10

Scatterplot showing restriction of range in X scores

In Figure 7.10, first consider the entire scatterplot showing the full (unrestricted) range of X and Y scores. We see a different batch of similar Y scores occurring as X increases, producing an elongated, relatively narrow ellipse that clearly slants upwards. Therefore, the correlation coefficient will be relatively large, and we will correctly conclude that there is a strong linear relationship between these variables.

However, say that instead we restricted the range of X when measuring the data, giving us only the scatterplot located between the lines labeled A and B in Figure 7.10. Now, we are seeing virtually the same batch of Y scores as these few X scores increase. This produces a scatterplot that looks relatively fat and more horizontal. Therefore, the correlation coefficient from these data will be very close to 0, so we will conclude that there is a very weak—if any—linear relationship here. This would be wrong, however, because without us restricting the range, we would have seen that nature actually produces a much stronger relationship. (Because either variable can be the X or Y variable, restricting the range of Y has the same effect.)

> **REMEMBER** *Restricting the range* of X or Y scores leads to an underestimate of the true strength of the relationship between the variables.

How do you avoid restricting the range? Generally, restriction of range occurs when researchers are too selective when obtaining participants. Thus, if you study the relationship between participants' high school grades and their subsequent salaries, don't restrict the range of grades by testing only honor students: Measure all students to get the entire range of grades. Or, if you're correlating personality types with degree of emotional problems, don't study only college students. People with severe emotional problems tend not to be in college, so you won't have their scores. Instead, include the full range of people from the general population. Likewise, any task you give participants should not be too easy (because then everyone scores in a narrow range of very high scores), nor should the task be too difficult (because then everyone obtains virtually the same low score). In all cases, the goal is to allow a wide range of scores to occur on both variables so that you have a complete description of the relationship.

STATISTICS IN PUBLISHED RESEARCH: CORRELATION COEFFICIENTS

In APA-style publications, the Pearson correlation coefficient is symbolized by r, and the Spearman coefficient is symbolized by r_S. Later we'll also see other coefficients that are designed for other types of scores, and you may find additional, advanced coefficients in published research. However, all coefficients are interpreted in the same ways that we have discussed: the coefficient will have an absolute value between 0 and 1, with 0 indicating no relationship and 1 indicating a perfectly consistent relationship.

In real research, however, a correlation coefficient near ± 1 simply does not occur. Recall from Chapter 2 that individual differences and extraneous environmental variables produce inconsistency in behaviors, which results in inconsistent relationships.

Therefore, adjust your expectations: Most research produces coefficients with absolute values in the neighborhood of only .30 to .50. Thus, coefficients below .20 tend to be considered very weak and often negligible. Coefficients around .40 are described as strong and coefficients above .60 are uncommon and considered very strong. A correlation near 1 is most likely a computational error.

PUTTING IT ALL TOGETHER

It should be obvious why you should compute a correlation coefficient whenever you have a relationship to summarize. It is the one number that allows you to envision and summarize the important information in a scatterplot. For example, in our study on nervousness and the amount of coffee consumed, say that I tell you that the r in the study equals $+.50$. *Without even seeing the data,* you know this is a positive linear relationship such that as coffee consumption increases, nervousness also tends to increase. Also, you know that it is a rather consistent relationship so there are similar Y scores paired with an X, producing a narrow, elliptical scatterplot that hugs the regression line. And, you know that coffee consumption is a reasonably good predictor of nervousness so, given someone's coffee score, you'll have considerable accuracy in predicting his or her nervousness score. No other type of statistic so directly summarizes a relationship. Therefore, as you'll see in later chapters, even when you conduct an experiment, always think "correlation coefficient" to describe the strength and type of relationship you've observed.

Using the SPSS Appendix As shown in Appendix B.4, the SPSS program will calculate the Pearson r, as well as computing the mean and standard deviation of the X scores and of the Y scores. SPSS will also compute the Spearman r_S (even if your data contains tied ranks.) Also, you may enter interval or ratio scores and the program will first convert them to ranks and then compute r_S.

CHAPTER SUMMARY

1. A *scatterplot* is a graph that shows the location of each pair of X–Y scores in the data. An *outlier* is a data point that lies outside of the general pattern in the scatterplot. It is produced when a participant has an unusual X or Y score.

2. The *regression line* summarizes a relationship by passing through the center of the scatterplot.

3. In a *linear relationship*, as the X scores increase, the Y scores tend to change in only *one* direction. In a *positive linear relationship,* as the X scores increase, the Y scores tend to increase. In a *negative linear relationship,* as the X scores increase, the Y scores tend to decrease. In a *nonlinear,* or *curvilinear, relationship,* as the X scores increase, the Y scores do not only increase or only decrease.

4. Circular or elliptical scatterplots that produce horizontal regression lines indicate no relationship. Scatterplots with regression lines sloping up as X increases indicate a positive linear relationship. Scatterplots with regression lines sloping down as X increases indicate a negative linear relationship. Scatterplots producing wavy regression lines indicate curvilinear relationships.

5. A *correlation coefficient* describes the *type* of relationship (the direction Y scores change) and the *strength* of the relationship (the extent to which one value of Y is consistently paired with one value of X).

6. A smaller absolute value of the correlation coefficient indicates a weaker, less consistent relationship, with greater variability in Y scores at each X, greater vertical spread in the scatterplot, and less accuracy in predicting Y scores based on correlated X scores.

7. The *Pearson correlation coefficient* (r) describes the type (either positive or negative) and the strength of the linear relationship between two interval and/or ratio variables.

8. The *Spearman rank-order correlation coefficient* (r_S) describes the type and strength of the linear relationship between two ordinal variables.

9. The *restriction of range problem* occurs when the range of scores from one or both variables is limited. Then the correlation coefficient underestimates the strength of the relationship that would be found if the range were not restricted.

10. Because a stronger relationship allows for greater accuracy in predicting Y scores, researchers say the X variable is a better *predictor* of Y scores, allowing us to *account for more variance in Y.*

KEY TERMS

ΣXY r r_S
correlation coefficient *136*
curvilinear relationship *141*
linear relationship *139*
negative linear relationship *141*
nonlinear relationship *141*
outlier *138*
Pearson correlation coefficient *147*

positive linear relationship *140*
regression line *140*
restriction of range *153*
scatterplot *138*
Spearman rank-order correlation
  coefficient *152*
strength of a relationship *142*
type of relationship *139*

REVIEW QUESTIONS

(Answers for odd-numbered questions are in Appendix D.)

1. What is the difference between an experiment and a correlational study in terms of how the researcher (a) collects the data? (b) examines the relationship?
2. (a) You have collected data that you think show a relationship. What do you do next? (b) What is the advantage of computing a correlation coefficient? (c) What two characteristics of a linear relationship are described by a correlation coefficient?
3. What are the two reasons why you can't conclude you have demonstrated a causal relationship based on correlational research?
4. (a) When do you compute a Pearson correlation coefficient? (b) When do you compute a Spearman coefficient?
5. (a) What is a scatterplot? (b) What is an outlier? (c) What is a regression line?
6. Why can't you obtain a correlation coefficient greater than ± 1?

7. (a) Define a positive linear relationship. (b) Define a negative linear relationship. (c) Define a curvilinear relationship.
8. As the value of r approaches ± 1, what does it indicate about the following? (a) The consistency in the X–Y pairs; (b) the variability of the Y scores at each X; (c) the closeness of Y scores to the regression line; (d) the accuracy with which we can predict Y if X is known.
9. What does a correlation coefficient equal to 0 indicate about the four characteristics in question 8?
10. (a) What is the restriction of range problem? (b) What produces a restricted range? (c) How is it avoided?
11. (a) What does a researcher mean when he states that a particular variable is a "a good predictor?" (b) What does a researcher mean when she says an X variable accounts for little of the variance in Y?

APPLICATION QUESTIONS

12. For each of the following, indicate whether it is a positive linear, negative linear, or nonlinear relationship: (a) Quality of performance (Y) increases with increased arousal (X) up to an optimal level; then quality of performance decreases with increased arousal. (b) Overweight people (X) are less healthy (Y). (c) As number of minutes of exercise increases each week (X), dieting individuals lose more pounds (Y). (d) The number of bears in an area (Y) decreases as the area becomes increasingly populated by humans (X).
13. Poindexter sees the data in question 12d and concludes, "We should stop people from moving into bear country so that we can preserve our bear population." What is the problem with Poindexter's conclusion?
14. For each of the following, give the symbol for the correlation coefficient you should compute. You measure (a) SAT scores and IQ scores; (b) taste rankings of tea by experts and those by novices; (c) finishing position in a race and amount of liquid consumed during the race.
15. Poindexter finds that $r = -.40$ between the variables of number of hours studied (X) and number of errors on a statistics test (Y). He also finds that $r = +.36$ between the variables of time spent taking the statistics test and the number of errors on the test. He concludes that the time spent taking a test forms a stronger relationship with the number of errors than does the amount of study time. (a) Describe the relative shapes of the two scatterplots. (b) Describe the relative amount of variability in Y scores at each X in each study. (c) Describe the relative closeness of Y scores to the regression line in each study. (d) Is Poindexter correct in his conclusion? If not, what's his mistake?
16. In question 15, (a) which variable is a better predictor of test errors and how do you know this? (b) Which variable accounts for more of the variance in test errors and how do you know this?
17. Foofy and Poindexter study the relationship between IQ score and high school grade average, measuring a large sample of students from PEST (the Program for Exceptionally Smart Teenagers), and compute $r = +.03$. They conclude that there is virtually no relationship between IQ and grade average. Should you agree or disagree with this conclusion? Is there a problem with their study?

18. A researcher measures the following scores for a group of people. The *X* variable is the number of errors on a math test, and the *Y* variable is the person's level of satisfaction with his/her performance. (a) With such ratio scores, what should the researcher conclude about this relationship? (*Hint:* Compute something!) (b) How well will he be able to predict satisfaction scores using this relationship?

Participant	Errors X	Satisfaction Y
1	9	3
2	8	2
3	4	8
4	6	5
5	7	4
6	10	2
7	5	7

19. You want to know if a nurse's absences from work in one month (*Y*) can be predicted by knowing her score on a test of psychological "burnout" (*X*). What do you conclude from the following ratio data?

Participant	Burnout X	Absences Y
1	2	4
2	1	7
3	2	6
4	3	9
5	4	6
6	4	8
7	7	7
8	7	10
9	8	11

20. In the following data, the *X* scores reflect participants' rankings in a freshman class, and the *Y* scores reflect their rankings in a sophomore class. To what extent do these data form a linear relationship?

Participant	Fresh. X	Soph. Y
1	2	3
2	9	7
3	1	2
4	5	9
5	3	1
6	7	8
7	4	4
8	6	5
9	8	6

21. A researcher observes the behavior of a group of monkeys in the jungle. He determines each monkey's relative position in the dominance hierarchy of the group (1 being most dominant) and also notes each monkey's relative weight (1 being the lightest). What is the relationship between dominance rankings and weight rankings in these data?

Participant	Dominance X	Weight Y
1	1	10
2	2	8
3	5	6
4	4	7
5	9	5
6	7	3
7	3	9
8	6	4
9	8	1
10	10	2

INTEGRATION QUESTIONS

22. In an experiment, (a) which variable is assumed to be the causal variable? (b) Which variable is assumed to be caused? (c) Which variable does the researcher manipulate? (d) Which variable occurs first? (Ch. 2)

23. In a correlational study, we measure participants' creativity and their intelligence. (a) Which variable does the researcher manipulate? (b) Which variable is the causal variable? (c) Which variable occurred first? (d) Which variable is called the independent variable? (Chs. 2, 7)

24. In question 23, (a) How would you determine which variable to call X? (b) In a different study, my title is "Creativity as a function of Intelligence." Which variable is my X variable? Why? (Ch. 2)

25. Indicate which of the following is a correlational design and the correlation coefficient to compute. (a) We measure participants' age and their daily cell phone usage. (b) We separate participants into three age groups, and then observe their cell phone usage during a one hour period. (c) A teacher uses students' grades on their first exam to predict their final exam grades. (d) We ask whether a website rated as most attractive has more visitors than one rated as second most attractive, and so on, for the top ten websites. (e) We compare performance on an attention test of people who were and were not given an energy drink. (Chs. 2, 7)

■ ■ SUMMARY OF FORMULAS

1. The formula for the Pearson r is

$$r = \frac{N(\Sigma XY) - (\Sigma X)(\Sigma Y)}{\sqrt{[N(\Sigma X^2) - (\Sigma X)^2][N(\Sigma Y^2) - (\Sigma Y)^2]}}$$

2. The formula for the Spearman r_S is

$$r_S = 1 - \frac{6(\Sigma D^2)}{N(N^2 - 1)}$$

8

Linear Regression

GETTING STARTED

To understand this chapter, recall the following:

- From Chapter 5, that when the mean is the predicted score, the variance reflects the "average error" in predictions.

- From Chapter 7, that the larger an r, the more consistent the relationship, so the closer the Y scores are to each other at an X, and the closer they are to the regression line. Also, the larger an r, the better we can predict Y scores and "account for variance."

Your goals in this chapter are to learn

- How a *regression line* summarizes a scatterplot.

- How the *regression equation* is used to predict the Y scores at a given X.

- How the *standard error of the estimate* measures the errors in prediction.

- How the strength of the relationship determines our accuracy in predicting Y scores.

- What the *proportion of variance accounted for* tells us and how to compute it.

Recall that, in a relationship, particular Y scores are naturally paired with certain X scores. Therefore, if we know an individual's X score and the relationship between X and Y, we can predict the individual's Y score. The statistical procedure for making such predictions is called *linear regression*. In the following sections, we'll examine the logic behind regression and see how to use it to predict scores. Then we'll look at ways of measuring the errors in prediction.

NEW STATISTICAL NOTATION

We use the following symbols for distinguishing participants' actual scores from the scores that we predict for them: As usual, Y stands for a participant's actual score. The symbol for a **predicted Y score** is Y'. The "'" is called prime, so this symbol is pronounced "Y prime."

Also, for a sample of Y scores, we will discuss the mean ($\overline{Y}$), sample variance (S_Y^2), and sample standard deviation (S_Y). These involve the same formulas we used previously, except now we plug in Y scores.

Thus,

$$\overline{Y} = \frac{\Sigma Y}{N} \quad \text{and} \quad S_Y^2 = \frac{\Sigma Y^2 - \dfrac{(\Sigma Y)^2}{N}}{N} \quad \text{and} \quad S_Y = \sqrt{\frac{\Sigma Y^2 - \dfrac{(\Sigma Y)^2}{N}}{N}}$$

WHY IS IT IMPORTANT TO KNOW ABOUT LINEAR REGRESSION?

A goal of research is to be able to predict when different behaviors will occur. This translates into predicting when someone has one score on a variable and when they have a different score. We use relationships to make these predictions. It's important that you know about linear regression because it is *the* statistical procedure for using a relationship to predict scores. Linear regression is commonly used in basic and applied research, particularly in educational, industrial and clinical settings. For example, the reason that students take the Scholastic Aptitude Test (SAT) when applying to some colleges is because, from previous research we know that SAT scores are somewhat positively correlated with college grades. Therefore, through regression techniques, the SAT scores of applying students are used to predict their future college performance. If the predicted grades are too low, the student is not admitted to the college. This approach is also used when people take a test when applying for a job so that the employer can predict who will be better workers, or when clinical patients are tested to identify those at risk of developing emotional problems.

REMEMBER   The importance of *linear regression* is that it is used to predict unknown Y scores based on the X scores from a correlated variable.

UNDERSTANDING LINEAR REGRESSION

Regression procedures center around drawing the linear regression line, the summary line drawn through a scatterplot. We use regression procedures in conjunction with the Pearson correlation. While r is the *statistic* that summarizes the linear relationship, the regression line is the *line* on the scatterplot that summarizes the relationship. Always compute r first to determine whether a relationship exists. If the correlation coefficient is not 0 and passes the inferential test, then perform linear regression to further summarize the relationship.

An easy way to understand a regression line is to compare it to a line graph of an experiment. In Chapter 4, we created a line graph by plotting the mean of the Y scores for each condition—each X—and then connecting adjacent data points with straight lines. The left-hand graph in Figure 8.1 shows the scatterplot and line graph of an experiment containing four conditions. Thus, for example, the arrows indicate that the mean of Y at X_3 is 3. Because the mean is the central score, we assume that those participants at X_3 scored *around* a Y of 3, so (1) 3 is our best single description of their scores, and (2) 3 is our best prediction for anyone else at that X.

It is difficult, however, to see the *linear* (straight-line) relationship in these data because the means do not fall on a straight line. Therefore, as in the right-hand graph in Figure 8.1, we summarize the linear relationship by drawing a regression line. Think of the regression line as a straightened-out version of the line graph: It is drawn so that it comes as close as possible to connecting the mean of Y at each X while still producing a straight line. Although not all means are on the line, the distance that some means are above the line averages out with the distance that other means are below the line. Thus, the regression line is called the *best-fitting* line because "on average" it passes through the center of the various Y means. Because each Y mean is located in the center of the corresponding Y scores, the regression line also passes through the center of the Y scores. Thus, the **linear regression line** is the straight line that summarizes the linear relationship in a scatterplot by, on average, passing through the center of the Y scores at each X.

FIGURE 8.1

Comparison of a line
graph and a regression
line

*Each data point is formed
by an X–Y pair. Each
asterisk (*) indicates the
mean Y score at an X.*

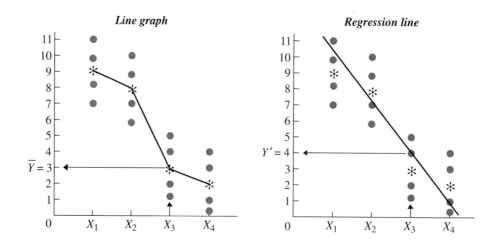

As usual, this is another descriptive procedure that allows us to summarize and envision data. Think of the regression line as reflecting the linear relationship hidden in the data. Because the actual Y scores fall above and below the line, the data only more or less fit this line. But we have no system for drawing a "more or less" linear relationship. Therefore, the regression line is how we envision what a perfect version of the linear relationship in the data would look like.

You should read the regression line in the same way that you read any graph: Travel vertically from an X until you intercept the regression line. Then travel horizontally until you intercept the Y axis. For example, the arrows in the right-hand graph of Figure 8.1 show that the value of Y at X_3 is now 4. The symbol for this value is Y' and it is our predicted Y score. A Y' is a summary of the Y scores for that X, based on the entire linear relationship. Therefore, considering the entire relationship in Figure 8.1, those participants at X_3 scored around 4, so 4 is our best prediction for anyone scoring that X. Likewise, any Y' is our best prediction of the Y scores at a corresponding X, based on the linear relationship that is summarized by the regression line.

Recognize that the Y' at any X is the value of Y falling *on* the regression line. The regression line therefore consists of the data points formed by pairing each possible value of X with its corresponding value of Y'. If you think of the line as reflecting a perfect version of the linear relationship hidden in the data, then each Y' is the Y score everyone would have at a particular X if a perfect relationship were present.

> **REMEMBER** The *linear regression line* summarizes the linear relationship in a sample and is used to obtain the Y' at any X.

Now you can see how regression techniques are used to predict unknown scores. First, we establish the relationship in a sample. Then we use the regression line to determine the Y' for each X. This is the Y *around* which everyone scored when at an X in our sample. For anyone else at that X, we'd assume they too would score around that Y'. Therefore, we can measure the X scores of individuals who were not in our sample, and the corresponding Y' is our best prediction of their Y scores.

The emphasis on prediction in correlation and regression leads to two important terms. We'll discuss using the X variable to predict Y scores. (There are procedures out there for predicting X scores from Y.) Therefore, as mentioned in the previous chapter, the X variable is called the **predictor variable**. The Y variable is called the **criterion variable**. Thus, when SAT scores are used to predict a student's future college grades,

SAT scores are the predictor variable, and college grade average is the criterion variable. (To remember *criterion,* remember that your predicted grades must meet a certain criterion for you to be admitted to the college.)

The first step in using regression techniques is to create the regression line. For that we use the *linear regression equation.*

THE LINEAR REGRESSION EQUATION

To draw a regression line, we don't simply eyeball the scatterplot and sketch in something that looks good. Instead, we use the linear regression equation. The **linear regression equation** is the equation for a straight line that produces the value of Y' at each X and thus defines the straight line that summarizes a relationship. When we plot the data points formed by the X–Y' pairs and draw a line connecting them, we have the regression line. The regression equation describes two characteristics of the regression line: its *slope* and its *Y intercept.*

The **slope** is a number that indicates how slanted the regression line is and the direction in which it slants. Figure 8.2 shows examples of regression lines having different slopes. When no relationship is present, the regression line is horizontal, such as line A, and the slope is zero. A positive linear relationship produces regression lines such as B and C; each of these has a slope that is a positive number. Because line C is steeper, its slope is a larger positive number. A negative linear relationship, such as line D, yields a slope that is a negative number.

The **Y intercept** is the value of Y at the point where the regression line intercepts, or crosses, the Y axis. In other words, the intercept is the value of Y' when X equals 0. In Figure 8.2, line B intercepts the Y axis at +2, so the Y intercept is +2. If we extended line C, it would intercept the Y axis at a point below the X axis, so its Y intercept is a negative Y score. Because line D reflects a negative relationship, its Y intercept is the relatively high Y score of 9. Finally, line A exhibits no relationship, and its Y intercept equals +8. Notice that here the predicted Y score for every X is always +8.

> **When there is no relationship, the regression line is flat and every Y' equals the Y intercept.**

The regression equation works like this: The slope indicates the *direction* in which the Ys change as X increases and the *rate* at which they change. In Figure 8.2, the steeply sloped line C reflects a relatively large increase in Y for each increase in X, as

FIGURE 8.2

Regression lines having different slopes and Y intercepts

Line A indicates no relationship, lines B and C indicate positive relationships having different slopes and Y intercepts, and line D indicates a negative relationship.

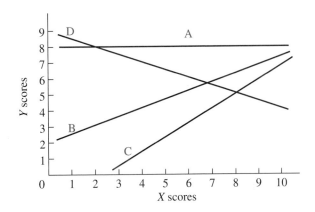

compared to, line B. The Y intercept indicates the starting point from which the Y scores begin to change. Thus, together, the slope and intercept describe how, starting at a particular Y score, the Y scores tend to change by a specific amount as the X scores increase. The summary of the new Y scores at each X is Y'.

The symbol for the slope of the regression line is b. The symbol for the Y intercept is a. Then

The linear regression equation is

$$Y' = bX + a$$

This formula says that to find the value of Y' for a given X, multiply the slope (b) times X and then add the Y intercept (a).

As an example, say that we have developed a test to identify (predict) those individuals who will be good or bad workers at a factory that makes "widgets." The first step is to determine the relationship between test scores and "widget-making." Therefore, say that we give the test to an unrealistically small N of 11 people and then measure the number of widgets each makes in an hour. Figure 8.3 shows the raw scores and resulting scatterplot. The predictor (X) variable is participants' scores on the widget test, and the criterion (Y) variable is the number of widgets they produced.

The first step is to find r:

$$r = \frac{N(\Sigma XY) - (\Sigma X)(\Sigma Y)}{\sqrt{[N(\Sigma X^2) - (\Sigma X)^2][N(\Sigma Y^2) - (\Sigma Y)^2]}}$$

so

$$r = \frac{11(171) - (29)(58)}{\sqrt{[11(89) - 841][11(354) - 3364]}}$$

The result is $r = +.74$. This is a very strong, positive linear relationship, and so the test will be what researchers call "a good predictor" of widget-making. Therefore, the next step is to compute the linear regression equation. To do that, we compute the slope and the Y intercept.

Compute the slope first.

Computing the Slope

The formula for the slope of the linear regression line is

$$b = \frac{N(\Sigma XY) - (\Sigma X)(\Sigma Y)}{N(\Sigma X^2) - (\Sigma X)^2}$$

FIGURE 8.3

Scatterplot and data for widget study

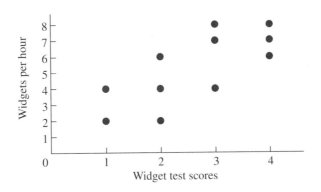

Participant	Widget Test Score: X	Widgets per Hour: Y	XY
1	1	2	2
2	1	4	4
3	2	4	8
4	2	6	12
5	2	2	4
6	3	4	12
7	3	7	21
8	3	8	24
9	4	6	24
10	4	8	32
11	4	7	28
$N = 11$	$\Sigma X = 29$	$\Sigma Y = 58$	$\Sigma XY = 171$
	$\Sigma X^2 = 89$	$\Sigma Y^2 = 354$	
	$(\Sigma X)^2 = 841$	$(\Sigma Y)^2 = 3364$	
	$\overline{X} = 29/11 = 2.64$	$\overline{Y} = 58/11 = 5.27$	

N is the number of pairs of scores in the sample, and X and Y are the scores in the sample. This is not a difficult formula because we typically compute the Pearson r first. The numerator of the formula for b is the same as the numerator in the formula for r, and the denominator of the formula for b is the left-hand quantity in the denominator of the formula for r. [An alternative formula for the slope is $b = (r)(S_Y/S_X)$.]

For the widget study, substituting the appropriate values into the formula gives

$$b = \frac{N(\Sigma XY) - (\Sigma X)(\Sigma Y)}{N(\Sigma X^2) - (\Sigma X)^2} = \frac{11(171) - (29)(58)}{11(89) - 841}$$

After multiplying and subtracting in the numerator,

$$b = \frac{199}{11(89) - 841}$$

After completing the denominator,

$$b = \frac{199}{138} = +1.44$$

Thus, the slope of the regression line for the widget study is $+1.44$. This positive slope indicates a positive relationship, which fits with the positive r of $+.74$. Had the relationship been negative, the formula would have produced a negative number here.

We are not finished yet. Now compute the Y intercept.

Computing the Y Intercept

The formula for the Y intercept of the linear regression line is

$$a = \bar{Y} - (b)(\bar{X})$$

First, multiply the mean of all X scores times the slope of the regression line. Then subtract that quantity from the mean of all Y scores.

For the widget study, b is $+1.44$, and from the data in Figure 8.3, $\bar{Y}$ is 5.27 and $\bar{X}$ is 2.64. Filling in the above formula gives

$$a = 5.27 - (+1.44)(2.64)$$

After multiplying,

$$a = 5.27 - (+3.80) = +1.47$$

Thus, the Y intercept of the regression line for the widget study is $+1.47$.

We're still not finished!

Describing the Linear Regression Equation

Once you have computed the Y intercept and the slope, rewrite the regression equation, substituting the computed values for a and b. Thus, for the widget study,

$$Y' = +1.44X + 1.47$$

This is the finished regression equation that describes the linear regression line for the relationship between widget test scores and widgets-per-hour scores.

We're still not finished. Next we plot the regression line.

Plotting the Regression Line

We use the finished regression equation to plot our linear regression line. To draw a line, we need at least two data points, so choose a low and high X score, insert each into the regression equation, and compute the Y' for that X. (Or, an easy low X to use is 0 because then Y' equals the Y intercept.) We'll use the widget test scores of 1 and 4. We begin with our finished regression equation:

$$Y' = +1.44X + 1.47$$

For $X = 1$, we have

$$Y' = +1.44(1) + 1.47 = 2.91$$

Multiplying $+1.44$ times 1 and adding 1.47 yields a Y' of 2.91. Likewise, for $X = 4$, $Y' = +1.44(4) + 1.47 = 5.76 + 1.47 = 7.23$.

FIGURE 8.4

Regression line for widget study

Widget Test Scores: X	Predicted Widgets per Hour: Y
1	2.91
2	4.35
3	5.79
4	7.23

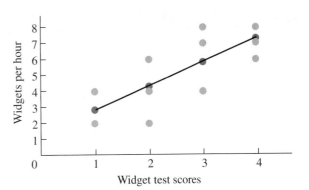

To graph the regression line, plot the data points for the previous X–Y′ pairs and draw the line. As shown in Figure 8.4, our widget regression line passes through the center of the original scatterplot, showing the straight-line relationship hidden in the data.

(We're almost there!)

Computing Predicted *Y* Scores

Remember that the regression line consists of all possible X–Y′ pairs. Therefore, we also use the finished regression equation to predict anyone's Y score if we know their X score. Thus, as we computed above, we will predict a Y′ of 2.91 for *anyone* scoring an X of 1, and a Y′ of 7.23 for *anyone* scoring an X of 4. Likewise, for the test scores of 2 and 3, we compute the Y′ scores of 4.35 and 5.79, respectively. As shown in Figure 8.4, these data points also fall on the regression. In fact, computing any Y′ using the equation is the equivalent of going to the graph and traveling vertically from the X score up to the regression line and then left to the value of Y′ on the Y axis.

We can compute Y′ for any value of X that falls *within* the range of Xs in our data, even if it's a score not found in the original sample: No one scored an X of 1.5, but entering this in the regression equation yields a predicted Y score of 3.63. Do not, however, make predictions using X scores beyond the original scores. Our regression equation is based only on widget test scores between 1 and 4, so we shouldn't predict a Y for an X of, for example, 6. This is because we can't be sure what the nature of the relationship is at 6—maybe it's curvilinear or has a steeper slope.

Now we are finished (really). Putting all of this together, the preceding computations are summarized in Table 8.1.

TABLE 8.1

Summary of steps in Linear Regression

1. Compute r.

2. Compute the slope, b, where $b = \dfrac{N(\Sigma XY) - (\Sigma X)(\Sigma Y)}{N(\Sigma X^2) - (\Sigma X)^2}$

3. Compute the Y intercept, a, where $a = \overline{Y} - (b)(\overline{X})$

4. Substitute the values of a and b into the formula for the regression equation:
$$Y' = (b)(X) + a$$

5. Plot the regression line.

6. Compute the predicted Y′ score for any X.

A QUICK REVIEW

- The linear regression equation is used to predict Y' scores based on X scores from a correlated variable and to draw the linear regression line.
- The regression equation is $Y' = bX + a$, where b is the slope and a is the Y intercept.

MORE EXAMPLES

To use X to predict Y in these scores,

X	Y
1	3
1	2
2	4
2	5
3	5
3	6

Compute b: $\Sigma X = 12$, $\Sigma Y = 25$, $\Sigma X^2 = 28$, $\Sigma XY = 56$, and $N = 6$.

$$b = \frac{N(\Sigma XY) - (\Sigma X)(\Sigma Y)}{N(\Sigma X^2) - (\Sigma X)^2} = \frac{6(56) - (12)(25)}{6(28) - (12)^2}$$

$$= \frac{336 - 300}{168 - 144} = +1.5$$

Compute a: $b = +1.5$, $\overline{X} = 2$, and $\overline{Y} = 4.167$.

$$a = \overline{Y} - (b)(\overline{X}) = 4.167 - (+1.5)(2) = 1.167$$

After rounding $Y' = +1.5X + 1.17$.

Say we want to predict Y for when $X = 2$. The Y' is

$$Y' = +1.5(2) + 1.17 = 3 + 1.17 = 4.17.$$

For Practice

1. We use linear regression when we want to _____.
2. The components of the regression equation to compute first are the _____ and _____.
3. Compute b for the following scores:

X	Y
1	5
1	3
2	4
2	3
3	2
4	1

---

4. Compute a for these scores.
5. What is the predicted Y score for an X of 2?

Answers

1. predict Y scores using the relationship with X
2. slope and Y intercept
3. $b = \dfrac{6(32) - (13)(18)}{6(35) - (13)^2} = -1.024$
4. $a = \overline{Y} - (b)(\overline{X}) = 3 - (-1.024)(2.167)$
   $= 3 - (-2.2.19) = +5.219$
5. $Y' - (-1.024)(2) + 5.219 = +3.171$

DESCRIBING ERRORS IN PREDICTION

Not all relationships are the same and so not all X variables are "good predictors." Therefore, a complete description of a relationship includes the descriptive statistics that summarize the error we have when using the relationship to predict Y scores. To describe the amount of prediction error we expect when predicting unknown scores, we first determine how well we can predict the actual Y scores in our sample: We pretend we don't know the scores, predict them, and then compare the predicted Y' scores to the actual Y scores.

The error in a single prediction is the amount that a participant's Y score differs from the corresponding predicted Y' score: In symbols this is $Y - Y'$, and it is literally the difference between the score a participant got and the score we predict he or she got. The predictions for some participants will be closer to their actual Y scores than for others, so we would like to compute something like the *average error* across all predictions.

To find the average error, we first compute Y' for everyone in the sample and subtract their Y' from their actual Y score. Statisticians equate errors with *deviations*, so

$Y - Y'$ equals the amount that Y *deviates* from Y'. To get the average error, we would like to simply sum these deviations and then find the average, but we cannot. Recall that the regression line is in the center of the scatterplot. Therefore, the Ys are equally spread out *around* their Y' scores, in the same way that previously we saw that Xs are spread out *around* their $\overline{X}$. Because of this, like with the mean, the positive and negative deviations with Y will cancel out, always producing a sum equal to zero. Therefore, the average deviation—the average error—will always equal zero.

To solve this problem, we *square* each deviation. The sum of the squared deviations of $Y - Y'$ is not necessarily zero, so neither is the average squared deviation. (Does this sound familiar?) When we find the average of the squared deviations, the answer is a type of *variance* that describes the "average" spread of the actual Y scores around—above and below—their predicted Y' scores.

Computing the Variance of the *Y* Scores Around *Y'*

The **variance of the *Y* scores around *Y'*** is the average squared difference between the actual Y scores and their corresponding predicted Y' scores. The symbol for this *sample* variance is $S_{Y'}^2$. The S^2 indicates sample variance or error, and the subscript Y' indicates that it is the error associated with using Y' to predict Y scores. The formula that defines the variance of the Y scores around Y' is

$$S_{Y'}^2 = \frac{\Sigma(Y - Y')^2}{N}$$

Like other definitional formulas we've seen, this formula is important because it shows the core calculation involved: We subtract the Y' predicted for each participant from his or her actual Y score giving us a measure of our error. Then we square each deviation, sum the squared deviations, and divide by N. The answer is one way to measure roughly the "average" amount of error we have when we use linear regression to predict Y scores.

Note: Among the approaches we might use, the regression procedures described in this chapter produce the smallest error in predictions possible, thereby producing the smallest sum of squared deviations possible. Researchers call this regression technique the "least-squares method." To get this name, they shorten "sum of squared deviations" to *squares,* and this method produces a sum that is the *least* it can be.

> **REMEMBER** The *variance of the Y scores around Y'* $(S_{Y'}^2)$ is one way to describe the average error when using linear regression to predict Y scores.

Using the above definitional formula for $S_{Y'}^2$ is very time consuming. Thankfully, there is a better way. In the defining formula, we can replace Y' with the formulas for finding Y' (for finding a, b, and so on). Among all of these formulas we'll find the components for the following computational formula.

The computational formula for the variance of the *Y* scores around *Y'* is

$$S_{Y'}^2 = S_Y^2(1 - r^2)$$

Much better! This formula uses r (which we compute before doing regression anyway) and, in computing r, we compute the ΣY and ΣY^2 needed for finding the variance in the Y scores. Therefore, finish the computations of S_Y^2 using the formula at the beginning of this chapter. Then this formula says to square r and subtract the answer from 1. Multiply the result times S_Y^2. The answer is $S_{Y'}^2$.

In the widget study, the data back in Figure 8.3 produced $r = +.736$, with $\Sigma Y = 58$, $\Sigma Y^2 = 354$, and $N = 11$. Therefore, we find that S_Y^2 is

$$S_Y^2 = \frac{\Sigma Y^2 - \frac{(\Sigma Y)^2}{N}}{N} = \frac{354 - \frac{(58)^2}{11}}{11} = \frac{354 - -305.818}{11} = 4.380$$

Then the above formula gives

$$S_{Y'}^2 = 4.380(1 - .736^2)$$

After squaring $+.736$ and subtracting the result from 1, we have

$$S_{Y'}^2 = 4.380(.458) = 2.01$$

Thus, we are "off" by something like an "average" of 2.01 when we predict participants' widget-per-hour scores (Y) based on their widget test scores (X).

Although this variance is a legitimate way to compute the error in our predictions, it is only somewhat like the "average" error, because of the usual problems when interpreting variance. First, squaring each difference between Y and Y' produces an unrealistically large number, inflating our error. Second, squaring produces error that is measured in squared units, so our predictions above are off by 2.01 *squared* widgets. (This *must* sound familiar!) The solution is to find the square root of the variance, and the result is a type of standard deviation. To distinguish the standard deviation found in regression, we call it the *standard error of the estimate*.

Computing the Standard Error of the Estimate

The **standard error of the estimate** is similar to a standard deviation of the Y scores around their Y' scores. It is the clearest way to describe the "average error" when using Y' to predict Y scores. The symbol for the standard error of the estimate is $S_{Y'}$. (Remember, S measures the *error* in the sample, and Y' is our *estimate* of a participant's Y score.) The definitional formula for the standard error of the estimate is

$$S_{Y'} = \sqrt{\frac{\Sigma(Y - Y')^2}{N}}$$

This is the same formula used previously for the variance of Y scores around Y', except with the added square root sign. By computing the square root, the answer is a more realistic number and we are no longer dealing with a squared variable. The core calculation, however, is still to find the error between participants' actual Y scores and their predicted Y' scores, and this is as close as we will come to computing the "average error" in our predictions.

To create a computational formula for the standard error of the estimate, we find the square root of each component of the previous computational formula for the variance of Y around Y' and have

The computational formula for the standard error of the estimate is

$$S_{Y'} = S_Y\sqrt{1 - r^2}$$

Here we first compute the standard deviation of all Y scores (S_Y) using the formula at the beginning of this chapter. Then we find the square root of the quantity $1 - r^2$ and then multiply it times the standard deviation of all Y scores.

For the widget study, we computed that r was $+.736$ and that the variance of all Y scores (S_Y^2) was 4.380. The standard deviation of the Y scores is the square root of the variance, so $S_Y = \sqrt{4.380} = 2.093$. Filling in the above formula gives

$$S_{Y'} = 2.093\sqrt{1 - .736^2}$$

Squaring $+.736$ yields .542, which subtracted from 1 gives .458. The square root of .458 is .677. Thus,

$$S_{Y'} = 2.093(.677) = 1.42$$

The standard error of the estimate is 1.42. Therefore, we conclude that when using the regression equation to predict the number of widgets produced per hour based on a person's widget test score, when we are wrong, we will be wrong by an "average" of about 1.42 widgets per hour.

> **REMEMBER** The *standard error of the estimate* $(S_{Y'})$ is interpreted as describing the "average error" in our predictions when we use Y' to predict Y scores.

It is appropriate to compute the standard error of the estimate anytime you compute a correlation coefficient, even if you do not perform regression—it's still important to know the average prediction error that your relationship would produce.

A QUICK REVIEW

- The variance of the Y scores around Y' $(S_{Y'}^2)$ and the standard error of the estimate $(S_{Y'})$ measure the errors in prediction when using regression, which are the differences between participants' Y and Y' scores.

MORE EXAMPLES

To compute $S_{Y'}$ first compute S_Y and r. Say that $S_Y = 3.6$ and $r = -.48$. Then $S_{Y'} = S_Y\sqrt{1 - r^2} = 3.6\sqrt{1 - (-.48^2)} = 3.6\sqrt{1 - .2304} = 3.6\sqrt{.7696} = (3.6)(.877) = 3.1572 = 3.16$. Using this relationship produces an "average error" of 3.16. The variance of the Y scores around Y' $S_{Y'}^2$ is $(3.16)^2$, or 9.986.

For Practice

1. The symbol for the variance of the Y scores around Y' is _____.
2. The symbol for the standard error of the estimate is _____.
3. The statistic we interpret as the "average error" in our predictions when using regression is _____.
4. If $S_Y = 2.8$ and $r = +.34$, then $S_{Y'}$ equals _____.

Answers

1. $S_{Y'}^2$
2. $S_{Y'}$
3. The standard error of the estimate, $S_{Y'}$
4. $S_{Y'} = 2.8\sqrt{1 - (.34^2)} = 2.8\sqrt{.8844} = 2.63$

Interpreting the Standard Error of the Estimate

In order for $S_{Y'}$ (and $S_{Y'}^2$) to accurately describe our prediction error, and for r to accurately describe the relationship, you should be able to assume that your data generally meet two requirements.

First, we assume homoscedasticity. **Homoscedasticity** occurs when the Y scores are spread out to the same degree at every X. The left-hand scatterplot in Figure 8.5 shows

FIGURE 8.5

Illustrations of homoscedastic and heteroscedastic data

On the left, the Ys have the same spread at each X; on the right they do not.

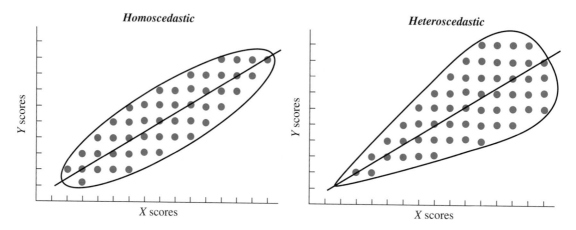

homoscedastic data. Because the vertical spread of the *Y* scores is constant at every *X*, the strength of the relationship is relatively constant at both low *X*s and at high *X*s, so *r* will accurately describe the relationship for all *X*s. Further, the vertical distance separating a data point above or below the regression line on the scatterplot is a way to visualize the difference between someone's *Y* and the *Y'* we predict. When the spread is constant, the standard error of the estimate ($S_{Y'}$) will accurately describe our average error, regardless of whether predictions are based on low or high *X*s.

Conversely, the right-hand scatterplot in Figure 8.5 shows an example of heteroscedastic data. **Heteroscedasticity** occurs when the spread in *Y* is not equal throughout the relationship. Now part of the relationship is very strong (forming a narrow ellipse) while part is much weaker (forming a fat ellipse). Therefore, *r* will not accurately describe the strength of the relationship for all *X*s. Likewise the $S_{Y'}$ will not accurately describe our average error for predictions from both low and high *X*s.

Second, we assume that the *Y* scores at each *X* form an approximately normal distribution. That is, if we constructed a frequency polygon of the *Y* scores at each *X*, we should have a normal distribution centered around *Y'*. Figure 8.6 illustrates this for the widget study. Meeting this assumption is important because $S_{Y'}$ is like a standard deviation. Recall that in a normal distribution approximately 68% of the scores fall between ±1 standard deviation from the mean. Therefore, approximately 68% of all *Y* scores

FIGURE 8.6

Scatterplot showing normal distribution of *Y* scores at each *X*

At each X, there is a normal distribution of Y scores centered around Y'.

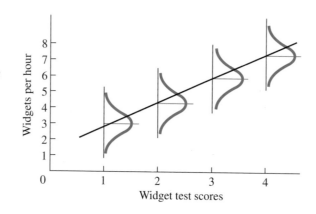

will be between $\pm 1 S_{Y'}$ from the regression line. In the widget study, $S_{Y'}$ is 1.42, so we expect approximately 68% of the actual Y scores to be between ± 1.42 from each value of Y'. Thus, we know the size of over two-thirds of our errors.

The Strength of a Relationship and Prediction Error

Finally, although the standard error of the estimate is the way to *quantify* our "average" prediction error, be sure you understand why this error is communicated by the size of r. A larger r indicates a stronger relationship and *the strength of a relationship determines the amount of prediction error that occurs.* This is because the strength of a relationship is the amount of variability—*spread*—in the Y scores at each X. For example, the left-hand scatterplot in Figure 8.7 shows a relatively strong relationship, with r close to -1. Thus, there is small vertical spread in the Ys at each $X,$ so the data points are close to the regression line. When the data points are close to the regression line it means that participants' actual Y scores are relatively close to their corresponding Y' scores. Therefore, we will find relatively small differences between the participants' Y scores and the Y' we predict for them, so we will have small error, and $S_{Y'}$ and $S_{Y'}^2$ will be small.

Conversely, the right-hand scatterplot in Figure 8.7 shows a weaker relationship, with r closer to 0. This indicates that the Y scores are more spread out vertically around the regression line. Therefore, more often, participants' actual Y scores are farther from their Y' scores, so we will have greater error, and $S_{Y'}$ and $S_{Y'}^2$ will be larger.

Thus, the size of $S_{Y'}$ and $S_{Y'}^2$ is *inversely* related to the size of r. This is why, as we saw in the previous chapter, the size of r allows us to describe the X variable as a good or poor "predictor" for predicting Y scores. When r is large, our prediction error, as measured by $S_{Y'}$ or $S_{Y'}^2$ is small, and so the X variable is a good predictor. However, when r is smaller, our error and $S_{Y'}$ or $S_{Y'}^2$ will be larger, so the X variable is a poorer predictor.

> *REMEMBER*   As the strength of the relationship increases, the actual Y scores are closer to their corresponding Y' scores, producing less prediction error and smaller values of $S_{Y'}$ and $S_{Y'}^2$.

FIGURE 8.7

Scatterplots of strong and week relationships

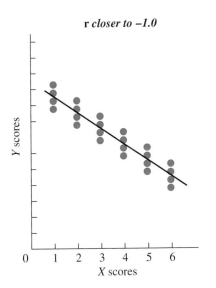

r *closer to –1.0*

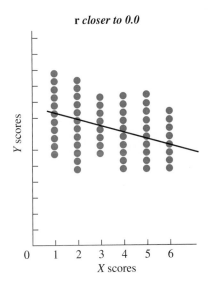

r *closer to 0.0*

In science we need to be more precise than using only terms such as *good* or *poor* predictor. The next section shows how we can *quantify* how effective a predictor variable is by computing the statistic with the strange name of the "proportion of variance accounted for."

COMPUTING THE PROPORTION OF VARIANCE ACCOUNTED FOR

The **proportion of variance accounted for** is the proportional improvement in the accuracy of our predictions produced by using a relationship to predict Y scores, compared to our accuracy when we do not use the relationship to predict Y scores. Understand that the term *proportion of variance accounted for* is a shortened version of "the proportion of variance in Y scores that is accounted for by the relationship with X." We "account" for variance in Y to the extent that the relationship allows us to accurately predict when different Y scores occur, resulting in less prediction error. Therefore, we will compute our "average" prediction error when we use regression and the relationship with X to predict Y scores as we've discussed. We will compare this error to our "average" error when we do not use regression and the relationship with X to predict Y.

To understand this, consider the scatterplots in Figure 8.8. In the graph on the left, we'll ignore that there is relationship with X for the moment. Without the relationship, our fall-back position is to compute the overall mean of all Y scores ($\overline{Y}$) and predict it as everyone's Y score. Here $\overline{Y} = 4$. On the graph, the mean is centered vertically among all Y scores, so it is as if we have the horizontal line shown: At any X, we travel vertically to the line and then horizontally to the predicted Y score, which in every case will be the $\overline{Y}$ of 4.

In Chapter 5 we saw that when we predict the mean score for everyone in a sample, our error in predictions is measured by computing the sample *variance*. Our error in one prediction is the difference between the actual Y score a participant obtains and the $\overline{Y}$ that we predict was obtained. In symbols, this error is $Y - \overline{Y}$. Then the sample variance of the Y scores (S_Y^2) is somewhat like the average error in these predictions. On the left-hand graph in Figure 8.8, each error is the vertical distance a data point is above or below the horizontal line. The distance that *all* Y scores are spread out above and below the horizontal line determines the size of S_Y^2. Researchers can always measure a sample of scores, compute the mean, and use it to predict scores. Therefore, the S_Y^2 is the largest error we are forced to accept. Because this variance is the worst that we

FIGURE 8.8

Scatterplots showing predictions when using and not using a relationship.

On the left, when a relationship is ignored, the mean of Y is always predicted; on the right, when using the relationship, the Y' scores are predicted.

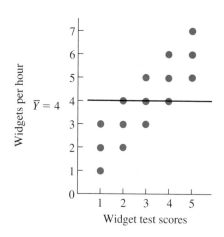

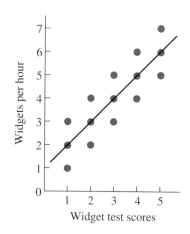

can do, anything that improves the accuracy of our predictions is measured relative to this variance. This is the variance we "account for."

> **REMEMBER** When we do *not* use the relationship to predict scores, our error is S_Y^2 which is computed by finding each $Y - \overline{Y}$, the difference between the Y score a participant actually obtained and the $\overline{Y}$ score we predict is obtained.

Now, let's use the relationship with X to predict scores, as in the right-hand scatterplot back in Figure 8.8. Here, we have the actual regression line and, for each X, we travel up to it and then over to the Y' score. Now our error is the difference between the actual Y scores that participants obtained and the Y' that we predict they obtained. In symbols, this is $Y - Y'$ for each participant. Based on this, as we saw earlier in this chapter, a way to measure our "average error" is the *variance of Y scores around Y'* or $S_{Y'}^2$. In the graph, our error will equal the distance the Y scores are vertically spread out around each Y' on the regression line.

> **REMEMBER** When we *do* use the relationship to predict scores, our error is $S_{Y'}^2$, which is computed by finding each $Y - Y'$, the difference between the Y score a participant actually obtained and the Y' we predict is obtained.

Notice that our error when using the relationship is always less than the error when we don't use the relationship. When we do not use the relationship, we cannot predict any of the differences among the Y scores, because we continuously predict the same $\overline{Y}$ for everyone. Our error is always smaller when we use the relationship because then we predict different scores for different participants: We can, at least, predict a lower Y score for those who tend to have lower Ys, a medium Y score for those scoring medium, and so on. Therefore, to some extent, we're closer to predicting when participants have one Y score and when they have different Y scores. You can see this in Figure 8.8 because most data points tend to be vertically closer to the actual regression line (and closer to their Y') than to the horizontal line that represents predicting the $\overline{Y}$ of 4 each time. Further, the stronger the relationship, the closer the Y scores will be to the regression line so the greater the advantage of using the relationship to predict scores. Therefore, the stronger the relationship, the greater the proportion of variance accounted for.

We compute the proportion of variance accounted for by comparing the error produced when using the relationship (the $S_{Y'}^2$) to the error produced when not using the relationship (the S_Y^2). First, we will do this using the definitional formula. The definitional formula for the proportion of variance accounted for is

$$\text{Proportion of variance accounted for} = 1 - \left(\frac{S_{Y'}^2}{S_Y^2} \right)$$

The formula says to first make a ratio of $S_{Y'}^2 / S_Y^2$. From the widget-making data back in Table 8.3, we know that when we predict the overall mean of Y for participants, our "average error" is the S_Y^2 of 4.38. But, when we predict Y' for participants, the "average error" is the $S_{Y'}^2$ of 2.01. Forming this ratio gives

$$\frac{S_{Y'}^2}{S_Y^2} = \frac{\text{error when using the relationship}}{\text{error when not using the relationship}} = \frac{2.01}{4.38} = .46$$

As shown, this is the ratio of our error when using the relationship to predict scores compared to our error when not using the relationship. The resulting proportion indicates how much of the error that occurs when not using the relationship is still present when we use the relationship. Here, .46 of the error remains when we use the relationship.

But, if .46 of the error remains, then using the relationship *eliminates* .54 of the error that occurs when not using the relationship. As in the formula, this proportion is found by subtracting the ratio from 1. Altogether,

$$1 - \frac{2.01}{4.38} = 1 - .46 = .54$$

Thus, using this relationship eliminates .54, or 54%, of the error that we'd have when not using the relationship, so we are 54% more accurate with it. Essentially, this tells us that, in the graphs back in Figure 8.8, the data points are, on average, 54% closer to the regression line (and Y') than they are to the horizontal line (and $\overline{Y}$). Therefore, if we know participants' X score and use this relationship to predict their Y scores, we are "on average" 54% closer to their actual Y scores than if we don't use this relationship. In statistical terms, we describe this as the proportion of variance accounted for, and so here we account for .54 of the variance in widget-making (Y) scores.

> **REMEMBER** The *proportion of variance accounted for* is the proportional improvement in accuracy when using the relationship with X to predict Y scores, compared to our accuracy when using the $\overline{Y}$ to predict Y scores.

Using *r* to Compute the Proportion of Variance Accounted For

Using the above definitional formula is rather time consuming. However, we've seen that the size of r is related to the error in our predictions by the formula for the standard error of the estimate:

$$S_{Y'}^2 = S_Y^2(1 - r^2)$$

In fact, this formula contains all of the components of the previous definitional formula for the proportion of variance accounted for. Rearranging them gives

$$r^2 = 1 - \frac{S_{Y'}^2}{S_Y^2}$$

On the right is the definitional formula for the proportion of variance accounted for, so

The computational formula for the proportion of variance accounted for is

Proportion of variance accounted for $= r^2$

Not too tough! All you do is compute r (which you would anyway) and then square it. This gives you the proportion of variance in Y scores that is accounted for by the relationship with X.

In the widget study, the r was $+.736$. Therefore, the proportion of variance accounted for is $(.736)^2$, which is again .54.

Thus, although r describes the overall consistency with which the Y scores are paired with the X scores, r^2 is slightly different. It reflects how much the *differences* among the Y scores match with the different X scores, showing how much closer we get to knowing each different Y score when we know a participant's X. The r^2 can be as low as 0 (when $r = 0$), indicating that the relationship in no way helps us to predict Y scores. Or, r^2 may be as high as 1 (when $r = \pm 1$), indicating that, whatever our errors might be without the relationship, 100% of that error is eliminated when using the relationship, because then we predict Y scores with 100% accuracy.

> **REMEMBER**   Compute r^2 whenever you find a relationship. This indicates the proportion of variance accounted for, which is the proportional improvement in accuracy achieved by using the relationship to predict Y scores, compared to not using the relationship.

Note: Sometimes r^2 is called the **coefficient of determination**. The proportion of variance *not* accounted for is called the **coefficient of alienation**.

> ## The computational formula for the proportion of variance not accounted for is
>
> Proportion of variance not accounted for $= 1 - r^2$

This is the proportion of the error we have without using the relationship that still remains when we use the relationship. In the widget study, $r^2 = .54$, so we still cannot account for $1 - .54 = .46$ of the variance in the Y scores.

Notice that r^2 describes the proportion of variance accounted for by the *sample* relationship. If the r passes the inferential statistical test, then r^2 is a *rough* estimate of the proportion of variance accounted for by the relationship in the population. Thus, we'd expect to be roughly 54% more accurate if we use the relationship and widget test scores to predict any other, unknown widget-making scores in the population.

Applying the Proportion of Variance Accounted For

The reason we make such a big deal out of the proportion of variance accounted for is that it is *the* statistical measure of how "important" a particular relationship is. Remember that scores reflect behavior, so when we accurately predict different Y score, we are actually predicting different *behaviors*. The goal of behavioral research is to understand and predict differences in behavior. Therefore, the greater the proportion of variance accounted for, the more that the relationship enables us to do this, so the more scientifically important and informative the relationship is.

Thus, we can now complete our evaluation of the widget-making relationship, where r was $+.74$ and $S_{Y'}$ was 1.42. With r^2 equal to .54, we are .54, or 54%, better off using this relationship than if we did not. And, our average prediction error of 1.42 is 54% less than we'd have without using this relationship. This would be deemed an especially useful relationship in real research because, as we discussed in Chapter 7, we usually find rs between $\pm.30$ to $\pm.50$. Squaring these translates into accounting for between only .09 and .25 of the variance. Therefore, .54 is *very* large, so this is an important relationship, and the widget test should prove valuable for identifying successful widget makers.

We also use r^2 when comparing *different* relationships to see which is more informative. Say that we find a relationship between the length of a person's hair and his or her creativity, but r is only $+.02$. Yes, this r indicates a relationship, but such a weak one is virtually useless. The fact that $r^2 = .0004$ indicates that knowing someone's hair length improves our knowledge about their creativity by only four-hundredths of *one* percent! However, say that we also find a relationship between age and creativity, and here r is $+.40$. This relationship is more important, at least in a statistical sense, because $r^2 = .16$. Age is the more important variable because knowing participants' ages gets us 16% closer to accurately predicting their creativity. Knowing their hair length gets us only .04% closer to predicting their creativity.

The logic of r^2 is applied to any relationship. For example, in the previous chapter, we discussed r_S. Squaring this coefficient also indicates the proportion of variance accounted for. (It is as if we performed the appropriate regression analysis, computed $S_{Y'}^2$ and so on.) Likewise, in later chapters, we will determine the proportion of variance accounted for in experiments. In all cases the answer indicates how useful the relationship is.

> *REMEMBER* Computing the proportion of variance accounted for is the way to evaluate the scientific importance of any relationship.

A QUICK REVIEW

- r^2 is the proportion of variance accounted for: the proportional improvement in accuracy produced by using a relationship to predict Y scores.

- The larger the proportion of variance accounted for, the greater the scientific importance of the relationship.

MORE EXAMPLES

We correlate students' scores on the first exam with their scores on the final exam: $r = +.40$, so $r^2 = (.40)^2 = .16$. To predict final exam scores, we can ignore the relationship and predict that everyone scored at the mean of the final exam. Or we can use this relationship and regression techniques to compute Y' for each student. By using the relationship, we will be "on average" 16% closer to each student's actual final exam score.

For Practice

Relationship A has $r = +.30$; relationship B has $r = -.60$.

1. The relationship with the scatterplot that hugs the regression line more is _____.

2. The relationship with Y scores closest to the Y' scores that we'll predict for them is _____.

3. As compared to predicting the $\overline{Y}$ for participants, relationship A produces Y' scores that are _____ closer to the Y scores.

4. As compared to predicting the $\overline{Y}$ for participants, relationship B produces Y' scores that are _____ closer to the Y scores.

5. Using relationship B to predict scores will improve our accuracy by _____ times as much as will using relationship A.

Answers
1. B
2. B
3. .09
4. .36
5. 4

A WORD ABOUT MULTIPLE CORRELATION AND REGRESSION

Sometimes we discover several X variables that each help us to more accurately predict a Y variable. For example, a positive correlation exists between height and ability to shoot baskets in basketball: The taller people are, the more baskets they tend to make. Also, a positive correlation exists between how much people practice basketball and their ability to shoot baskets: The more they practice, the more baskets they tend to make. Obviously, to be as accurate as possible in predicting how well people will shoot baskets, we should consider both how tall they are and how much they practice. This example has two predictor variables (height and practice) that predict one criterion variable (basket shooting). When we wish to simultaneously consider *multiple predictor variables* for *one criterion variable,* we use the statistical procedures known as multiple correlation and multiple regression. Although the computations involved in these procedures are beyond this text, understand that the **multiple correlation coefficient**, called the *multiple R*, indicates the strength of the relationship between the multiple predictors taken together, and the criterion variable. The **multiple regression equation** allows us to predict someone's Y score by simultaneously considering his or her scores on all X variables. The squared multiple R is the proportion of variance in the Y variable accounted for by using the relationship to predict Y scores.

STATISTICS IN PUBLISHED RESEARCH: LINEAR REGRESSION

In addition to multiple correlation, you may encounter studies that use other, advanced versions of correlation and regression. Understand that the basic approach in these procedures is also to summarize the strength and type of relationship that is present, and to use an X score to predict a central, summary Y score.

In reports of a regression analysis, you will sometimes see our Y', but you may also see the symbol $\hat{Y}$. Our other symbols are generally also found in publications, but a variation of the slope—b—may be referred to as "beta" and "β." Also, when researchers discuss measures of prediction error, they often use the term *residual.* Finally, you may not always see scatterplots in published reports. Instead a graph showing only the regression line may be included, and from the size of r and $S_{Y'}$, you must envision the general scatterplot.

PUTTING IT ALL TOGETHER

This and the previous chapter introduced many new symbols and concepts. However, they boil down to four major topics:

1. *The correlation coefficient:* The correlation coefficient communicates the *type* and *strength* of a relationship. The larger the coefficient, the stronger is the relationship: the more consistently one value of Y is paired with one value of X and the closer the data come to forming a perfect straight-line relationship.

2. *The regression equation:* The regression equation allows you to draw the regression line through the scatterplot and to use the relationship with X to predict any individual's Y score.

3. *Errors in prediction:* The standard error of the estimate indicates the "average" amount your predictions will be in error when using a particular relationship.

4. The proportion of variance accounted for: By squaring a correlation coefficient, you know how much smaller the errors in predicting Y scores are when you use the relationship, compared to if you do not use the relationship.

Using the SPSS Appendix Appendix B.4 explains how to compute the components of the linear regression equation as well as the standard error of the estimate (although it is an estimated population version—as if in our defining formula for $S_{Y'}$ we divided by $N - 2$ instead of N).

CHAPTER SUMMARY

1. *Linear regression* is the procedure for predicting unknown Y scores based on correlated X scores. It produces the *linear regression line,* which is the best-fitting straight line that summarizes a linear relationship.

2. The *linear regression equation* includes the *slope,* indicating how much and in what direction the regression line slants, and the Y *intercept,* indicating the value of Y when the line crosses the Y axis.

3. For each X, the regression equation produces Y', which is the predicted Y score for that X.

4. The *standard error of the estimate* $(S_{Y'})$ is interpreted as the "average error" when using Y' to predict Y scores. This error is also summarized by the *variance of the Y scores around* $Y'(S_{Y'}^2)$.

5. With regression we assume that (1) the Y scores are *homoscedastic,* meaning that the spread in the Y scores around all Y' scores is the same, and (2) the Y scores at each X are normally distributed around their corresponding value of Y'.

6. The stronger the relationship, the smaller are the values of $S_{Y'}$ and $S_{Y'}^2$ because then the Y scores are closer to Y' and so the smaller the difference between Y and Y'.

7. The *proportion of variance accounted for* is the proportional improvement in accuracy that is achieved by using the relationship to predict Y scores, compared to using $\overline{Y}$ to predict scores. This *coefficient of determination* is computed by squaring the correlation coefficient.

8. The proportion of variance not accounted for—the *coefficient of alienation*—is the proportion of the prediction error that is not eliminated when Y' is the predicted score instead of $\overline{Y}$.

9. The proportion of variance accounted for indicates the statistical importance of a relationship.

10. *Multiple correlation and multiple regression* are procedures for describing the relationship when multiple predictor (X) variables are simultaneously used to predict scores on one criterion (Y) variable.

KEY TERMS

Y' b a $S_{Y'}^2$ $S_{Y'}$ r^2
coefficient of alienation *177*
coefficient of determination *177*
criterion variable *162*
heteroscedasticity *172*
homoscedasticity *171*
linear regression equation *163*
linear regression line *161*
multiple correlation coefficient *179*

multiple regression equation *179*
predicted Y score *161*
predictor variable *162*
proportion of variance accounted *174*
slope *163*
standard error of the estimate *170*
variance of the Y scores around Y' *169*
Y intercept *163*

REVIEW QUESTIONS

(Answers for odd-numbered questions are in Appendix D.)

1. What is the linear regression line?
2. What is the linear regression procedure used for?
3. What is Y' and how do you obtain it?
4. What is the general form of the linear regression equation? Identify its component symbols.
5. (a) What does the Y intercept indicate? (b) What does the slope indicate?
6. Distinguish between the *predictor variable* and the *criterion variable* in linear regression.
7. (a) What is the name for $S_{Y'}$? (b) What does $S_{Y'}$ tell you about the spread in the Y scores? (c) What does $S_{Y'}$ tell you about your errors in prediction?
8. (a) What two assumptions must you make about the data in order for the standard error of the estimate to be accurate, and what does each mean? (b) How does heteroscedasticity lead to an inaccurate description of the data?
9. How is the value of $S_{Y'}$ related to the size of r? Why?
10. When are multiple regression procedures used?
11. (a) What are the two statistical names for r^2? (b) How do you interpret r^2?

APPLICATION QUESTIONS

12. What research steps must you go through to use the relationship between a person's intelligence and grade average in high school so that, if you know a person's IQ, you can more accurately predict the person's grade average?
13. We find that the correlation between math ability (X) and musical aptitude scores (Y) is $r = +.44$. The standard error of the estimate is $S_{Y'} = 3.90$. Bubbles has a math score of 60 and Foofy a score of 72. (a) Based on their math scores, who is predicted to have the higher music score and why? (b) What procedure would we use to predict their music scores? (c) If our predictions are wrong, what is the "average" amount we expect to be "off"? (d) How much smaller is this error than if we don't use math scores to predict music scores? (e) What is you answer in part (d) called?

14. (a) Explain conceptually why the proportion of variance accounted for equals 1.0 with a perfect correlation. (b) Why should you expect most relationships to account for only about 9% to 25% of the variance?

15. What do you know about a research project when you read that it employed multiple correlation and regression procedures?

16. Poindexter finds $r = -.80$ when correlating number of hours studied and number of errors made on a statistics test. He also finds $r = +.40$ between speed of taking the test and number of errors on the test. He concludes that hours studied forms twice as strong a relationship and is therefore twice as important as the speed of taking the test. (a) Why is he correct or incorrect? (b) Compare these relationships and draw the correct conclusion.

17. (a) In question 16 what advanced statistical procedures can Poindexter employ to improve his predictions about test errors even more? (b) Say that the resulting correlation coefficient is .67. Using the proportion of variance accounted for, explain what this means.

18. A researcher finds that the correlation between variable A and variable B is $r = +.20$. She also finds that the correlation between variable C and variable B is $r = -.40$. Which relationship is scientifically more useful and by how much?

19. You measure how much people are initially attracted to a person of the opposite sex and how anxious they become during their first date. For the following ratio data, answer the questions below.

Participant	Attraction X	Anxiety Y
1	2	8
2	6	14
3	1	5
4	3	8
5	6	10
6	9	15
7	6	8
8	6	8
9	4	7
10	2	6

(a) Compute the statistic that describes the relationship here. (b) Compute the linear regression equation. (c) What anxiety score do you predict for a person who has an attraction score of 9? (d) When using this relationship, what is the "average" amount of error you should expect in your predictions?

20. (a) For the relationship in question 19, what is the proportion of variance accounted for? (b) What is the proportion of variance not accounted for? (c) Why is or is not this a valuable relationship?

21. In question 19 of the Application Questions in Chapter 7, we correlated "burnout" scores (X) with absenteeism scores (Y). Using those data: (a) Compute the linear regression equation. (b) Compute the standard error of the estimate. (c) For a burnout of 4, what absence score is predicted? (d) How useful is knowing burnout scores for predicting absenteeism?

22. A researcher measures how positive a person's mood is and how creative he or she is, obtaining the following interval scores:

Participant	Mood X	Creativity Y
1	10	7
2	8	6
3	9	11
4	6	4
5	5	5
6	3	7
7	7	4
8	2	5
9	4	6
10	1	4

(a) Compute the statistic that summarizes this relationship. (b) What is the predicted creativity score for anyone scoring 3 on mood? (c) If your prediction is in error, what is the amount of error you expect to have? (d) How much smaller will your error be if you use the regression equation than if you merely used the overall mean creativity score as the predicted score for all participants?

23. Dorcas complains that it is unfair to use SAT scores to determine college admittance because she might do much better in college than predicted. (a) What statistic(s) will indicate whether her complaint is likely to be correct? (b) In reality, the positive correlation coefficient between SAT scores and college performance is only moderate. What does this tell you about how useful the SAT is for predicting college success?

INTEGRATION QUESTIONS

24. (a) How are experiments and correlational designs similar in their purpose?
(b) What is the distinguishing characteristic between an experiment and a correlational study? (Chs. 2, 7)

25. In the typical experiment, (a) Do we group the scores to summarize them, and if so how? (b) What are the two descriptive statistics we compute for the summary? (Chs. 4, 5)

26. In the typical correlational design, (a) Do we group the scores to summarize them, and if so how? (b) What are the three descriptive statistics we compute for the summary? (Chs. 7, 8)

27. In a typical experiment, a researcher says she has found a good predictor (a) What name do we give to this variable? (b) Scores on which variable are being predicted?
(c) For particular participants, what score will be used as the predicted score?
(d) What statistic describes the "average" error in predictions? (Chs. 4, 5)

28. In a typical correlational study, a researcher says he has found a good predictor
(a) What name do we give to this variable? (b) Scores on which variable are being predicted? (c) For a particular participant, what score will be the predicted score?
(d) What statistic describes the "average" error in predictions? (Chs. 7, 8)

29. You know that a relationship accounts for a substantial amount of variance. What does this tell you about (a) its strength? (b) the consistency that a particular Y is paired with only one X, (c) the variability in Y scores at each X? (d) how closely the scatterplot hugs the regression line? (e) the size of the correlation coefficient? (f) how well we can predict Y scores? (Chs. 7, 8)

■ ■ SUMMARY OF FORMULAS

1. The formula for the linear regression equation is

$$Y' = bX + a$$

2. The formula for the slope of the linear regression line is

$$b = \frac{N(\Sigma XY) - (\Sigma X)(\Sigma Y)}{N(\Sigma X^2) - (\Sigma X)^2}$$

3. The formula for the Y intercept of the linear regression line is

$$a = \overline{Y} - (b)(\overline{X})$$

4. The formula for the variance of Y scores around Y' is

$$S_{Y'}^2 = S_Y^2(1 - r^2)$$

5. The formula for the standard error of the estimate is

$$S_{Y'} = S_Y\sqrt{1 - r^2}$$

6. The formula for the proportion of variance accounted for is

$$\text{Coefficient of determination} = r^2$$

7. The formula for the proportion of variance not accounted for is

$$\text{Coefficient of determination} = 1 - r^2$$

9 Using Probability to Make Decisions about Data

GETTING STARTED

To understand this chapter, recall the following:

- From Chapter 3, that relative frequency is the proportion of time that scores occur.
- From Chapter 6, that by using z-scores we can determine the *proportion of the area under the normal curve* for particular scores and thereby determine their relative frequency.
- Also from Chapter 6, that by using a *sampling distribution of means* and z-scores we can determine the proportion of the area under the normal curve for particular sample means and thereby determine their relative frequency.

Your goals in this chapter are to learn

- What *probability* is and how it is computed.
- How to compute the probability of raw scores and sample means using z-scores.
- How *random sampling* and *sampling error* may or may not produce *representative* samples.
- How to use a sampling distribution of means to determine whether a sample represents a particular population.

You now know the common descriptive statistics used in behavioral research. Therefore, you are ready to learn the other type of procedures, called *inferential statistics*. Later chapters will show you the different procedures that are used with different research designs. This chapter sets the foundation for all inferential procedures by introducing you to the wonderful world of probability. Don't worry, though. The discussion is rather simple, and there is little in the way of formulas. However, you do need to understand the basics of probability. In the following sections, we discuss (1) what probability is, (2) how we determine probability, and (3) how to use probability to draw conclusions about samples.

NEW STATISTICAL NOTATION

In daily conversation, the words *chances, odds,* and *probability* are used interchangeably. In statistics, however, they have different meanings. Odds are expressed as fractions or ratios ("The odds of winning are 1 in 2"). Chance is expressed as a percentage ("There is a 50% chance of winning"). Probability is expressed as a decimal ("The probability of winning is .50"). For inferential procedures, always express the answers you compute as probabilities.

The symbol for probability is p. The probability of a particular event—such as event A—is $p(A)$, which is pronounced "p of A" or "the probability of A."

WHY IS IT IMPORTANT TO KNOW ABOUT PROBABILITY?

Recall that ultimately we want to draw inferences about relationships in nature—in what we call the population. However, even though we may find a relationship in our sample data, we can never "prove" that this relationship is found in the population. We must measure the entire population to *know* what occurs, but if we could measure the (infinite) population, we wouldn't need samples to begin with. Thus, we will always have some uncertainty about whether the relationship in our sample is found in the population. The best that we can do is to describe what we think is *likely* to be found. This is where inferential statistics come into play. Recall from Chapter 2 that **inferential statistics** are for deciding whether our sample data accurately represent the relationship found in the population (in nature). Essentially, we use probability and inferential procedures to make an intelligent bet about whether we would find the sample's relationship if we could study everyone in the population. Therefore, it is important that you understand the basics of probability so that you understand how we make these bets. This is especially so, because we always have this uncertainty about the population, so we perform inferential statistics in *every* study. Therefore, using probability—and making bets—is an integral part of *all* behavioral research.

THE LOGIC OF PROBABILITY

Probability is used to describe *random,* or chance, events. People often mistakenly believe that the definition of *random* is that events occur in a way that produces no pattern. However, the events may form a pattern, so, for example, a lottery might produce the random sequence of 1, 2, 3, 4, 5. Rather, an event is random if it is *uncontrolled,* so that the events occur naturally without any influence from us. That is, we allow nature to be fair, with no bias toward one event over another (no rigged roulette wheels or loaded dice). Thus, a random event occurs or does not occur merely because of the luck of the draw. Probability is our way of mathematically describing how luck operates to produce an event.

Using statistical terminology, the event that *does* occur in a given situation is a *sample.* The larger collection of all possible events that *might* occur in this situation is the *population.* Thus, the event might be drawing a particular playing card from the population of a deck of cards, or tossing a coin and obtaining a particular sample of heads and tails from the population of possible heads and tails. In research, the event is obtaining a particular sample that contains particular participants or scores.

Because probability deals only with random events, we compute the probability only of samples obtained through random sampling. **Random sampling** is selecting a sample in such a way that all elements or individuals in the population have an equal chance of being selected. Thus, in research, random sampling is anything akin to blindly drawing names from a large hat. In all cases, a particular random sample occurs or does not occur simply because of the luck of the draw.

But how can we describe an event that happens only by chance? By paying attention to how *often* the event occurs when chance is operating. The probability of any event is based on how often the event occurs *over the long run.* Intuitively, we use this logic all the time: If event A happens frequently over the long run, then we think it is likely to happen again now, and we say that it has a high probability. If event B happens infrequently, then we think it is unlikely, and we say that it has a low probability.

When we decide that an event happens frequently, we are making a *relative* judgment, describing the event's *relative frequency*. This is the proportion of time that the event occurs out of all events that might occur from the population. This is also the event's probability. The **probability** of an event is equal to the event's relative frequency in the population of possible events that can occur. An event's relative frequency is a number between 0 and 1, so the event's probability is from 0 to 1.

> *REMEMBER* The *probability* of an event equals the event's relative frequency in the population.

Probability is essentially a system for expressing our *confidence* that a particular random event will occur. First, we assume that an event's past relative frequency will continue over the long run in the future. Then we express our confidence that the event will occur in any *single* sample by expressing the relative frequency as a probability between 0 and 1. For example, I am a rotten typist, and I randomly make typos 80% of the time. This means that in the population of my typing, typos occur with a relative frequency of .80. We expect the relative frequency of my typos to continue at a rate of .80 in anything I type. This expected relative frequency is expressed as a probability: the probability is .80 that I will make a typo when I type the next woid.

Likewise, all probabilities communicate our confidence. If event A has a relative frequency of zero in a particular situation, then the probability of event A is zero. This means that we do not expect A to occur in this situation because it never does. But if event A has a relative frequency of .10 in this situation, then A has a probability of .10: Because we expect it to occur in only 10% of our samples, we have some—but not much—confidence that A will occur in the next sample. On the other hand, if A has a probability of .95, we are confident that it will occur: It occurs 95% of the time in the population, we expect it in 95% of our samples, and so our confidence is at .95 that it will occur now. At the most extreme, an event's relative frequency can be 1: It is 100% of the population, so its probability is 1. Here, we are positive it will occur in this situation because it always does.

All possible events together constitute 100% of the population. This means that the relative frequencies of all events must add up to 1, so the probabilities must also add up to 1. Thus, if the probability of my making a typo is .80, then because $1 - .80 = .20$, the likelihood that I will type error-free words is $p = .20$.

Understand that except when p equals either 0 or 1, it is up to chance whether a particular sample contains the event. For example, even though I make typos 80% of the time, I may go for quite a while without making one. That 20% of the time I make no typos has to occur sometime. Thus, although the probability is .80 that I will make a typo in each word, it is only over the long run that we expect to see precisely 80% typos.

People who fail to understand that probability implies *over the long run* fall victim to the "gambler's fallacy." For example, after observing my errorless typing for a while, the fallacy is thinking that errors "must" occur now, essentially concluding that errors have become more likely. Or, say we are flipping a coin and get seven heads in a row. The fallacy is thinking that a head is now less likely to occur because it's already occurred too often (as if the coin says, "Hold it. That's enough heads for a while!"). The mistake of the gambler's fallacy is failing to recognize that whether an event occurs or not in a sample does not alter its probability because probability is determined by what happens "over the long run."

COMPUTING PROBABILITY

Computing the probability of an event is simple: We need only determine its relative frequency in the population. When we know the relative frequency of all events, we have a probability distribution. A **probability distribution** indicates the probability of all events in a population.

Creating Probability Distributions

We have two ways to create a probability distribution. One way is to observe the relative frequency of the events, creating an *empirical probability distribution.* Typically, we cannot observe the entire population, so we observe samples from the population. We assume that the relative frequencies in the samples represent the relative frequencies in the population.

For example, say that Dr. Fraud is sometimes very cranky, and apparently his crankiness is random. We observe him on 18 days and he is cranky on 6 of them. Relative frequency equals f/N, so the relative frequency of Dr. Fraud's crankiness is 6/18, or .33. We expect him to continue to be cranky 33% of the time, so $p = .33$ that he will be cranky today. Conversely, he was not cranky on 12 of the 18 days, which is 12/18, or .67. Thus, $p = .67$ that he will not be cranky today. Because his cranky days plus his noncranky days constitute all possibilities here, we have the complete probability distribution for his crankiness.

Statistical procedures usually rely on the other way to create a probability distribution. A *theoretical probability distribution* is a theoretical model of the relative frequencies of events in a population, based on how we *assume* nature distributes the events. From such a model, we determine the expected relative frequency of each event, which is then the probability of each event.

For example, when tossing a coin, we assume that nature has no bias toward heads or tails, so over the long run we expect the relative frequency of heads to be .50 and the relative frequency of tails to be .50. Because relative frequency in the population *is* probability, we have a theoretical probability distribution for coin tosses: The probability of a head on any toss is $p = .50$, and the probability of a tail is $p = .50$.

Or, consider a deck of 52 playing cards. The deck is actually the population, so each card occurs once out of 52 draws. Therefore, the probability of you drawing any particular card from a full deck is $1/52 = .019$. Likewise, with 4 "Kings" in a full deck, the probability of you selecting one is $4/52 = .077$.

Finally, if the numbers you select for your state's lottery drawing have a 1 in 17 million chance of winning, it's because there are 17 million different possible combinations of numbers to select from. We expect to draw all 17 million combinations equally often over the long run. Therefore, we'll draw your selection at a rate of once out of every 17 million draws, so your chance of winning on today's draw is 1 in 17 million.

And that is the logic of probability. First, we either theoretically or empirically devise a model of the expected relative frequency of each event in the population. Then, an event's relative frequency equals its probability (our confidence) that it will occur in a particular sample.

Factors Affecting the Probability of an Event

Not all random events are the same, and their characteristics influence their probability. First, events may be either independent or dependent. Two events are **independent events** when the probability of one is *not* influenced by the occurrence of the other.

For example, contrary to popular belief, washing your car does *not* make it rain. These are independent events, so the probability of rain does not change when you wash your car. On the other hand, two events are **dependent events** when the probability of one *is* influenced by the occurrence of the other. For example, whether you pass an exam usually depends on whether you study: The probability of passing increases or decreases *depending* on whether studying occurs.

An event's probability is also affected by the type of sampling we perform. When **sampling with replacement**, any previously selected individuals or events are *replaced* back into the population before drawing additional ones. For example, say we will select two playing cards. Sampling with replacement occurs if, after drawing the first card, we return it to the deck before drawing the second card. Notice that the probabilities on each draw are based on 52 possible outcomes, and so they stay constant. On the other hand, when **sampling without replacement**, previously selected individuals or events are *not* replaced into the population before selecting again. Thus, sampling without replacement occurs if, after a card is drawn, it is discarded. Now the probability of selecting a particular card on the first draw is based on 52 possible outcomes, but the probability of selecting a card on the second draw is different because it is based on only 51 outcomes.

A QUICK REVIEW

- An event's probability equals its relative frequency in the population.
- A probability distribution indicates all probabilities for a population, and is influenced by whether the events are independent or dependent and whether sampling is with or without replacement.

MORE EXAMPLES

One hundred raffle tickets are sold each week. Assuming that all tickets are equally likely to be drawn, each should be selected at a rate of 1 out of 100 draws over the long run. Therefore, the probability that you hold the winning ticket this week is $p = 1/100 = .01$.

For Practice

1. The probability of any event equals its _____ in the _____.
2. As the p of an event decreases, the event's relative frequency in this situation _____.
3. As the p of an event increases, our confidence that the event will occur _____.
4. Tossing a coin (heads or tails) is sampling _____ replacement.

Answers

1. relative frequency; population.
2. decreases
3. increases
4. with

OBTAINING PROBABILITY FROM THE STANDARD NORMAL CURVE

The reason we discuss probability is not because we have an uncontrollable urge to flip coins and draw cards. In research, the random events that we are interested in are scores. For our discussions, pretend that we've already measured everyone's score in a particular situation so that we can randomly select directly from the population of *scores*. Further, we will assume that scores are independent (whether someone else scores high on a test does not influence the p that you'll score high) and sampled with replacement (the selection of a particular score does not remove it from the population). Then our theoretical probability distribution is usually based on the standard normal curve. Here's how it works.

Determining the Probability of Individual Scores

In Chapter 6, you used *z*-scores to find *the proportion of the total area under the normal curve* in any part of a distribution. This proportion corresponds to the relative frequency of the scores in that part of the population. Now, however, you know that the relative frequency of scores in the population *is* their probability. Therefore, the proportion of the total area under the curve for particular scores equals the probability of those scores.

For example, Figure 9.1 shows the distribution for a set of scores. Say that we seek the probability of randomly selecting a score below the mean of 59. To answer this, think first in terms of *z*-scores. Raw scores below the mean produce negative *z*-scores, so the question becomes "What is the probability of randomly selecting a negative *z*-score?" Negative *z*-scores constitute 50% of the area under the curve and thus have a relative frequency of .50. Therefore, the probability is .50 that we will select a negative *z*-score. (Using our "parking lot" view of the curve, 50% of the parking lot holds people with negative *z*-scores. If we put the names of everyone in the lot in a hat and started selecting, we'd expect to draw someone with a negative *z*-score 50% of the time: therefore, *p* = .5 of getting one on a single draw.) Now think raw scores: Because negative *z*-scores correspond to raw scores below 59, the probability is also .50 that we will select a raw score below 59.

Likewise, say that we seek the *p* of selecting a score between 59 and 69. In Figure 9.1, a raw score of 69 has a *z* = +1. From column B of the *z*-tables in Appendix C, *z*-scores between the mean and a *z* of +1 occur .3413 of the time. Thus, the probability is .3413 that we will select one of these *z*-scores, so the probability is also .3413 that we will select a raw score between 59 and 69.

Or, say that we seek the probability of selecting a raw score above 79. Figure 9.1 shows that a raw score of 79 is at a *z*-score of +2. From column C of the *z*-table, the relative frequency of scores beyond a *z* of +2 is .0228. Therefore, the probability is .0228 that we will select a raw score above 79.

Finally, understand what we mean when a score is beyond a *z* of *plus or minus* some amount. For example, beyond *z*-scores of ±2 means that we seek scores in the tail above +2 or in the tail below −2. We saw that beyond *z* = +2 is .0228 of the curve and so, beyond *z* = −2 is also .0228. The word *or* indicates that we don't distinguish between the tails, so we *add* the two areas together. In total, .0228 + .0228, or .0456, of the area under the curve contains the scores we seek. In Figure 9.1, a raw score of 39 is at *z* = −2, and 79 is at *z* = +2. Thus, *p* = .0456 that we will select a raw score below 39 or above 79.

Researchers seldom determine the probability of an individual score. However, by understanding the preceding, you can understand a major part of inferential statistics, which is to determine the probability of obtaining particular sample means.

FIGURE 9.1

z-distribution showing the area for scores below the mean, between the mean and *z* = +1, and above *z* = +2

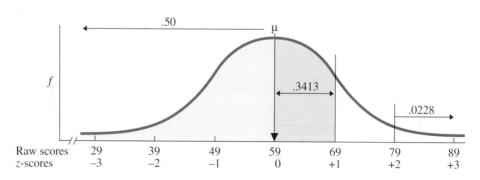

Raw scores	29	39	49	59	69	79	89
z-scores	−3	−2	−1	0	+1	+2	+3

Determining the Probability of Sample Means

In Chapter 6, we conceptualized the *sampling distribution of means* as the frequency distribution of all possible sample means—the population—that would result if our bored statistician randomly sampled a raw score population an infinite number of times using a particular N. For example, Figure 9.2 shows the sampling distribution of means produced from the population of SAT scores when $\mu = 500$. Recognize that the different values of $\overline{X}$ occur here simply because of the luck of the draw of which scores are selected for a sample. Because many raw scores are near 500, the bored statistician frequently selects samples with means at 500. But sometimes a sample mean somewhat higher than 500 occurs because, by chance, she happened to select scores that are predominantly above 500. Sample means that are *far* above 500, however, occur less frequently because chance seldom produces only scores very far above 500. Likewise, sometimes she obtains scores that produce means somewhat below 500, while less frequently will luck produce means far below 500. Thus, the sampling distribution provides a picture of how often different sample means occur due to chance when sampling this underlying SAT raw score population. Therefore, the sampling distribution is a theoretical probability distribution.

We use the sampling distribution to determine the probability of sample means in the same way that we previously determined the probability of raw scores. First, as in Chapter 6, we compute the mean's z-score. To do this, recall that we first compute the *standard error of the mean* using the formula

$$\sigma_{\overline{X}} = \frac{\sigma_X}{\sqrt{N}}$$

Then we compute a z-score for the sample mean using the formula

$$z = \frac{\overline{X} - \mu}{\sigma_{\overline{X}}}$$

Then, by applying the standard normal curve model and z-table, we can determine the probability of particular sample means.

For example, in Figure 9.2, our $N = 25$ and $\sigma_X = 100$, so the standard error of the mean is 20. Say that we're interested in the sample means between the μ of 500 and the $\overline{X}$ of 520. First, the mean's z-score is $(520 - 500)/20 = +1$. Thus, in other words, we're interested in means having z-scores between 0 and $+1$. The relative frequency of such z-scores is .3413. Therefore, $p = .3413$ that we will randomly select a sample mean with a z-score between 0 and $+1$. Likewise, the probability is .3413 that we will select a sample mean that is between 500 and 520.

Think about this: Randomly selecting a sample mean is the same as selecting a sample of raw scores that produce that mean. Also, randomly selecting a sample of raw scores is the same as selecting a sample of participants who then produce these scores. Therefore, when we use the sampling distribution to determine the probability of selecting particular

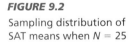

FIGURE 9.2

Sampling distribution of SAT means when $N = 25$

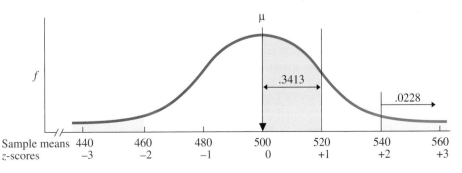

sample means, we are actually finding the probability that we will select a sample of participants whose scores will produce those means. Thus, we can rephrase our finding above: When we randomly select 25 participants from the SAT population, the probability is .3413 that we will select a sample that produces a mean between 500 and 520.

Likewise, we can find probabilities in other parts of the sampling distribution. For example, back in Figure 9.2, a mean of 540 has $z = +2$, and we've seen that the relative frequency of scores beyond this z is .0228. Therefore, $p = .0228$ that we will select a sample of SAT scores that produce a mean above 540. Or, we might seek the probability of means below 460 *or* above 540. In Figure 9.2, these means translate into z-scores of ± 2. With .0228 of the area under the curve beyond each z, a total of .0456 of the curve is beyond ± 2. Thus, the probability is only .0456 that we will select a sample that produces a mean below 460 or above 540.

> *REMEMBER* Determine the probability of particular sample means by computing their z-score on the sampling distribution.

As with individual raw scores, computing the exact probability of sample means will not be a big part of what we do. Instead, you should understand the general logic of how z-scores and a sampling distribution indicate the likelihood of various sample means. In particular, as in Figure 9.2, see how a small z-score indicates that we are generally close to the center of the sampling distribution, where samples having such means are relatively frequent. Therefore, *samples that produce means near μ are relatively likely to occur.* Conversely, the larger a z-score, the farther into the tail of the sampling distribution we are, so samples having these means are relatively infrequent. *Samples that produce means in the tails of the sampling distribution are unlikely to occur.* Thus, for example, an SAT mean of 560 has a z-score of $+3$. We—or the bored statistician—are extremely unlikely to randomly select a sample having such a mean because such samples (and means) hardly ever occur, even over the long run.

> *REMEMBER* The larger the z-score, the less likely the corresponding sample mean is to occur.

A QUICK REVIEW

- To find the probability of particular sample means, we envision the sampling distribution, compute the z-score, and apply the z-tables.
- The farther into the tail of the sampling distribution that a sample mean falls, the less likely it is to occur.

MORE EXAMPLES

In a population, $\mu = 35$ and $\sigma_X = 8$. What is the probability of obtaining a sample $(N = 16)$ with a mean above $\overline{X} = 38.3$? To compute the z-score, first compute the standard error of the mean: $\sigma_{\overline{X}} = \sigma_X / \sqrt{N} = 8/\sqrt{16} = 2$. Then $z = (\overline{X} - \mu)/\sigma_{\overline{X}} = (38.3 - 35)/2 = +1.65$. The means above 38.3 are in the upper tail of the distribution, so from column C of the z-table, sample means above 38.3 have a $p = .0495$.

For Practice

1. With $\mu = 500$, $\sigma_X = 100$, and $N = 25$, what is the probability of selecting a $\overline{X}$ above 530?

2. Approximately, what is the probability of selecting an SAT sample mean having a z-score between ± 1?

3. If $\mu = 100$, are we more likely to obtain a sample mean that is close to 100 or a mean that is very different from 100?

4. The farther that sample means are into the tail of the sampling distribution, the *lower/higher* their probability.

Answers

1. $\sigma_{\overline{X}} = 100/\sqrt{25} = 20$; $z = (530 - 500)/20 = +1.5$; $p = .0668$
2. With about 68% of the distribution, $p = .68$.
3. A mean close to 100 is more likely.
4. lower

RANDOM SAMPLING AND SAMPLING ERROR

Computing the probability of sample means as we've described forms the basis for all inferential statistics. Why do we need such procedures? Recall that in research we want to conclude that the way our sample behaves is also the way the entire population would behave. However, we need inferential statistics because there is no guarantee that the sample *accurately* reflects the population. In other words, we are never certain that a sample is representative. In a **representative sample**, the characteristics of the individuals and scores in the sample accurately reflect the characteristics of individuals and scores found in the population. Thus, if 55% of the population is female, then a sample will be representative if it is also 55% female. If 20% of the population scored 475, then a sample is representative if 20% of the sample's scores are 475. And so on, so that the proportions of the sample made up by the various individuals and their scores equal the proportions found in the population. Thus, to put it simply, a representative sample is a miniature version of the population. This is why, if the μ in the SAT population is 500, then the $\overline{X}$ in a representative sample will be 500.

To produce representative samples, researchers select participants using random sampling. A random sample *should* be representative because, by being unselective in choosing participants, we allow the characteristics of the population to occur naturally in the sample, in the same ways that they occur in the population. Thus, if 55% of the population is female, then 55% of a random sample should be female because that is how often we will encounter females. In the same way, random sampling should produce a sample having all of the characteristics of the population.

At least we hope it works that way! A random sample "should" be representative, but nothing forces this to occur. The problem is that, *just by the luck of the draw,* we can obtain a sample whose characteristics do not match those of the population. However, representativeness is not all or nothing. Depending on the individuals and scores selected, a sample can be somewhat representative, only somewhat matching the population. For example, 20% of the population may score at 475, but simply through the luck of who is selected, this score might occur 10% or 30% of the time in our sample. If so, the sample will have characteristics that are only somewhat similar to those of the population, and although μ may be 500, the sample mean will not be 500. In the same way, depending on the scores we happen to select, any sample may not be perfectly representative of the population from which it is selected, so the sample mean will not equal the population mean it is representing.

The statistical term for communicating that chance produced an unrepresentative sample is to say that the sample reflects sampling error. **Sampling error** occurs when random chance produces a sample statistic (such as $\overline{X}$) that is not equal to the population parameter it represents (such as μ). Sampling error conveys that the reason a sample mean is different from μ is because, by chance, the sample is unrepresentative of the population. That is, because of the luck of the draw, the sample contains too many high scores or too many low scores relative to the population, so the sample is in *error* in representing the population.

> **REMEMBER** *Sampling error* results when, by chance, the scores that are selected produce a sample statistic that is different from the population parameter it represents.

Here then, is the central problem for researchers and the reason for inferential statistics: When sampling error produces a sample that is different from the population that it comes from and represents, it has the characteristics of some *other* population. The problem is that then the sample appears to come from and represent that other population. Thus, although a sample always represents some population, we are never sure *which* population it represents: Through sampling error the sample may poorly represent one population although it doesn't look like it represents that one, or the sample may accurately represent some other population altogether.

> *REMEMBER* Any sample may poorly represent one population, or it may accurately represent a different population.

For example, say that we return to the SAT scores of Prunepit University and find that a random sample obtains a mean score of 550. This is surprising because the ordinary, national population of SAT scores has a μ of 500. Therefore, we should have obtained a sample mean of 500 if our sample was perfectly representative of this population. How do we explain a sample mean of 550? On the one hand, maybe we simply have sampling error. Maybe because of the luck of the draw, we selected too many students with high scores and not enough with low scores so that the sample mean came out to be 550 instead of 500. Thus, it's possible that chance produced a less than perfectly representative sample, but the population being represented is still that ordinary population where μ is 500. On the other hand, perhaps the sample does not come from and represent the ordinary population of SAT scores. After all, these *are* Prunepit students, so they may belong to a very different population of students, having some other μ. For example, maybe Prunepit students belong to the population where μ is 550, and their sample is perfectly representing this population.

The solution to this dilemma is to use inferential statistics to make a decision about the population being represented by our sample. The next chapter puts all of this into a research context, but in the following sections we'll examine the basics of deciding whether a sample represents a particular population.

DECIDING WHETHER A SAMPLE REPRESENTS A POPULATION

We deal with the possibility of sampling error in this way: Because we rely on random sampling, how representative a sample is depends on random chance—the luck of the draw of which individuals and scores are selected. Therefore, we can determine whether our sample is *likely* to come from and thus represent a particular population. If chance is *likely* to produce our sample from the population, then we decide that our sample *does* come from and represent that population, although maybe with a little sampling error. However, if chance is *unlikely* to produce our sample from the population, then we decide that the sample does *not* represent that population, and instead represents some other population.

Here's a non-math example. You obtain a paragraph of someone's typing, but you don't know whose. Is it mine? Does it represent the population of my typing? Say there are zero typos in the paragraph. It's possible that some quirk of chance produced such an unrepresentative sample, but it's not likely: I type errorless words only 20% of the time, so the probability of an errorless *paragraph* is extremely small. Thus, because chance is unlikely to produce such a sample from the population of my typing, you should conclude that the sample represents the population of a competent typist where such a sample is more likely.

On the other hand, say that there are typos in 75% of the words in the paragraph. This is consistent with what you would expect if the sample represents my typing, but we have a little sampling error. Although you expect 80% typos from me over the long run, you would not expect precisely 80% typos in every sample. Rather, a sample with 75% errors seems likely to occur simply by chance when the population of my typing is sampled. Therefore, you can accept that this paragraph represents my typing, albeit somewhat poorly.

We will use the same logic to decide whether our Prunepit sample represents the population of SAT scores where μ is 500: we will determine the probability of obtaining a sample mean of 550 from this population. As you've seen, we determine the probability of a sample mean by computing its z-score on the sampling distribution of means. Thus, we first envision the sampling distribution showing the different means that the bored statistician would obtain if, using our N, she randomly sampled the ordinary SAT population an infinite number of times. This is shown in Figure 9.3.

Notice, the statistician is definitely representing the SAT population where μ is 500, so whether she obtained a particular mean that was high, low, or in-between depends purely on the luck of the draw of which scores she happened to select. Therefore, think of a sampling distribution as a "picture of chance," showing how often chance produces different sample means when we sample a particular raw score population. Essentially, it shows how often different degrees of sampling error occur.

The next step is to calculate the z-score for our sample mean of 550 so that we can determine its likelihood. In reality we would not always expect a *perfectly* representative sample, so we would not expect a sample mean of precisely 500 every time. Instead, if our sample is representing this population, then the sample mean should be *close* to 500. For example, say that the z-score for our mean is at location A in Figure 9.3. Read what the frequency distribution indicates by following the dotted line: This mean has a very high relative frequency and thus is very *likely* when we are drawing a sample from the ordinary SAT population. By being "close" to μ, this tells us that samples are often unrepresentative to this extent, and so the bored statistician frequently encountered this mean when she *was* representing the SAT population. Thus, this is a mean that we'd expect to see if *we* are representing this population. In fact, to put it simply, we obtained an expected mean that happens often with this population. Therefore, we will conclude it is a good bet that our sample comes from and represents the ordinary SAT population, even though it doesn't look like it represents

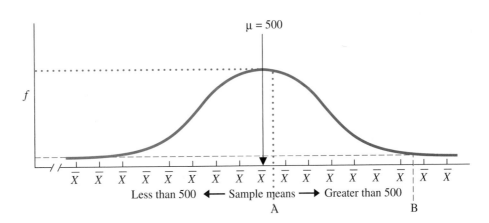

FIGURE 9.3

Sampling distribution of SAT means showing two possible locations of our sample mean that population. We assume the discrepancy is due to sampling error where, by chance, we obtained a few too many high scores so our $\overline{X}$ turned out to equal 550 instead of 500.

However, say that, instead, our sample has a z-score at location B back in Figure 9.3. Following the dashed line shows that this is a very infrequent and *unlikely* mean. By being "far" from μ, this tells us that samples are seldom unrepresentative to this extent, and so the bored statistician almost never encountered such an extreme case of sampling error when she *was* representing this SAT population. Thus, this is a mean we would not expect to see if *we* are representing this population. To put it simply, we obtained an unexpected mean that almost never happens with this population! Therefore, because it is a bad bet that our sample comes from and represents the ordinary SAT population, we will *reject* that we are representing this population, rejecting that the discrepancy between our $\overline{X}$ and 500 reflects sampling error. Instead, it makes more sense to conclude that the sample represents and comes from some other raw score population (having some other μ), where this sample is more likely.

Be sure you understand the above logic before proceeding, because it is used in all inferential procedures. We will always have a known, underlying raw score population that a sample may or may not represent. From that raw score population we envision the sampling distribution of means that would be produced. Then we determine the location of our sample mean on the sampling distribution. The farther into the tail of the sampling distribution the sample mean is, the less likely that the sample comes from and represents the original underlying raw score population.

A QUICK REVIEW

- If the z-score shows that a sample mean is *unlikely* in the sampling distribution, reject that the sample is merely poorly representing the underlying raw score population.

- If the z-score shows that a sample mean is *likely* in the sampling distribution, conclude that the sample represents the underlying raw score population, albeit somewhat poorly.

MORE EXAMPLES

On the sampling distribution created from body weights in the United States, an $\overline{X}$ produces $z = +5$! Such a mean is so unlikely when representing this population that we reject that our sample represents this population. However, another $\overline{X}$ produced $z = -.02$. Such a mean is close to μ and thus very likely, so this sample is likely to represent this population.

For Practice

1. _____ communicates that a sample mean is different from the μ it represents.

2. Sampling error occurs because of _____.

3. A sample mean has $z = +1$ on the sampling distribution created from the population of psychology majors. Is this likely to be a sample of psychology majors?

4. A sample mean has $z = -4.0$ on the above sampling distribution. Is this likely to be a sample of psychology majors?

Answers

1. Sampling error
2. Random chance
3. Yes
4. No

Setting Up the Sampling Distribution

To decide if our Prunepit sample represents the ordinary SAT population (with $\mu = 500$), we must perform two tasks: (1) Determine the probability of obtaining our sample from the ordinary SAT population and (2) decide whether the sample is unlikely to be representing this population. We perform both tasks simultaneously by setting up the sampling distribution.

In Figure 9.4, we have the sampling distribution of means from the ordinary SAT population. (The X axis is labeled twice, showing the sample means and their corresponding z-scores.) The first step in setting up the sampling distribution is to create the shaded areas in the tails of the distribution. We saw in the previous section that, at some point, an SAT sample mean could be so far above 500 in the upper tail of the sampling distribution that we could not believe that it represents the underlying raw score population. Recognize that any sample mean lying *beyond* that point, farther into the tail, would also be unbelievable. Therefore, we will draw a line in the upper tail of the sampling distribution creating the shaded area that encompasses all of these means. Likewise, an SAT sample mean could be so far *below* 500 in the lower tail of the sampling distribution that we would also not believe that it represents the underlying population. And, any samples beyond that point, farther into the tail, would also be unbelievable. We draw a line in the lower tail of the distribution to create the shaded area that encompasses these means. In statistical terms, the shaded areas are each called the *region of rejection*. As shown, very infrequently are samples so poor at representing the SAT population that they have means lying in the region of rejection. In fact,

> **Samples with means in the region of rejection are *so* unrepresentative of the underlying raw score population that it's a better bet they represent some *other* population.**

Thus, the **region of rejection** contains means that are so unlikely to be representing the underlying population, that if ours is one of them, we *reject* that it represents that population. Essentially, we "shouldn't" get an SAT sample mean that lies in the region of rejection if we're representing the ordinary SAT population because such

FIGURE 9.4

Setup of sampling distribution of SAT means showing the region of rejection and critical values

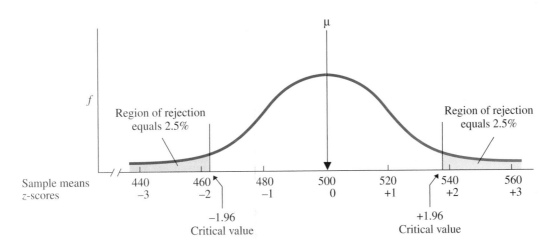

means almost never occur with this population. Therefore, if we do get such a mean, we probably are *not* representing this population: We *reject* that the sample represents the underlying raw score population and decide that it represents some other population.

Conversely, if our sample mean is not in the region of rejection, then the sample is not unlikely to be representing the ordinary SAT population. In fact, by our definition, samples not in the region of rejection are likely to represent this population, although with some sampling error. In such cases, we *retain* the idea that the sample is simply poorly representing this population of SAT scores.

How do we know where to draw the line that starts the region of rejection? By defining our *criterion probability*. The **criterion probability** is the probability that defines samples as too unlikely for us to accept as representing a particular population. Researchers usually use .05 as their default criterion probability. By this criterion, those sample means that together would occur only 5% of the time when representing the ordinary SAT population are so unlikely that if we get any one of them, we'll reject that our sample represents this population.

The criterion that we select determines the size of the region of rejection. The sample means that occur 5% of the time are those that make up the extreme 5% of the sampling distribution. However, we consider the means that are above *or* below 500 that together are a *total* of 5% of the curve, so we divide the 5% in half. Therefore, as Figure 9.4 showed, the extreme 2.5% of the curve in each tail of the sampling distribution forms our region of rejection.

> REMEMBER The *criterion probability* that defines samples as unlikely—and also determines the size of the *region of rejection*—is usually $p = .05$.

Now the task is simply to determine if our sample mean falls into the region of rejection. To do this, we will compare the sample's z-score to the *critical value*.

Identifying the Critical Value

Figure 9.4 also showed our *critical values*. These are the z-scores at the lines that mark the beginning of the upper and lower regions of rejection. Because z-scores get larger as we go farther into the tails, if the z-score for our sample mean is *greater* than the critical value, then we know that our sample mean lies in the region of rejection. Thus, a **critical value** marks the inner edge of the region of rejection and therefore defines the value required for a sample to fall into the region of rejection. Essentially, it is the minimum z-score that defines a sample as too unlikely.

> REMEMBER The *critical value* of z defines the minimum value of z a sample must have in order to be in the region of rejection.

How do we determine the critical value? By considering our criterion. With a criterion of .05, we set up the region of rejection in Figure 9.4 so that in each tail is the extreme 2.5%, or .025, of the total area under the curve. Then from the z-table in Appendix C, we see that .025 of the curve lies beyond the z-score of 1.96. Therefore, in each tail, the region of rejection begins at 1.96, so ± 1.96 is our critical value of z.

Thus, back in Figure 9.4, labeling the inner edges of the region of rejection with ± 1.96 completes how you should set up the sampling distribution. (*Note:* In the next chapter, using both tails like this is called a "two-tailed test.") We'll use Figure 9.4 to determine whether our Prunepit sample mean lies in the region of rejection by comparing the sample's z-score to the critical value. We will use this rule:

A sample mean lies in the region of rejection only if its z-score is *beyond* the critical value.

Thus, if our Prunepit mean has a z-score that is *larger* than ±1.96, then the sample lies *in* the region of rejection. If the z-score is *smaller than* or *equal to* the critical value, then the sample is *not* in the region of rejection.

Deciding If the Sample Represents the Population

Now, at long last, we can evaluate our sample mean of 550 from Prunepit U. First, we compute the sample's z-score on the sampling distribution created from the ordinary SAT population. With $\sigma_X = 100$ and $N = 25$, the standard error of the mean is

$$\sigma_{\bar{X}} = \frac{\sigma_X}{\sqrt{N}} = \frac{100}{\sqrt{25}} = 20$$

Then the z-score is

$$z = \frac{\bar{X} - \mu}{\sigma_{\bar{X}}} = \frac{550 - 500}{20} = +2.5$$

Think about this z-score. If the sample represents the ordinary SAT population, it's doing a very poor job of it. With a population mean of 500, a perfectly representative sample would have a mean of 500 and thus have a z-score of 0. Good old Prunepit produced a z-score of $+2.5$!

To confirm our suspicions, we compare the sample's z-score to the critical value of ±1.96. Locating the sample's z-score on the sampling distribution gives us the complete picture, which is shown in Figure 9.5. (When performing this procedure yourself, you should draw the complete picture, too.) The sample's z of $+2.5$—and the underlying sample mean of 550—lies in the region of rejection. This tells us that a

FIGURE 9.5

Completed sampling distribution of SAT means showing location of the Prunepit U. sample relative to the critical value

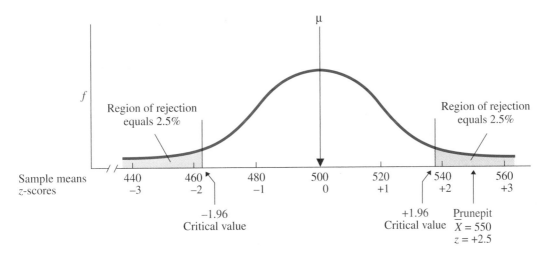

sample mean of 550 is among those means that are extremely unlikely when someone is representing the ordinary population of SAT scores. In other words, very seldom does chance—the luck of the draw—produce such unrepresentative samples from this population, so it is not a good bet that chance produced *our* sample from this population. Therefore, we say that we "reject" that our sample represents the population of SAT scores having a μ of 500.

Notice that we make a definitive, yes-or-no decision. Because our sample is so unlikely to represent the SAT raw score population where μ is 500, we decide that no, it does not represent this population.

We wrap up our conclusions in this way: If the sample does not represent the ordinary SAT population, then it must represent some other population. For example, perhaps the Prunepit students obtained the high mean of 550 because they lied about their scores, so they may represent the population of students who lie about the SAT. Whatever the reason, having rejected that the sample represents the population where μ is 500, we use the sample mean to estimate the μ of the population that the sample *does* represent. A sample having a mean of 550 is most likely to come from a population having a μ of 550. Therefore, our best estimate is that the Prunepit sample represents an SAT population that has a μ of 550.

On the other hand, say that our original sample mean had been 474, resulting in a z-score of $(474 - 500)/20 = -1.30$. Because -1.30 does *not* lie beyond the critical value of ± 1.96, this sample mean is *not* in the region of rejection. Looking back at Figure 9.5, we see that when sampling the underlying SAT population, this sample mean is relatively frequent and thus likely. Because of this, we say that we "retain" the idea that random chance produced a less than perfectly representative sample but that it probably represents and comes from the SAT population where μ is 500.

> *REMEMBER* When a sample's z-score is beyond the critical value, *reject* that the sample represents the underlying raw score population. When the z-score is not beyond the critical value, *retain* the idea that the sample represents the underlying raw score population.

Other Ways to Set Up the Sampling Distribution

Previously, the region of rejection was in both tails of the distribution because we wanted to identify unrepresentative sample means that were either too far above or too far below 500. Instead, however, we can place the region of rejection in only one tail of the distribution. (In the next chapter, you'll find out why you would want to use this "one-tailed" test.)

Say that we are interested only in sample means that are *less* than 500, having negative z-scores. Our criterion is still .05, but now we place the *entire* region of rejection in the lower, left-hand tail of the sampling distribution, as shown in Figure 9.6. *This produces a different critical value.* From the z-table (and using the interpolation procedures described in Appendix A.2), the extreme lower 5% of a distribution lies beyond a z-score of -1.645. Therefore, the z-score for our sample must lie beyond the critical value of -1.645 for it to be in the region of rejection. If it does, we will again conclude that the sample is so unlikely to be representing the SAT population where $\mu = 500$ that we'll reject that the sample represents this population. However, if the z-score is anywhere else on the sampling distribution, even far into the upper tail, we will *not* reject that the sample represents the SAT population where $\mu = 500$.

FIGURE 9.6

Setup of SAT sampling distribution to test negative *z*-scores

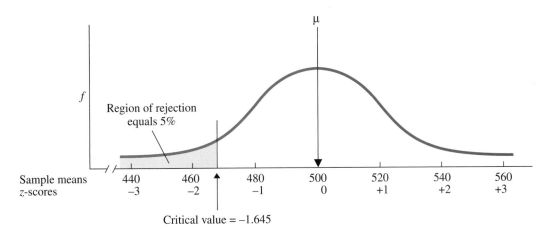

On the other hand, say that we're interested only in sample means *greater* than 500, having positive *z*-scores. Here, we place the entire region of rejection in the upper, right-hand tail of the sampling distribution, as shown in Figure 9.7. Now the critical value is *plus* 1.645, so only if our sample's *z*-score is beyond +1.645 does the sample mean lie in the region of rejection. Only then do we reject the idea that the sample represents the underlying raw score population.

REMEMBER When using one tail of the distribution and a criterion of .05, we use only the critical value of +1.645 or only the critical value of −1.645.

FIGURE 9.7

Setup of SAT sampling distribution to test positive *z*-scores

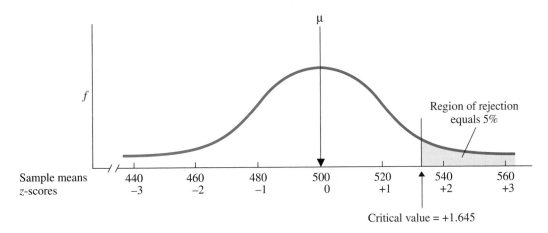

A QUICK REVIEW

- To decide if a sample represents a particular raw score population, compute the sample mean's z-score and compare it to the critical value on the sampling distribution.

MORE EXAMPLES

A sample of SAT scores ($N = 25$) produces $\overline{X} = 460$. Does the sample represent the SAT population where $\mu = 500$ and $\sigma_X = 100$? Compute z: $\sigma_{\overline{X}} = \sigma_X/\sqrt{N} = 100/\sqrt{25} = 20$; $z = (\overline{X} - \mu)/\sigma_{\overline{X}} = (460 - 500)/20 = -2.0$. With a criterion of .05 and the region of rejection in both tails, the critical value is ± 1.96. The sampling distribution is like Figure 9.5. The z of -2 is beyond -1.96, so it is in the region of rejection. Conclusion: The sample does not represent this SAT population.

For Practice

1. The region of rejection contains those samples considered to be *likely/unlikely* to represent the underlying raw score population.

2. The ____ defines the z-score that is required for a sample to be in the region of rejection.

3. For a sample to be in the region of rejection, its z-score must be *smaller/larger* than the critical value.

4. On a test, $\mu = 60$ and $\sigma_X = 18$. A sample ($N = 100$) produces $\overline{X} = 65$. Using the .05 criterion and both tails, does this sample represent this population?

Answers

1. unlikely
2. critical value
3. larger (beyond)
4. $\sigma_{\overline{X}} = 18/\sqrt{100} = 1.80$; $z = (65 - 60)/1.80 = +2.78$. this z is beyond ± 1.96, so reject that the sample represents this population; it's likely to represent the population with $\mu = 65$.

PUTTING IT ALL TOGETHER

The decision-making process discussed in this chapter is the essence of all inferential statistics. The basic question is always "Do the sample data represent a specified raw score population?" The sample may not look like it represents the population, but that may simply be sampling error or the sample may represent some other population. To decide, we

1. Draw a sampling distribution for the specified underlying raw score population.
2. Select the criterion probability, determine the critical value, and label your distribution, showing the region of rejection.
3. Compute a z-score for the sample mean.
4. If z is beyond the critical value, it is in the region of rejection. Therefore, the sample is unlikely to represent the underlying raw score population, so *reject* that it does. Conclude that the sample represents another population that is more likely to produce such data.
5. If z is not beyond the critical value, then the sample is not in the region of rejection and is likely to merely reflect sampling error. Therefore, *retain* the idea that the sample represents the specified population, although somewhat poorly.

With these steps, we make an intelligent bet about the population that our sample represents. Recognize that, as with any bet, our decisions might be wrong. (We'll discuss this in the next chapter.) Such errors are not likely, however, and that is why we perform inferential procedures. By incorporating probability into our decision making, we are confident that *over the long run* we will correctly identify the population that a sample represents. In the context of research, therefore, we have greater confidence that we are interpreting our data correctly.

CHAPTER SUMMARY

1. *Probability* (p) indicates the likelihood of an event when random chance is operating.

2. *Random sampling* is selecting a sample so that all elements or individuals in the population have an equal chance of being selected.

3. The probability of an event is equal to its relative frequency in the population.

4. Two events are *independent* if the probability of one event is not influenced by the occurrence of the other. Two events are *dependent* if the probability of one event *is* influenced by the occurrence of the other.

5. *Sampling with replacement* is replacing individuals or events back into the population before selecting again. *Sampling without replacement* is *not* replacing individuals or events back into the population before selecting again.

6. The standard normal curve is a theoretical probability distribution. The proportion of the area under the curve for particular z-scores is also the probability of the corresponding raw scores or sample means.

7. In a *representative sample*, the individuals and scores in the sample accurately reflect the types of individuals and scores found in the population.

8. *Sampling error* results when chance produces a sample statistic (such as $\overline{X}$) that is different from the population parameter (such as μ) that it represents.

9. *The region of rejection* is in the extreme tail or tails of a sampling distribution. Sample means here are unlikely to represent the underlying raw score population.

10. The *criterion probability* is the probability (usually .05) that defines samples as unlikely to represent the underlying raw score population.

11. The *critical value* is the minimum z-score needed for a sample mean to lie in the region of rejection.

KEY TERMS

p
criterion probability *198*
critical value *198*
dependent event *189*
independent event *188*
inferential statistics *186*
probability *187*

probability distribution *188*
random sampling *186*
region of rejection *197*
representative sample *193*
sampling error *193*
sampling with replacement *189*
sampling without replacement *189*

REVIEW QUESTIONS

(Answers for odd-numbered questions are in Appendix D.)

1. (a) What does probability convey about an event's occurence in a sample? (b) What is the probability of a random event based on?

2. What is random sampling?

3. (a) What is sampling with replacement? (b) What is sampling without replacement? (c) How does sampling without replacement affect the probability of events, compared to sampling with replacement?

4. (a) When are events independent? (b) When are they dependent?

5. What does the term sampling error indicate?

6. When testing the representativeness of a sample mean, (a) What is the criterion probability? (b) What is the region of rejection? (c) What is the critical value?

7. What does comparing a sample's *z*-score to the critical value indicate?

8. What is the difference between using both tails versus one tail of the sampling distribution in terms of (a) the size of the region of rejection? (b) the critical value?

APPLICATION QUESTIONS

9. Poindexter's uncle is building a house on land that has been devastated by hurricanes 160 times in the past 200 years. However, there hasn't been a major storm there in 13 years, so his uncle says this is a safe investment. His nephew argues that he is wrong because a hurricane must be due soon. What are the fallacies in the reasoning of both men?

10. Four airplanes from different airlines have crashed in the past two weeks. This terrifies Bubbles, who must travel on a plane. Her travel agent claims that the probability of a plane crash is minuscule. Who is correctly interpreting the situation? Why?

11. Foofy conducts a survey to learn who will be elected class president and concludes that Poindexter will win. It turns out that Dorcas wins. What is the statistical explanation for Foofy's erroneous prediction?

12. (a) Why does random sampling produce representative samples? (b) Why does random sampling produce unrepresentative samples?

13. In the population of typical college students, $\mu = 75$ on a statistics final exam $(\sigma_{\overline{X}} = 6.4)$. For 25 students who studied statistics using a new technique, $\overline{X} = 72.1$. Using two tails of the sampling distribution and the .05 criterion: (a) What is the critical value? (b) Is this sample in the region of rejection? How do you know? (c) Should we conclude that the sample represents the population of typical students? (d) Why?

14. In a population, $\mu = 100$ and $\sigma_X = 25$. A sample $(N = 150)$ has $\overline{X} = 102$. Using two tails of the sampling distribution and the .05 criterion: (a) What is the critical value? (b) Is this sample in the region of rejection? How do you know? (c) What does this indicate about the likelihood of this sample occurring in this population? (d) What should we conclude about the sample?

15. The mean of a population of raw scores is 33 $(\sigma_X = 12)$. Use the criterion of .05 and the upper tail of the sampling distribution to test whether a sample with $\overline{X} = 36.8$ $(N = 30)$ represents this population. (a) What is the critical value? (b) Is the sample in the region of rejection? How do you know? (c) What does this indicate about the likelihood of this sample occurring in this population? (d) What should we conclude about the sample?

16. We obtain a $\overline{X} = 46.8$ ($N = 15$) which may represent the population where $\mu = 50$ ($\sigma_X = 11$). Using the criterion of .05 and the lower tail of the sampling distribution: (a) What is our critical value? (b) Is this sample in the region of rejection? How do you know? (c) What should we conclude about the sample? (d) Why?

17. The mean of a population of raw scores is 28 ($\sigma_X = 9$). Your $\overline{X}$ is 34 (with $N = 35$). Using the .05 criterion with the region of rejection in both tails of the sampling distribution, should you consider the sample to be representative of this population? Why?

18. The mean of a population of raw scores is 48 ($\sigma_X = 16$). Your $\overline{X}$ is 44 (with $N = 40$). Using the .05 criterion with the region of rejection in both tails of the distribution, should you consider the sample to be representative of this population? Why?

19. A couple with eight daughters decides to have one more baby, because they think this time they are sure to have boy! Is this reasoning accurate?

20. On a standard test of motor coordination, a sports psychologist found that the population of average bowlers had a mean score of 24, with a standard deviation of 6. She tested a random sample of 30 bowlers at Fred's Bowling Alley and found a sample mean of 26. A second random sample of 30 bowlers at Ethel's Bowling Alley had a mean of 18. Using the criterion of $p = .05$ and both tails of the sampling distribution, what should she conclude about each sample's representativeness of the population of average bowlers?

21. (a) In question 20, if a particular sample does not represent the population of average bowlers, what is your best estimate of the μ of the population it does represent? (b) Explain the logic behind this conclusion.

22. Foofy computes the $\overline{X}$ from data that her professor says is a random sample from population Q. She correctly computes that this mean has a z-score of $+41$ on the sampling distribution for population Q. Foofy claims she has proven that this could not be a random sample from population Q. Do you agree or disagree? why?

INTEGRATION QUESTIONS

23. In a study you obtain the following data representing the aggressive tendencies of some football players:

    40   30   39   40   41   39   31   28   33

    (a) Researchers have found that in the population of nonfootball players, μ is 30 ($\sigma_X = 5$.) Using both tails of the sampling distribution, determine whether your football players represent a different population. (b) What do you conclude about the population of football players and its μ? (Chs. 4, 6, 9)

24. We reject that a sample, with $\overline{X} = 95$, is merely poorly representing the population where $\mu = 100$. (a) What is our best estimate of the population μ that the sample is representing? (b) Why can we claim this value of μ? (Ch. 4)

25. For a distribution in which $\overline{X} = 43$ and $S_X = 8$, using z-scores, what is the relative frequency of (a) scores below 27? (b) Scores above 51? (c) A score between 42 and 44? (d) A score below 33 or above 49? (e) For each of the questions above, what is the probability of randomly selecting participants who have these scores? (Chs. 6, 9)

26. When we compute the z-score of a sample mean, (a) What must you compute first? (b) What is its formula? (c) What is the formula for the z-score of a sample mean? (d) It describes the mean's location among other means on what distribution? (e) What does this distribution show? (Chs. 6, 9)

27. The mean of a population of raw scores is 50 ($\sigma_X = 18$). (a) Using the z-table, what is the relative frequency of sample means below 46 when $N = 40$? (b) What is the probability of randomly selecting a sample of 40 scores having a $\overline{X}$ below 46? (Chs. 6, 9)

28. The mean of a population of raw scores is 18 ($\sigma_X = 12$). (a) Using the z-table, what is the relative frequency of sample means above 24 when $N = 30$? (b) What is the probability of randomly selecting a sample of 30 participants whose scores produce a mean above 24? (Chs. 6, 9)

■ ■ SUMMARY OF FORMULAS

1. The formula for transforming a sample mean into a z-score is

$$z = \frac{\overline{X} - \mu}{\sigma_{\overline{X}}}$$

where the standard error of the mean is

$$\sigma_{\overline{X}} = \frac{\sigma_X}{\sqrt{N}}$$

10 Introduction to Hypothesis Testing

GETTING STARTED

To understand this chapter, recall the following:

- From Chapter 2, what the conditions of an independent variable are and what the dependent variable is.
- From Chapter 4, that a relationship in the population occurs when different means from the conditions represent different μs and thus different distributions of dependent scores.
- From Chapter 9, that when a sample's z-score falls into the region of rejection, the sample is unlikely to represent the underlying raw score population.

Your goals in this chapter are to learn

- Why the possibility of sampling error causes researchers to perform inferential statistical procedures.
- When *experimental hypotheses* lead to either a *one-tailed* or a *two-tailed test*.
- How to create the *null* and *alternative hypotheses*.
- How to perform the *z-test*.
- How to interpret *significant* and *nonsignificant* results.
- What *Type I errors, Type II errors,* and *power* are.

From the previous chapter, you know the basic logic of all inferential statistics. Now we will put these procedures into a research context and present the statistical language and symbols used to describe them. Until further notice, we'll be talking about experiments. This chapter shows (1) how to set up an inferential procedure, (2) how to perform the z-test, (3) how to interpret the results of an inferential procedure, and (4) the way to describe potential errors in our conclusions.

NEW STATISTICAL NOTATION

Five new symbols will be used in stating mathematical relationships.

1. The symbol for *greater than* is $>$. Thus, $A > B$ means that A is greater than B.

2. The symbol for *less than* is $<$, so $B < A$ means that B is less than A.

3. The symbol for *greater than or equal to* is $\geq$, so $B \geq A$ indicates that B is greater than or equal to A.

4. The symbol for *less than or equal to* is $\leq$, so $B \leq A$ indicates that B is less than or equal to A.

5. The symbol for *not equal to* is $\neq$, so $A \neq B$ means that A is different from B.

Also, the introduction of each inferential statistical procedure will include a checklist of the procedure's *assumptions*. The assumptions are the rules for *when* to use it: they tell you that a procedure is appropriate *assuming* that your design and your data meet certain requirements. (*Note*: Select your inferential procedure and check its assumptions *prior* to actually collecting data. Otherwise, your data may fit no procedure, in which case the study is useless.)

WHY IS IT IMPORTANT TO KNOW ABOUT THE *z*-TEST?

The *z*-test is one of the simplest inferential statistics around, so it is a good starting point for learning these procedures. Also, when reading behavioral research, you may encounter a study that employs it, so you should understand how it works. Most importantly, the discussion will introduce the formal system researchers use in *all* inferential procedures. Therefore, understand the general steps and terminology involved here because you'll see them again and again.

THE ROLE OF INFERENTIAL STATISTICS IN RESEARCH

As you saw in the previous chapter, a random sample may be more or less representative of a population because, just by the luck of the draw, the sample may contain too many high scores or too many low scores relative to the population. Because the sample is not perfectly representative, it reflects *sampling error*, and so the sample mean does not equal the population mean.

Here is how sampling error can impact on an experiment. Recall that in experiments, we hope to see a relationship in which, as we change the conditions of the independent variable, scores on the dependent variable change in a consistent fashion. Therefore, if the means for the conditions are different from each other, we infer that if we measured the entire population, we would find a different population of scores located around a different μ for each condition. But here is where sampling error comes in. Maybe the sample means for the conditions differ because of sampling error, and actually they all poorly represent the *same* population. If so, then the relationship does not exist: We'd find the same population of scores, having the same μ, in each condition. Or because of sampling error, perhaps the actual relationship in the population is different from the relationship in our sample data.

For example, say that we compare men and women on the dependent variable of creativity. In nature, men and women don't really differ on this variable, so their μs are equal. However, through sampling error—the luck of the draw—we might end up with some female participants who are more creative than our male participants, or vice versa. Then sampling error will mislead us into thinking that this relationship exists, even though it really does not. Or, say that we measure the heights of some men and women and, by chance, obtain a sample of relatively short men and a sample of tall women. If we didn't already know that men are taller, sampling error would mislead us into concluding that women are taller.

Researchers perform inferential statistics in every study, because it is always possible that we are being misled by sampling error so that the relationship we see in our sample data is not the relationship found in nature.

Previously we've said that **inferential statistics** are used to decide if sample data represent a particular relationship in the population. Using the process discussed in the previous chapter, the decision boils down to this: (1) Should we believe that the relationship we see in the sample data is generally the same as the relationship we would find if we tested the entire population? or (2) Should we conclude that the relationship in the sample is a coincidence produced by sampling error, and that the sample does not accurately represent the populations and relationship found in nature?

The specific inferential procedure employed in a given research situation depends upon the *research design* and on the *scale of measurement* used when measuring the *dependent variable*. We have, however, two general categories of inferential statistics: **Parametric statistics** are procedures that require specific assumptions about the characteristics of our populations. (Remember *parametric* by recalling that a population's characteristics are called *parameters*.) Two assumptions common to all parametric procedures are (1) the population of dependent scores forms a normal distribution, and (2) the scores are interval or ratio scores. Thus, parametric procedures are used when it is appropriate to compute the mean in each condition. In this and upcoming chapters, we'll focus on parametric procedures.

The other category is **nonparametric statistics,** which are inferential procedures that do not require stringent assumptions about our populations. These procedures are used with nominal or ordinal scores or with skewed interval or ratio distributions (when it is appropriate to calculate the median or mode). Chapter 15 presents nonparametric procedures.

> **REMEMBER** *Parametric and nonparametric inferential statistics* are for deciding if the data accurately represent a relationship in nature, or if sampling error is misleading us into thinking there is this relationship.

As we'll see, parametric procedures are often preferable, so typically we use nonparametric procedures only when the data clearly violate the assumptions of parametric procedures. Instead, we can use a parametric procedure if the data come close to meeting its assumptions. For example, if our population is approximately normally distributed, we can still use a parametric procedure.

As you'll see, both parametric and nonparametric procedures are performed in the same way. The first step is setting up the procedure.

SETTING UP INFERENTIAL PROCEDURES

Researchers follow four steps when setting up an experiment: Create the experimental hypotheses, design the experiment to test these hypotheses, translate the experimental hypotheses into statistical hypotheses, and select and set up the appropriate statistical procedure to test the statistical hypotheses.

Creating the Experimental Hypotheses

Recognize that the purpose of all experiments is to obtain data that will help us to resolve the simplest of debates: maybe my independent variable works as I think it does versus maybe it does not. From this, we first create two experimental hypotheses. **Experimental hypotheses** describe the predicted relationship we may or may not find. One hypothesis states that we will demonstrate the predicted relationship (manipulating the independent variable will work as expected). The other hypothesis states that we will not demonstrate the predicted relationship (manipulating the independent variable will not work as expected).

However we can predict a relationship in one of two ways. Sometimes we expect a relationship, but we are not sure whether scores will increase or decrease as we change the independent variable. This leads to a two-tailed statistical procedure. A **two-tailed test** is used when we predict a relationship but do not predict the direction in which scores will change. Notice that a two-tailed test occurs when we predict that one group will produce *different dependent scores* than the other group, without saying which group will score higher. For example, we have a two-tailed test if we propose that "men and women *differ* in creativity" or that "higher anxiety levels will *alter* participants' test scores."

At other times, we do predict the direction in which the dependent scores will change. A **one-tailed test** is used when we predict the *direction* in which scores will change. We may predict that as we change the independent variable, the dependent scores will only increase, or we may predict that they will only decrease. Notice that a one-tailed test occurs when we predict which group will have the *higher dependent scores*. For example, we have a one-tailed test if we predict that "men are *more* creative than women" or that "higher anxiety levels will *lower* test scores."

> *REMEMBER* A *two-tailed test* is used when you do not predict the direction that scores will change. A *one-tailed test* is used when you do predict the direction that scores will change.

Let's first examine a study involving a two-tailed test. Say that we've discovered a chemical that is related to intelligence, which we are ready to test on humans in an "IQ pill." The amount of the pill is our independent variable, and a person's resulting IQ is the dependent variable. We believe that this pill will affect IQ, but we are not sure whether it will make people smarter or dumber. Therefore, here are our two-tailed experimental hypotheses:

1. We will demonstrate that the pill works by either increasing or decreasing IQ scores.

2. We will not demonstrate that the pill works, because IQ scores will not change.

Remember, however, that ultimately researchers want to describe what occurs in nature, in what we call the *population*. Therefore, although we must first see that the independent variable works as predicted in our sample, the real issue is whether we can conclude that it works in the population.

Designing a One-Sample Experiment

There are many ways we might design a study to test our pill, but the simplest way is as a *one-sample experiment*. We will randomly select one sample of participants and give each person, say, one pill. Then we'll give participants an IQ test. The sample will represent the population of people who have taken one pill, and the sample $\overline{X}$ will represent the population μ.

To demonstrate a relationship, however, we must demonstrate that *different* amounts of the pill produce *different* populations of IQ scores, having different μs. Therefore, we must compare the population represented by our sample to some other population receiving some other amount of the pill. *To perform a one-sample experiment, we must already know the population mean under some other condition of the independent variable.* Here our independent variable is the amount of the pill taken, and one amount that we already know about is zero amount. The IQ test has been given to many people over the years who have *not* taken the pill, and let's say this population has a μ of 100.

We will compare this population that has not taken the pill to the population that has taken the pill that is represented by our sample. If the population without the pill has a different μ than the population with the pill, then we will have demonstrated a relationship.

Creating the Statistical Hypotheses

So that we can apply statistical procedures, we translate our experimental hypotheses into *statistical hypotheses*. We are still debating whether our independent variable works, but now we state this in terms of the corresponding statistical outcomes. **Statistical hypotheses** describe the population parameters that the sample data represent if the predicted relationship does or does not exist. The two statistical hypotheses are the *alternative hypothesis* and the *null hypothesis*.

The Alternative Hypothesis   It is easier to create the alternative hypothesis first because it corresponds to the experimental hypothesis that the experiment *does* work as predicted. The **alternative hypothesis** describes the population parameters that the sample data represent if the predicted relationship exists. The alternative hypothesis is always the hypothesis of a difference; it says that changing the independent variable produces the predicted difference in the populations.

For example, Figure 10.1 shows the populations we'd find if the pill *increases* IQ. This shows a relationship because everyone's IQ score is increased so that the distribution moves to the right, over to the higher scores. We don't know how much scores will increase, so we do not know the value of the new μ with the pill. But we do know that the μ of the population with the pill will be *greater* than 100 because 100 is the μ of the population without the pill.

On the other hand, Figure 10.2 shows the populations if the pill *decreases* IQ. Here, the pill moves the distribution to the left, over to the lower scores. Again, we don't know how much the pill will decrease scores, but we do know that the μ of the population with the pill will be *less than* 100.

The alternative hypothesis is a shorthand way of communicating all of the above. If the pill works as predicted, then the population with the pill will have a μ that is either greater than or less than 100. In other words, the population mean with the pill will *not equal* 100. The symbol for the alternative hypothesis is H_a. (The H stands for hypothesis and the subscript a stands for alternative.) Our alternative hypothesis in this study is

$$H_a: \mu \neq 100$$

FIGURE 10.1

Relationship in the population if the IQ pill increases IQ scores

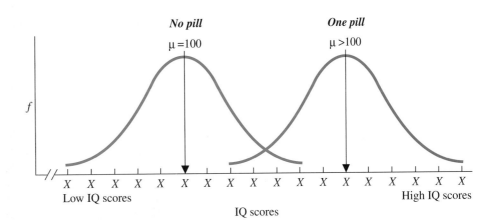

FIGURE 10.2

Relationship in the population if the IQ pill decreases IQ scores

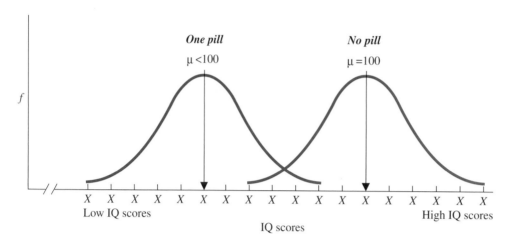

This proposes that the sample mean produced by our pill represents a μ not equal to 100. Because the μ without the pill is 100, H_a implies that a relationship exists in the population. Thus, we can interpret H_a as stating that our independent variable works as predicted.

The Null Hypothesis   The statistical hypothesis corresponding to the experimental hypothesis that the independent variable does *not* work as predicted is called the null hypothesis. The **null hypothesis** describes the population parameters that the sample data represent if the predicted relationship does *not* exist. It is the hypothesis of "no difference," saying that changing the independent variable does *not* produce the predicted difference in the population.

   If the IQ pill does not work, then it would be as if the pill were not present. We already know that the population of IQ scores without the pill has a μ of 100. Therefore, if the pill does not work, the population of scores will be unchanged and μ will still be 100. Thus, if we measured the population with and without the pill, we would have one population of scores, located at the μ of 100, as shown in Figure 10.3.

FIGURE 10.3

Population of scores if the IQ pill does not affect IQ scores

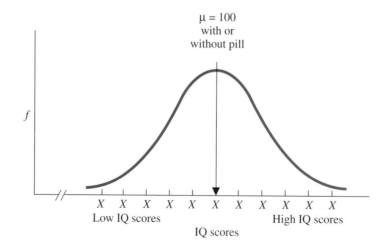

The null hypothesis is a shorthand way of communicating the above. The symbol for the null hypothesis is H_0. (The subscript is 0 because *null* means zero, as in zero relationship.) The null hypothesis for the IQ pill study is

$$H_0: \mu = 100$$

This proposes that our sample comes from and represents the population where μ is 100. Because this is the same population found without the pill, H_0 implies that the predicted relationship does not exist in nature (that the independent variable does not work).

> **REMEMBER** The *alternative hypothesis* (H_a) says the sample data represent a μ and population that reflects the predicted relationship. The *null hypothesis* (H_0) says the data represent the μ and population that is found when the predicted relationship does not occur in nature.

A QUICK REVIEW

- The null hypothesis shows the value of μ that our $\overline{X}$ represents if the predicted relationship does not exist.
- The alternative hypothesis shows the value of μ that our $\overline{X}$ represents if the predicted relationship does exist.

MORE EXAMPLES

In an experiment, we compare a sample of men to the population of women who have a μ of 75. We predict simply that men are different from women, so this is a two-tailed test. The alternative hypothesis is that our men represent a different population, so their μ is not 75; thus, $H_a: \mu \neq 75$. The null hypothesis is that men are the same as women, so the men's μ is also 75, so $H_0: \mu = 75$.

For Practice

1. A ____ test is used when we do *not* predict the direction that scores will change; a ____ test is

used when we *do* predict the direction that scores will change.

2. The ____ hypothesis says that the sample data represent a population where the predicted relationship exists. The ____ hypothesis says that the sample data represent a population where the predicted relationship does not exist.

3. The μ for adults on a personality test is 140. We test a sample of children to see if they are different from adults. What are H_a and H_0?

4. The μ for days absent among workers is 15.6. We train a sample of new workers and ask whether the training changes worker absenteeism. What are H_a and H_0?

Answers

1. two-tailed; one-tailed
2. alternative; null
3. $H_a: \mu \neq 140$; $H_0: \mu = 140$
4. $H_a: \mu \neq 15.6$; $H_0: \mu = 15.6$

The Logic of Statistical Hypothesis Testing

The statistical hypotheses for the IQ pill study are $H_0: \mu = 100$ and $H_a: \mu \neq 100$. Remember, these are hypotheses—guesses—about the population that our sample may represent. Notice that, together, H_0 and H_a include all possibilities because the μ after everyone has taken the pill would either equal or not equal 100. We use inferential procedures to test (choose between) these hypotheses. (Inferential procedures are also called *statistical hypothesis testing*.)

Say that we randomly selected a sample of 36 people, gave them the pill, measured their IQ, and found that their mean score was 105. On the one hand, the obvious interpretation is this: People who have not taken this pill have a mean IQ of 100, so if the pill did not work, then the sample mean "should" have been 100. Therefore, a sample mean of 105 suggests that the pill does work, raising IQ scores about 5 points. If the pill does this for the sample, it should do this for the population. Therefore, our results appear to support our alternative hypothesis, H_a: $\mu \neq 100$: If we measured everyone in the population with and without the pill, we would have the two distributions shown back in Figure 10.1, with the population that received the pill located at the μ of 105. Conclusion: It seems that the pill works. We appear to have evidence of a relationship in nature where increased amounts of the pill are associated with increased IQ scores.

But hold on! Remember sampling error? We just assumed that our sample is *perfectly* representative of the population it represents. But what if there was sampling error? Maybe we obtained a mean of 105 not because the pill works, but because we inaccurately represented the situation where the pill does *not* work. Maybe the pill does nothing, but by chance we happened to select too many participants who *already* had an above-average IQ and too few with a low IQ, so that our mean is 105 instead of 100. Thus, maybe the null hypothesis is correct. Even though it doesn't look like it, maybe our sample actually represents the population where μ is 100. Maybe we have not demonstrated that the pill works.

In fact, we can never know whether our pill works based on the results of one study. Whether the sample mean is 105, 1050, or 105,000, it is still possible that the sample mean is different from 100 simply because of sampling error. As this illustrates, one side of the debate (that we're calling the null hypothesis) is to always argue that the independent variable does not work as predicted, regardless of what our sample data seem to show. Instead, it is always possible that the data poorly represent the situation where the predicted relationship does not occur in nature.

> **REMEMBER** The null hypothesis always implies that if our sample data show the predicted relationship, we are being misled by sampling error and there really is not that relationship in nature.

Thus, we cannot automatically infer that the relationship exists in the population when our sample data show the predicted relationship because two things can produce such data: sampling error or our independent variable. Maybe H_0 is correct because sampling error produced our sample data, the independent variable really does not work as predicted, and thus the μ we're representing is 100. Or maybe H_a is correct because a relationship in nature produced our sample data, so we can believe that the independent variable does work as predicted, and thus the μ we're representing is not 100.

The only way to resolve this dilemma for certain would be to give the pill to the entire population and see whether μ was 100 or 105. We cannot do that so we can never prove whether the null hypothesis is true. However, we can determine how *likely* it is to be true. That is, we can determine the probability that sampling error would produce a sample mean of 105 when the sample actually comes from and represents the population where μ is 100. If such a mean is very unlikely, we'll *reject* the H_0 that our sample represents this population.

If this sounds familiar, it's because it is the procedure discussed in the previous chapter. In fact, that procedure is a parametric inferential procedure called the *z-test*.

PERFORMING THE z-TEST

The **z-test** is the procedure for computing a z-score for a sample mean on the sampling distribution of means. The formula for the z-test is the formula we used in Chapters 6 and 9 (and we'll see it again in a moment). The z-test is used in a one-sample experiment when we can meet these four assumptions:

1. We have randomly selected one sample.
2. The dependent variable is at least approximately normally distributed in the population, and involves an interval or ratio scale.
3. We *know* the mean of the population of raw scores under some other condition of the independent variable.
4. We *know* the true standard deviation (σ_X) of the above population.

Say that from past research, we know that in the IQ population where μ is 100, the standard deviation is 15. Therefore, the z-test is appropriate for our study.

> *REMEMBER*   The *z-test* is used only if the raw score population's σ_X is known.

Setting Up the Sampling Distribution for a Two-Tailed Test

We always test H_0 by examining the sampling distribution created from the raw score population that H_0 says we are representing. Here H_0 says that the sample represents a population with $\mu = 100$. Therefore, it is as if we again hired our (very) bored statistician. Using the N of 36 that we used, she infinitely samples the raw score population of IQ scores without the pill where μ is 100. This produces a sampling distribution of means with a μ of 100, as in Figure 10.4. Notice, the μ of the sampling distribution always equals the value of μ given in the null hypothesis.

> *REMEMBER*   The mean of the sampling distribution always equals the μ of the raw score population that H_0 says we are representing.

A sampling distribution always describes the situation *when H_0 is true*. Here it shows the sample means that occur when we *are* drawing samples from the IQ population where μ *is* 100. Any sample mean not equal to 100 occurs *solely* because of sampling error—the luck of the draw that produced an unrepresentative sample. Thus, you can

FIGURE 10.4

Sampling distribution of IQ means for a two-tailed test

A region of rejection is in each tail of the distribution, marked by the critical values of ±1.96

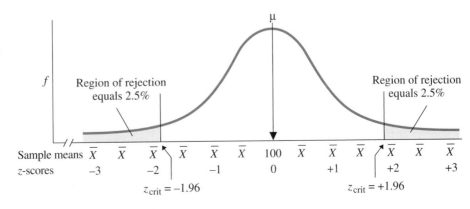

think of the sampling distribution as showing the frequency of all $\overline{X}$s we might get through sampling error when the pill doesn't work. (Always add the phrase "when H_0 is true" to any information you get from a sampling distribution.)

Next we set up the sampling distribution as we did in the previous chapter: Determine the size and location of the region of rejection and then identify the critical value. However, we have some new symbols and terms.

1. *Choose alpha:* Recall that the *criterion* probability defines sample means as being too unlikely to represent the underlying raw score population, which in turn defines the size of the region of rejection. The symbol for the criterion probability is α, the Greek letter **alpha.** Usually the criterion is .05, so in code, a $\alpha = .05$.

2. *Locate the region of rejection:* Recall that the region of rejection may be in both tails or only one tail of the sampling distribution. To decide, consider your hypotheses. With our pill, we created *two-tailed* hypotheses for a *two-tailed test* because we predicted the pill might raise or lower IQ scores. We will be correct if our sample mean is either above 100 or below 100 and we can reject that it represents the no-pill population. Thus, back in Figure 10.4, with a two-tailed test, we placed a region of rejection in each tail.

3. *Determine the critical value:* We'll abbreviate the critical value of z as z_{crit}. With $\alpha = .05$, the total region of rejection is .05, of the curve, so the region in each tail is .025 of the curve. From the z-table, a z-score of 1.96 demarcates this region, and so we complete Figure 10.4 by adding that z_{crit} is ± 1.96.

Now the test of H_0 boils down to comparing the z-score for our sample mean to the z_{crit} of ± 1.96. Therefore, it's time to compute the z-score for the sample.

Computing *z*

Here is some more code. The z-score we compute is "obtained" from the data, so we'll call it *z obtained*, which we abbreviate as z_{obt}. You know how to compute this from previous chapters.

The formula for the *z*-test is

$$z_{obt} = \frac{\overline{X} - \mu}{\sigma_{\overline{X}}}$$

where

$$\sigma_{\overline{X}} = \frac{\sigma_X}{\sqrt{N}}$$

First, we compute the standard error of the mean ($\sigma_{\overline{X}}$) In the formula, N is the number of scores in the sample and σ_X is the true population standard deviation. For our IQ pill study, σ_X is 15 and N is 36, so

$$\sigma_{\overline{X}} = \frac{\sigma_X}{\sqrt{N}} = \frac{15}{\sqrt{36}} = \frac{15}{6} = 2.50$$

Next we compute z_{obt}. In the formula, the value of μ is the μ of the sampling distribution, which is also the μ of the underlying raw score population that H_0 says the sample represents. The $\overline{X}$ is computed from our sample. The $\sigma_{\overline{X}}$ is the standard error of the mean. Thus, our $\overline{X} = 105$, $\mu = 100$, and $\sigma_{\overline{X}} = 2.50$, so

$$z_{obt} = \frac{\overline{X} - \mu}{\sigma_{\overline{X}}} = \frac{105 - 100}{2.5} = \frac{+5}{2.5} = +2.00$$

Our z_{obt} is $+2.00$. The final step is to interpret this z_{obt} by comparing it to z_{crit}.

Comparing the Obtained *z* to the Critical Value

Figure 10.5 shows the location of our z_{obt} (and sample mean) on the sampling distribution. Remember that the sampling distribution describes the situation *when H_0 is true*: Here it shows all possible means that occur when, as our H_0 claims happened to us, samples are drawn from the population where μ is 100. If we are to believe H_0, the sampling distribution should show that a mean of 105 occurs relatively frequently and is thus likely in this situation. However, Figure 10.5 shows just the opposite.

A z_{obt} of $+2$ tells us that the bored statistician hardly ever obtained a sample mean of 105 when drawing samples from the population where μ is 100. This makes it difficult to believe that our sample came from this population. In fact, because a z_{obt} of $+2$ is beyond the z_{crit} of ±1.96, our sample is in the region of rejection. Therefore, we conclude that our sample is unlikely to have come from and represent the population where $\mu = 100$, rejecting that our sample is poorly representing this population.

In statistical terms, we have "rejected" the null hypothesis. If we reject H_0, then we are left with H_a, and so we "accept H_a." Here, H_a is $\mu \neq 100$, so we accept that our sample represents a population where μ is not 100. Thus, in sum, we have determined that the sample is unlikely to represent the population where μ is 100, so we conclude that it is likely to represent a population where μ is not 100.

> **REMEMBER** When a sample statistic falls beyond the critical value, the statistic lies in the region of rejection, so we *reject H_0* and *accept H_a*.

Once we have made a decision about the statistical hypotheses (H_0 and H_a), we then make a decision about the corresponding original experimental hypothesis. We rejected H_0, so we will also reject the experimental hypothesis that our independent variable does

FIGURE 10.5

Sampling distribution of IQ means

The sample mean of 105 is located at $z_{obt} = +2.00$

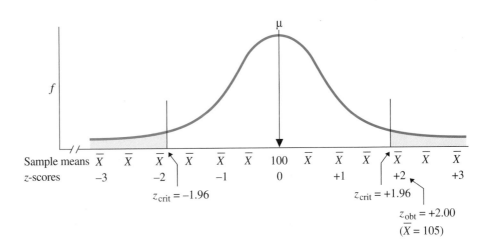

not work as predicted. Therefore, we will reject that our pill does not work. If this makes your head spin, it may be because the logic actually involves a "double negative." When our sample falls in the region of rejection, we say "no" to the H_0 that says we're representing the population with $\mu = 100$. But H_0 says this as a way of saying there is "no relationship" involving our pill. By rejecting H_0, we are saying no to "no relationship." This is actually saying yes, there *is* a relationship involving our pill, which is what H_a says. Therefore, by rejecting H_0 and accepting H_a, we also accept the corresponding experimental hypothesis that the independent variable does work as predicted. Here, it appears that we have demonstrated a relationship in nature such that the pill would change the population of IQ scores. In fact, we can be more specific: A sample mean of 105 is most likely to represent the population where μ is 105. Thus, without the pill, the population μ is 100, but with the pill, we expect that scores would increase to a μ of around 105.

INTERPRETING SIGNIFICANT RESULTS

The shorthand way to communicate that we have rejected H_0 and accepted H_a is to say that the results are *significant*. (Statistical hypothesis testing is sometimes called "significance testing.") *Significant* does *not* mean important or impressive. **Significant** indicates that our results are unlikely to occur if the predicted relationship does not exist in the population. Therefore, we imply that the relationship found in the experiment is "believable," representing a "real" relationship found in nature, and that it was not produced by sampling error from the situation in which the relationship does not exist.

> **REMEMBER** The term *significant* means that we have rejected the null hypothesis and believe that the data reflect a relationship found in nature.

Notice that your decision is simply either yes, reject H_0, or no, do not reject H_0. All z-scores in the region of rejection are treated the same, so one z_{obt} cannot be "more significant" than another. Likewise, there is no such thing as "very significant" or "highly significant." (That's like saying "very yes" or "highly yes.") If z_{obt} is beyond z_{crit}, regardless of how far it is beyond, the results are simply significant, period!

Although we accept that a relationship exists, we have three very important restrictions on how far we can go when interpreting significant results in any experiment.

First, *we did not prove that H_0 is false.* With our pill, the only thing we have "proven" is that a sample mean of 105 is *unlikely* to come from a population where $\mu = 100$. However, the sampling distribution shows that means of 105 *do* occur once in a while when *we are* representing this population. Maybe our sample was one of them. Maybe the pill did not work, and our sample was very unrepresentative of this.

Second, *we did not prove it was our independent variable that caused the scores to change.* Although our pill *might* have caused the higher IQ scores, some other, hidden variable also might have produced them. Maybe our participants cheated on the IQ test, or there was something in the air that made them smarter, or there were sunspots, or who-knows-what! If we've performed a good experiment and can eliminate such factors, then we can *argue* that it is our independent variable that changed the scores.

Finally, *the μ represented by our sample may not equal our $\overline{X}$.* Even assuming that our pill does increase IQ, the population μ is probably not *exactly* 105. Our sample may reflect (you guessed it) sampling error! That is, the sample may accurately reflect that the pill increases IQ, but it may not perfectly represent *how much* the pill increases scores. Therefore, if we gave the pill to the population, we might find a μ of 104, or 106, or *any* other value. However, a sample mean of 105 is most likely when the population μ is 105, so we would conclude that the μ resulting from our pill is probably *around* 105.

Bearing these qualifications in mind, we interpret the $\overline{X}$ of 105 the way we wanted to several pages back: Apparently, the pill increases IQ scores by about 5 points. But now, because the results are significant, we are confident that we are not being misled by sampling error. Therefore, we are more confident that we have discovered a relationship in nature. (But stay tuned, we could be wrong.) At this point, we return to being behavioral researchers and interpret the results "psychologically": We describe how the ingredients in the pill affect intelligence, what brain mechanisms are involved, and so on.

INTERPRETING NONSIGNIFICANT RESULTS

Let's say that the IQ pill had instead produced a sample mean of 99. Now the z-score for the sample is

$$z_{\text{obt}} = \frac{\overline{X} - \mu}{\sigma_{\overline{X}}} = \frac{99 - 100}{2.5} = \frac{-1}{2.5} = -.40$$

As in Figure 10.6, a z_{obt} of $-.40$ is *not* beyond the z_{crit} of ± 1.96, so the sample is not in the region of rejection. This indicates that we will frequently obtain a sample mean of 99 when sampling the population where $\mu = 100$. Therefore, the null hypothesis is reasonable: our sample is likely to be a poor representation of the population where μ is 100. So we will not reject H_0. Sampling error from this population can explain our results just fine, thank you, so we will not reject this explanation. In such situations, we say that we have "failed to reject H_0" or that we "retain H_0."

When we retain H_0, we also retain the experimental hypothesis that our independent variable does not work as predicted. We've found that our sample mean was likely to occur if we were representing the situation where the pill is not present. Therefore, it makes no sense to conclude that the pill works if our results were likely to occur *without* the pill. Likewise, we never conclude that an independent variable works if the results were likely to be due to sampling error from the situation where it does not work.

FIGURE 10.6

Sampling distribution of IQ means

The sample mean of 99 has a z_{obt} of -40.

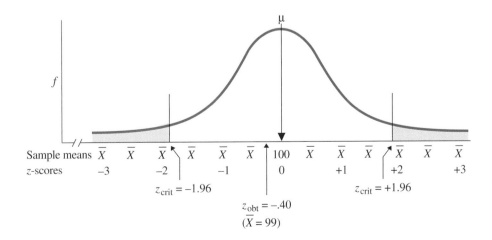

The shorthand way to communicate all of this is to say that the results are *not significant* or that they are *nonsignificant*. (Don't say *insignificant*.) **Nonsignificant** indicates that the results are likely to reflect chance, sampling error, without there being a relationship in nature.

> **REMEMBER** *Nonsignificant* indicates that we have failed to reject H_0 because the results are not in the region of rejection and are thus likely to occur when there is not the predicted relationship in nature.

When we retain H_0, we also retain the experimental hypothesis that the independent variable does not work as predicted. However, we have not *proven* that H_0 is true, so we have not *proven* that our independent variable does not work. We have simply failed to find convincing evidence that it *does* work. The only thing we're sure of is that sampling error *could* have produced our data. Therefore, we *still* have two hypotheses that are both viable: (1) H_0, that the data do not really represent a relationship, and (2) H_a, that the data do represent a relationship. Thus, maybe in fact the pill does not work. Or maybe the pill does work, but our sample poorly represents this. We simply don't know whether the pill works or not.

Thus, with nonsignificant results, you should not say anything about whether the independent variable influences behavior or not, and do not even begin to interpret the results "psychologically." All that you can say is that you failed to demonstrate that the predicted relationship exists

> **REMEMBER** Nonsignificant results provide no convincing evidence—one way or the other—as to whether a relationship exists in nature.

For this reason, you cannot design a study to show that a relationship does not exist. For example, you could not set out to show that the IQ pill does not work. At best, you'll end up retaining both H_0 and H_a, and at worst, you'll end up rejecting H_0, showing that it does work.

SUMMARY OF THE *z*-TEST

Altogether, the preceding discussion can be summarized as the following four steps. For a one-sample experiment that meets the assumptions of the *z*-test:

1. *Determine the experimental hypotheses and create the statistical hypothesis*: Predict the relationship the study will or will not demonstrate. Then H_0 describes the μ that the $\overline{X}$ represents if the predicted relationship does not exist. H_a describes the μ that the $\overline{X}$ represents if the relationship does exist.

2. *Compute $\overline{X}$, compute $\sigma_{\overline{X}}$, and then compute z_{obt}.* In the formula for z, the value of μ is the μ of the sampling distribution, which is also the μ of the raw score population that H_0 says is being represented.

3. *Set up the sampling distribution:* Select α, locate the region of rejection, and determine the critical value.

4. *Compare z_{obt} to z_{crit}:* If z_{obt} lies beyond z_{crit}, then reject H_0, accept H_a, and the results are "significant." Then interpret the relationship. If z_{obt} does not lie beyond z_{crit}, do not reject H_0 and the results are "nonsignificant." Do not draw any conclusions about the relationship.

- If z_{obt} lies beyond z_{crit}, reject H_0, the results are significant, and conclude there is evidence for the predicted relationship. Otherwise, the results are not significant, and we make no conclusion about the relationship.

MORE EXAMPLES

We test a new technique for teaching reading. Without it, the μ on a reading test is 220, with $\sigma_X = 15$. An N of 25 participants has $\overline{X} = 211.55$. Then:

1. With a two-tailed test, $H_0: \mu = 220$; $H_a: \mu \neq 220$.
2. Compute z_{obt}: $\sigma_{\overline{X}} = \sigma_X/\sqrt{N} = 15/\sqrt{25} = 3$; $z_{obt} = (\overline{X} - \mu)/\sigma_{\overline{X}} = (211.55 - 220)/3 = -2.817$.
3. With $\alpha = .05$, z_{crit} is ± 1.96, and the sampling distribution is like Figure 10.6.
4. The z_{obt} of -2.817 is beyond the z_{crit} of -1.96, so the results are significant: the data reflect a relationship, with the μ of the population using the technique at around 211.55, while for those not using it at $\mu = 220$.

Above a different mean produced $Z = -1.83$. This z_{obt} is not beyond the z_{crit} so the results are not signifi-

cant: Make no conclusion about the influence of the technique on reading.

For Practice

We test whether a sample of 36 successful dieters are more or less satisfied with their appearance than in the population of nondieters, where $\mu = 40$ ($\sigma_X = 12$).

1. What are H_0 and H_a?
2. The $\overline{X}$ for dieters is 45. Compute z_{obt}.
3. Set up the sampling distribution.
4. What should we conclude?

Answers

1. $H_0: \mu = 40$; $H_a: \mu \neq 40$
2. $\sigma_{\overline{X}} = 12/\sqrt{36} = 2$; $z_{obt} = (45 - 40)/2 = +2.50$
3. With $\alpha = .05$ the sampling distribution has a region of rejection in each tail, with $z_{crit} = \pm 1.96$ (as in Figure 10.6).
4. The z_{obt} of $+2.50$ is beyond z_{crit} of ± 1.96, so the results are significant: The population of dieters are more satisfied (at a μ around 45) than the population of nondieters (at $\mu = 40$).

THE ONE-TAILED TEST

Recall that a *one-tailed test* is used when we predict the *direction* in which scores will change. The statistical hypotheses and sampling distribution are different in a one-tailed test.

The One-Tailed Test for Increasing Scores

Say that we had developed a "smart" pill, so the experimental hypotheses are (1) the pill makes people smarter by increasing IQ scores, or (2) the pill does not make people smarter. For the statistical hypotheses, start with the alternative hypothesis: People without the pill produce $\mu = 100$, so if the pill makes them smarter, their μ will be greater than 100. Therefore, our alternative hypothesis is that our sample represents this population, so $H_a: \mu > 100$. On the other hand, if the pill does not work as predicted, either it will leave IQ scores *unchanged* or it will *decrease* them (making people dumber). Then μ will either equal 100 or be less than 100. Therefore, our null hypothesis is that our sample represents one of these populations, so $H_0: \mu \leq 100$.

We again test H_0, and we do so by testing whether the sample represents the raw score population in which μ *equals* 100. This is because first the pill *must* make the sample smarter, producing a $\overline{X}$ above 100, or we have no evidence that the "smart" pill works. If we then conclude that the population μ is above 100, then it is automatically above any value less than 100.

> **REMEMBER** A one-tailed null hypothesis always includes a population with μ equal to some value. Test H_0 by testing whether the sample data represent that population.

Thus, as in Figure 10.7, the sampling distribution again shows the means that occur when sampling the population where μ is 100. We again set $\alpha = .05$, but because we have a one-tailed test, the region of rejection is in *one tail* of the sampling distribution. You can identify which tail by identifying the result you must see to claim that your independent variable works as predicted (to support H_a). For us to believe that the smart pill works, we must conclude that the $\overline{X}$ is *significantly larger* than 100. On the sampling distribution, the means that are significantly larger than 100 are in the region of rejection in the *upper* tail of the sampling distribution. Therefore, the entire region is in the upper tail of the distribution. Then, as in the previous chapter, the region of rejection is 5% of the curve, so z_{crit} is $+1.645$.

Say that after testing the pill ($N = 36$) we find $\overline{X} = 106.58$. The sampling distribution is still based on the IQ population with $\mu = 100$ and $\sigma_X = 15$, so $\sigma_{\overline{X}} = 15/\sqrt{36} = 2.5$. Then $z_{\text{obt}} = (106.58 - 100)/2.5 = +2.63$. As in Figure 10.7, this z_{obt} is beyond z_{crit}, so it is in the region of rejection. Therefore, the sample mean is unlikely to represent the population having $\mu = 100$. If the sample is unlikely to represent the population where μ is 100, it is even less likely to represent a population where μ is below 100. Therefore, we reject the null hypothesis that $\mu \leq 100$, and accept the alternative hypothesis that $\mu > 100$. We conclude that the pill produces a *significant increase* in IQ scores and estimate that μ would equal about 106.58 (keeping in mind all of the cautions and qualifications for interpreting significant results that we discussed previously).

Notice that a one-tailed z_{obt} is significant only if it lies beyond z_{crit} and has the same *sign*. Thus, if z_{obt} had not been in our region of rejection, we would retain H_0 and have no evidence whether the pill works or not. This would be the case even if we had obtained very low scores producing a very large negative z-score. We have no region of rejection in the lower tail for this study and, no, you cannot move the region of rejection to make the results significant. Remember, which tail you use is determined by your experimental hypothesis. After years of developing a "smart pill," it would make no sense to suddenly say, "Whoops, I meant to call it a dumb pill." Likewise, you

FIGURE 10.7

Sampling distribution of IQ means for a one-tailed test of whether scores increase

The region of rejection is entirely in the upper tail.

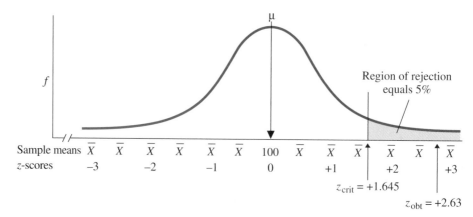

cannot switch from a one-tailed to a two-tailed test. Therefore, use a one-tailed test only when confident of the direction in which the dependent scores will change. When in doubt, use a two-tailed test.

The One-Tailed Test for Decreasing Scores

Say that we had created a pill to lower IQ scores. If the pill works, then μ would be *less than* 100, so H_a: $\mu < 100$. But, if the pill does not work, it would produce the same scores as no pill (with $\mu = 100$), or it would make people smarter (with $\mu > 100$). So H_0: $\mu \geq 100$.

We again test H_0 using the previous sampling distribution. Now, however, to conclude that the pill lowers IQ, our sample mean must be significantly *less* than 100. Therefore, the region of rejection is in the lower tail of the distribution, as in Figure 10.8.

With $\alpha = .05$, z_{crit} is now *minus* 1.645. If the sample produces a *negative* z_{obt} beyond -1.645 (for example, $z_{obt} = -1.69$), then we reject the H_0 that the sample mean represents a μ equal to or greater than 100 and accept the H_a that the sample represents a μ less than 100. However, if z_{obt} does not fall in the region of rejection (for example, if $z_{obt} = -1.25$), we do not reject H_0, and we have no evidence as to whether the pill works or not.

FIGURE 10.8

Sampling distribution of IQ means for a one-tailed test of whether scores decrease

The region of rejection is entirely in the lower tail.

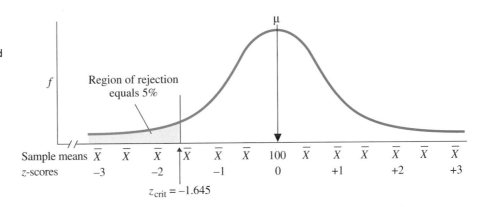

A QUICK REVIEW

- Perform a one-tailed test when predicting the direction the scores will change.
- When predicting that $\overline{X}$ will be higher than μ, the region of rejection is in the upper tail of the sampling distribution. When predicting that $\overline{X}$ will be lower than μ, the region of rejection is in the lower tail.

MORE EXAMPLES

We predict that learning statistics will increase a student's IQ. Those not learning statistics have $\mu = 100$ and $\sigma_X = 15$. For 25 statistics students, $\overline{X} = 108.6$.

1. With a one-tailed test, H_a: $\mu > 100$; H_0: $\mu \leq 100$.
2. Compute z_{obt}: $\sigma_{\overline{X}} = \sigma_X/\sqrt{N} = 15/\sqrt{25} = 3$; $z_{obt} = (\overline{X} - \mu)/\sigma_{\overline{X}} = (108.6 - 100)/3 = +2.87$.
3. With $\alpha = .05$, z_{crit} is $+1.645$. The sampling distribution is as in Figure 10.7.
4. The z_{obt} of $+2.87$ is beyond z_{crit}, so the results are significant: Learning statistics gives a μ around 108.6, whereas people not learning statistics have $\mu = 100$.

Say that a different mean produced $z_{obt} = +1.47$. This is not beyond z_{crit}, so it is not significant. We'd have no evidence that learning statistics raises IQ.

continued

For Practice

You test the effectiveness of a new weight-loss diet.

1. Why is this a one-tailed test?

2. For the population of nondieters, $\mu = 155$. What are H_a and H_0?

3. In which tail is the region of rejection?

4. With $\alpha = .05$, the z_{obt} for the sample of dieters is -1.86. What do you conclude?

Answers

1. Because a successful diet *lowers* weight scores
2. H_a: $\mu < 155$ and H_0: $\mu \geq 155$
3. The left-hand tail
4. The z_{obt} is beyond z_{crit} of -1.645, so it is significant: The μ for dieters will be less than the μ of 155 for nondieters.

STATISTICS IN PUBLISHED RESEARCH: REPORTING SIGNIFICANCE TESTS

Every study must indicate whether the results are significant or nonsignificant. With our IQ pill, a report might say, "The pill produced a significant difference in IQ scores." This indicates that the difference between the sample mean and the μ without the pill is too large to accept as being due to sampling error. Or a report might say that we obtained a "significant z": The z_{obt} is beyond the z_{crit}. Or we might observe a "significant effect of the pill": The change in IQ scores reflected by the sample mean is unlikely to be caused by sampling error, so presumably it is the effect of—caused by—changing the conditions of the independent variable.

Whether any result is significant depends on how we have defined *unlikely*. We do not always use $\alpha = .05$, so you must always report the α used. The APA format for reporting a result is to indicate the symbol for the statistic, the obtained value, and then the alpha level. For example to report a significant z_{obt} of $+2.00$, we write: $z = +2.00$, $p < .05$. Notice that instead of using α we use p (for probability), and with significant results, we say that p is *less than* .05. (We'll discuss the reason for this shortly.) For a nonsignificant z_{obt} of say, $-.40$, we report $z = -.40$, $p > .05$. Notice, with nonsignificant results, p is *greater than* .05.

ERRORS IN STATISTICAL DECISION MAKING

We have one other issue to consider, and it involves potential errors in our decisions: Regardless of whether we conclude that the sample does or does not represent the predicted relationship, we may be wrong.

Type I Errors: Rejecting H_0 When H_0 Is True

Sometimes, the variables we investigate are *not* related in nature, so H_0 is really true. When in this situation, if we obtain data that cause us to reject H_0, then we make an error. A **Type I error** is defined as rejecting H_0 when H_0 is true. In other words, we conclude that the independent variable works when it really doesn't.

Thus, when we rejected H_0 and claimed that the pill worked, it's possible that it did not work and we made a Type I error. How could this happen? Because our sample was exactly what the sampling distribution indicated it was: a very unlikely and unrepresentative sample from the population having a μ of 100. In fact, the sample so poorly represented the situation where the pill did not work, we mistakenly thought that the pill did work. *In a Type I error, there is so much sampling error that we—and our statistical procedures—are fooled into concluding that the predicted relationship exists when it really does not.*

Any time researchers discuss Type I errors, it is a *given* that H_0 is true. Think of it as being in the "Type I situation" whenever you discuss the situation in which the predicted relationship does not exist. If you *reject* H_0 in this situation, then you've made a Type I error. If you *retain* H_0 in this situation, then you've avoided a Type I error: By not concluding that the pill works, you've made the correct decision because, in reality, the pill doesn't work.

We never know if we're making a Type I error because only nature knows if the variables are related. However, we do know that the theoretical probability of a Type I error equals our α. Here's why. Assume that the IQ pill does not work, so we're in the Type I situation. Therefore, we can only obtain IQ scores from the population where μ is 100. If we repeated this experiment many times, then the sampling distribution in Figure 10.9 shows the different means we'd obtain over the long run. With a $\alpha = .05$, the total region of rejection is 5% of the distribution, so sample means in the region of rejection would occur 5% of the time in this situation. These means would cause us to *reject* H_0 even though H_0 is true. Rejecting H_0 when it is true is a Type I error, so over the long run, the relative frequency of Type I errors would be .05. Therefore, anytime we reject H_0, the theoretical probability that we've just made a Type I error is .05. (The same is true in a one-tailed test.)

You either will or will not make the correct decision when H_0 is true, so the probability of avoiding a Type I error, is $1 - \alpha$. This is because, if 5% of the time samples are in the region of rejection when H_0 is true, then 95% of the time they are not in the region of rejection when H_0 is true. Therefore, 95% of the time we will not obtain sample means that cause us to erroneously reject H_0: Anytime you retain H_0, the theoretical probability is .95 that you've avoided a Type I error.

Although the *theoretical* probability of a Type I error equals α, the *actual* probability is slightly less than α. This is because the region of rejection includes the critical value. Yet to reject H_0, the z_{obt} must be *larger* than z_{crit}. We cannot determine the precise area under the curve at z_{crit}, so we can't remove it from our 5%. We can only say that the region of rejection is slightly less than 5% of the curve. Therefore, the actual probability of a Type I error is also slightly less than α.

Thus, in our examples when we rejected H_0, the probability that we made a Type I error was slightly less than .05. That is why we report a significant result using $p < .05$. This is code for "the probability of a Type I error is less than .05." The reason you must always report your alpha level is so that you indicate the probability of making a Type I error.

FIGURE 10.9

Sampling distribution of sample means showing that 5% of all sample means fall into the region of rejection when H_0 is true

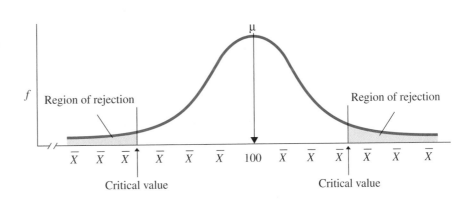

On the other hand, we reported a nonsignificant result using $p > .05$. This communicates that we did not call this result significant because to do so would require a region greater than 5% of the curve. But then the probability of a Type I error would be greater than our α of .05, and that's unacceptable.

Typically, researchers do not use an α larger than .05 because then it is too likely that they will make a Type I error. This may not sound like a big deal, but the next time you fly in an airplane, consider that the designer's belief that the wings will stay on may actually be a Type I error: He's been misled by sampling error into *erroneously* thinking the wings will stay on. A 5% chance of this is scary enough—we certainly don't want more than a 5% chance that the wings will fall off. In science, we are skeptical and careful, so we want to be convinced that sampling error did not produce our results. Having only a 5% chance that it did is reasonably convincing.

Type I errors are the reason a study must meet the assumptions of a statistical procedure. If we violate the assumptions, then the true probability of a Type I error will be *larger* than our α (so it's larger than we think it is). Thus, if we severely violate a procedure's assumptions, we may think that α is .05 when in fact it is, say, .20! But recall that with parametric tests we can violate the assumptions somewhat. This is allowed because the probability of a Type I error will still be close to α (it will be only, say, .051 when we've set α at .050).

Sometimes making a Type I error is so dangerous that we want to reduce its probability even further. Then we usually set alpha at .01. For example, say that the smart pill had some dangerous side effects. We would not want to needlessly expose the public to such dangers, so $\alpha = .01$ would make us even less likely to conclude that the pill works when it does not. When α is .01, the region of rejection is the extreme 1% of the sampling distribution, so the probability of making a Type I error is now $p < .01$.

However, we use the term *significant* in an all-or-nothing fashion: A result is *not* "more" significant when $\alpha = .01$ than when $\alpha = .05$. If z_{obt} lies in the region of rejection that was used to define significant, then the result is significant, period! The *only* difference is that when $\alpha = .01$ the probability that we've made a Type I error is smaller.

Finally, computer programs such as SPSS compute the exact probability of a Type I error. For example, we might see $p = .02$. This indicates that the z_{obt} lies in the extreme 2% of the sampling distribution, and thus the probability of a Type I error here is .02. If our α is .05, then this result is significant. However, we might see $p = .07$, which indicates that to call this result significant we'd need a region of rejection that is the extreme 7% of the sampling distribution. This implies an α of .07, which is greater than .05, and thus this result is not significant.

> **REMEMBER** When H_0 is true: Rejecting H_0 is a Type I error, and its probability is α; retaining H_0 is avoiding a Type I error, and its probability is $1 - \alpha$.

Type II Errors: Retaining H_0 When H_0 Is False

It is also possible to make a totally different kind of error. Sometimes the variables we investigate really *are* related in nature, and so H_0 really is false. When in this situation, if we obtain data that cause us to retain H_0, then we make a Type II error. A **Type II error** is defined as retaining H_0 when H_0 is false (and H_a is true). In other words, here we fail to identify that the independent variable really does work.

Thus, when our IQ sample mean of 99 caused us to retain H_0 and not claim the pill worked, it's possible that the pill did work and we made a Type II error. How could this happen? Because the sample mean of 99 was so close to 100 (the μ without the pill) that the difference could easily be explained as sampling error, so we weren't convinced the pill worked. Or perhaps the pill would actually *increase* IQ greatly, say to a μ of 105, but we obtained an unrepresentative sample of this. Either way, *in a Type II error, the sample mean is so close to the μ described by H_0 that we—and our statistics—are fooled into concluding that the predicted relationship does not exist when it really does*.

Anytime we discuss Type II errors, it's a *given* that H_0 is false and H_a is true. That is, we're in the "Type II situation," in which the predicted relationship does exist. If we *retain* H_0 in this situation, then we've made a Type II error. If we *reject* H_0 in this situation, then we've avoided a Type II error: We've made the correct decision because we concluded that the pill works and it does work.

We never know when we make a Type II error, but we can determine its probability. The computations of this are beyond the introductory level, but you should know that the symbol for the theoretical probability of a Type II error is β, the Greek letter beta. Whenever you retain H_0, β is the probability that you've made a Type II error. On the other hand, $1 - \beta$ is the probability of avoiding a Type II error. Thus, anytime you reject H_0, the probability is $1 - \beta$ that you've made the correct decision and rejected a false H_0.

> **REMEMBER** When H_0 is false: Retaining H_0 is a Type II error, and its probability is β; rejecting H_0 is avoiding a Type II error, and its probability is $1 - \beta$.

Comparing Type I and Type II Errors

You probably think that Type I and Type II errors are two of the most confusing inventions ever devised. So, first recognie that if there's a possibility you've made one type of error, then there is no chance that you've made the other type of error. Remember: In the Type I *situation*, H_0 is really true (the variables are not related in nature). In the Type II *situation*, H_0 is really false (the variables are related in nature). You can't be in both situations simultaneously. Second, if you don't make one type of error, then you are *not* automatically making the other error because you might be making a correct decision. Therefore, look at it this way: The type of error you can *potentially* make is determined by your situation—what nature "says" about whether there is a relationship. Then, whether you actually make the error depends on whether you agree or disagree with nature.

Thus, four outcomes are possible in any study. Look at Table 10.1. As in the upper row of the table, sometimes H_0 is really true: Then if we reject H_0, we make a Type I error (with a $p = \alpha$). If we retain H_0, we avoid a Type I error and make the correct

TABLE 10.1
Possible Results of Rejecting or Retaining H_0

		Our Decision	
		We Reject H_0	*We Retain* H_0
The truth about H_0	Type I situation: H_0 *is true* (*no relationship exists*)	We make a Type I error ($p = \alpha$)	We are correct, a voiding a Type I error ($p = 1 - \alpha$)
	Type II situation: H_0 *is false* (*a relationship exists*)	We are correct, avoiding a Type II error ($p = 1 - \beta$)	We make a Type II error ($p = \beta$)

decision (with $p = 1 - \alpha$). But, as in the lower row of the table, sometimes H_0 is really false: Then if we retain H_0, we make a Type II error (with $p = \beta$), and if we reject H_0, we avoid a Type II error and make the correct decision (with $p = 1 - \beta$).

In any experiment, the results of your inferential procedure will place you in one of the columns of Table 10.1. If you reject H_0, then either you've made a Type I error, or you've made the correct decision and avoided a Type II error. If you retain H_0, then either you've made a Type II error or you've made the correct decision and avoided a Type I error.

The most serious error is a Type I, concluding that an independent variable works when really it does not. For example, concluding that new drugs, surgical techniques, or engineering procedures work when they really do not can cause untold damage. For this reason, researchers always use a small α to minimize the likelihood of these errors. On the other hand, a Type II error is not as harmful because it is merely failing to identify an independent variable that works. We have faith that future research will eventually discover the variable. However, we still prefer to avoid Type II errors, and for that we need *power*.

Power

Of the various outcomes back in Table 10.1, the goal of research is to reject H_0 when it is false: We conclude that the pill works, and the truth is that the pill does work. Not only have we avoided any errors, but we have learned about a relationship in nature. This ability has a special name: **Power** is the probability that we will reject H_0 when it is false, correctly concluding that the sample data represent a relationship. In other words, power is the probability of not making a Type II error, so power equals $1 - \beta$.

Power is important because, after all, why bother to conduct a study if we're unlikely to reject the null hypothesis even when there is a relationship present? Therefore, power is a concern anytime we do not reject H_0 because we wonder, "Did we just miss a relationship?" For example, previously, when we did not find a significant effect of the pill, maybe the problem was that we lacked power: Maybe we were unlikely to reject H_0 *even if the pill really worked.*

To avoid this doubt, we strive to *maximize* the power of a study (maximizing the size of $1 - \beta$). Then we'll have confidence in our decision if we do ultimately retain null. Essentially, the idea is to do everything we can to ensure that in case we end up in the Type II situation where there is a relationship in nature, we—and our statistics—will not miss the relationship. If we still end up retaining H_0, we know that it's not for lack of trying. We're confident that if the relationship was there, we would have found it, so it must be that the relationship is *not* there. Therefore, we are confident in the decision to retain H_0, and, in statistical lingo, we say that we're confident we have avoided a Type II error.

> **REMEMBER** We seek to maximize *power* so that, if we retain H_0, we are confident we are not making a Type II error.

The time to build in power is when we design a study. We're talking about being in the Type II situation here, so it's a given that the relationship exists in nature. We can't do anything to ensure that we're in this situation (that's up to nature), but assuming we are, then the goal is to have *significant results*. Therefore, we increase power by increasing the likelihood that our results will be significant. Results are significant if z_{obt} is larger than z_{crit}, so anything that increases the size of the obtained value relative to the critical value increases power.

We influence power first through the statistics we use. It is better to design a study so that you can use parametric procedures because parametric procedures are more powerful than nonparametric ones: Analyzing data using a parametric test is more likely to produce significant results than analyzing the same data using a nonparametric test. Then, in case we're in the situation where H_0 is false, we won't miss the relationship.

Also, when we can predict the direction of the relationship, using a one-tailed test is more powerful than a two-tailed test. This is because the z_{crit} for a one-tailed test (1.645) is smaller than the z_{crit} for a two-tailed test (1.96). All other things being equal, a z_{obt} is more likely to be beyond 1.645, so it's more likely to be significant.

In later chapters, you'll see additional ways to maximize power. Do not think that we are somehow "rigging" the decisions here. We are simply protecting ourselves against errors. Setting α at .05 or less protects us if we end up in the situation where H_0 is true (limiting Type I errors). Maximizing power protects us if we end up in the situation where H_0 is false (limiting Type II errors). Together, these strategies minimize our errors, regardless of whether or not there is really a relationship.

> *REMEMBER*   When discussing power, it is a given that H_0 is false. Power is increased by increasing the size of the obtained value relative to the critical value so that the results are more likely to be significant.

A QUICK REVIEW

- A Type I error is rejecting a true H_0. A Type II error is retaining a false H_0.
- Power is the probability of not making a Type II error.

MORE EXAMPLES

When H_0 is true, there is no relationship: If the data cause us to reject H_0, we make a Type I error. To decrease the likelihood of this, we keep alpha small. If the data cause us to retain H_0, we avoid this error. When H_0 is false, there is a relationship: If the data cause us to retain H_0, we make a Type II error. If the data cause us to reject H_0, we avoid this error. To increase the likelihood of this, we increase power.

For Practice

1. Claiming that an independent variable works although in nature it does not is a _____ error.
2. Failing to conclude that an independent variable works although in nature it does is a _____ error.
3. If we reject H_0, we cannot make a _____ error.
4. If we retain H_0, we cannot make a _____ error.
5. To be confident in a decision to retain μ, our power should be _____.

Answers

1. Type I
2. Type II
3. Type II
4. Type I
5. high

**PUTTING IT
ALL TOGETHER**

Essentially, the purpose of inferential statistics is to minimie the probability of making Type I and Type II errors. If we had not performed the z-test in our initial IQ pill study, we might have made a Type I error: We might have erroneously concluded that the pill raises IQ to around 105 when, in fact, we were being misled by sampling error. We would have no idea if this had occurred, nor even the chances that it had occurred. After finding a significant result, however, we are confident that we did not make a Type I error because the probability of doing so is less than .05. Likewise, if the results were not significant, through power we minimize the probability of a Type II error, so we'd be confident that we did not miss a pill that actually works.

All parametric and nonparametric inferential procedures follow the logic described here: H_0 is the hypothesis that says your data represent the populations you would find if the predicted relationship does not exist; H_a says that your data represent the predicted relationship. You then compute something like a z-score for your data on the sampling distribution when H_0 is true. If the z-score is larger than the critical value, the results are unlikely to represent the populations described by H_0, so we reject H_0 and accept H_a. The results are called significant, meaning essentially that they are "believable": The relationship depicted in the sample data can be believed as existing in nature rather than being a chance pattern resulting from sampling error. That's it! That's inferential statistics (well, not quite).

CHAPTER SUMMARY

1. *Inferential statistics* are procedures for deciding whether sample data represent a particular relationship in the population.

2. *Parametric* inferential procedures require assumptions about the raw score populations being represented. They are performed when we compute the mean.

3. *Nonparametric* inferential procedures do not require stringent assumptions about the populations being represented. They are performed when we compute the median or mode.

4. *The alternative hypothesis* (H_a) is the statistical hypothesis that describes the population μs being represented if the predicted relationship exists.

5. *The null hypothesis* (H_0) is the statistical hypothesis that describes the population μs being represented if the predicted relationship does not exist.

6. A *two-tailed test* is used when we do not predict the direction in which the dependent scores will change. A *one-tailed test* is used when the direction of the relationship is predicted.

7. The *z-test* is the parametric procedure used in a one-sample experiment if (a) the population contains normally distributed interval or ratio scores and (b) the standard deviation of the population (σ_X) is *known*.

8. If z_{obt} lies beyond z_{crit}, then the corresponding sample mean is unlikely to occur when sampling from the population described by H_0. Therefore, we *reject* H_0 and *accept* H_a. This is a *significant* result and is evidence of the predicted relationship in the population.

9. If z_{obt} does not lie beyond z_{crit}, then the corresponding sample mean is likely to occur when sampling the population described by H_0. Therefore, we *retain* H_0. This is a *nonsignificant* result and is not evidence for or against the predicted relationship.

10. A *Type I error* occurs when a true H_0 is rejected. Its theoretical probability equals α. If a result is significant, the probability of a Type I error is $p < \alpha$. The theoretical probability of avoiding a Type I error when retaining H_0 is $1 - \alpha$.

11. A *Type II error* occurs when a false H_0 is retained. Its theoretical probability is β. The theoretical probability of avoiding a Type II error when rejecting H_0 is $1 - \beta$.

12. *Power* is the probability of rejecting a false H_0, and it equals $1 - \beta$.

KEY TERMS

$>$ $<$ $\geq$ $\leq$ $\neq$ H_a H_0 α z_{crit} z_{obt} β

alpha *216*
alternative hypothesis *211*
beta *227*
experimental hypotheses *209*
inferential statistics *208*
nonparametric statistics *209*
nonsignificant *220*
null hypothesis *212*

one-tailed test *210*
parametric statistics *209*
power *228*
significant *218*
statistical hypotheses *211*
two-tailed test *210*
Type I error *224*
Type II error *226*
z-test *215*

REVIEW QUESTIONS

(Answers for odd-numbered questions are in Appendix D.)

1. Why does the possibility of sampling error present a problem to researchers when inferring a relationship in the population?
2. What are inferential statistics used for?
3. What does α stand for, and what two things does it determine?
4. (a) What are the two major categories of inferential procedures? (b) What characteristics of your data determine which you should use? (c) What happens if you seriously violate the assumptions of a procedure? (d) What is a statistical reason to design a study so you can use parametric procedures?
5. What are experimental hypotheses?
6. (a) What does H_0 communicate? (b) What does H_a communicate?
7. (a) When do you use a one-tailed test? (b) When do you use a two-tailed test?
8. (a) What does "significant" convey about the results of an experiment? (b) Why is obtaining significant results a goal of behavioral research? (c) Why is declaring the results significant not the final step in a study?
9. (a) What is power? (b) Why do researchers want to maximize power? (c) What result makes us worry whether we have sufficient power? (d) Why is a one-tailed test more powerful than a two-tailed test?
10. (a) What are the advantage and disadvantage of two-tailed tests? (b) What are the advantage and disadvantage of one-tailed tests?

APPLICATION QUESTIONS

11. Describe the experimental hypotheses and the independent and dependent variables when we study: (a) whether the amount of pizza consumed by college students during finals week increases relative to the rest of the semester, (b) whether breathing exercises alter blood pressure, (c) whether sensitivity to pain is affected by increased hormone levels, and (d) whether frequency of day-dreaming decreases as a function of more light in the room.
12. For each study in question 11, indicate whether a one- or a two-tailed test should be used and state the H_0 and H_a. Assume that $\mu = 50$ when the amount of the independent variable is zero.
13. Listening to music while taking a test may be relaxing or distracting. We test 49 participants while listening to music, and they produce an $\overline{X} = 54.36$. The mean

of the population taking this test without music is 50 ($\sigma_X = 12$). (a) Is this a one-tailed or two-tailed test? Why? (b) What are our H_0 and H_a? (c) Compute z_{obt}. (d) With $\alpha = .05$, what is z_{crit}? (e) Do we have evidence of a relationship in the population? If so, describe the relationship.

14. We ask whether attending a private school leads to higher or lower performance on a test of social skills. A sample of 100 students from a private school produces a mean of 71.30 on the test, and the national mean for students from public schools is 75.62 ($\sigma_X = 28.0$). (a) Should we use a one-tailed or a two-tailed test? Why? (b) What are H_0 and H_a? (c) Compute z_{obt}. (d) With $\alpha = .05$, what is z_{crit}? (e) What should we conclude about this relationship?

15. (a) In question 13, what is the probability that we made a Type I error? What would be the error in terms of the independent and dependent variables? (b) What is the probability that we made a Type II error? What would be the error in terms of the independent and dependent variables?

16. (a) In question 14, what is the probability that we made a Type I error? What would be the error in terms of the independent and dependent variables? (b) What is the probability that we made a Type II error? What would be the error in terms of the independent and dependent variables?

17. Foofy claims that a one-tailed test is cheating because we use a smaller z_{crit}, and therefore it is easier to reject H_0 than with a two-tailed test. If the independent variable doesn't work, she claims, we are more likely to make a Type I error. Why is she correct or incorrect?

18. Poindexter claims that the real cheating occurs when we increase power by increasing the likelihood that results will be significant. He reasons that if we are more likely to reject H_0, then we are more likely to do so when H_0 is true. Therefore, we are more likely to make a Type I error. Why is he correct or incorrect?

19. Bubbles reads that in study A the $z_{obt} = +1.97, p < .05$. She also reads that in study B the $z_{obt} = +14.21, p < .0001$. (a) She concludes that the results of study B are way beyond the critical value used in study A, falling into a region of rejection containing only .0001 of the sampling distribution. Why is she correct or incorrect? (b) She concludes that the results of study B are more significant than those of study A, both because the z_{obt} is so much larger and because α is so much smaller. Why is she correct or incorrect? (c) In terms of their conclusions, what is the difference between the two studies?

20. Researcher A finds a significant relationship between increasing stress level and ability to concentrate. Researcher B replicates this study but finds a nonsignificant relationship. Identify the statistical error that each researcher may have made.

21. A report indicates that brand X toothpaste significantly reduced tooth decay relative to other brands, with $p < .44$. (a) What does "significant" indicate about the researcher's decision about brand X? (b) What makes you suspicious of the claim that brand X works better than other brands?

22. We ask if the attitudes toward fuel costs of 100 owners of hybrid electric cars ($\overline{X} = 76$) are different from those on a national survey of owners of non-hybrid cars ($\mu = 65, \sigma_X = 24$). Higher scores indicate a more positive attitude. (a) Is this a one- or two-tailed test? (b) In words what is H_0 and H_a? (c) Perform the z-test. (d) What do you conclude about attitudes here? (e) Report your results in the correct format.

23. We ask if visual memory ability for a sample of 25 art majors ($\overline{X} = 49$) is better than that of engineers who, on a nationwide test, scored $\mu = 45$ and $\sigma_X = 14$. Higher scores indicate a better memory. (a) Is this a one- or two-tailed test? (b) In words what is H_0 and H_a? (c) Perform the z-test. (d) What do you conclude about memory ability here? (e) Report your results in the correct format.

INTEGRATION QUESTIONS

24. We measure the self-esteem scores of a sample of statistics students, reasoning that this course may lower their self-esteem relative to that of the typical college student ($\mu = 55$ and $\sigma_X = 11.35$). We obtain these scores:

44   55   39   17   27   38   36   24   36

(a) Summarize your sample data. (b) Is this a one-tailed or two-tailed test? Why? (c) What are H_0 and H_a? (d) Compute z_{obt}. (e) With $\alpha = .05$, what is z_{crit}? (f) What should we conclude about the relationship here? (Chs. 4, 5, 10)

25. (a) What is the difference between the independent variable and the dependent variable in an experiment? (b) When the assumptions of a procedure require normally distributed interval/ratio scores, are we referring to scores on the independent or dependent variable? (c) What distinguishes an interval and ratio variable from nominal or ordinal variables? (d) What distinguishes a skewed versus a normal distribution? (Chs. 2, 3, 4, 10)

26. For the following, identify the independent variable and the dependent variable and explain why we should use a parametric or nonparametric procedure?
(a) When ranking the intelligence of a group of people given a smart pill.
(b) When comparing the median income for a group of college professors to that of the national population of all incomes. (c) When comparing the mean reading speed for a sample of hearing-impaired children to the average reading speed in the population of hearing children. (d) When measuring interval scores from a personality test given to a group of emotionally troubled people and comparing them to the population μ for emotionally healthy people. (Chs. 2, 4, 10)

27. We have a $\overline{X}$ of 40 under the condition of people tested in the morning versus a $\overline{X}$ of 60 for people tested in the evening. Assuming they accurately represent their populations, how do you envision this relationship in the population? (Chs. 4, 10)

28. (a) What does a sampling distribution of means show? (b) A mean having a z beyond ± 1.96 is where? (c) How often do means in the region of rejection occur when dealing with a particular raw score population? (d) What does this tell you about your mean? (Chs. 6, 9, 10)

29. (a) Why do researchers want to discover relationships? (b) What is the difference between a real relationship and one produced by sampling error? (c) What does a relationship produced by sampling error tell us about nature? (Chs. 2, 10)

30. (a) Why can no statistical result prove that changing the independent variable *causes* the dependent scores to change? (b) What one thing does a significant result prove? (Chs. 2, 10)

■ ■ ■ SUMMARY OF FORMULAS

To perform the z-test,

$$\sigma_{\overline{X}} = \frac{\sigma_X}{\sqrt{N}}$$

$$z_{obt} = \frac{\overline{X} - \mu}{\sigma_{\overline{X}}}$$

where μ is the mean of the population described by H_0.

11

Performing the One-Sample *t*-Test and Testing Correlation Coefficients

GETTING STARTED

To understand this chapter, recall the following:

- From Chapter 5, that s_X is the *estimated* population standard deviation, that s_X^2 is the *estimated* population variance, and that both involve degrees of freedom, or *df*, which equals $N - 1$.
- From Chapter 7, the uses and interpretation of r and r_S.
- From Chapter 10, the basics of significance testing, including one- and two-tailed tests, H_0 and H_a, Type I and Type II errors, and power.

Your goals in this chapter are to learn

- When and how to perform the t-*test*.
- How the *t-distribution* and *degrees of freedom* are used.
- What is meant by the *confidence interval for* μ and how it is computed.
- How to perform significance testing of r and r_S.
- How to increase the power of a study.

The logic of hypothesis testing discussed in the previous chapter is common to all inferential statistical procedures. Your goal now is to learn how slightly different procedures are applied to different research designs. This chapter begins the process by introducing the *t*-test, which is very similar to the *z*-test. The chapter presents (1) when and how to perform the *t*-test, (2) how to use a similar procedure to test correlation coefficients, and (3) a new procedure—called the *confidence interval*—that is used to estimate μ.

WHY IS IT IMPORTANT TO KNOW ABOUT *t*-TESTS?

The *t*-test is important because, like the *z*-test, the *t*-test is used for significance testing in a one-sample experiment. In fact, the *t*-test and the "*t*-distribution" are used more often in behavioral research. That's because with the *z*-test we must *know* the standard deviation of the raw score population (σ_X). However, usually researchers do *not* know such things because they're exploring uncharted areas of behavior. Instead, we usually *estimate* σ_X by using the sample data to compute the unbiased, estimated population standard deviation (s_X). Then we compute something *like* a *z*-score for our sample mean. However, because we are estimating, we are computing *t*. The **one-sample *t*-test** is the parametric inferential procedure for a one-sample experiment when the standard deviation of the raw score population must be estimated.

> *REMEMBER* Use the *z*-test when σ_X is known; use the *t*-test when σ_X is not known.

Further, *t*-tests are important because they form the basis for several other procedures that we'll see in this and later chapters.

PERFORMING THE ONE-SAMPLE *t*-TEST

The one-sample *t*-test is applied when we have a one-sample experiment. Here's an example: Say that one of those "home-and-gardening/good-housekeeper" magazines describes a test of housekeeping abilities. The magazine is targeted at women, and it reports that the national average score for women is 75 (so their μ is 75), but it does not report the standard deviation. Our question is, "How do men score on this test?" To answer this, we'll give the test to a random sample of men and use their $\overline{X}$ to estimate the μ for the population of all men. Then we can compare the μ for men to the μ of 75 for women. If we can conclude that men produce one population of scores located at one μ, but women produce a different population of scores at a different μ, then we've found a relationship in which, as gender changes, test scores change.

As usual, we first set up the statistical test.

1. *The statistical hypotheses:* Say that we're being open minded and look for any kind of difference, so we have a two-tailed test. If men are different from women, then the μ for men will not equal the μ for women of 75, so H_a is $\mu \neq 75$. If men are not different, then their μ will equal that of women, so H_0 is $\mu = 75$.

2. *Alpha:* We select alpha; .05 sounds good.

3. *Check the assumptions:* The one-sample *t*-test is appropriate if we can assume the following about the dependent variable:

   a. We have one random sample of interval or ratio scores.

   b. The raw score population forms a normal distribution.

   c. The standard deviation of the raw score population is estimated by computing s_X.

Based on similar research that we've read, our test scores meet these assumptions, so we proceed. For simplicity, we test nine men. (For *power,* you should never collect so few scores.) Say that the sample produces a $\overline{X} = 65.67$. Based on this, we might conclude that the population of men has a μ of 65.67, while women have a μ of 75. On the other hand, maybe we are being misled by sampling error: Maybe by chance we selected some exceptionally sloppy men for our *sample,* but men in the population are not different from women, and so our sample actually poorly represents that the male population also has a $\mu = 75$.

To test this null hypothesis, we use the logic that we've used previously: H_0 says that the men's mean represents a population where μ is 75, so we will create a sampling distribution showing the means that occur by chance when representing this population. Then using the formula for the *t*-test, we will compute t_{obt}, which will locate our sample mean on this sampling distribution in the same way that *z*-scores did. The larger the absolute value of t_{obt}, the farther our sample mean is into the tail of the sampling distribution. Therefore, we will compare t_{obt} to the critical value, called t_{crit}. If t_{obt} is beyond t_{crit}, our sample mean lies in the region of rejection, so we'll reject that the sample poorly represents the population where μ is 75.

The only novelty here is that t_{obt} is calculated differently than z_{obt} and t_{crit} comes from the *t*-distribution.

Computing t_{obt}

The computation of t_{obt} consists of three steps that parallel the three steps in the *z*-test. After computing our sample mean, the first step in the *z*-test was to determine the standard deviation (σ_X) of the raw score population. For the *t*-test, we could compute the estimated standard deviation (s_X), but to make your computations simpler, we'll use the estimated population variance (s_X^2). Recall that the formula for the estimated population variance is

$$s_X^2 = \frac{\sum X^2 - \dfrac{(\sum X)^2}{N}}{N - 1}$$

The second step of the *z*-test was to compute the standard error of the mean $(\sigma_{\overline{X}})$, which is like the "standard deviation" of the sampling distribution. However, because now we are estimating the population variability, we compute the **estimated standard error of the mean,** which is an estimate of the "standard deviation" of the sampling distribution of means. The symbol for the estimated standard error of the mean is $s_{\overline{X}}$. (The lowercase *s* stands for an estimate of the population, and the subscript $\overline{X}$ indicates that it is for a population of means.)

Previously we computed $\sigma_{\overline{X}}$ by dividing σ_X by $\sqrt{N}$. We could use a similar formula here so that $s_{\overline{X}}$ equals s_X divided by $\sqrt{N}$. However, recall that to get any standard deviation we first compute the variance and then find its square root. Thus, buried in the above is the extra step of finding the square root of the variance that we then divide by $\sqrt{N}$. Instead, by using the variance, we only need to take the square root once.

> **The formula for the estimated standard error of the mean is**
>
> $$s_{\overline{X}} = \sqrt{\frac{s_X^2}{N}}$$

This says to divide the estimated population variance by the *N* of our sample and then find the square root.

The third step in the *z*-test was to compute z_{obt} using the formula

$$z_{obt} = \frac{\overline{X} - \mu}{\sigma_{\overline{X}}}$$

This says to find the difference between our sample mean and the μ in H_0 and then divide by the standard error. Likewise, the final step in computing t_{obt} is to use this formula:

> **The formula for the one-sample *t*-test is**
>
> $$t_{obt} = \frac{\overline{X} - \mu}{s_{\overline{X}}}$$

Here, the $\overline{X}$ is our sample mean, μ is the mean of the H_0 sampling distribution (which equals the value of μ described in the null hypothesis), and $s_{\overline{X}}$ is the estimated standard error of the mean. The t_{obt} is like a *z*-score, however, indicating how far our sample mean is from the μ of the sampling distribution, when measured in estimated standard error units.

For our housekeeping study, say that we obtained the data in Table 11.1. First, compute s_X^2. Substituting the data into the formula gives

$$s_X^2 = \frac{\sum X^2 - \dfrac{(\sum X)^2}{N}}{N - 1} = \frac{39,289 - \dfrac{349,281}{9}}{9 - 1} = 60.00$$

Thus, the estimated variance of the population of housekeeping scores is 60.

Next, compute the estimated standard error of the mean. With $s_X^2 = 60.00$ and $N = 9$,

$$s_{\overline{X}} = \sqrt{\frac{s_X^2}{N}} = \sqrt{\frac{60.00}{9}} = \sqrt{6.667} = 2.582$$

Finally, compute t_{obt}. Our sample mean is 65.67, the μ that H_0 says we're representing is 75, and the estimated standard error of the mean is 2.582. Therefore,

$$t_{obt} = \frac{\overline{X} - \mu}{s_{\overline{X}}} = \frac{65.67 - 75}{2.582} = \frac{-9.33}{2.582} = -3.61$$

Our t_{obt} is -3.61.

TABLE 11.1

Test Scores of Nine Men

Subject	Grades (X)	X^2
1	50	2,500
2	75	5,625
3	65	4,225
4	72	5,184
5	68	4,624
6	65	4,225
7	73	5,329
8	59	3,481
9	64	4,096
$N = 9$	$\sum X = 591$	$\sum X^2 = 39,289$
	$(\sum X)^2 = 349,281$	
	$\overline{X} = 65.67$	

A QUICK REVIEW

- Perform the one-sample *t*-test in a one-sample experiment when you do not know the population standard deviation.

MORE EXAMPLES

In a study, H_0 is that $\mu = 60$. To compute t_{obt}, say that $\overline{X} = 62$, $s_X^2 = 25$, and $N = 36$.

$$s_{\overline{X}} = \sqrt{\frac{s_X^2}{N}} = \sqrt{\frac{25}{36}} = \sqrt{.694} = .833$$

$$t_{obt} = \frac{\overline{X} - \mu}{s_{\overline{X}}} = \frac{62 - 60}{.833} = \frac{+2}{.833} = +2.40$$

(continued)

For Practice

In a study, H_0 is that $\mu = 10$. The data are 6, 7, 9, 8, and 8.

1. To compute the t_{obt}, what two descriptive statistics are computed first?

2. What do you compute next?

3. Compute the t_{obt}.

Answers

1. $\overline{X}$ and s_X^2
2. $s_{\overline{X}}$
3. $\overline{X} = 7.6$, $s_X^2 = 1.30$, $N = 5$; $s_{\overline{X}} = \sqrt{1.3/5} = .51$;
   $t_{obt} = (7.6 - 10)/.51 = -4.71$

The *t*-Distribution and Degrees of Freedom

In our housekeeping study, the sample mean produced a t_{obt} of -3.61 on the sampling distribution of means in which $\mu = 75$. The question is, "Is this t_{obt} significant?" To answer this, the final step is to compare t_{obt} to t_{crit}, and for that we examine the *t-distribution.*

Think of the *t*-distribution in the following way. One last time we hire our *very* bored statistician. She infinitely draws samples having our N from the raw score population described by H_0. For each sample she computes $\overline{X}$, but she also computes s_X^2 and, ultimately, t_{obt}. Then she plots the usual sampling distribution—a frequency distribution of the sample means—but also labels the X axis using each t_{obt}. Thus, the **t-distribution** is the distribution of all possible values of t computed for random sample means selected from the raw score population described by H_0. For our example, the *t*-distribution essentially shows all sample means—and their corresponding values of t_{obt}—that occur when men and women belong to the same population of housekeeping scores.

You can envision the *t*-distribution as in Figure 11.1. As with *z*-scores, increasing positive values of t_{obt} are located farther to the right of μ, and increasing negative values of t_{obt} are located farther to the left of μ. If t_{obt} places our mean close to the center of the distribution, then this mean is frequent and thus likely when H_0 is true. (So in our example, our sample of men is likely to be representing the population where μ is 75, the same population as for women.) But, if t_{obt} places our sample mean far into a tail of the sampling distribution, then this mean is infrequent and thus unlikely when H_0 is true. (Our sample is unlikely to represent the population where μ is 75, so it is unlikely that men and women have the same population of scores.) To determine if our mean is far enough into a tail, we find t_{crit} and create the region of rejection.

FIGURE 11.1

Example of a *t*-distribution of random sample means

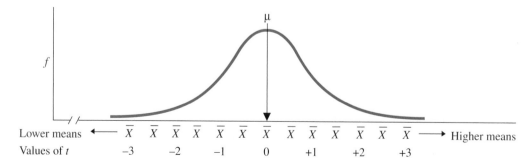

But we have one important novelty here: There are actually *many* versions of the *t*-distribution, each having a slightly different shape. The shape of a particular distribution depends on the sample size that is used when creating it. If the statistician uses small samples, the *t*-distribution will be only a rough approximation to the normal curve. This is because small samples will often contain large sampling error, so often each estimate of the population variability (s_X^2) will be very different from the next and from the true population variability. This inconsistency produces a *t*-distribution that is only approximately normal. However, large samples are more representative of the population, so each estimate of the population variability will be very close to the true population variability. As we saw when computing the *z*-test, using the true population variability (σ_X) produces a sampling distribution that forms a normal curve. In-between, as sample size increases, each *t*-distribution will be a successively closer approximation to the normal curve.

However, in this context, the size of a sample is determined by the quantity $N - 1$, what we call the *degrees of freedom*, or *df* . Because we compute the estimated population standard deviation using $N - 1$, it is our *df* that determines how close we are to the true population variability, and thus it is the *df* that determines the shape of the *t*-distribution. The larger the *df*, the closer the *t*-distribution comes to forming a normal curve. However, a tremendously large sample is not required to produce a perfect normal *t*-distribution. When *df* is greater than 120, the *t*-distribution is virtually identical to the standard normal curve. But when *df* is between 1 and 120 (which is often the case in research), a differently shaped *t*-distribution will occur for each *df*.

The fact that *t*-distributions are differently shaped is important for one reason: Our region of rejection should contain precisely that portion of the area under the curve defined by our α. If α = .05, then we want to mark off precisely 5% of the area under the curve. On distributions that are shaped differently, we mark off that 5% at different locations. Because the location of the region of rejection is marked off by the critical value, *with differently shaped* t-*distributions we will have different critical values*. For example, Figure 11.2 shows two *t*-distributions. Notice the size of the (blue) region of rejection in a tail of Distribution A. Say that this corresponds to the extreme 5% of Distribution A and is beyond the t_{crit} of ±2.5. However, if we also use ±2.5 as t_{crit} on Distribution B, the region of rejection is larger, containing *more* than 5% of the

FIGURE 11.2

Comparison of two *t*-distributions based on different sample *N*s

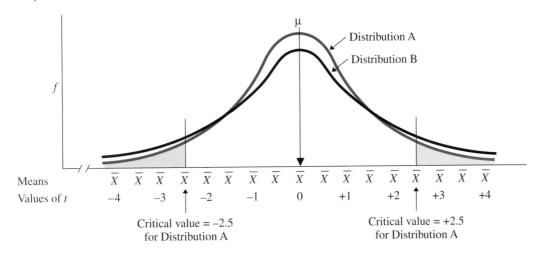

distribution. Conversely, the t_{crit} marking off 5% of Distribution B will mark off *less* than 5% of Distribution A. (The same problem exists for a one-tailed test.)

This issue is important because α is not only the size of the region of rejection, but it is also the probability of a Type I error. Unless we use the appropriate t_{crit}, the actual probability of a Type I error will not equal our α and that's not supposed to happen! Thus, there is only one version of the *t*-distribution to use when testing a particular t_{obt}: the one that the bored statistician would create by using the *same df* as in our sample. Therefore, we are no longer automatically using the critical values of 1.96 or 1.645. Instead, when your *df* is between 1 and 120, use the *df* to first identify the appropriate sampling distribution for your study. The t_{crit} on that distribution will accurately mark off the region of rejection so that the probability of a Type I error equals your α. Thus, in the housekeeping study with an *N* of 9, we will use the t_{crit} from the *t*-distribution for *df* = 8. In a different study, however, where *N* might be 25, we would use the different t_{crit} from the *t*-distribution for *df* = 24. And so on.

> **REMEMBER**   The appropriate t_{crit} for the one-sample *t*-test comes from the *t*-distribution that has *df* equal to *N* − 1, where *N* is the number of scores in the sample.

Using the *t*-Tables

We obtain the different values of t_{crit} from Table 2 in Appendix C, entitled "Critical Values of *t*." In these "*t*-tables," you'll find separate tables for two-tailed and one-tailed tests. Table 11.2 contains a portion of the two-tailed table.

To find the appropriate t_{crit}, first locate the appropriate column for your α (either .05 or .01). Then find the value of t_{crit} in the row at the *df* for your sample. For example, in the housekeeping study, *N* is 9, so *df* is *N* − 1 = 8. For a two-tailed test with α = .05 and *df* = 8, t_{crit} is 2.306.

Here's another example: In a different study, *N* is 61. Therefore, the *df* = *N* − 1 = 60. Look in Table 2 of Appendix C to find the two-tailed t_{crit} with α = .05. It is 2.000. The one-tailed t_{crit} here is 1.671.

The table contains no positive or negative signs. In a two-tailed test, you add the "$\pm$," and, in a one-tailed test, you supply the appropriate "+" or "−." Also, the table uses the symbol for infinity (∞) for *df* greater than 120. With this *df*, using the estimated population standard deviation is virtually the same as using the true population standard deviation. Therefore, the *t*-distribution matches the standard normal curve, and the critical values are those of the *z*-test.

TABLE 11.2

A Portion of the *t*-Tables

	Alpha Level	
df	α = .05	α = .01
1	12.706	63.657
2	4.303	9.925
3	3.182	5.841
4	2.776	4.604
5	2.571	4.032
6	2.447	3.707
7	2.365	3.499
8	2.306	3.355

Interpreting the *t*-Test

Once you've calculated t_{obt} and identified t_{crit}, you can make a decision about your results. In our housekeeping study, we must decide whether or not the men's mean of 65.67 represents the same population of scores that women have. Our t_{obt} is −3.61, and t_{crit} is ± 2.306, producing the sampling distribution in Figure 11.3. Remember, this can be interpreted as showing the frequency of all means that occur by chance when H_0 is true. Essentially, here the distribution shows all sample means that occur when the men's population of scores is the same as the women's population, with a μ of 75. Our t_{obt} lies beyond t_{crit}, so the results are significant: Our

FIGURE 11.3

Two-tailed *t*-distribution for *df* = 8 when H_0 is true and $\mu = 75$

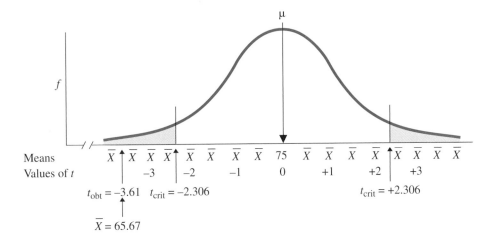

sample mean is so unlikely to occur if it had been representing the population in which μ is 75, that we reject the H_0 that it represents this population.

We interpret significant results using the same rules as discussed in the previous chapter. Thus, although we consider whether we've made a Type I error, with a sample mean of 65.67, our best estimate is that the μ for men is around 65.67. Because we expect a different population for women located at 75, we conclude that the results demonstrate a relationship in the population between gender and test scores. Then we return to being a researcher and interpret the relationship in psychological or sociological terms: What do the scores and relationship indicate about the underlying behaviors and their causes? Are men really more ignorant about housekeeping than women, and if so, why? Do men merely pretend to be ignorant, and if so, why? And so on.

Conversely, if t_{obt} was not beyond t_{crit} (for example, if $t_{obt} = +1.32$), it would not be significant. Then we would have no evidence for a relationship between gender and test scores, *one way or the other.* We would, however, consider whether we had sufficient *power* so that we were not making a Type II error.

The One-Tailed *t*-Test

As usual, we perform one-tailed tests when we predict the direction of the difference between our conditions. Thus, if we had predicted that men score *higher* than women H_a would be that the sample represents a population with μ greater than 75 ($H_a: \mu > 75$). H_0 would be that μ is less than or equal to 75 ($H_0: \mu \leq 75$). We then examine the sampling distribution that occurs when $\mu = 75$ (as we did in the two-tailed test). We find the one-tailed t_{crit} from the *t*-tables for our *df* and α. To decide in which tail of the sampling distribution to put the region of rejection, we determine what's needed to support H_a. Here, for the sample to represent a population of higher scores, the $\overline{X}$ must be *greater* than 75 and be *significant.* As shown in the left-hand graph in Figure 11.4, such means are in the upper tail, which is where we place the region of rejection, and t_{crit} is positive.

However, predicting that men score *lower* than women would produce the sampling distribution on the right in Figure 11.4. Now H_a is that μ is less than 75, and H_0 is that μ is greater than or equal to 75. Because we seek a $\overline{X}$ that is *significant* and *lower* than 75, the region of rejection is in the lower tail, and t_{crit} is negative.

In either case, calculate t_{obt} using the previous formulas. If the absolute value of t_{obt} is larger than t_{crit} and has the same sign, then the $\overline{X}$ is unlikely to be representing a μ described by H_0. Therefore, reject H_0, accept H_a, and the results are significant.

FIGURE 11.4

H_0 sampling distributions of *t* for one-tailed tests

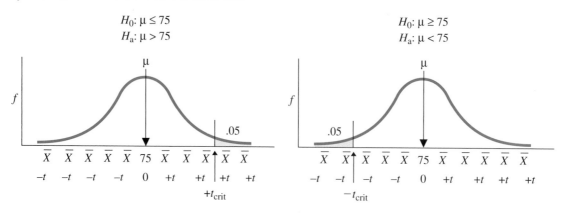

Some Help When Using the *t*-Tables

In the *t*-tables, you will *not* find a critical value for every *df* between 1 and 120. When the *df* of your sample does not appear in the table, you can take one of two approaches.

First, remember that all you need to know is whether t_{obt} is beyond t_{crit}, but you do not need to know how far beyond it is. Often you can determine this by examining the critical values given in the table for the *df* immediately above and below your *df*. For example, say that we perform a one-tailed *t*-test at $\alpha = .05$ with 49 *df*. The *t*-tables give a t_{crit} of 1.684 for 40 *df* and a t_{crit} of 1.671 for 60 *df*. Arrange the values like this:

$$df \qquad 60 \qquad 49 \qquad 40$$
$$t_{crit} \qquad 1.671 \qquad ? \qquad 1.684$$

nonsignificant ← less than 1.671   greater than 1.684 → significant

Because 49 *df* lies between 40 *df* and 60 *df*, our unknown t_{crit} is a number larger than 1.671 but smaller than 1.684. Therefore, any t_{obt} that is beyond 1.684 is already beyond the smaller unknown t_{crit}, and so it is significant. On the other hand, any t_{obt} that is not beyond 1.671 will not be beyond the larger unknown t_{crit}, and so it is not significant.

The second approach is used when t_{obt} falls *between* the two critical values given in the tables. Then you must compute the exact t_{crit} by performing "linear interpolation," as described in Appendix A.2.

A QUICK REVIEW

- Perform the one-sample *t*-test when σ_X is unknown.

MORE EXAMPLES

In a study, μ is 40. We predict our condition will change scores relative to this μ. $N = 25$. This is a two-tailed test, so $H_0: \mu = 40$; $H_a: \mu \neq 40$. Then compute t_{obt}.

The scores produce $\overline{X} = 46$ and $s_X^2 = 196$.

$$s_{\overline{X}} = \sqrt{\frac{s_X^2}{N}} = \sqrt{\frac{196}{25}} = 2.80$$

$$t_{obt} = \frac{\overline{X} - \mu}{s_{\overline{X}}} = \frac{46 - 40}{2.80} = +2.14$$

Find t_{crit}: With $\alpha = .05$ and $df = 24$, $t_{crit} = \pm 2.064$. The t_{obt} lies beyond the t_{crit}. Conclusion: The independent variable significantly increases scores from a μ of 40 to a μ around 46.

(continued)

For Practice

We test if artificial sunlight during the winter months *lowers* one's depression. Without the light, a depression test has $\mu = 8$. With the light, our sample scored 4, 5, 6, 7, and 8.

1. What are the hypotheses?

2. Compute t_{obt}.

3. What is t_{crit}?

4. What is the conclusion?

Answers

1. To "lower" is a one-tailed test: $H_a: \mu < 8$; $H_0: \mu \geq 8$.
2. $\overline{X} = 6$; $s_X^2 = 2.5$; $s_{\overline{X}} = \sqrt{2.5/5} = .707$;
   $t_{obt} = (6 - 8)/.707 = -2.83$
3. With $\alpha = .05$ and $df = 4$, $t_{crit} = -2.132$.
4. t_{obt} is beyond t_{crit}. Conclusion: Artificial sunlight significantly lowers depression scores from a μ of 8 to a μ around 6.

ESTIMATING μ BY COMPUTING A CONFIDENCE INTERVAL

As you've seen, after rejecting H_0, we estimate the population μ that the sample mean represents. There are two ways to estimate μ.

The first way is **point estimation,** in which we describe a point on the variable at which the μ is expected to fall. Earlier we estimated that the μ of the population of men is located on the variable of housekeeping scores at the *point* identified as 65.67. However, if we actually tested the entire population, μ would probably not be *exactly* 65.67. The problem with point estimation is that it is extremely vulnerable to sampling error. Our sample probably does not *perfectly* represent the population of men, so we can say only that the μ is *around* 65.67.

The other, better way to estimate μ is to include the possibility of sampling error and perform interval estimation. With **interval estimation,** we specify a range of values within which we expect the population parameter to fall. You often encounter such intervals in real life, although they are usually phrased in terms of "plus or minus" some amount (called the **margin of error**). For example, the evening news may report that a sample survey showed that 45% of the voters support the president, with a margin of error of plus or minus 3%. This means that the pollsters expect that, if they asked the *entire* population, the result would be within ±3% of 45%: They believe that the true portion of the population that supports the president is inside the *interval* that is between 42% and 48%.

We will perform interval estimation in a similar way by creating a *confidence interval*. Confidence intervals can be used to describe various population parameters, but the most common is for a single μ. The **confidence interval for a single μ** describes a range of values of μ, one of which our sample mean is likely to represent. Thus, when we say that our sample of men represents a μ *around* 65.67, a confidence interval is the way to define *around*. To do so, we'll identify those values of μ above and below 65.67 that the sample mean is likely to represent, as shown here:

$$\mu_{low} \cdots \mu \ \mu \ \mu \ \mu \ 65.67 \ \mu \ \mu \ \mu \ \mu \cdots \mu_{high}$$

values of μ, one of which is likely to be
represented by our sample mean

The μ_{low} is the lowest value of μ that our sample mean is likely to represent, and μ_{high} is the highest value of μ that the mean is likely to represent. When we compute these two values, we will have the confidence interval.

When is a sample mean likely to represent a particular μ? It depends on sampling error. For example, intuitively we know that sampling error is unlikely to produce a sample mean of 65.67 if μ is, say, 500. In other words, 65.67 is significantly different from 500. But sampling error *is* likely to produce a sample mean of 65.67 if, for example, μ is 65 or 66. In other words, 65.67 is not significantly different from these μs. Thus, a sample mean is likely to represent any μ that the mean is *not* significantly different from. The logic behind a confidence interval is to compute the highest and lowest values of μ that are not significantly different from the sample mean. All μs between these two values are also not significantly different from the sample mean, so the mean is likely to represent one of them.

> **REMEMBER** A *confidence interval* describes the highest and lowest values of μ that are not significantly different from our sample mean, and so the mean is likely to represent one of them.

We usually compute a confidence interval only after finding a significant t_{obt}. This is because we must be sure that our sample is not representing the μ described in H_0 before we estimate any other μ that it might represent.

Computing the Confidence Interval

The *t*-test forms the basis for the confidence interval, and here's what's behind the formula for it. We seek the highest and lowest values of μ that are not significantly different from the sample mean. The most that μ can differ from a sample mean and still not be significant is when t_{obt} *equals* t_{crit}. We can state this using the formula for the *t*-test:

$$t_{crit} = \frac{\overline{X} - \mu}{s_{\overline{X}}}$$

To find the largest and smallest values of μ that do not differ significantly from our sample mean, we determine the values of μ that we can put into this formula along with our $\overline{X}$ and $s_{\overline{X}}$. Because we are describing the μ above and below the sample mean, we use the two-tailed value of t_{crit}. Then by rearranging the above formula, we create the formula for finding the value of μ to put in the *t*-test so that the answer equals $-t_{crit}$. We also rearrange this formula to find the value of μ to put in so that the answer equals $+t_{crit}$. Our sample mean represents a μ *between* these two μs, so we combine these rearranged formulas to produce:

> **The formula for the confidence interval for a single μ is**
>
> $$(s_{\overline{X}})(-t_{crit}) + \overline{X} \le \mu \le (s_{\overline{X}})(+t_{crit}) + \overline{X}$$

The symbol μ stands for the unknown value represented by the sample mean. The $\overline{X}$ and $s_{\overline{X}}$ are computed from your data. Find the two-tailed value of t_{crit} in the *t*-tables at your α for $df = N - 1$, where N is the sample N.

REMEMBER Use the *two-tailed* critical value when computing a confidence interval, even if you performed a one-tailed *t*-test.

For our housekeeping study, the $\overline{X} = 65.67$ and $s_{\overline{X}} = 2.582$. The two-tailed t_{crit} for $df = 8$ and $\alpha = .05$ is ± 2.306. Filling in the formula, we have

$$(2.582)(-2.306) + 65.67 \le \mu \le (2.582)(+2.306) + 65.67$$

After multiplying 2.582 times -2.306 and times $+2.306$, we have

$$-5.954 + 65.67 \le \mu \le +5.954 + 65.67$$

The formula at this point tells us that our mean represents a μ of 65.67, *plus or minus* 5.954.

After adding ± 5.954 to 65.67, we have

$$59.72 \le \mu \le 71.62$$

This is the finished confidence interval. Returning to our previous diagram, we replace the symbols μ_{low} and μ_{high} with the numbers 59.72 and 71.62, respectively.

$$59.72 \ldots \mu \; \mu \; \mu \; \mu \; 65.67 \; \mu \; \mu \; \mu \; \mu \ldots 71.62$$

values of μ, one of which is likely to be
represented by our sample mean

As shown, our sample mean probably represents a μ around 65.67, meaning that μ is greater than or equal to 59.72, but less than or equal to 71.62.

Because we created this interval using the t_{crit} for an α of .05, there is a 5% chance that we are in error and the μ being represented by our mean is outside of this interval. On the other hand, there is a 95% chance $[(1 - \alpha)(100)]$ that the μ being represented is *within* this interval. Therefore, we have created what is called the *95% confidence interval:* We are 95% confident that the interval between 59.72 and 71.62 contains the μ represented by our sample mean.

For greater confidence, we could have used the t_{crit} for $\alpha = .01$. Then we would create the 99% confidence interval of $57.01 \le \mu \le 74.33$. Notice, however, that greater confidence comes at the cost of less precision: This interval spans a wider range of values than did the 95% interval, so we have less precisely identified the value of μ. Usually, researchers compromise between precision and confidence by creating the 95% confidence interval.

Thus, we conclude our one-sample *t*-test by saying, with 95% confidence, that our sample of men represents a μ between 59.72 and 71.62. Because the center of the interval is at 65.67, we still communicate that μ is *around* 65.67, but we have much more information than if we merely said that μ is somewhere around 65.67. In fact, therefore, you should compute a confidence interval anytime you are describing the μ represented by the mean of a condition in any significant experiment.[1]

---

[1] You can also compute a confidence interval when performing the *z*-test. Use the formula above, except use the critical values of *z*. If $\alpha = .05$, then $z_{\text{crit}} = \pm 1.96$. If $\alpha = .01$, then $z_{\text{crit}} = \pm 2.575$.

A QUICK REVIEW

- A confidence interval for μ provides a range of μs, any one of which our $\overline{X}$ is likely to represent.

MORE EXAMPLES

A t_{obt} is significant with $\overline{X} = 50$, $N = 20$, and, $s_{\overline{X}} = 4.7$. To compute the 95% confidence interval, $df = 19$, so the two-tailed $t_{crit} = \pm 2.093$. Then,

$$(s_{\overline{X}})(-t_{crit}) + \overline{X} \le \mu \le (s_{\overline{X}})(+t_{crit}) + \overline{X}$$
$$(4.7)(-2.093) + 50 \le \mu \le (4.7)(+2.093) + 50$$
$$(-9.837) + 50 \le \mu \le (+9.837) + 50$$
$$40.16 \le \mu \le 59.84$$

For Practice

1. What does this 95% confidence indicate:
   $15 \le \mu \le 20$?

2. With $N = 22$, you perform a one-tailed test ($\alpha = .05$.) What is t_{crit} for computing the confidence interval?

3. The t_{obt} is significant when $\overline{X} = 35$, $s_{\overline{X}} = 3.33$, and $N = 22$. Compute the 95% confidence interval.

Answers

1. We are 95% confident that our $\overline{X}$ represents a μ between 15 and 20.
2. With $df = 21$, the *two-tailed* $t_{crit} = \pm 2.080$
3. $(3.33)(-2.080) + 35 \le \mu \le (3.33)(+2.080) + 35 = 28.07 \le \mu \le 41.93$

Summary of the One-Sample *t*-Test

All of the preceding boils down to the following steps for the *t*-test.

1. *Check that the experiment meets the assumptions.*
2. *Create the hypotheses:* Create either the two-tailed or one-tailed H_0 and H_a.
3. *Compute t_{obt}:* From the sample data, compute s_X^2, compute $s_{\overline{X}}$, and then compute t_{obt}.
4. *Find the appropriate t_{crit}:* Use the *df* that equal $N - 1$.
5. *If t_{obt} is beyond t_{crit}:* Reject H_0, the results are significant, and so interpret the relationship "psychologically."
6. *If t_{obt} is not beyond t_{crit}:* the results are not significant. Draw no conclusion.
7. *Compute the confidence interval:* For significant results, use the two-tailed t_{crit} to describe the μ represented by your $\overline{X}$.

STATISTICS IN PUBLISHED RESEARCH: REPORTING THE *t*-TEST

Report the results of a *t*-test in the same way that you reported the *z*-test, but also include the *df*. For our housekeeping study, we had 8 *df*, the t_{obt} was -3.47, and it was significant. We always include the descriptive statistics too, so in a report you might read: "the national average for women is 75, although this sample of men scored lower ($M = 65.67$, $SD = 7.75$). This difference was significant, with $t(8) = -3.47, p < .05$."

First, note the *df* in parentheses. Also note that the results are significant because the probability is *less* than our alpha of .05 that we are making a Type I error. (Had these results not been significant, we'd report $t(8) = -3.47, p > .05$.)

In the *t*-tables, when α is .01, the t_{crit} is ± 3.355, so our t_{obt} of -3.47 would also be significant if we had used the .01 level. Therefore, instead of saying $p < .05$, we would provide more information by reporting that $p < .01$, because this communicates that the probability of a Type I error is not in the neighborhood of .04, .03, or .02. For this reason, researchers usually report the smallest values of alpha at which a result is significant.

Usually, confidence intervals are reported in sentence form (and not symbols), but we always indicate the confidence level used. Thus, in a report you might see: "The 95% confidence interval for this mean was between 59.72 and 71.62."

SIGNIFICANCE TESTS FOR CORRELATION COEFFICIENTS

It's time to shift mental gears and consider that other type of one-sample study—a correlational study. Here's a new example: We examine the relationship between a man's age and his housekeeping score in a correlational design. We measure the test scores and the ages of a sample of 25 men and determine that the Pearson correlation coefficient is appropriate. Say that, using the formula from Chapter 7, we compute an $r = -.45$, indicating that the older a man is, the lower his housekeeping score is.

Although this correlation coefficient describes the relationship in the *sample,* ultimately, we want to describe the relationship in the population. That is, we seek the correlation coefficient that would be produced if we could measure everyone's X and Y scores in the population. Of course, we cannot do that, so instead we use the sample coefficient to estimate the correlation that we'd expect to find if we could measure the entire population.

Recall that symbols for population parameters involve the Greek alphabet, so we need a new symbol:

> **The symbol for the Pearson correlation coefficient in the population is the Greek letter called "rho," which looks like this: ρ.**

We interpret ρ in the same way as r: It is a number between 0 and ± 1.0, indicating either a positive or a negative linear relationship in the population. The larger the absolute value of ρ, the stronger the relationship: The more that one value of Y is associated with each X, the more closely the scatterplot for the population hugs the regression line, and the better we can predict unknown Y scores by using X scores.

Thus, using the sample r in our study of $-.45$, we might estimate that ρ would equal $-.45$ if we measured the ages and housekeeping scores of the entire population of men.

But hold on, there's a problem here: That's right, sampling error. The problem of sampling error applies to *all* statistics. Here the idea is that, because of the luck of the draw of who was selected for the sample, their scores happened to produce this correlation. But, in the population—in nature—there is no relationship, and so ρ is really zero. So, here we go again! For any correlation coefficient you compute, you must determine whether it is significant.

> *REMEMBER*   Never accept that a sample correlation coefficient reflects a real relationship in nature unless it is significant.

Testing the Pearson *r*

As usual, the first step is to make sure that a study meets the assumptions of the statistical procedure. The Pearson correlation coefficient has three assumptions:

1. We have a random sample of pairs of X and Y scores, and each variable is an interval or ratio variable.

2. The *X* and *Y* scores each represent a normal distribution. Further, they represent a *bivariate normal distribution*. This means that the *Y* scores at each *X* form a normal distribution and the *X* scores at each *Y* form a normal distribution. (If *N* is larger than 25, however, violating this assumption is of little consequence.)

3. The null hypothesis is that in the population there is zero correlation. This is the most common approach and the one that we'll use. (However, you can also test the H_0 that your sample represents a nonzero ρ. Consult an advanced statistics book for the details.)

Our housekeeping and age scores meet these assumptions, so we set α at .05 and test *r*. First, we create the statistical hypotheses. You can perform either a one- or a two-tailed test. Use a two-tailed test if you do not predict the direction of the relationship. For example, let's say that we are unsure whether men produce higher or lower scores as they age. This is a two-tailed test because we're predicting either a positive or a negative correlation. For our alternative hypothesis, if the correlation in the population is either positive or negative, then ρ does not equal zero. Therefore we have

$$H_a: \rho \neq 0$$

On the other hand, the null hypothesis is always that the predicted relationship does not exist, so here it says that the correlation in the population is zero. Thus,

$$H_0: \rho = 0$$

This implies that if *r* does not equal zero, it's because of sampling error. You can understand this by looking at Figure 11.5. It shows the scatterplot in the population that H_0 says we would find: There is no relationship here, so ρ equals 0. Recall, however, that a slanting elliptical scatterplot reflects an *r* that is not equal to zero. Thus, H_0 implies that, by chance, we selected an elliptical sample scatterplot from this population plot. Therefore, it says, although age and housekeeping scores are not really related, the scores in our sample happen to pair up so that it looks like they're related. Conversely, H_a implies that the population's scatterplot would not look like Figure 11.5, but rather it would be similar to our sample's scatterplot.

FIGURE 11.5

Scatterplot of a population for which ρ = 0, as described by H_0

Any r *is a result of sampling error when selecting a sample from this scatterplot.*

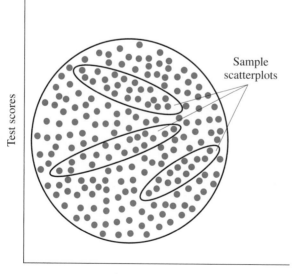

The Sampling Distribution of *r* We test H_0 by determining the likelihood of obtaining our sample *r* from the population where ρ is zero. To do so, we envision the sampling distribution of *r*.

The bored statistician quit! By now, however, you could create the sampling distribution yourself: Using the same *N* as in our study, you would select an infinite number of samples of $X-Y$ pairs from the population where ρ = 0 (as if you pulled each sample from the scatterplot in Figure 11.5). Each time, you would compute *r*. If you then plotted the frequency of the various values of *r,* you would have the sampling distribution of *r.* The **sampling distribution of *r*** is a frequency distribution showing all possible values of *r* that occur by chance when samples are drawn from a population in which ρ is zero. Such a distribution is shown in Figure 11.6.

The only novelty here is that along the *X* axis are now different values of *r*. When ρ = 0, the most frequent sample *r* is also 0, so the mean of the sampling distribution—the average *r*—is 0. Because of sampling error, however, sometimes we'll obtain a positive *r* and sometimes a negative *r*. Most often the *r* will be relatively small and close to 0. But, less frequently, we'll obtain a larger *r* that falls into a tail of the distribution. Thus, the larger the *r* (whether positive or negative), the less likely it is to occur when the sample actually represents a population in which ρ = 0.

To test H_0, we determine where our *r* lies on this distribution. To do so, we could perform a variation of the *t*-test, but luckily that is not necessary. Instead, *r* directly communicates its location on the sampling distribution. The mean of the sampling distribution is always zero, so, for example, our *r* of −.45 is a distance of .45 below the mean. Therefore, we test H_0 simply by examining our obtained *r*, which is r_{obt}. To determine whether r_{obt} is in the region of rejection, we compare it to r_{crit}.

As with the *t*-distribution, the shape of the sampling distribution of *r* is slightly different for each *df,* so there is a different value of r_{crit} for each *df*. *But,* here's a new one: With the Pearson correlation coefficient, the degrees of freedom equals $N - 2$, where *N* is the number of *pairs* of scores in the sample.

> *REMEMBER* For the Pearson *r*, the degrees of freedom equals $N - 2$, where *N* is the number of pairs of scores.

Table 3 in Appendix C gives the critical values of the Pearson correlation coefficient. Use these "*r*-tables" in the same way that you've used the *t*-tables: Find r_{crit} for either a one- or a two-tailed test at the appropriate α and *df*. For the housekeeping correlation, *N* was 25, so *df* = 23, and, for a two-tailed test with α = .05, r_{crit} is ±.396. We set up

FIGURE 11.6

Sampling distribution of *r*

It is an approximately normal distribution, with values of r *plotted along the* X *axis.*

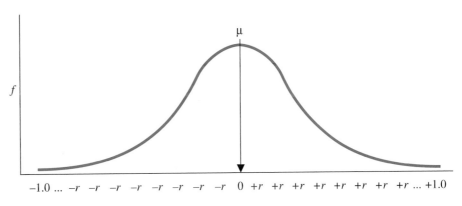

f

μ

−1.0 ... −*r* −*r* −*r* −*r* −*r* −*r* −*r* −*r* 0 +*r* +*r* +*r* +*r* +*r* +*r* +*r* +*r* ... +1.0

Values of *r*

the sampling distribution in Figure 11.7. Our r_{obt} of $-.45$ is beyond the r_{crit} of $\pm.396$, so it is in the region of rejection. As usual, this means that the results are significant: This *r* is so unlikely to occur if we had been representing the population where ρ is 0, that we reject the H_0 that we were representing this population. We conclude that the r_{obt} is "significantly different from 0."

Interpreting *r* The rules for interpreting a significant result here are the same as before. Remember that rejecting H_0 does not prove anything. In particular, this was a correlational study, so we have not proven that changes in age *cause* test scores to change. In fact, we have not even proven that the relationship exists because we may have made a Type I error. Here, a Type I error is rejecting the H_0 that there is zero correlation in the population, when in fact there *is* zero correlation in the population. As usual, though, with $\alpha = .05$, the probability that we have made a Type I error is slightly less than .05.

Report the Pearson correlation coefficient using the same format as with previous statistics. Our r_{obt} of $-.45$ was significant with 23 *df*, so in a published report we'd say that $r(23) = -.45$, $p < .05$. As usual, our *df* is in parentheses and because $\alpha = .05$ and the coefficient is significant, we indicate that the probability is *less* than .05 that we've made a Type I error.

Because the sample r_{obt} is $-.45$, our best estimate is that, in this population, ρ equals $-.45$. However, recognizing that the sample may contain sampling error, we expect that ρ is probably *around* $-.45$. (We could more precisely describe ρ by computing a confidence interval. However, this is computed using a very different procedure from the one discussed previously.)

In Chapter 8, you saw that we further describe a relationship by computing the regression equation and the proportion of variance accounted for. However, do this only when r_{obt} is significant! Only then are we confident that we're describing a "real" relationship. Thus, for the housekeeping study, we would now compute the linear regression equation for predicting test scores if we know a man's age. We would also compute r^2, which is $(-.45)^2$ or .20. Recall, this is the *proportion of variance in Y* scores that is *accounted for* by the relationship with *X*. Here, an r^2 of .20 tells us that we are 20% more accurate when we use the relationship with age to predict housekeeping scores than when we do not use the relationship.

Remember that it is r^2 and not "significance" that determines how important a relationship is. *Significant* indicates only that the sample relationship is unlikely to be a fluke of chance. The r^2 indicates the importance of a relationship because it indicates the extent to which knowing participants' *X* scores improves our accuracy in predicting and understanding differences in their *Y* scores. This allows us to understand

FIGURE 11.7

H_0 sampling distribution of *r*

For the two-tailed test, there is a region of rejection for positive values of r_{obt} *and for negative values of* r_{obt}.

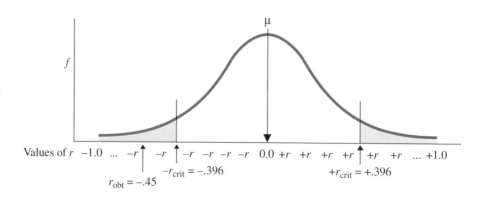

differences in their *behavior,* and so the larger the r^2, the greater is the importance of the relationship.

Thus, a relationship must be significant to be even potentially important (because it must first be believable). But, a significant relationship is not necessarily important. For example, an r_{obt} of $+.10$ might be significant, but it is not statistically important: $+.10^2$ is only .01, so this relationship accounts for only 1% of the variance and so is virtually useless in explaining differences in Y scores. (We're only 1% better off with it than without it.) Thus, although this relationship is unlikely to occur through sampling error, it is also an unimportant relationship.

After describing the relationship, as usual the final step is to interpret it in terms of behaviors. For example, perhaps our correlation coefficient reflects socialization processes, with older men scoring lower on the housekeeping test because they come from generations in which wives typically did the housekeeping, while men were the "breadwinners."

Of course, if r_{obt} does not lie beyond r_{crit}, then you would retain H_0 and conclude that the correlation is not significant. In this case, make no claims about the relationship that may or may not exist, and do not compute the regression equation or r^2.

One-Tailed Tests of r   If we had predicted only a positive correlation or only a negative correlation, then we would have performed a one-tailed test. When we predict a positive relationship, we are predicting a positive ρ (a number greater than 0) so our alternative hypothesis is H_a: $\rho > 0$. The H_0 says that we're wrong (that ρ is 0 or less than 0), so H_0: $\rho \leq 0$. On the other hand, when we predict a negative relationship, we are predicting a negative ρ (a number less than 0) so we have H_a: $\rho < 0$. The H_0 says we're wrong (that ρ is 0 or greater than 0) so H_0: $\rho \geq 0$.

We test each H_0 by again testing whether the sample represents a population in which there is zero relationship—so again we examine the sampling distribution for $\rho = 0$. From the *r*-tables in Appendix C, find the one-tailed critical value for *df* and α. Then set up one of the sampling distributions shown in Figure 11.8. When predicting a positive correlation, use the left-hand distribution: r_{obt} is significant if it is positive and falls beyond the positive r_{crit}. When predicting a negative correlation, use the right-hand distribution: r_{obt} is significant if it is negative and falls beyond the negative r_{crit}.

FIGURE 11.8

H_0 sampling distribution of *r* where $\rho = 0$ for one-tailed test

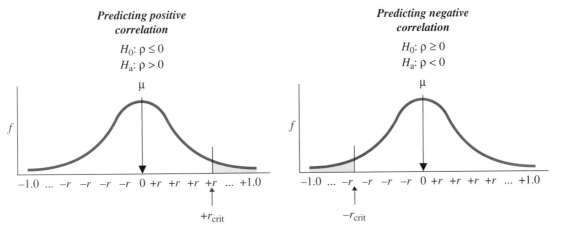

Testing the Spearman r_S

We also use the above logic when testing the Spearman correlation coefficient (r_S). Recall that r_S describes the linear relationship in a sample when X and Y are both ordinal (ranked) scores. Again our ultimate goal is to use the sample coefficient to estimate the correlation coefficient we would see if we could measure everyone in the population.

The symbol for the Spearman correlation coefficient in the population is ρ_S.

However, before we can use r_S to estimate ρ_S, we must first deal with the usual problem: That's right, maybe our r_S merely reflects sampling error. Perhaps if we measured the population, we'd find that ρ_S is 0. Therefore, before we can conclude that the correlation reflects a relationship in nature, we must perform hypothesis testing. So, we

1. *Set alpha:* how about .05?
2. *Consider the assumptions of the test:* The r_S requires a random sample of pairs of *ranked* (ordinal) scores. (*Note:* Because of the data involved and the lack of parametric assumptions, r_S is technically a *nonparametric* procedure.)
3. *Create the statistical hypotheses:* You can test the one- or two-tailed hypotheses that we saw previously with ρ, except now use the symbol ρ_S.

The only new aspect of testing r_S is the sampling distribution. The **sampling distribution of r_S** is a frequency distribution showing all possible values of r_S that occur when samples are drawn from a population in which ρ_S is zero. This creates a new family of sampling distributions and a different table of critical values. Table 4 in Appendix C, entitled "Critical Values of the Spearman Rank-Order Correlation Coefficient," contains the critical values for one- and two-tailed tests of r_S. Obtain critical values as in previous tables, except here use N, *not* degrees of freedom. (Note the instructions at the top of this table for when your N is not listed.)

> *REMEMBER* The critical value of r_S is obtained using N, the number of pairs of scores in the sample.

Here's an example. In Chapter 7, we correlated the aggressiveness rankings given to nine children by two observers and found that $r_S = +.85$. We had assumed that the observers' rankings would agree, predicting a positive correlation. Therefore, we have a one-tailed test with the hypotheses H_0: $\rho_S \leq 0$ and H_a: $\rho_S > 0$. From Table 4 in Appendix C, with $\alpha = .05$ and $N = 9$, the one-tailed critical value is $+.600$. This produces the sampling distribution shown in Figure 11.9.

FIGURE 11.9

One-tailed H_0 sampling distribution of values of r_s when H_0 is $\rho_s = 0$

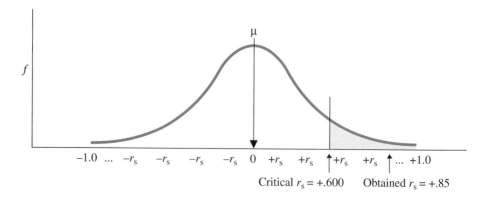

f

−1.0 ... $-r_s$ $-r_s$ $-r_s$ $-r_s$ 0 $+r_s$ $+r_s$ $+r_s$ $+r_s$ ... +1.0

Critical $r_s = +.600$ Obtained $r_s = +.85$

Because an r_S of $+.85$ is beyond the critical value of $+.600$, we reject H_0. Thus, our r_S is significantly different from zero, and we estimate that ρ_S in the population of such rankings is around $+.85$. In a publication, this would be reported as $r_S(9) = +.85$, $p < .05$. Note that the N of the sample is given in parentheses. We would also compute the squared r_S to determine the proportion of variance accounted for. Then interpret these results as we have done previously. (With different predictions, we might have performed the other one-tailed test or a two-tailed test.)

Summary of Testing a Correlation Coefficient

All of the preceding boil down to the following steps:

1. *Check the assumptions of r or r_S.*
2. *Create the hypotheses: Create either the two-tailed or one-tailed H_0 and H_a.*
3. *Compute the correlation coefficient.*
4. *Obtain the critical value from Appendix C: The critical value for r is in Table 3, using $df = N - 2$. The critical value for r_S is in Table 4, using $df = N$.*
5. *Compare the obtained to the critical value: If the obtained coefficient is beyond the critical value, the results are significant. If the coefficient is not beyond the critical value, the results are not significant.*
6. *For significant results, compute the proportion of variance accounted for by squaring the obtained coefficient. For r, compute the linear regression equation.*

- Always perform hypothesis testing on a correlation coefficient so that we are confident it is not due to sampling error.

MORE EXAMPLES

We compute an $r = +.32$ ($N = 42$). We predicted some kind of relationship, so H_0: $\rho = 0$; H_a: $\rho \neq 0$. With $\alpha = .05$ and $df = 42 - 2 = 40$, the two-tailed $r_{crit} = \pm.304$. The r_{obt} is beyond r_{crit}, so it is significant: We expect the population correlation coefficient (ρ) to be around $+.32$.

We compute an $r_S = -.22$, using $N = 24$. With $\alpha = .05$, the two-tailed critical value is $\pm.409$. The r_S is not in the region of rejection and is not significant.

For Practice

We predict a negative relationship and obtain $r_{obt} = -.44$.

1. What are H_0 and H_a?
2. With $\alpha = .05$ and $N = 10$, what is r_{crit}?
3. What is the conclusion about r_{obt}?
4. What is the conclusion about the relationship in the population and r^2?

Answers

1. H_0: $\rho \geq 0$; H_a: $\rho < 0$
2. $df = 8$, $r_{crit} = -.549$
3. Not significant
4. Make no conclusion and do not compute r^2.

MAXIMIZING THE POWER OF STATISTICAL TESTS

The *t*-test is our first design in which we—the researcher—have some control over all components of our data, because we are not stuck with a given σ_X. Therefore, it is appropriate to revisit the topic of *power*, so that you can understand how researchers use this control to increase the power of a study. Recall that **power** is the probability of *not* committing a Type II error. We're talking about those times when H_0 really is false, and the error would be to retain H_0. Instead, we should reject H_0, correctly concluding that the predicted relationship exists in nature. Essentially, *power is the probability that we will not miss a relationship that really exists in nature.* We maximize power by doing everything we can to reject H_0 so that we don't miss the relationship. If we still end up retaining H_0, we can be confident that we did not do so incorrectly and miss a relationship that exists, but rather that the relationship does not exist.

We maximize power by the way that we design an experiment or correlational study.[2] We're talking about those times when we *should* reject the null hypothesis, so maximizing power boils down to maximizing the probability that our results will be significant. This translates into designing the study to maximize the size of our obtained statistic relative to the critical value, so that the obtained will be significant.

For the one-sample *t*-test, three aspects of the design produce a relatively larger t_{obt} and thus increase power. (These also apply to other types of experiments that we will discuss.) Look at the formulas:

$$t_{obt} = \frac{\overline{X} - \mu}{s_{\overline{X}}} \quad s_{\overline{X}} = \sqrt{\frac{s_X^2}{N}}$$

First, *larger differences produced by changing the independent variable increase power.* In the housekeeping study, the greater the difference between the sample mean for men and the μ for women, the greater the power. Logically, the greater the difference between men and women, the less likely we are to miss that a difference exists. Statistically, in the formula this translates to a larger difference between $\overline{X}$ and μ that produces a larger numerator, which results in a larger t_{obt} that is more likely to be significant. Therefore, when designing any experiment, the rule is to select conditions that are substantially different from one another, so that we produce a big difference in dependent scores *between* the conditions.

Second, *smaller variability in the raw scores increases power.* Recall that *variability* refers to the differences among the scores. Logically, smaller variability indicates more consistent behavior and a more consistent, stronger relationship. This makes a clearer pattern that we are less likely to miss. Statistically, in the formula, smaller variability produces a smaller estimated variance (s_X^2), which produces a smaller standard error $(s_{\overline{X}})$. Then in the *t*-test, dividing by a smaller denominator produces a larger t_{obt}. We will see smaller variability in scores the more that all participants experience the study in the same way. Therefore, the rule is to conduct any study in a consistent way that minimizes the variability of scores *within* each condition.

Third, *a larger N increases power.* Logically, a larger N provides a more accurate representation of the population, so we are less likely to make *any* type of error. Statistically, dividing s_X^2 by a larger N produces a smaller $s_{\overline{X}}$, which results in a larger t_{obt}. Also, a larger N produces larger *df*, which produces a smaller t_{crit}. Then our t_{obt} is more likely to be significant. Therefore, the rule is to design any experiment with the largest practical N. However, this is for *small* samples. Generally, an N of 30 per condition is needed for minimal power, and increasing N up to 121 adds substantially to it. However, an N of, say, 500 is not substantially more powerful than an N of, say, 450.

> *REMEMBER* Increase power in an experiment by maximizing differences in dependent scores *between* conditions, minimizing differences among scores *within* conditions, and testing a larger N.

---

[2]More advanced textbooks contain procedures for determining the amount of power that is present in a given study.

Likewise, we maximize the power of a correlational study by maximizing the size of the correlation coefficient relative to the critical value. Three aspects will increase the power of a correlation coefficient. First, *avoiding a restricted range increases power.* Recall from Chapter 7 that having a small range of scores on the *X* or *Y* variable produces a coefficient that is smaller than it would be without a restricted range. Therefore, always measure the full range of possible *X* and *Y* scores. Second, *minimizing the variability of the* Y *scores at each* X *increases power.* Recall that the smaller the variability in Y scores at each X, the larger the correlation coefficient. Therefore, always test participants in a consistent fashion to minimize the variability in Y scores at each X. Third, *increasing N increases power.* With a larger *N*, the *df* are larger, so the critical value is smaller, and thus a given coefficient is more likely to be significant.

> *REMEMBER* Increase power in a correlation coefficient by avoiding a restricted range, minimizing the variability in *Y* scores, and increasing *N*.

PUTTING IT ALL TOGETHER

In one sense, I hope that you found this chapter rather boring—not because it *is* boring, but because, for each statistic, we performed virtually the same operations. In testing *any* statistic, we ultimately do and say the same things. In all cases, if the obtained statistic is out there far enough in the sampling distribution, it is too unlikely for us to accept as representing the H_0 situation, so we reject H_0. Any H_0 implies that the sample does not represent the predicted relationship, so rejecting H_0 increases our confidence that the data *do* represent the predicted relationship. We're especially confident because the probability is less than α that we've made an error in this decision. If we fail to reject H_0, then hopefully we have sufficient power, so that we're unlikely to have made an error here, too. These are the fundamentals of *all* inferential statistics.

Using the SPSS Appendix As described in Appendix B.5, SPSS will compute the one-sample t_{obt}. It also indicates the smallest *two-tailed* region of rejection (and alpha level) for which your t_{obt} is significant. Further, it computes the $\overline{X}$ and s_X for the sample and it computes the 95% confidence interval.

As part of computing the Pearson *r* or the Spearman r_S, SPSS automatically performs a one- or two-tailed significance test. This includes indicating the smallest alpha level at which the coefficient is significant.

CHAPTER SUMMARY

1. The *one-sample t-test* is for testing a one-sample experiment when the standard deviation of the raw score population is not known.

2. A *t-distribution* is a theoretical sampling distribution of all possible values of *t* when a raw score population is infinitely sampled using a particular *N*. A *t-distribution* that more or less forms a perfect normal curve will occur depending on the *degrees of freedom* (*df*) of the samples used to create it.

3. In *point estimation,* a μ is assumed to be at a point on the variable equal to $\overline{X}$. Because the sample probably contains sampling error, a point estimate is likely to be incorrect. In *interval estimation,* a μ is assumed to lie within a specified interval. Interval estimation is performed by computing a *confidence interval*.

4. The *confidence interval for a single* μ describes a range of μs, one of which the sample mean is likely to represent. The interval contains the highest and lowest values of μ that are not significantly different from the sample mean.

5. The symbol for the Pearson correlation coefficient in the population is ρ (called rho). The symbol for the Spearman correlation coefficient in the population is ρ_S.

6. The *sampling distribution of the Pearson r* is a frequency distribution showing all possible values of *r* that occur when samples are drawn from a population in which ρ is zero.

7. The *sampling distribution of the Spearman r_S* is a frequency distribution showing all possible values of r_S that occur when samples are drawn from a population in which ρ_S is zero.

8. Only when a correlation coefficient is significant is it appropriate to compute the linear regression equation and the proportion of variance accounted for.

9. Maximize the power of experiments by (a) creating large differences in scores between the conditions of the independent variable, (b) minimizing the variability of the scores within each condition, and (c) increasing the *N* of small samples.

10. Maximize the power of a correlation coefficient by (a) avoiding a restricted range, (b) minimizing the variability in *Y* at each *X,* and (c) increasing the *N* of small samples.

KEY TERMS

t_{obt} t_{crit} $s_{\overline{X}}$ df H_0 H_a r_{obt} r_{crit} ρ
confidence interval for a single μ 243
estimated standard error of the
   mean 236
interval estimation 243
margin of error 243

one-sample *t*-test 234
point estimation 243
sampling distribution of *r* 249
sampling distribution of r_S 252
t-distribution 238

REVIEW QUESTIONS

(Answers for odd-numbered questions are in Appendix D.)

1. A scientist has conducted a one-sample experiment. (a) What two parametric procedures are available to her? (b) What is the deciding factor for selecting between them? (c) What are the other assumptions of the *t*-test?

2. In this chapter, you learned how to perform four different statistical procedures. List them.

3. (a) What is the difference between $s_{\overline{X}}$ and $\sigma_{\overline{X}}$? (b) How is their use the same?

4. (a) Why are there different values of t_{crit} when samples have different *N*s? (b) What must you compute in order to find t_{crit}?

5. (a) What is the symbol for the Pearson correlation coefficient in the population? (b) What is the symbol for the Spearman correlation coefficient in the population? (c) Summarize the steps involved in analyzing a Pearson correlational study.

6. Summarize the steps involved in analyzing the results of a one-sample experiment.

7. What is the final step when results are significant in any study?

8. Say that you have a sample mean of 44 in a study. (a) Estimate the corresponding μ using point estimation. (b) What would a confidence interval for this μ tell you? (c) Why is computing a confidence interval better than using a point estimate? (d) What is the difference between reporting an estimate of μ using a margin of error versus using a confidence interval?

9. (a) What is power? (b) What outcome should cause you to worry about having sufficient power? (c) Why? (d) At what stage do you build in power?

10. (a) What are the three aspects of maximizing the power of a t-test? (b) What are the three aspects of maximizing the power of a correlation coefficient?

APPLICATION QUESTIONS

11. We ask whether a new version of our textbook is beneficial or detrimental to students learning statistics. On a national statistics exam, $\mu = 68.5$ for students using other textbooks. A sample of students using this book has the following scores:

   64   69   92   77   71   99   82   74   69   88

   (a) What are H_0 and H_a for this study? (b) Compute t_{obt}. (c) With $\alpha = .05$, what is t_{crit}? (d) What do you conclude about the use of this book? (e) Compute the confidence interval for μ.

12. A researcher predicts that smoking cigarettes decreases a person's sense of smell. On a test of olfactory sensitivity, the μ for nonsmokers is 18.4. A sample of people who smoke a pack a day produces these scores:

   16   14   19   17   16   18   17   15   18   19   12   14

   (a) What are H_0 and H_a for this study? (b) Compute t_{obt}. (c) With $\alpha = .05$, what is t_{crit}? (d) What should the researcher conclude about this relationship? (e) Compute the confidence interval for μ.

13. Foofy studies if hearing an argument in favor of an issue alters participants' attitudes toward the issue one way or the other. She presents a 30-second argument to 8 people. In a national survey about this issue, $\mu = 50$. She obtains $\overline{X} = 53.25$ and $s_X^2 = 569.86$. (a) What are H_0 and H_a? (b) What is t_{obt}? (c) With $\alpha = .05$, what is t_{crit}? (d) If appropriate, compute the confidence interval for μ. (e) What conclusions should Foofy draw about the relationship?

14. In question 13, (a) What statistical principle should Foofy be concerned with? (b) Identify three problems with her study from a statistical perspective. (c) Why would correcting these problems improve her study?

15. Poindexter examined the relationship between ratio scores measuring the quality of sneakers worn by volleyball players and their average number of points scored per game. Studying 10 people who owned sneakers of good to excellent quality, he computed $r = +.21$. Without further ado, he immediately claimed to have support for the notion that better-quality sneakers are related to better performance on a somewhat consistent basis. He then computed r^2 and the regression equation. Do you agree or disagree with his approach? Why?

16. Eventually, for the study in question 15, Poindexter reported that $r(8) = +.21$, $p > .05$. (a) What should he conclude about this relationship? (b) What other

computations should he perform to describe the relationship in these data?
(c) What statistical goal described in this chapter should he be concerned with?
(d) Identify three problems with his study. (e) Why will correcting these problems improve his study?

17. A scientist suspects that as a person's stress level changes, so does the amount of his or her impulse buying. With $N = 72$, his r is $+.38$. (a) What are H_0 and H_0? (b) With $\alpha = .05$, what is r_{crit}? (c) Report these results using the correct format. (d) What conclusions should he draw? (e) What other calculations should be performed to describe the relationship in these data?

18. Foofy computes the correlation between an individual's physical strength and his or her college grade point average. Using a computer, the correlation for a sample of 2000 people is $r(1998) = +.08, p = < .0001$. She claims this is a useful tool for predicting which college applicants are likely to succeed academically. Do you agree or disagree? Why?

19. We study the influence of wearing uniforms in middle school on attitudes toward achieving good grades. On a national survey, the average attitude score for students who do not wear uniforms is $\mu = 12$. A sample of students who wear uniforms has scores of 8, 12, 10, 9, 6, and 7. Perform all parts of the *t*-test and draw the appropriate conclusions.

20. A newspaper article claims that the academic rank of a college is negatively related to the rank of its football team. From a sample of 28 colleges, you obtain a correlation coefficient of $-.32$. (a) Which type of correlation coefficient did you compute? (b) What are H_0 and H_a? (c) With $\alpha = .05$, what is the critical value? (d) What should you conclude about the accuracy of the newspaper's claim? (e) In predicting a particular school's academic ranking in your sample, how important is it that you look at the school's football ranking?

21. (a) How would you report your results if $\alpha = .05, N = 43$, and $t_{obt} = +6.72$ is significant? (b) How would you report your results if $\alpha = .05, N = 6$, and $t_{obt} = -1.72$ is not significant?

22. While reading a published research report, you encounter the following statements. For each, identify the N, the procedure performed and the outcome, the relationship, and the type of error possibly being made. (a) "When we examined the perceptual skills data, the mean of 55 for the sample of adolescents differed significantly from the population mean of 70 for adults, $t(45) = 3.76, p < .01$." (b) "The correlation between personality type and emotionality, however, was not significantly different from zero, with $r(25) = +.42, p > .05$."

23. You wish to compute the 95% confidence interval for a sample with a df of 80. Using interpolation, determine the t_{crit} that you should use.

24. In a two-tailed test, N is 35. (a) Is the t_{obt} of $+2.019$ significant? (b) Is the t_{obt} of $+4.0$ significant?

INTEGRATION QUESTIONS

25. (a) Why must a relationship be significant to be important? (b) Why can a relationship be significant and still be unimportant? (Chs. 8, 10, 11)

26. (a) What is the difference between the purpose of descriptive and inferential statistics? (b) When should you should use a parametric versus a nonparametric inferential procedure? (Chs. 2, 10)

27. What is the design of the study when we compute the *z*-test and *t*-test versus when we compute a correlation coefficient? (Chs. 2, 7, 10, 11)

28. (a) What does a correlation coefficient tell you? (b) When do you compute r? (c) When do you compute r_S? (d) When is linear regression used? (Chs. 7, 8, 11)

29. (a) Study A reports a result with $p = .031$. Study B reports results with $p < .001$. What is the difference between the results of A and B in terms of (a) how significant they are? (b) Their critical values? (c) Their region of rejection? (d) The probability of a statistical error? (Chs. 9, 10, 11)

30. (a) What do we mean by the *restriction of range*? (b) Why is it a problem for the size of correlation coefficient? (c) Why is it a problem for the power of a correlation coefficient? (Chs. 7, 11)

31. For the following, specify which descriptive and inferential procedures should be performed, explain what is being compared, and identify the key to answering the researcher's question. (a) A researcher measures a group of participants using standard tests of "social nervousness" and "introversion" to determine if introversion is a good predictor of nervousness. (Scores are interval scores.) (b) The average worker at a calculator plant can assemble 106 calculators in his or her first hour of work. During their final hour, a sample of workers produced an average of only 97.4 calculators ($s_X = 17.3$). Should we conclude that performance decreases during the final hour for all workers? (c) For 20 years the basketball coach recorded his team's performance when making free throws, with $\mu = 71.1$ and $\sigma_X = 14.2$. A sports psychologist trained 20 players on this year's team to visualize each shot beforehand. They shot an average of 77.6 for the year. Is visualization a way to improve the performance of all players? (Chs. 4, 5, 7, 8, 10, 11)

■ ■ ■ SUMMARY OF FORMULAS

1. To perform the one-sample t-test,

$$s_X^2 = \frac{\Sigma X^2 - \dfrac{(\Sigma X)^2}{N}}{N - 1}$$

$$s_{\bar{X}} = \sqrt{\frac{s_X^2}{N}}$$

$$t_{obt} = \frac{\bar{X} - \mu}{s_{\bar{X}}}$$

Values of t_{crit} are found in Table 2 of Appendix C, for $df = N - 1$.

2. The formula for a confidence interval for a single μ is

$$(s_{\bar{X}})(-t_{crit}) + \bar{X} \le \mu \le (s_{\bar{X}})(+t_{crit}) + \bar{X}$$

where t_{crit} is the two-tailed value for $df = N - 1$.

3. Critical values of the Pearson r are found in Table 3 of Appendix C, using $df = N - 1$, where N is the number of pairs.

4. Critical values of the Spearman r_S are found in Table 4 of Appendix C, using N, the number of pairs.

12 The Two-Sample *t*-Test

GETTING STARTED

To understand this chapter, recall the following:

- From Chapter 2, what the terms *condition, independent variable,* and *dependent variable refer to.*
- From Chapter 4, how to graph the results of experiments.
- From Chapter 8, how to interpret the proportion of variance accounted for.
- From Chapter 11, what a confidence interval indicates and what you've learned about inferential statistics.

Your goals in this chapter are to learn

- The logic of a *two-sample experiment.*
- The difference between *independent samples* and *related samples.*
- How to perform the *independent-samples* t-*test* and *related-samples* t-*test.*
- How to compute a *confidence interval for the difference between two* μs and for the μ *of difference scores.*
- How r_{pb}^2 is used to describe *effect size* in a two-sample experiment.

This chapter presents the two-sample *t*-test, which is the major parametric procedure used when an experiment involves *two* samples. As the name implies, this test is similar to the one-sample *t*-test you saw in Chapter 11, except that a two-sample design requires that we use slightly different formulas. This chapter discusses (1) one version of this *t*-test, called the *independent-samples t*-test, and its confidence interval; (2) the other version of this *t*-test, called the *related-samples t-test*, and its confidence interval; and (3) procedures for summarizing the results of any two-sample experiment.

NEW STATISTICAL NOTATION

So far, N has stood for the number of scores in a sample. Actually, N indicates the total number of scores in the *study*, but with only one condition, N was also the number of scores in the sample. However, now we will be discuss experiments with two conditions, so the lowercase n with a subscript will stand for the number of scores in each sample. Thus, n_1 is the number of scores in condition 1, and n_2 is the number of scores in condition 2. N is the total number of scores in the experiment, so adding the ns together equals N.

> **REMEMBER** N stands for the total number of scores in an experiment; n stands for the number of scores in a condition.

WHY IS IT IMPORTANT TO KNOW ABOUT THE TWO-SAMPLE *t*-TEST?

The one-sample experiments discussed in previous chapters are not often found in real research, because they require that we know μ under one condition of the independent variable. Usually, however, researchers explore new behaviors and variables, so they do not know any μs beforehand. Instead, the usual approach is to measure a sample of participants under each condition of the independent variable and to use the sample mean to estimate the corresponding population μ that would be found. Often we test only two conditions, and then our inferential procedures involve two-sample *t*-tests. Thus, it is important for you to know about these procedures because they apply to a more realistic and common way of conducting experiments that you'll often encounter. Further, by understanding studies with two conditions, you will understand the more complicated designs and analyses that we'll discuss in the remaining chapters and that also are common in the literature.

UNDERSTANDING THE TWO-SAMPLE EXPERIMENT

In a two-sample experiment, we measure participants' scores under two conditions of the independent variable. Condition 1 produces $\overline{X}_1$ that represents μ_1, the μ we would find if we tested everyone in the population under condition 1. Condition 2 produces $\overline{X}_2$ that represents μ_2, the μ we would find if we tested everyone in the population under condition 2. A possible outcome from such an experiment is shown in Figure 12.1. If each sample mean represents a different population and μ for each condition, then the experiment has demonstrated a relationship in nature.

However, there is the usual problem of sampling error. Even though we may have different sample means, the relationship may not exist in the population. Instead, if we tested the population, we might find the *same* population of scores under both conditions. In Figure 12.1 we might find only the lower or upper distribution, or we might find one in-between. Then there would be only one value of μ: Call it μ_1 or μ_2, it wouldn't matter because it's the *same* μ. Therefore, before we make any conclusions about the experiment, we must determine whether the difference between the sample means reflects sampling error.

FIGURE 12.1

Relationship in the population in a two-sample experiment

As the conditions change, scores in the population tend to change.

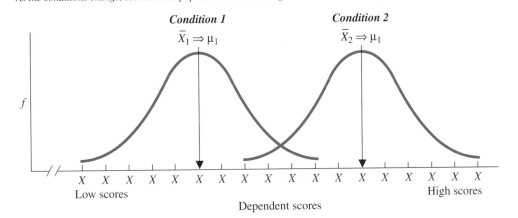

The parametric statistical procedure for determining whether the results of a two-sample experiment are significant is the two-sample *t*-test. However, we have two different ways to create the samples, so we have two different versions of the *t*-test: One is called the *independent-samples* t-*test* and the other is called the *related-samples* t-*test*.

> REMEMBER The two ways to calculate the two-sample *t*-test are the independent-samples *t*-test or the related-samples *t*-test.

THE INDEPENDENT-SAMPLES *t*-TEST

The **independent-samples *t*-test** is the parametric procedure for testing two sample means from **independent samples.** Two samples are independent when we randomly select participants for a sample, without regard to who else has been selected for either sample. Then the scores are *independent events,* which, as in Chapter 9, means that the probability of a particular score occurring in one sample is not influenced by the scores that occur in the other sample. You can recognize independent samples by the absence of anything fancy when selecting participants, such as creating pairs of participants or repeatedly testing the same participants in both conditions.

Here is a study that calls for the independent-samples *t*-test. People who witness a crime or other event may recall the event differently when they are hypnotized. We'll select two samples of participants who watch a videotape of a supposed robbery. Later, one group will be hypnotized and then answer 30 questions about the event. The other group will answer the questions without being hypnotized. Thus, the conditions of the independent variable are the presence or absence of hypnosis, and the dependent variable is the amount of information correctly recalled. This design is shown in Table 12.1. After replacing the *X*s with the actual recall scores, we will compute the mean of each condition (each column). If the means differ, we'll have evidence of a relationship where, as amount of hypnosis changes, recall scores also change.

First, as always, we check that the study meets the assumptions of the statistical test. In addition to requiring independent samples, this *t*-test also requires

1. The dependent scores measure an interval or ratio variable.

2. The populations of raw scores form at least roughly normal distributions.

3. And here's a new one: The populations have homogeneous variance. **Homogeneity of variance** means that the variances of the populations being represented are equal. That is, we assume that if we computed σ_X^2 for each population, we would have the same answer each time.

4. It is not required that each condition have the same *n*, but the *n*s should not be massively unequal—a difference in the neighborhood of 10 to 20 is best. (The more the *n*s differ from each other, the more important it is to have homogeneity of variance.)

You'll know if you meet these assumptions by seeing how the variables are treated in previously published research related to your study.

Statistical Hypotheses for the Independent-Samples *t*-Test

As usual, we may have a one- or a two-tailed test. For now, say that we don't predict whether hypnosis will increase or decrease recall scores so we have a two-tailed test.

TABLE 12.1

Diagram of Hypnosis Study using an Independent-Samples Design

The independent variable is amount of hypnosis, and the dependent variable is recall.

	No Hypnosis	*Hypnosis*
Recall Scores	X	X
	X	X
	X	X
	X	X
	X	X
	$\overline{X}$	$\overline{X}$

First, the alternative hypothesis: A relationship exists if one population mean (μ_1) is larger or smaller than the other (μ_2), producing two distributions, similar to that back in Figure 12.1. In other words, μ_1 should not equal μ_2. We could state this as H_a: $\mu_1 \neq \mu_2$, but there is a better way. If the two μs are not equal, then their *difference* does not equal zero. Thus, the two-tailed alternative hypothesis for our study is

$$H_a\text{: } \mu_1 - \mu_2 \neq 0$$

H_a implies that the means from our two conditions each represent a different population of recall scores having a different μ, so a relationship is present.

Of course, there's also our old nemesis, the null hypothesis. Perhaps there is no relationship, so if we tested everyone under the two conditions, we would find the same population and μ. In other words, μ_1 *equals* μ_2. We could state this as H_0: $\mu_1 = \mu_2$, but, again, there is a better way. If the two μs are equal, then their *difference* is zero. Thus, our two-tailed null hypothesis is

$$H_0\text{: } \mu_1 - \mu_2 = 0$$

H_0 implies that both samples represent the same population of scores, having the same μ, so a relationship is not present. If our sample means differ, it's because of sampling error in representing that one μ.

Notice that these hypotheses do not contain a specific value of μ. Therefore, these are the two-tailed hypotheses for *any* independent-samples *t*-test, when you are testing an H_0 that says there is *zero* difference between the populations. This is the most common approach and the one that we'll use. (Consult an advanced statistics book to test for nonzero differences.)

As usual, we test the null hypothesis, and to do that we examine the sampling distribution.

The Sampling Distribution for the Independent-Samples *t*-Test

To understand the sampling distribution here, say that we find a mean recall score of 20 in the no-hypnosis condition and a mean of 23 in the hypnosis condition. We can summarize these results by looking at the *difference* between the means: Changing from no hypnosis to hypnosis results in a difference in mean recall of 3 points. We always test H_0 by finding the probability of obtaining our results when there is not a relationship, so here we will determine the probability of obtaining a *difference* of 3 between two $\overline{X}$s when they both actually represent the same μ.

> **REMEMBER** The independent-samples *t*-test determines the probability of obtaining our *difference* between $\overline{X}$s when H_0 is true.

You can think of the sampling distribution as follows. Using the same *n*s as in our study, we select *two* random samples from *one* raw score population. (Just like H_0 says we did in our study.) We compute the two sample means and subtract one from the other. The result is the *difference between the means,* which we symbolize by $\overline{X}_1 - \overline{X}_2$. We do this an infinite number of times and plot a frequency distribution of these differences, producing the **sampling distribution of differences between means.** This is the distribution of all possible differences between two means when they are drawn from one raw score population. You can envision this sampling distribution as shown in Figure 12.2. On the *X* axis, each score is the *difference* between two randomly selected sample means. The axis is labeled twice, first using the symbols of $\overline{X}_1 - \overline{X}_2$ and, beneath them,

FIGURE 12.2

Sampling distribution of differences between means when H_0: $\mu_1 - \mu_2 = 0$

Each $\overline{X}_1 - \overline{X}_2$ symbolizes the difference between two sample means. Larger differences are less likely when H_0 is true.

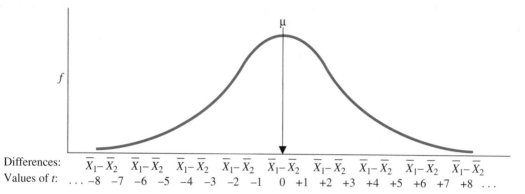

actual differences we might find. The mean of the sampling distribution is zero because, most often, both sample means will equal the μ of the population of raw scores, so their difference will be zero. However, sometimes $\overline{X}_1$ or $\overline{X}_2$ is larger, so the difference will be a positive or negative amount. Small negative or positive differences will occur relatively frequently, but larger differences occur less frequently. The larger the difference between the means, the farther into the tail of the distribution it lies.

To test H_0, we determine where our difference between means lies on this sampling distribution. To do so, we compute a new version of t_{obt} but it provides information similar to previous *t*-tests: A difference of zero between $\overline{X}_1$ and $\overline{X}_2$, located at the μ of the distribution, produces a t_{obt} of zero. A positive difference produces a positive t_{obt} and a negative difference produces a negative t_{obt}. Larger differences between the means are further into a tail of the distribution and have a larger t_{obt}. Therefore, if the difference between our sample means produces a t_{obt} close to the center of the distribution, then our difference occurs frequently when H_0 is true: In our example, our two samples are likely to represent the same population of recall scores. But, if t_{obt} places our difference beyond t_{crit}, far into a tail of the sampling distribution, then this difference is unlikely when H_0 is true: Our two samples are unlikely to represent the same population of recall scores.

So now we compute t_{obt}.

Computing the Independent-Samples *t*-Test

In the previous chapter, you computed t_{obt} by computing $\overline{X}$ and then performing three steps: (1) estimating the variance of the raw score population, (2) computing the estimated standard error of the sampling distribution, and (3) computing t_{obt}. For the two-sample *t*-test, after computing $\overline{X}_1$ and $\overline{X}_2$, you perform three similar steps.

Estimating the Population Variance   First, calculate s_X^2 for *each* condition, using the formula

$$s_X^2 = \frac{\Sigma X^2 - \dfrac{(\Sigma X)^2}{n}}{n - 1}$$

Each time, use the Xs from only one condition, and n is the number of scores in that condition.

This will give us s_1^2 and s_2^2, and each is an estimate of the population variance. However, each may contain sampling error. (Because of this, if s_1^2 does not equal s_2^2, we have not necessarily violated the assumption of homogeneity of variance.) To obtain the best estimate, we compute a "weighted average" of the two variances. The "weight" we give to each variance is based on the number of participants in a sample, using each sample's *df*. This weighted average is called the **pooled variance,** and its symbol is s_{pool}^2.

The formula for the pooled variance is

$$s_{pool}^2 = \frac{(n_1 - 1)s_1^2 + (n_2 - 1)s_2^2}{(n_1 - 1) + (n_2 - 1)}$$

This says to multiply the s_X^2 from each sample times $n - 1$ for that sample. Then add the results together and divide by the sum of $(n_1 - 1) + (n_2 - 1)$.

For example, say that the hypnosis study produced the results shown in Table 12.2. Let's label the hypnosis condition as condition 1, so it produces $\overline{X}_1$, s_1^2 and n_1. The no-hypnosis condition produces $\overline{X}_2$, s_2^2, and n_2. Filling in the above formula, we have

$$s_{pool}^2 = \frac{(17 - 1)9.0 + (15 - 1)7.5}{(17 - 1) + (15 - 1)}$$

In the numerator, 16 times 9 is 144, and 14 times 7.5 is 105. In the denominator, 16 plus 14 is 30, so

$$s_{pool}^2 = \frac{144 + 105}{30} = \frac{249}{30} = 8.30$$

Thus, we estimate that the variance of the population of recall scores represented by our samples is 8.30.

Computing the Standard Error of the Difference The next step is to use s_{pool}^2 to compute the standard error of the sampling distribution. It is called the standard error of the difference. The **standard error of the difference** is the estimated "standard deviation" of the sampling distribution of differences between the means. The symbol for the standard error of the difference is $s_{\overline{X}_1 - \overline{X}_2}$. (The subscript indicates that we are dealing with differences between pairs of means.)

In the previous chapter, we computed the standard error by dividing the variance by N and then taking the square root. However, instead of dividing by N, we can multiply

TABLE 12.2

Data from the Hypnosis Study

	Condition 1: *Hypnosis*	*Condition 2:* *No Hypnosis*
Mean details recalled	$\overline{X}_1 = 23$	$\overline{X}_2 = 20$
Number of participants	$n_1 = 17$	$n_2 = 15$
Estimated variance	$s_1^2 = 9.0$	$s_2^2 = 7.5$

by $1/N$. Then, for the two-sample *t*-test, we substitute the pooled variance and our two *ns*, producing this formula:

The formula for the standard error of the difference is

$$s_{\overline{X}_1 - \overline{X}_2} = \sqrt{(s_{\text{pool}}^2)\left(\frac{1}{n_1} + \frac{1}{n_2}\right)}$$

To use this formula, first reduce the fractions $1/n_1$ and $1/n_2$ to decimals. Then add them together and multiply the sum times s_{pool}^2. Then find the square root.

For the hypnosis study, s_{pool}^2 is 8.30, n_1 is 17, and n_2 is 15. Thus:

$$s_{\overline{X}_1 - \overline{X}_2} = \sqrt{8.3\left(\frac{1}{17} + \frac{1}{15}\right)}$$

First, $1/17$ is .059 and $1/15$ is .067. Their sum is .126. Then

$$s_{\overline{X}_1 - \overline{X}_2} = \sqrt{8.3(.126)} = \sqrt{1.046} = 1.023$$

Thus, $s_{\overline{X}_1 - \overline{X}_2}$ is 1.023.

Computing t_{obt}   In previous chapters we found how far the result of the study $(\overline{X})$ was from the mean of the H_0 sampling distribution (μ), measured in standard error units. In general, this formula is

$$t_{\text{obt}} = \frac{(\text{result of the study}) - (\text{mean of } H_0 \text{ sampling distribution})}{\text{standard error}}$$

Now the "result of the study" is the *difference between* the two sample means, so in the formula we will put in $\overline{X}_1 - \overline{X}_2$. Likewise, instead of one μ we have the *difference* described by H_0, so we put in $\mu_1 - \mu_2$. Finally, we replace "standard error" with $s_{\overline{X}_1 - \overline{X}_2}$. All together we have

---

The formula for the independent-samples t_{obt} is

$$t_{\text{obt}} = \frac{(\overline{X}_1 - \overline{X}_2) - (\mu_1 - \mu_2)}{s_{\overline{X}_1 - \overline{X}_2}}$$

Here, $\overline{X}_1$ and $\overline{X}_2$ are our sample means, $s_{\overline{X}_1 - \overline{X}_2}$ is computed as above, and the value of $\mu_1 - \mu_2$ is the difference specified by the null hypothesis. This value is always 0 (unless you are testing for a nonzero difference.)

For the hypnosis study, our sample means were 23 and 20, the difference between μ_1 and μ_2 is 0, and $s_{\overline{X}_1 - \overline{X}_2}$ is 1.023. Therefore,

$$t_{\text{obt}} = \frac{(23 - 20) - 0}{1.023} = \frac{(+3.0) - 0}{1.023} = \frac{+3.0}{1.023} = +2.93$$

Our t_{obt} is $+2.93$. Thus, the difference of $+3.0$ between our sample means is located at something like a *z*-score of $+2.93$. on the sampling distribution of differences produced when both samples represent the same population.

A QUICK REVIEW

To compute the independent-samples t_{obt}:

- Compute $\overline{X}_1$, s_1^2 and n_1; $\overline{X}_2$, s_2^2 and n_2.
- Then compute the pooled variance (s_{pool}^2).
- Then compute the standard error of the difference $(s_{\overline{X}_1 - \overline{X}_2})$.
- Then compute t_{obt}.

MORE EXAMPLES

An independent-samples study produced the following data: $\overline{X}_1 = 27$, $s_1^2 = 36$, $n_1 = 11$, $\overline{X}_2 = 21$, $s_2^2 = 33$, and $n_2 = 11$.

$$s_{\text{pool}}^2 = \frac{(n_1 - 1)s_1^2 + (n_2 - 1)s_2^2}{(n_1 - 1) + (n_2 - 1)}$$

$$= \frac{(10)36 + (10)33}{10 + 10} = 34.5$$

$$s_{\overline{X}_1 - \overline{X}_2} = \sqrt{s_{\text{pool}}^2 \left(\frac{1}{n_1} + \frac{1}{n_2} \right)}$$

$$= \sqrt{34.5 \left(\frac{1}{11} + \frac{1}{11} \right)} = 2.506$$

$$t_{\text{obt}} = \frac{(\overline{X}_1 - \overline{X}_2) - (\mu_1 - \mu_2)}{s_{\overline{X}_1 - \overline{X}_2}} = \frac{(27 - 21) - 0}{2.506}$$

$$= +2.394$$

For Practice

We find $\overline{X}_1 = 33$, $s_1^2 = 16$, $n_1 = 21$, $\overline{X}_2 = 27$, $s_2^2 = 13$ and $n_2 = 21$.

1. Compute the pooled variance (s_{pool}^2).
2. Compute the standard error of the difference $(s_{\overline{X}_1 - \overline{X}_2})$.
3. Compute t_{obt}.

Answers

1. $s_{\text{pool}}^2 = \dfrac{(20)16 + (20)13}{20 + 20} = 14.5$

2. $s_{\overline{X}_1 - \overline{X}_2} = \sqrt{14.5 \left(\dfrac{1}{21} + \dfrac{1}{21} \right)} = 1.18$

3. $t_{\text{obt}} = \dfrac{(33 - 27) - 0}{1.18} = +5.08$

Interpreting the Independent-Samples *t*-Test

To determine if t_{obt} is significant, we compare it to t_{crit}, which is found in the *t*-tables (Table 2 in Appendix C). As usual, we obtain t_{crit} using degrees of freedom, but with two samples, the *df* are computed differently: Now the degrees of freedom equals $(n_1 - 1) + (n_2 - 1)$.

> **REMEMBER**   Critical values of *t* for the independent-samples *t*-test are found for $df = (n_1 - 1) + (n_2 - 1)$.

Another way of expressing this is $df = (n_1 + n_2) - 2$.

For the hypnosis study, $n_1 = 17$ and $n_2 = 15$, so $df = (17 - 1) + (15 - 1) = 30$. With alpha at .05, the two-tailed t_{crit} is ± 2.042. Figure 12.3 locates these values on the sampling distribution of differences.

The sampling distribution shows the frequency of various differences between sample means that occur when the samples really represent no difference in the population. Our H_0 says that the difference between our sample means is merely a poor representation of no difference. But, looking at the sampling distribution, we see that our difference of $+3$ hardly ever occurs when the samples represent no difference. In fact our t_{obt}

FIGURE 12.3

H_0 sampling distribution of differences between means when $\mu_1 - \mu_2 = 0$. *A larger* t *indicates a larger difference between means.*

The t_{obt} *shows the location of a difference of +3.0*

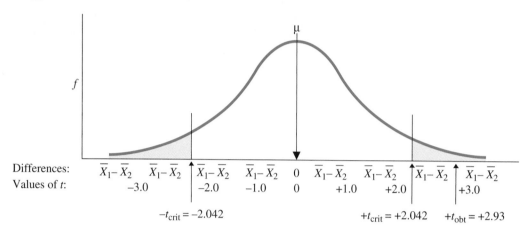

Differences:	$\overline{X}_1 - \overline{X}_2$	$\overline{X}_1 - \overline{X}_2$	$\overline{X}_1 - \overline{X}_2$	$\overline{X}_1 - \overline{X}_2$	0	$\overline{X}_1 - \overline{X}_2$	$\overline{X}_1 - \overline{X}_2$	$\overline{X}_1 - \overline{X}_2$	$\overline{X}_1 - \overline{X}_2$
Values of *t*:		−3.0	−2.0	−1.0	0	+1.0	+2.0		+3.0

$-t_{crit} = -2.042$ $\qquad$ $+t_{crit} = +2.042$ $\qquad$ $+t_{obt} = +2.93$

of $+2.93$ lies beyond t_{crit}, so the results are significant: Our difference of $+3$ is so unlikely to occur if our samples were representing no difference in the population that we reject that this is what they represent. Therefore, we reject H_0 and accept the H_a that we are representing a difference between μs that is *not* zero.

We can say that our difference of $+3$ is significantly different from 0. Or we can say that our two means differ significantly *from each other*. Here, the mean for hypnosis (23) is larger than the mean for no hypnosis (20), so we can conclude that increasing the amount of hypnosis leads to significantly higher recall scores.

If t_{obt} was not beyond t_{crit}, we would not reject H_0, and we would have no evidence *for or against* a relationship between hypnosis and recall. Then we would consider if we had sufficient *power* to prevent a Type II error (retaining a false H_0). As in the previous chapter, we maximize power here by designing the study to (1) maximize the size of the difference between the means, (2) minimize the variability of scores within each condition, and (3) maximize the size of N. These steps will maximize the size of t_{obt} relative to t_{crit} so that we are unlikely to miss the relationship if it really exists.

Because we did find a significant result, we describe and interpret the relationship. First, from our sample means, we expect the μ for no hypnosis to be around 20 and the μ for hypnosis to be around 23. To more precisely describe these μs, we could compute a *confidence interval* for each. To do so, we would use the formula for a confidence interval in the previous chapter, looking at only one condition at a time, using only one $s_{\overline{X}}^2$ and $\overline{X}$, and computing a new standard error and t_{crit}. Then we'd know the range of values of μ likely to be represented by each of our means.

However, another way to describe the populations represented by our samples is to create a *confidence interval for the difference between the μs.*

Confidence Interval for the Difference between Two μs

Above we found a difference of $+3$ between our sample means, so if we could examine the corresponding μ_1 and μ_2, we'd expect their difference would be *around* $+3$. To more precisely define "around," we can compute a confidence interval for this difference. We will compute the largest and smallest difference between μs that our difference between sample means is likely to represent. Then we will have a range of differences between the population μs that our difference between $\overline{X}$s may represent. The **confidence interval for the difference between two μs** describes a range of *differences* between two μs, one of which is likely to be represented by the *difference* between our two sample means.

The formula for the confidence interval for the difference between two μs is

$$(s_{\overline{X}_1-\overline{X}_2})(-t_{\text{crit}}) + (\overline{X}_1 - \overline{X}_2) \leq \mu_1 - \mu_2 \leq (s_{\overline{X}_1-\overline{X}_2})(+t_{\text{crit}}) + (\overline{X}_1 - \overline{X}_2)$$

Here, $\mu_1 - \mu_2$ stands for the unknown difference we are estimating. The t_{crit} is the *two-tailed* value found for the appropriate α at $df = (n_1 - 1) + (n_2 - 1)$. The values of $s_{\overline{X}_1-\overline{X}_2}$ and $(\overline{X}_1 - \overline{X}_2)$ are computed in the *t*-test.

In the hypnosis study, the two-tailed t_{crit} for $df = 30$ and $\alpha = .05$ is ± 2.042, $s_{\overline{X}_1-\overline{X}_2} = 1.023$, and $\overline{X}_1 - \overline{X}_2 = +3$. Filling in the formula gives

$$(1.023)(-2.042) + (+3) \leq \mu_1 - \mu_2 \leq (1.023)(+2.042) + (+3)$$

Multiplying 1.023 times ± 2.042 gives

$$-2.089 + (+3) \leq \mu_1 - \mu_2 \leq +2.089 + (+3)$$

So, finally,

$$.911 \leq \mu_1 - \mu_2 \leq 5.089$$

Because $\alpha = .05$, this is the 95% confidence interval: We are 95% confident that the interval between .911 and 5.089 contains the difference we'd find between the μs for no hypnosis and hypnosis. In essence, if someone asked us how big a difference hypnosis makes for everyone in the population when recalling information in our study, we'd be 95% confident that the difference is, on average, between about .91 and 5.09 correct answers.

Performing One-Tailed Tests with Independent Samples

As usual, we perform a one-tailed test whenever we predict the specific direction in which the dependent scores will change. Thus, we would have performed a one-tailed test if we had predicted that hypnosis would increase recall scores. Everything discussed previously applies here, but to prevent confusion, we'll use the subscript h for hypnosis and n for no hypnosis. Then follow these steps:

1. Decide which $\overline{X}$ and corresponding μ is expected to be larger. (We predict the μ_h for hypnosis is larger.)

2. Arbitrarily decide which condition to subtract from the other. (We'll subtract no hypnosis *from* hypnosis.)

3. Decide whether the predicted difference will be positive or negative. (Subtracting the smaller μ_n from the larger μ_h should produce a positive difference, *greater* than zero.)

4. Create H_a and H_0 to match this prediction. (Our H_a is that $\mu_h - \mu_n > 0$; H_0 is that $\mu_h - \mu_n \leq 0$.)

5. Envision the same sampling distribution that we used in the two-tailed test. Obtain the one-tailed t_{crit} from Table 2 in Appendix C. Locate the region of rejection based on your prediction. If we expect a positive difference, it is in the right-hand tail of the sampling distribution, so t_{crit} is positive. If we predict a negative difference, it is in the left-hand tail and t_{crit} is negative.

6. Compute t_{obt} as we did previously, but be sure to subtract the $\overline{X}$s in the same way as in H_a. (We used $\mu_h - \mu_n$, so we'd compute $\overline{X}_h - \overline{X}_n$.)

Conversely, if we had predicted that hypnosis would decrease scores, subtracting $h - n$, we would have H_a: $\mu_1 - \mu_2 < 0$ and H_0: $\mu_1 - \mu_2 \geq 0$, and t_{crit} would be negative.

SUMMARY OF THE INDEPENDENT-SAMPLES t-TEST

After checking that the study meets the assumptions, the independent-samples t-test involves the following:

1. *Create either the two-tailed or one-tailed* H_0 *and* H_a.
2. *Compute* t_{obt}:
   a. Compute $\overline{X}_1$, s_1^2, and n_1; $\overline{X}_2$, s_2^2, and n_2.
   b. Compute the pooled variance (s_{pool}^2).
   c. Compute the standard error of the difference $(s_{\overline{X}_1 - \overline{X}_2})$.
   d. Compute t_{obt}.
3. *Find* t_{crit}: In the t-tables, use $df = (n_1 - 1) + (n_2 - 1)$.
4. *Compare* t_{obt} *to* t_{crit}: If t_{obt} is beyond t_{crit}, the results are significant; describe the relationship. If t_{obt} is not beyond t_{crit}, the results are not significant; make no conclusion about the relationship.
5. *Compute the confidence interval*: Describe the μ represented by each condition and/or the difference between the μs.

A QUICK REVIEW

- Perform the independent-samples t-test in experiments that test two independent samples.

MORE EXAMPLES

We perform a two-tailed experiment, so H_0: $\mu_1 - \mu_2 = 0$ and H_a: $\mu_1 - \mu_2 \neq 0$. The $\overline{X}_1 = 24$, $s_1^2 = 9$, $n_1 = 14$, $\overline{X}_2 = 21$, $s_2^2 = 9.4$, and $n_2 = 16$. Then

$$s_{pool}^2 = \frac{(n_1 - 1)s_1^2 + (n_2 - 1)s_2^2}{(n_1 - 1) + (n_2 - 1)}$$

$$= \frac{(13)9 + (15)9.4}{13 + 15} = 9.214$$

$$s_{\overline{X}_1 - \overline{X}_2} = \sqrt{s_{pool}^2 \left(\frac{1}{n_1} + \frac{1}{n_2}\right)}$$

$$= \sqrt{9.214\left(\frac{1}{14} + \frac{1}{16}\right)} = 1.111$$

$$t_{obt} = \frac{(\overline{X}_1 - \overline{X}_2) - (\mu_1 - \mu_2)}{s_{\overline{X}_1 - \overline{X}_2}} = \frac{(24 - 21) - 0}{1.111}$$

$$= +2.70$$

With $\alpha = .05$ and $df = (n_1 - 1) + (n_2 - 1) = 28$, $t_{crit} = \pm 2.048$. The t_{obt} is significant: We expect μ_1

(continued)

to be around 24 and μ_2 to be around 21. The confidence interval for the difference between the μs is

$$(s_{\overline{X}_1-\overline{X}_2})(-t_{crit}) + (\overline{X}_1 - \overline{X}_2) \le \mu_1 - \mu_2 \le$$
$$(s_{\overline{X}_1-\overline{X}_2})(+t_{crit}) + (\overline{X}_1 - \overline{X}_2)$$
$$(1.111)(-2.048) + 3 \le \mu_1 - \mu_2 \le$$
$$(1.111)(+2.048) + 3$$
$$0.725 \le \mu_1 - \mu_2 \le 5.275$$

For Practice

We test whether "cramming" for an exam is harmful to grades. Condition 1 crams for a pretend exam, but condition 2 does not. Each $n = 31$, the cramming $\overline{X}$ is 43 ($s_X^2 = 64$), and the no-cramming $\overline{X}$ is 48 ($s_X^2 = 83.6$).

1. Subtracting cramming from no cramming, what are H_0 and H_a?
2. Will t_{crit} be positive or negative?
3. Compute t_{obt}.
4. What do you conclude about this relationship?
5. Compute the confidence interval between μs.

Answers

1. H_a: $\mu_{nc} - \mu_c > 0$; H_0: $\mu_{nc} - \mu_c \le 0$
2. Positive
3. $s_{pool}^2 = (1920 + 2508)/60 = 73.80$;
   $s_{\overline{X}_1-\overline{X}_2} = \sqrt{73.80(.065)} = 2.190$;
   $t_{obt} = (5)/2.190 = +2.28$
4. With $\alpha = .05$ and $df = 60$, $t_{crit} = +1.671$, t_{obt} is significant: μ_c is around 43; μ_{nc} is around 48.
5. $(2.19)(-2.00) + 5 \le \mu_1 - \mu_2 \le (2.19)(+2.000) + 5$
   $= 0.62 \le \mu_1 - \mu_2 \le 9.38$

THE RELATED-SAMPLES *t*-TEST

Now we will discuss the other version of the two-sample *t*-test. The **related-samples *t*-test** is the parametric procedure used with two related samples. **Related samples** occur when we pair each score in one sample with a particular score in the other sample. Researchers create related samples to have more equivalent and thus more comparable samples. Two types of research designs produce related samples. They are *matched-samples designs* and *repeated-measures designs*.

In a **matched-samples design,** we match each participant in one condition with a participant in the other condition. For example, say that we want to measure how well people shoot baskets when using either a standard basketball or a new type of ball (one with handles). If, however, by luck, one condition contained taller people than the other, then differences in basket shooting could be due to the differences in height instead of the different balls. The solution is to create two samples containing people who are the same height. We do this by matching pairs of people who are the same height and assigning a member of the pair to each condition. Thus, if two participants are 6 feet tall, one will be assigned to each condition. Likewise, a 4-foot person in one condition is matched with a 4-footer in the other condition, and so on. This will produce two samples that, overall, are equivalent in height, so any differences in basket shooting between them cannot be due to differences in height. Likewise, we might match participants using age, or physical ability, or we might use naturally occurring pairs such as roommates or identical twins.

The other, more common, way of producing related samples is called repeated measures. In a **repeated-measures design,** each participant is tested under all conditions of the independent variable. For example, we might first test people when they use the standard basketball and then measure the same people again when they use the new ball. (Although we have one sample of participants, we have two samples of scores.) Here, any differences in basket shooting between the samples cannot be due to differences in height or to any other attribute of the participants.

Matched-groups and repeated-measures designs are analyzed in the same way because both produce related samples. Related samples are also called *dependent samples*. In Chapter 9, two events were dependent when the probability of one is influenced by the occurrence of the other. Related samples are dependent because the probability that a score in a pair is a particular value is influenced by the paired score. For example, if I make zero baskets in one condition, I'll probably make close to zero in the other condition. This is not the case with independent samples: In the hypnosis study, whether someone scores 0 in the no-hypnosis condition will not influence the probability of anyone scoring 0 in the hypnosis condition.

We cannot use the independent-samples *t*-test in such situations because its sampling distribution describes the probability of differences between means from *independent* samples. With related samples, we must compute this probability differently, so we create the sampling distribution differently and we compute t_{obt} differently. However, except for requiring related samples, the assumptions of the related-samples *t*-test are the same as those for the independent-samples *t*-test: (1) The dependent variable involves an interval or ratio scale, (2) the raw score populations are at least approximately normally distributed, and (3) the populations have homogeneous variance. Because related samples form pairs of scores, the *n* in the two samples must be equal.

The Logic of Hypotheses Testing in the Related-Samples *t*-Test

Let's say that we are interested in phobias (irrational fears of objects or events). We have a new therapy we want to test on spider-phobics—people who are overly frightened by spiders. From the local phobia club, we randomly select the unpowerful *N* of five spider-phobics and test our therapy using repeated measures of two conditions: before therapy and after therapy. Before therapy we measure each person's fear response to a picture of a spider by measuring heart rate, perspiration, and so on. Then we compute a "fear" score between 0 (no fear) and 20 (holy terror!). After providing the therapy, we again measure the person's fear response to the picture. (A before-and-after, or *pretest/posttest*, design such as this always uses the related-samples *t*-test.)

Say that we obtained the raw scores shown on the left side of Table 12.3. First, compute the mean of each condition (each column). Before therapy the mean fear score is 14.80, but after therapy the mean is 11.20. Apparently, the therapy reduced fear scores by an average of $14.80 - 11.20 = 3.6$ points. But, on the other hand, maybe therapy does nothing; maybe this difference is solely the result of sampling error from the *one* population of fear scores we'd have with or without therapy.

TABLE 12.3

Finding the Difference Scores in the Phobia Study

Each D = Before − After

Participant	*Before Therapy*	−	*After Therapy*	=	**D**	**D²**
1 (Foofy)	11	−	8	=	+3	9
2 (Biff)	16	−	11	=	+5	25
3 (Cleo)	20	−	15	=	+5	25
4 (Attila) -	17	−	11	=	+6	36
5 (Slug)	10	−	11	=	−1	1
	$\overline{X} = 14.80$		$\overline{X} = 11.20$		$\Sigma D = +18$	$\Sigma D^2 = 96$
$N = 5$					$\overline{D} = +3.6$	

To test these hypotheses, we first transform the data and then perform a *t*-test on the transformed scores. As shown in Table 12.3, we transform the data by first finding the difference between the two raw scores for each participant. This *difference score* is symbolized by *D*. Here, we subtracted after therapy from before therapy. You could subtract in the reverse order, but subtract all scores in the same way. If this were a matched-samples design, we'd subtract the scores from each pair of matched participants.

Next, compute the *mean difference,* symbolized as $\overline{D}$. Add the positive and negative differences to find the sum of the differences, symbolized by ΣD. Then divide by *N*, the number of difference scores. In Table 12.3, $\overline{D}$ equals 18/5, which is +3.6. Notice that this is also the difference between our original means of 14.80 and 11.20. Anyway you approach it, the before scores were, on average, 3.6 points higher than the after scores. (As in the far right-hand column of Table 12.3, later we'll need to square each difference and then find the sum, finding ΣD^2.)

Now here's the strange part: Forget about the before and after scores for the moment and consider only the difference scores. We have one sample mean from one random sample of scores. As in the previous chapter, with one sample we perform the one-sample *t*-test! The fact that we have difference scores is irrelevant, so we create the statistical hypotheses and test them in virtually the same way that we did with the one-sample *t*-test.

> **REMEMBER**   The *related-samples* t-*test* is performed by applying the one-sample *t*-test to the difference scores.

STATISTICAL HYPOTHESES FOR THE RELATED-SAMPLES *t*-TEST

Our sample of difference scores represents the population of difference scores that would result if we could measure the population's fear scores under each condition and then subtract the scores in one population from the corresponding scores in the other population. The population of difference scores has a μ that we identify as μ_D. To create the statistical hypotheses, we determine the predicted values of μ_D in H_0 and H_a.

In reality, we expect the therapy to reduce fear scores, but let's first perform a two-tailed test. H_0 always says no relationship is present, so it says the population of before-scores is the same as the population of after-scores. However, when we subtract them as we did in the sample, not every *D* will equal zero because, due to random physiological or psychological fluctuations, some participants will not score identically when tested before and after. Therefore, we will have a *population* of different *D*s, as shown on the left in Figure 12.4.

On average, the positive and negative differences should cancel out to produce a $\mu_D = 0$. This is the population that H_0 says that our sample of *D*s represents, and that our $\overline{D}$ somewhat poorly represents this μ_D. Therefore, $H_0: \mu_D = 0$.

For the alternative hypothesis, if the therapy alters fear scores in the population, then either the before scores or the after scores will be consistently higher. Then, after subtracting them, the population of *D*s will tend to contain only positive or only negative scores. Therefore, the average difference (μ_D) will be a positive or negative number, and not zero. Thus, $H_a: \mu_D \neq 0$.

We test H_0 by examining the sampling distribution, which here is the *sampling distribution of mean differences*. Shown on the right side of Figure 12.4, it is as if we infinitely sampled the population of *D*s on the left that H_0 says our sample represents.

FIGURE 12.4
Population of difference scores described by H_0 and the resulting sampling distribution of mean differences

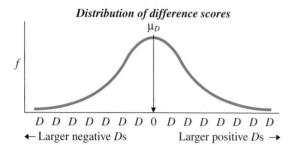
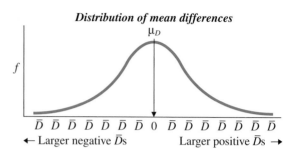

The **sampling distribution of mean differences** shows all possible values of $\overline{D}$ that occur when samples are drawn from a population of difference scores where $\mu_D = 0$. For the phobia study, it essentially shows all values of $\overline{D}$ we might get by chance when the therapy does not work. The $\overline{D}$s that are farther into the tails of the distribution are less likely to occur if H_0 was true and the therapy did not work.

Notice that the hypotheses H_0: $\mu_D = 0$ and H_a: $\mu_D \neq 0$ and the above sampling distribution are appropriate for the two-tailed test for *any* study when testing whether the data represent *zero* difference between your conditions. This is the most common approach and the one that we'll discuss. (Consult an advanced statistics book to test for nonzero differences.)

We test our H_0 by determining where on the sampling distribution our $\overline{D}$ is located. To do so, we compute t_{obt}.

Computing the Related-Samples *t*-Test

Computing t_{obt} here is identical to computing the one-sample *t*-test discussed in Chapter 11—only the symbols have been changed from X to D There, we first computed the estimated population variance (s_X^2), then the standard error of the mean $(s_{\overline{X}})$, and then t_{obt}. We perform the same three steps here.

First, find s_D^2, which is the estimated population variance of the difference scores.

> **The formula for s_D^2 is**
>
> $$s_D^2 = \frac{\Sigma D^2 - \dfrac{(\Sigma D)^2}{N}}{N - 1}$$

(*Note:* For all computations in this *t*-test, *N* equals the number of *difference* scores.)

Using the data from the phobia study in Table 12.3, we have

$$s_D^2 = \frac{\Sigma D^2 - \dfrac{(\Sigma D)^2}{N}}{N - 1} = \frac{96 - \dfrac{(18)^2}{5}}{4} = 7.80$$

Second, find $s_{\overline{D}}$. This is the **standard error of the mean difference,** or the "standard deviation" of the sampling distribution of $\overline{D}$. (Just as $s_{\overline{X}}$ was the standard deviation of the sampling distribution when we called the mean $\overline{X}$.)

The formula for the standard error of the mean difference is

$$s_{\overline{D}} = \sqrt{\frac{s_D^2}{N}}$$

Divide s_D^2 by N and then find the square root. For the phobia study, $s_D^2 = 7.80$ and $N = 5$, so

$$s_{\overline{D}} = \sqrt{\frac{s_D^2}{N}} = \sqrt{\frac{7.80}{5}} = \sqrt{1.56} = 1.249$$

Third, find t_{obt}.

The formula for the related-samples *t*-test is

$$t_{obt} = \frac{\overline{D} - \mu_D}{s_{\overline{D}}}$$

Here, $\overline{D}$ is the mean of your difference scores, $s_{\overline{D}}$ is computed as above, and μ is the value given in H_0: It is always zero (unless you are testing a nonzero difference). Then, as usual, t_{obt} is like a *z*-score, indicating how far our $\overline{D}$ is from the μ_D of the sampling distribution when measured in standard error units.

For the phobia study, $\overline{D} = 3.6$, $s_{\overline{D}} = 1.249$ and $\mu_D = 0$. Filling in the formula, we have

$$t_{obt} = \frac{\overline{D} - \mu_D}{s_{\overline{D}}} = \frac{+3.6 - 0}{1.249} = +2.88$$

Thus, $t_{obt} = +2.88$.

Interpreting the Related-Samples *t*-Test

Interpret t_{obt} by comparing it to t_{crit} from the *t*-tables in Appendix C. Here, $df = N - 1$.

REMEMBER   The degrees of freedom in the related-samples *t*-test are $df = N - 1$, where N is the number of difference scores.

For the phobia study, with $\alpha = .05$ and $df = 5-1 = 4$, the two-tailed t_{crit} is ± 2.776. The completed sampling distribution is shown in Figure 12.5. The t_{obt} is in the region of rejection, so the results are significant: Our sample with $\overline{D} = +3.6$ is so unlikely to be representing the population of *D*s where $\mu_D = 0$ that we reject the H_0 that our

FIGURE 12.5

Two-tailed sampling distribution of $\bar{D}$s when $\mu_D = 0$

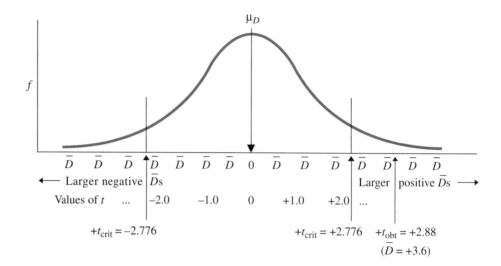

sample represents this population. Because this population has a $\mu = 0$, we conclude that our $\bar{D}$ of +3.6 is significantly different from zero. Therefore, we accept H_a, concluding that the sample represents a population of Ds having a μ_D that is not zero, with μ_D probably around +3.6.

Now we work backwards to our original fear scores. Recall that our $\bar{D}$ of +3.6 also equals the difference between the original mean fear score before therapy (14.80) and the mean fear score after therapy (11.20): Any way you approach it, the therapy reduced fear scores by an average of +3.6. Because we have determined that this reduction is significant using $\bar{D}$, we can also conclude that this reduction is significant using our original fear scores. Therefore, we conclude that the means of 14.80 and 11.20 differ significantly from each other, and are unlikely to represent the same population of fear scores. Instead, we conclude that our therapy works, with the sample data representing a relationship in the population of spider-phobics such that fear scores go from a μ around 14.80 before therapy to a μ around 11.20 after therapy.

If t_{obt} had not been beyond t_{crit}, the results would not be significant. Then we'd want to have maximized our power in the same ways as discussed previously: We maximize the differences between the conditions, minimize the variability in the scores within the conditions, and maximize N. *Note:* A related-samples *t*-test is intrinsically more powerful than an independent-samples *t*-test because the Ds will be less variable than the original raw scores. For example, back in Table 12.3, Biff and Cleo show variability between their before scores and between their after scores, but they have the *same* difference scores. Thus, by designing a study that uses related samples, we will tend to have greater power than when we design a similar study that uses independent samples.

> ***REMEMBER*** A related-samples *t*-test is intrinsically more powerful than an independent-samples *t*-test.

With significant results, we use the sample means to estimate the μ of the fear scores for each condition as described above. It would be nice to compute a confidence interval for each μ, as in the previous chapter, but we cannot do that. That procedure assumes each mean comes from an *independent* sample. We can, however, compute a *confidence interval for* μ_D.

Computing the Confidence Interval for μ_D

Because our $\overline{D}$ is +3.6, we assume that if we measured the entire population before and after therapy, the population of difference scores would have a μ_D *around* +3.6. To better define "around," we compute a confidence interval. The **confidence interval for μ_D** describes a range of values of μ_D, one of which our sample mean is likely to represent.

> **The formula for the confidence interval for μ_D is**
>
> $$(s_{\overline{D}})(-t_{\text{crit}}) + \overline{D} \leq \mu_D \leq (s_{\overline{D}})(+t_{\text{crit}}) + \overline{D}$$

This is the same formula used in Chapter 11, except that the symbol $\overline{X}$ has been replaced by $\overline{D}$. The t_{crit} is the *two-tailed* value for $df = N - 1$, where N is the number of difference scores, $s_{\overline{D}}$ is the standard error of the mean difference computed as above, and $\overline{D}$ is the mean of the difference scores.

In the phobia study, $s_{\overline{D}} = 1.25$ and $\overline{D} = +3.6$, and with $\alpha = .05$ and $df = 4$, t_{crit} is ± 2.776. Filling in the formula gives

$$(1.25)(-2.776) + 3.6 \leq \mu_D \leq (1.25)(+2.776) + 3.6$$

which becomes

$$0.13 \leq \mu_D \leq 7.07$$

Thus, we are 95% confident that our $\overline{D}$ of +3.6 represents a population μ_D within this interval. In other words, we would expect the average difference in before and after scores in the population to be between 0.13 and 7.07.

Performing One-Tailed Tests with Related Samples

As usual, we perform a one-tailed test when we predict the direction of the difference between our two conditions. Realistically, in the phobia study, we would predict we'd find *lower* scores in the after-therapy condition. Then to create H_a, first arbitrarily decide which condition to subtract from which and what the differences should be. We subtracted the predicted lower after-scores from the predicted higher before-scores, so this should produce Ds that are positive. Then $\overline{D}$ should be positive, representing a population that has a positive μ_D. Therefore, $H_a: \mu_D > 0$. Then $H_0: \mu_D \leq 0$.

We again examine the sampling distribution that occurs when $\mu_D = 0$. Obtain the one-tailed t_{crit} from Table 2 in Appendix C. Then locate the region of rejection based on your prediction: Our $\overline{D}$ should be positive and, as in Figure 12.6, the positive values of $\overline{D}$ are in the right-hand tail, and so t_{crit} is positive.

Had we predicted *higher* scores in the after-therapy condition then, by subtracting before from after, the Ds and $\overline{D}$ should be negative, representing a negative μ_D. Thus, $H_a: \mu_D < 0$ and $H_0: \mu_D \geq 0$. Now the region of rejection is in the lower tail of the sampling distribution, and t_{crit} is negative.

Compute t_{obt} using the previous formula. But subtract to get your Ds in the same way as when you created your hypotheses.

FIGURE 12.6

One-Tailed Sampling Distribution of $\overline{D}$s When $\mu_D = 0$

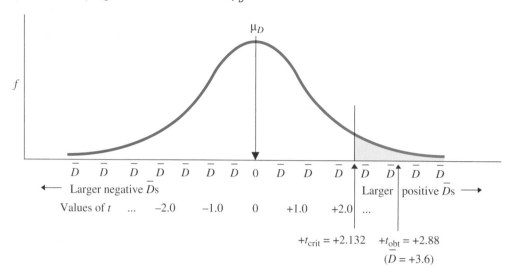

SUMMARY OF THE RELATED-SAMPLES *t*-TEST

After checking that the design is matched samples or repeated measures and meets the assumptions, the related-samples *t*-test involves the following:

1. *Create either the two-tailed or one-tailed H_0 and H_a.*
2. *Compute t_{obt}:*
   a. Compute the *difference score* for each pair of scores.
   b. Compute $\overline{D}$ and s_D^2.
   c. Compute $s_{\overline{D}}$.
   d. Compute t_{obt}.
3. *Find t_{crit}: In the t-tables, use $df = N - 1$.*
4. *Compare t_{obt} to t_{crit}: If t_{obt} is beyond t_{crit}, the results are significant; describe the populations of raw scores and interpret the relationship. If t_{obt} is not beyond t_{crit}, the results are not significant; make no conclusion about the relationship.*
5. *Compute the confidence interval for μ_D.*

A QUICK REVIEW

- Perform the related-samples *t*-test with a matched-groups or repeated-measures design.

MORE EXAMPLES

In a two-tailed study, we compare husband-and-wife pairs, with H_0: $\mu_D = 0$ and H_a: $\mu_D \neq 0$. Subtracting wife – husband produces

Wife	Husband	D
4	6	−2
5	8	−3
3	9	−6
5	8	−3
$\overline{X} = 4.25$	$\overline{X} = 7.75$	$\overline{D} = -3.5$

(continued)

$$\overline{D} = -14/4 = -3.5$$

$$s_D^2 = \frac{\Sigma D^2 - \dfrac{(\Sigma D)^2}{N}}{N - 1} = \frac{58 - \dfrac{(-14)^2}{4}}{3} = 3.0$$

$$s_{\overline{D}} = \sqrt{\frac{s_D^2}{N}} = \sqrt{3/4} = .866$$

$$t_{obt} = \frac{\overline{D} - \mu_D}{s_{\overline{D}}} = (-3.5 - 0)/.866 = -4.04$$

With $\alpha = .05$ and $df = 3$, t_{crit} is ±3.182 The t_{obt} is significant. For wives, we expect μ is 4.25, and for husbands, we expect μ is 7.75. For the confidence interval of μ_D,

$$(s_{\overline{D}})(-t_{crit}) + \overline{D} \le \mu_D \le (s_{\overline{D}})(-t_{crit}) + \overline{D}$$
$$(.866)(-3.182) + -3.5 \le \mu_D \le$$
$$(.866)(+3.182) + -3.5 =$$
$$-6.256 \le \mu_D \le -0.744$$

For Practice

A two-tailed study with repeated measures gives

A	B
8	7
10	5
9	6
8	5
11	6

1. What are H_0 and H_a?
2. Subtracting A – B, perform the t-test.
3. Compute the confidence interval of μ_D.
4. Subtracting A – B, what are H_0 and H_a if we predicted that B would produce lower scores?

Answers

1. $H_0: \mu_D = 0$; $H_a: \mu_D \ne 0$
2. $\overline{D} = 17/5 = +3.4$; $s_D^2 = 2.8$: $s_{\overline{D}} = \sqrt{2.8/5} = .748$; $t_{obt} = (3.4 - 0)/.748 = +4.55$. With $\alpha = 0.5$, $t_{crit} = \pm2.776$ and t_{obt} is significant.
3. $(.748)(-2.776) + 3.4 \le \mu_D \le (.748)(+2.776) + 3.4 = 1.324 \le \mu_D \le 5.476$
4. $H_0: \mu_D \le 0$; $H_a: \mu_D > 0$

DESCRIBING THE RELATIONSHIP IN A TWO-SAMPLE EXPERIMENT

In either two-sample t-test, the fact that t_{obt} is significant is not the end of the story. If you stop after hypothesis testing, then you've found a relationship, but you have not described it. Instead, whenever (and only) when you have significant results, you should fully describe the relationship in your sample data. This involves two steps—graphing the results and computing "effect size."

Graphing the Results of a Two-Sample Experiment

From Chapter 4, you know that we plot the mean of each condition on the Y axis and the conditions of the independent variable on the X axis. The results of our previous studies are shown in Figure 12.7. Notice that for the phobia study the means of the original fear scores from the before and after conditions are plotted, not the Ds.

FIGURE 12.7

Line graphs of the results of the hypnosis study and the phobia study

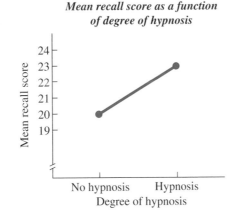

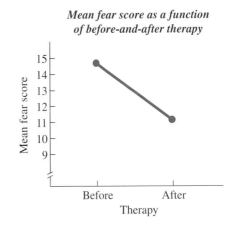

From each graph, you can first discern the *type*—direction—of the relationship: The hypnosis graph shows a positive linear relationship, and the phobia graph shows a negative relationship. Further, recall that the *regression line* summarizes a relationship by running through the center of the *scatterplot*. With only two conditions, the line graph *is* the regression line. Therefore, we can envision the scatterplot in each graph as being around the line, with participants' data points located above and below each mean's data point. Finally, recall that we use the regression line to predict *Y* scores based on *X*. Therefore, for participants in a particular condition, we travel vertically to the line and then horizontally to *Y*, predicting that they scored at the mean score for that condition.

What's missing is that we don't know the nature of each relationship. Recall that some relationships are stronger (more consistent) than others. Likewise, some independent variables have a greater impact on a behavior than others. Researchers address these issues by computing a measure of *effect size.*

Measuring Effect Size in the Two-Sample Experiment

An important statistic for describing the results of an experiment is called a *measure of effect size.* The "effect" is from *cause and effect*, because in an experiment we assume that changing the independent variable "causes" the dependent scores to change. (The quotes are used because there's always a chance that something else was the cause.) However, not all independent variables will cause scores to change to the same degree, so we need to know the influence that a particular variable has. **Effect size** indicates the amount of influence that changing the conditions of the independent variable had on dependent scores. Thus, for example, the extent to which changing hypnosis influenced recall scores is the effect size of hypnosis.

The larger the effect size, the greater is the independent variable's impact in determining participants' scores. We want to study those variables that most influence the behavior measured by these scores, so **the larger the effect size, the more scientifically important the independent variable is.** Remember that *significant* does not mean important, but only that the sample relationship is unlikely to reflect sampling error. Although a relationship must be significant to be potentially important, it can be significant and still be unimportant. Thus, you should always compute a measure of effect size for any significant result, because this is the only way to determine whether your independent variable is important in influencing a behavior. In fact, the American Psychological Association requires published research to report effect size.

> *REMEMBER* The larger the *effect size*, the greater the influence that an independent variable has on dependent scores and thus the more important the variable is.

We will discuss two methods for measuring effect size. The first is to compute *Cohen's* d.

Effect Size Using Cohen's *d*   One way to describe the impact of an independent variable is in terms of how *big* a difference we see between the means of our conditions. For example, we saw that the presence/absence of hypnosis produced a difference in recall scores of 3. However, the problem is that we don't know whether, in the grand scheme of things, 3 is large, small, or in between. We need a frame of reference, and here we use the estimated population standard deviation. Recall that the standard deviation reflects the "average" amount that scores differ from the mean and from each other. Individual scores always differ much more than their means, but this still provides a frame of reference. For example, if individual scores differ by an "average" of 20, then we know that many large differences among scores occur in this situation. Therefore, a difference of 3 between two samples of such scores is not all that impressive. However, say that scores differ by an "average" of only 5. Because smaller differences occur in this situation, a difference between conditions of 3 is more impressive.

Thus, we *standardize* the difference between our sample means by comparing it to the population standard deviation. This is the logic behind the measure of effect size known as **Cohen's *d***: It measures effect size as the magnitude of the difference between the conditions, relative to the population standard deviation. We have two versions of how it is computed.

The formulas for Cohen's *d* are:

Independent-samples *t*-test	Related-samples *t*-test
$d = \dfrac{\overline{X}_1 - \overline{X}_2}{\sqrt{s^2_{pool}}}$	$d = \dfrac{\overline{D}}{\sqrt{s^2_D}}$

For the independent-samples *t*-test, the difference between the conditions is measured as $\overline{X}_1 - \overline{X}_2$ and the standard deviation comes from the square root of the pooled variance. For our hypnosis study, the means were 23 and 20, and s^2_{pool} was 8.3, so

$$d = \frac{\overline{X}_1 - \overline{X}_2}{\sqrt{s^2_{pool}}} = \frac{23 - 20}{\sqrt{8.3}} = \frac{+3}{2.88} = 1.04$$

This tells us that the effect of changing our conditions was to change scores by an amount that is slightly larger than one standard deviation.

For the related-samples *t*-test, the difference between the conditions is measured by $\overline{D}$ and the standard deviation comes from finding the square root of the estimated variance (s^2_D.) In our phobia study, $\overline{D} = +3.6$ and $s^2_D = 7.8$, so

$$d = \frac{\overline{D}}{\sqrt{s^2_D}} = \frac{+3.6}{\sqrt{7.8}} = \frac{+3.6}{2.79} = 1.29$$

Thus, the effect size of the therapy was 1.29.

We can interpret the above *d*s in two ways. First, the larger the *absolute* size of *d*, the larger the impact of the independent variable. In fact, Cohen[1] proposed the following interpretations when *d* is the neighborhood of the following amounts:

Values of d	Interpretation of Effect Size
d = .2	Small effect
d = .5	Medium effect
d = .8	Large effect

---

[1]Cohen, J. (1988) *Statistical power analysis for the behavioral sciences.* Hillsdale, NJ: Lawrence Erlbaum Associates.

Thus, above we found two *very* large effects. Second, we can compare the *relative* size of different *ds* to determine the relative impact of a variable. Above, the *d* for hypnosis was 1.04, but for therapy it was 1.29. Therefore, the therapy manipulation had a slightly larger impact. (*Note:* The difference between the conditions may produce a + or − that some researchers use to indicate the direction the scores change. Others think of *d* as the *amount* of impact the independent variable has, which cannot be negative.)

Another way to measure effect size is by computing the *proportion of variance accounted for.*

Effect Size Using Proportion of Variance Accounted For This approach measures effect size, not in terms of the *size* of the changes in scores but in terms of how *consistently* the scores change. Here, a variable has a greater impact, the more it "causes" everyone to behave in the same way, producing virtually the same score for everyone in a particular condition. This then is an important variable, because *by itself*, it pretty much controls the score (and behavior) that everyone exhibits. A variable is more minor if it exhibits less control of a behavior and scores.

We measure this effect by measuring the "proportion of variance accounted for." Recall from Chapter 5 that variance reflects differences in scores and that when we predict scores, we "account for variance." In Chapter 8, we saw that the proportion of variance accounted for was the proportional improvement achieved when we use a relationship to predict scores compared to when we do not use the relationship to predict scores. In an experiment, the scores we predict are the means of the conditions. Thus, in an experiment, the *proportion of variance accounted for* is the proportional improvement achieved when we use the mean of a condition as the predicted score of participants tested in that condition compared to when we do not use this approach. Put simply it is the extent to which individual scores in each condition are close to the mean of the condition, so if we predict the mean for someone, we are close to his or her actual score. When the independent variable has more control of a behavior, everyone in a condition will score more consistently. Then scores will be closer to the mean, so we will have a greater improvement in accurately predicting the scores, producing a larger proportion of variance accounted for. On the other hand, when the variable produces very different, inconsistent scores in each condition, our ability to predict them is not improved by much, and so little of the variance will be accounted for.

> *REMEMBER* The larger the proportion of variance accounted for, the greater the effect size of the independent variable in terms of consistently changing scores, so the more important the variable is.

In Chapter 8, we saw that the computations for the proportion of variance accounted for are performed by computing the squared correlation coefficient. For the two-sample experiment, we compute a new correlation coefficient and then square it. The **squared point-biserial correlation coefficient** indicates the proportion of variance accounted for in a two-sample experiment. Its symbol is r^2_{pb}. This can produce a proportion as a low as 0 (when the variable has no effect) to as high as 1.0 (when the variable perfectly controls scores so that we can accurately predict 100% of them). In real research, however, a variable typically accounts for between about 10% and 30% of the variance, with more than 30% being a *very* substantial amount.

> ### The formula for computing r_{pb}^2 is
>
> $$r_{pb}^2 = \frac{(t_{obt})^2}{(t_{obt})^2 + df}$$

This formula is used with either the independent-samples or related-samples t-test. In either case, start with your significant t_{obt} and square it. Then, for independent samples, $df = (n_1 - 1) + (n_2 - 1)$ For related samples, $df = N - 1$.

   In our hypnosis study, $t_{obt} = +2.93$ with $df = 30$, so,

$$r_{pb}^2 = \frac{(t_{obt})^2}{(t_{obt})^2 + df} = \frac{(2.93)^2}{(2.93)^2 + 30} = \frac{8.585}{8.585 + 30} = \frac{8.585}{38.585} = .22$$

Thus, on average, we are 22% closer to predicting participants' recall scores when we predict the mean of each hypnosis condition for them, compared to when we ignore this relationship. Hypnosis is not of major importance here, because scores are not consistently very close to the mean in each condition. Further, these inconsistent scores have a cause, so other, hidden variables must be causing them (perhaps IQ, memory ability, or motivation are operating). Therefore, hypnosis is only one of a number of variables that play a role here, and, thus, it is only somewhat important in determining recall.

   On the other hand, in the phobia study, $t_{obt} = +2.88$ and $df = 4$, so

$$r_{pb}^2 = \frac{(t_{obt})^2}{(t_{obt})^2 + df} = \frac{(2.88)^2}{(2.88)^2 + 4} = \frac{8.294}{12.294} = .67$$

Thus, the presence/absence of therapy accounts for 67% of the variance in fear scores. This variable plays a substantial role in determining these scores. Further, fewer other variables need to be considered in order to completely predict scores, so this is an important relationship for understanding phobias and the therapy.

   We also use the proportion of variance accounted for to compare the relationships from different studies. Thus, the role of therapy in determining fear scores (at 67%) is about three times larger than the role of hypnosis in determining recall scores (which was only 22%).

> *REMEMBER*   Compute *effect size* by computing d or r_{pb}^2 to determine the size of the impact of an independent variable on dependent scores.

STATISTICS IN PUBLISHED RESEARCH: THE TWO-SAMPLE EXPERIMENT

Report the results of either two-sample t-test using the same format used previously, but, remember, we also report the mean and standard deviation of each group, the confidence interval, and effect size. Thus, a published report of our independent-samples hypnosis study might say,

> **"The hypnosis group ($M = 23.00$, $SD = 3.00$) produced significantly higher recall scores than did the control group ($M = 20.00$, $SD = 2.74$), with $t(30) = +2.93$, $p < .05$. The 95% confidence interval for the difference is 0.91 to 5.09. The effect size of the hypnosis manipulation was $r_{pb}^2 = .22$."**

   For the phobia study, you would see a similar statement that included the means of the raw fear scores in each condition.

PUTTING IT ALL TOGETHER

Remember that one of your goals in this course is to learn *when* to use different statistical procedures. Obviously, you perform the independent-samples *t*-test if you've created two independent samples and the related-samples *t*-test if you've created two related samples.

In both procedures, if t_{obt} is not significant, consider whether you have sufficient power. If t_{obt} is significant, then focus on the means from each condition so that you summarize the typical score—and typical behavior—found in each condition. Use effect size to gauge how big a role the independent variable plays in determining the behaviors. Finally, interpret the relationship in terms of the underlying behaviors and causes that it reflects. (This last step will become easier as you learn more about the behavioral sciences in your other courses.)

Using the SPSS Appendix See Appendix B.6 to use SPSS to perform the independent-samples *t*-test or the related-samples *t*-test (but it is called the "paired-samples" *t*-test). For either, the program indicates the α at which t_{obt} is significant, but for a two-tailed test only. It also computes the descriptive statistics for each condition and automatically computes the confidence interval for either $\mu_1 - \mu_2$ or μ_D. The program does not compute *d* or r_{pb}^2.

CHAPTER SUMMARY

1. Two samples are *independent* when participants are randomly selected for each, without regard to who else has been selected, and each participant is in only one condition.

2. The *independent-samples* t-test requires (a) two independent samples, (b) normally distributed interval or ratio scores, and (c) homogeneous variance.

3. *Homogeneity of variance* means that the variances in the populations being represented are equal.

4. The *confidence interval for the difference between two μs* contains a range of differences between two μs, one of which is likely to be represented by the difference between our two sample means.

5. Two samples are *related* either when we *match* each participant in one condition to a participant in the other condition, or when we use *repeated measures* of one group of participants tested under both conditions.

6. The *confidence interval for* μ_D contains a range of values of μ_D, any one of which is likely to be represented by the sample's $\overline{D}$.

7. The power of a two-sample t-test increases with (a) larger differences in scores *between* the conditions, (b) smaller variability of scores *within* each condition, and (c) larger *n*s The related-samples t-test is more powerful than the independent-samples t-test.

8. *Effect size* indicates the amount of influence that changing the conditions of the independent variable had on the dependent scores.

9. *Cohen's d* measures effect size as the magnitude of the difference between the conditions.

10. The *proportion of variance accounted for* (computed as r_{pb}^2) measures effect size as the *consistency* of scores produced within each condition. The larger the proportion, the more accurately the mean of a condition predicts individual scores in that condition.

KEY TERMS

n N s_{pool}^2 $s_{\overline{X}_1 - \overline{X}_2}$ $\overline{D}$ s_D^2 $s_{\overline{D}}$ d r_{pb}^2

Cohen's *d* 281
Confidence interval for μ_D 277
confidence interval for the difference
   between two μs 269
effect size 280
homogeneity of variance 262
independent samples 262
independent-samples *t*-test 262
matched-samples design 271
pooled variance 265
related samples 271

related-samples *t*-test 271
repeated-measures design 271
sampling distribution of differences
   between the means 263
sampling distribution of mean
   differences 274
standard error of the difference 265
standard error of the mean
   difference 275
squared point-biserial correlation
   coefficient 282

REVIEW QUESTIONS

(Answers for odd-numbered questions are in Appendix D.)

1. A scientist has conducted a two-sample experiment. (a) What two parametric procedures are available to him? (b) What is the deciding factor for selecting between them?

2. How do you create independent samples?

3. (a) What are the two ways to create related samples? (b) What other assumptions must be met before using either two-sample *t*-test?

4. What is homogeneity of variance?

5. (a) What is $s_{\overline{X}_1 - \overline{X}_2}$? (b) What is $s_{\overline{D}}$? (c) What is the difference between n and N?

6. All other things being equal, should you create a related-samples or an independent-samples design? Why?

7. What does the confidence interval for μ_D indicate?

8. What does a confidence interval for the difference between two μs indicate?

9. (a) What does effect size indicate? (b) What does *d* indicate? (c) What does r_{pb}^2 indicate?

10. (a) What is the final step when completing an experiment? (b) Why is effect size useful at this stage?

11. Foofy obtained a statistically significant two-sample t_{obt}. What three things should she do to complete her analysis?

APPLICATION QUESTIONS

12. For the following, which type of *t*-test is required? (a) Studying the effects of a memory drug on Alzheimer's patients, testing a group of patients before and after administration of the drug. (b) Studying whether men and women rate the persuasiveness of an argument delivered by a female speaker differently. (c) The study described in part (b), but with the added requirement that for each man of a particular age, there is a woman of the same age.

13. We study the relationship between hot or cold baths and the amount of relaxation they produce. The relaxation scores from two independent samples are

    Sample 1 (hot): $\bar{X} = 43, s_X^2 = 22.79, n = 15$
    Sample 2 (cold): $\bar{X} = 39, s_X^2 = 24.6, n = 15$

    (a) What are H_0 and H_a? (b) Compute t_{obt} (c) With $\alpha = .05$, what is t_{crit}? (d) What should we conclude about this relationship? (e) Compute the confidence interval for the difference between the μs. (f) Using our two approaches, how big of an effect does bath temperature have on relaxation? (g) Describe how you would graph these results.

14. We investigate if a period of time feels longer or shorter when people are bored compared to when they are not bored. Using independent samples, we obtain these estimates of the time period (in minutes):

    Sample 1 (bored): $\bar{X} = 14.5, s_X^2 = 10.22, n = 28$
    Sample 2 (not bored): $\bar{X} = 9.0, s_X^2 = 14.6, n = 34$

    (a) What are H_0 and H_a? (b) Compute t_{obt} (c) With $\alpha = .05$, what is t_{crit}? (d) What should the researcher conclude about this relationship? (e) Compute the confidence interval for the difference between the μs. (f) Using our two approaches, how important is boredom in determining how quickly time seems to pass?

15. A researcher asks if people score higher or lower on a questionnaire measuring their well-being when they are exposed to much sunshine compared to when they're exposed to little sunshine. A sample of 8 people is measured under both levels of sunshine and produces these well-being scores:

    Low: 14   13   17   15   18   17   14   16
    High: 18   12   20   19   22   19   19   16

    (a) Subtracting low from high, what are H_0 and H_a? (b) Compute t_{obt}. (c) With $\alpha = .05$, what do you conclude about this study? (d) Compute the appropriate confidence interval. (e) What is the predicted well-being score for someone when tested under low sunshine? Under high sunshine? (f) On average, how much more accurate are these predictions than if you did not know how much sunshine people experience? (g) How scientifically important are these results?

16. A researcher investigates whether classical music is more or less soothing to air-traffic controllers than modern music. She plays a classical selection to one group and a modern selection to another. She gives each person an irritability questionnaire and obtains the following:

    Sample A (classical): $n = 6, \bar{X} = 14.69, s_X^2 = 8.4$
    Sample B (modern): $n = 6, \bar{X} = 17.21, s_X^2 = 11.6$

(a) Subtracting $A - B$, what are H_0 and H_a? (b) What is t_{obt}? (c) With $\alpha = .05$, are the results significant? (d) Report the results using the correct format. (e) What should she conclude about the relationship in nature between type of music and irritability? (f) What other statistics should be computed? (g) What statistical flaw is likely in the experiment?

17. We predict that children exhibit more aggressive acts after watching a violent television show. The scores for ten participants before and after watching the show are

Sample 1 (After)	Sample 2 (Before)
5	4
6	6
4	3
4	2
7	4
3	1
2	0
1	0
4	5
3	2

(a) Subtracting before from after, what are H_0 and H_a? (b) Compute t_{obt} (c) With $\alpha = .05$, what is t_{crit}? (d) What should the researcher conclude about this relationship? (e) Compute the appropriate confidence interval. (f) How large is the effect of violence in terms of the *difference* it produces in aggression scores?

18. You investigate whether the older or younger male in pairs of brothers tends to be more extroverted. You obtain the following extroversion scores:

Sample 1 (Younger)	Sample 2 (Older)
10	18
11	17
18	19
12	16
15	15
13	19
19	13
15	20

(a) What are H_0 and H_a? (b) Compute t_{obt} (c) With $\alpha = .05$, what is t_{crit}? (d) What should you conclude about this relationship? (e) Which of our approaches should we use to determine if this a scientifically important relationship?

19. A rather dim student proposes testing the conditions of "male" and "female" using a repeated-measures design. What's wrong with this idea?

20. With $\alpha = .05$ and $df = 40$, a significant independent-samples t_{obt} was $+4.55$. How would you report this in the literature?

21. An experimenter investigated the effects of a sensitivity training course on a policeman's effectiveness at resolving domestic disputes (using independent

samples who had or had not completed the course). The dependent variable was the ability to resolve a domestic dispute. These success scores were obtained:

No Course	Course
11	13
14	16
10	14
12	17
8	11
15	14
12	15
13	18
9	12
11	11

(a) Should a one-tailed or a two-tailed test be used? (b) What are H_0 and H_a? (c) Subtracting course from no course, compute t_{obt} and determine whether it is significant. (d) Compute the confidence interval for the difference between the μs. (e) What conclusions can the experimenter draw from these results? (f) Using our two approaches, compute the effect size and interpret it.

22. When reading a research article, you encounter the following statements. For each, identify the N, the design, the statistical procedure performed and the result, the relationship, and if a Type I or Type II error is possibly being made. (a) "The *t*-test indicated a significant difference between the mean for men ($M = 5.4$) and for women ($M = 9.3$), with $t(58) = 7.93, p < .01$. Unfortunately, the effect size was only .08." (b) "The *t*-test indicated that participants' weights after three weeks of dieting were significantly reduced relative to the pretest measure, with $t(40) = 3.56, p < .05$, and $r_{pb}^2 = .16$."

INTEGRATION QUESTIONS

23. How do you distinguish the independent variable from the dependent variable? (Ch. 2)

24. What is the difference between an experiment versus a correlational study in terms of (a) the design? (b) How we examine the relationship? (c) How sampling error might play a role? (Chs. 2, 4, 7, 11, 12)

25. (a) When do you perform parametric inferential procedures in experiments? (b) What are the four parametric inferential procedures for experiments that we have discussed and what is the design in which each is used? (Chs. 10, 11, 12)

26. In recent chapters, you have learned about three different versions of a confidence interval. (a) What are they called? (b) How are all three similar in terms of what they communicate? (c) What are the differences between them? (Chs. 11, 12)

27. (a) What does it mean to "account for variance"? (b) How do we predict scores in an experiment? (c) Which variable in an experiment is potentially the good predictor and important? (d) When does that occur? (Chs. 5, 7, 8, 12)

28. (a) In an experiment, what are the three ways to try to maximize power? (b) What does maximizing power do in terms of our errors? (c) For what outcome is it most important for us to have maximum power and why? (Chs. 10, 11, 12)

29. You have performed a one-tailed t-test. When computing a confidence interval, should you use the one-tailed or two-tailed t_{crit}? (Chs. 11, 12)

30. For the following, identify the inferential procedure to perform and the key information for answering the research question. *Note*: If no inferential procedure is appropriate, indicate why. (a) Ten students are tested for accuracy of distance estimation when using one or both eyes. (b) We determine that the average number of cell phone calls in a sample of college students is 7.2 per hour. We want to describe the likely national average (μ) for this population. (c) We compare children who have siblings to those who do not, rank ordering the children in terms of their willingness to share their toys. (d) Two gourmet chefs have rank ordered the 10 best restaurants in town. How consistently do they agree? (e) We measure the influence of sleep deprivation on driving performance for groups having 4 or 8 hours sleep. (f) We test whether wearing black uniforms produces more aggression by comparing the mean number of penalties a hockey team receives per game when wearing black to the league average for teams with non-black uniforms. (Chs. 11, 12)

■ ■ ■ SUMMARY OF FORMULAS

1. To perform the independent samples t-test:

$$s_X^2 = \frac{\Sigma X^2 - \dfrac{(\Sigma X)^2}{N}}{N - 1}$$

$$s_{pool}^2 = \frac{(n_1 - 1)s_1^2 + (n_2 - 1)s_2^2}{(n_1 - 1) + (n_2 - 1)}$$

$$s_{\overline{X}_1 - \overline{X}_2} = \sqrt{s_{pool}^2 \left(\frac{1}{n_1} + \frac{1}{n_2} \right)}$$

$$t_{obt} = \frac{(\overline{X}_1 - \overline{X}_2) - (\mu_1 - \mu_2)}{s_{\overline{X}_1 - \overline{X}_2}}$$

$$df = (n_1 - 1) + (n_2 - 1)$$

2. The formula for the confidence interval for the difference between two μs is

$$(s_{\overline{X}_1 - \overline{X}_2})(-t_{crit}) + (\overline{X}_1 - \overline{X}_2) \le \mu_1 - \mu_2 \le$$
$$(s_{\overline{X}_1 - \overline{X}_2})(+t_{crit}) + (\overline{X}_1 - \overline{X}_2)$$

3. To perform the related samples t-test:

$$s_D^2 = \frac{\Sigma D^2 - \dfrac{(\Sigma D)^2}{N}}{N - 1}$$

$$s_{\overline{D}} = \sqrt{\frac{s_D^2}{N}}$$

$$t_{obt} = \frac{\overline{D} - \mu_D}{s_{\overline{D}}}$$

$$df = N - 1$$

4. The formula for the confidence interval for μ_D is

$$(s_{\overline{D}})(-t_{crit}) + \overline{D} \le \mu_D \le (s_{\overline{D}})(+t_{crit}) + \overline{D}$$

5. The formula for Cohen's d for independent samples is

$$d = \frac{\overline{X}_1 - \overline{X}_2}{\sqrt{s_{pool}^2}}$$

6. The formula for Cohen's d for related samples is

$$d = \frac{\overline{D}}{\sqrt{s_D^2}}$$

7. The formula for r_{pb}^2 is

$$r_{pb}^2 = \frac{(t_{obt})^2}{(t_{obt})^2 + df}$$

With independent samples, $df = (n_1 - 1) + (n_2 - 1)$ With related samples, $df = N - 1$.

13

The One-Way Analysis of Variance

GETTING STARTED

To understand this chapter, recall the following:

- From Chapter 5, variance indicates variability, which is the differences between scores. Also, it is called "error variance."
- From Chapter 10, why we limit the probability of a Type I error to .05.
- From Chapter 12, what independent and related samples are, how we "pool" the sample variances to estimate the variance in the population, and what effect size is and why it is important.

Your goals in this chapter are to learn

- The terminology of *analysis of variance*.
- When and how to compute F_{obt}.
- Why F_{obt} should equal 1 if H_0 is true, and why it is greater than 1 if H_0 is false.
- When to compute *Fisher's protected* t-*test* or *Tukey's HSD*.
- How *eta squared* describes effect size.

Believe it or not, we have only one more common inferential procedure to learn and it is called the *analysis of variance*. This is the parametric procedure used in experiments involving more than two conditions. This chapter will show you (1) the general logic behind the analysis of variance, (2) how to perform this procedure for one common design, and (3) how to perform an additional analysis called *post hoc comparisons*.

NEW STATISTICAL NOTATION

The analysis of variance has its own language that is also commonly used in research publications:

1. Analysis of variance is abbreviated as **ANOVA.**
2. An independent variable is also called a **factor.**
3. Each condition of the independent variable is also called a **level,** or a **treatment,** and differences in scores (and behavior) produced by the independent variable are a **treatment effect.**
4. The symbol for the number of levels in a factor is k.

WHY IS IT IMPORTANT TO KNOW ABOUT ANOVA?

It is important to know about analysis of variance because it is *the* most common inferential statistical procedure used in experiments. Why? Because there are actually many versions of ANOVA, so it can be used with many different designs: It can be applied to an experiment involving independent samples or related samples, to an independent variable involving any number of conditions, and to a study involving any number of independent variables. Such complex designs are common because, first, the hypotheses of the study may require comparing more than two conditions of an independent variable. Second, researchers often add more conditions because, after all of the time and effort involved in creating two conditions, little more is needed to test additional conditions. Then we learn even more about a behavior (which *is* the purpose of research). Therefore, you'll often encounter the ANOVA when conducting your own research or when reading about the research of others.

AN OVERVIEW OF ANOVA

Because different versions of ANOVA are used depending on the design of a study, we have important terms for distinguishing among them. First, a **one-way ANOVA** is performed when only one independent variable is tested in the experiment (a two-way ANOVA is used with two independent variables, and so on). Further, different versions of the ANOVA depend on whether participants were tested using independent or related samples. However, in earlier times participants were called "subjects," and in ANOVA, they still are. Therefore, when an experiment tests a factor using independent samples in all conditions, it is called a **between-subjects factor** and requires the formulas from a **between-subjects ANOVA.** When a factor is studied using related samples in all levels, it is called a **within-subjects factor** and involves a different set of formulas called a **within-subjects ANOVA.** In this chapter, we'll discuss the one-way, between-subjects ANOVA. (The slightly different formulas for a one-way, within-subjects ANOVA are presented in Appendix A.3.)

As an example of this type of design, say we conduct an experiment to determine how well people perform a task, depending on how difficult they believe the task will be (the "perceived difficulty" of the task). We'll create three conditions containing the unpowerful n of five participants each and provide them with the same easy ten math problems. However, we will tell participants in condition 1 that the problems are easy, in condition 2 that the problems are of medium difficulty, and in condition 3 that the problems are difficult. Thus, we have three *levels* of the *factor* of perceived difficulty. Our dependent variable is the number of problems that participants then correctly solve within an allotted time. If participants are tested under only one condition and we do not match them, then this is a one-way, between-subjects design.

The way to diagram a one-way ANOVA is shown in Table 13.1. Each column is a level of the factor, containing the scores of participants tested under that condition (here symbolized by X). The symbol n stands for the number of scores in a condition, so here $n = 5$ per level. The mean of each level is the mean of the scores from that column. With three levels in this factor, $k = 3$. (Notice that the general format is to label the factor as factor A, with levels A_1, A_2, A_3, and so on.) The total number of scores in the experiment is N, and here $N = 15$. Further, the overall mean of all scores in the experiment is the mean of all 15 scores.

TABLE 13.1

Diagram of a Study
Having Three Levels
of One Factor

*Each column represents
a condition of the
independent variable.*

	Factor A: Independent Variable of Perceived Difficulty		
Level A₁: Easy	*Level A₂: Medium*	*Level A₃: Difficult*	← *Conditions* $k = 3$
X	X	X	
X	X	X	
X	X	X	
X	X	X	
X	X	X	
$\overline{X}_1$	$\overline{X}_2$	$\overline{X}_3$	Overall $\overline{X}$
$n_1 = 5$	$n_2 = 5$	$n_3 = 5$	$N = 15$

Although we now have three conditions, our purpose is still to demonstrate a relationship between the independent variable and the dependent variable. Ideally, we'll find a different mean for each condition, suggesting that if we tested the entire population under each level of difficulty, we would find three different populations of scores, located at three different μs. However, it's possible that we have the usual problem: Maybe the independent variable really does nothing to scores, the differences between our means reflect sampling error, and actually we would find the same population of scores, having the same μ, for all levels of difficulty. Therefore, as usual, before we can conclude that a relationship exists, we must eliminate the idea that our sample means poorly represent that no relationship exists. The **analysis of variance** is the parametric procedure for determining whether significant differences occur in an experiment containing two or more conditions. Thus, when you have only two conditions, you can use either a two-sample *t*-test or the ANOVA: You'll reach exactly the same conclusions, and both have the same probability of making Type I and Type II errors. However, you *must* use ANOVA when you have more than two conditions (or more than one independent variable).

Otherwise, the ANOVA has the same assumptions as previous parametric procedures. Performing a one-way, between-subjects ANOVA is appropriate when

1. The experiment has only one independent variable and all conditions contain independent samples.

2. The dependent variable measures normally distributed interval or ratio scores.

3. The variances of the populations are homogeneous.

Although the *n*s in all conditions need not be equal, violations of the assumptions are less serious when the *n*s are equal. Also, the procedures are *much* easier to perform with equal *n*s.

How ANOVA Controls the Experiment-Wise Error Rate

You might be wondering why we even need ANOVA. Couldn't we use the independent-samples *t*-test to test for significant differences among the three means above? That is, we might test whether $\overline{X}_1$ differs from $\overline{X}_2$, then whether $\overline{X}_2$ differs from $\overline{X}_3$, and finally whether $\overline{X}_1$ differs from $\overline{X}_3$. We cannot use this approach because of the resulting probability of making a Type I error (rejecting a true H_0).

To understand this, we must first distinguish between making a Type I error *when comparing a pair of means* as in the *t*-test, and making a Type I error *somewhere in the experiment* when there are more than two means. In our example, we have three means so we can make a Type I error when comparing $\overline{X}_1$ to $\overline{X}_2$, $\overline{X}_2$ to $\overline{X}_3$, or $\overline{X}_1$ to $\overline{X}_3$. Technically, when we set $\alpha = .05$, it is the probability of making a Type I error when we make *one* comparison. However, it also defines what we consider to be an acceptable probability of making a Type I error *anywhere* in an experiment. The probability of making a Type I error *anywhere* among the comparisons in an experiment is called the **experiment-wise error rate.**

We can use the *t*-test when comparing the only two means in an experiment because with only one comparison, the experiment-wise error rate equals α. But with more than two means in the experiment, performing multiple *t*-tests results in an experiment-wise error rate that is much larger than α. Because of the importance of avoiding Type I errors, however, we do not want the error rate to be larger than we think it is, and it should never be larger than .05. Therefore, we perform ANOVA because then our actual experiment-wise error rate will equal the alpha we've chosen.

> *REMEMBER* The reason for performing ANOVA is that it keeps the *experiment-wise error rate* equal to α.

Statistical Hypotheses in ANOVA

ANOVA tests only two-tailed hypotheses. The null hypothesis is that there are no differences between the populations represented by the conditions. Thus, for our perceived difficulty study, with the three levels of easy, medium, and difficult, we have

$$H_0: \mu_1 = \mu_2 = \mu_3$$

In general, for any ANOVA with k levels, the null hypothesis is $H_0: \mu_1 = \mu_2 = \cdots = \mu_k$. The "$\cdots = \mu_k$" indicates that there are as many μs as there are levels.

You might think that the alternative hypothesis would be $\mu_1 \neq \mu_2 \neq \mu_3$. However, a study may demonstrate differences between *some* but not *all* conditions. For example, perceived difficulty may only show differences in the population between our easy and difficult conditions. To communicate this idea, the alternative hypothesis is

$$H_a: \text{not all } \mu\text{s are equal}$$

H_a implies that a relationship is present because two or more of our levels represent different populations.

As usual, we test H_0, so ANOVA always tests whether all of our level means represent the same μ.

The Order of Operations in ANOVA: The *F* Statistic and Post Hoc Comparisons

The statistic that forms the basis for ANOVA is *F*. We compute F_{obt}, which we compare to F_{crit}, to determine whether any of the means represent different μs.

When F_{obt} is not significant, it indicates no significant differences between any means. Then the experiment has failed to demonstrate a relationship, we are finished with the statistical analyses, and it's back to the drawing board.

When F_{obt} is significant, it indicates only that *somewhere* among the means *two* or more of them differ significantly. However, F_{obt} does not indicate *which* specific means differ significantly. Thus, if F_{obt} for the perceived difficulty study is significant, we will know that we have significant differences somewhere among the means of the easy, medium, and difficult levels, but we won't know where they are.

Therefore, when F_{obt} is significant we perform a second statistical procedure, called post hoc comparisons. **Post hoc comparisons** are like *t*-tests in which we compare all possible *pairs* of means from a factor, one pair at a time, to determine which means differ significantly. Thus, for the difficulty study, we'll compare the means from easy and medium, from easy and difficult, and from medium and difficult. Then we will know which of our level means actually differ significantly from each other. By performing post hoc comparisons after F_{obt} is significant, we ensure that the experiment-wise probability of a Type I error equals our α.

> *REMEMBER* If F_{obt} is significant, then perform *post hoc comparisons* to determine which specific means differ significantly.

The one exception to this rule is when you have only two levels in the factor. Then the significant difference indicated by F_{obt} must be between the only two means in the study, so it is unnecessary to perform post hoc comparisons.

A QUICK REVIEW

- The one-way ANOVA is performed when testing two or more conditions from one independent variable.
- A significant F_{obt} followed by post hoc comparisons indicates which level means differ significantly, with the experiment-wise error rate equal to α.

MORE EXAMPLES

We measure the mood of participants after they have won $0, $10, or $20 in a rigged card game. With one independent variable, a *one-way* design is involved, and the *factor* is the amount of money won. The *levels* are $0, $10, or $20. If independent samples receive each *treatment,* we perform the *between-subjects ANOVA.* (Otherwise, perform the *within-subjects ANOVA.*) A significant F_{obt} will indicate that at least two of the conditions produced significant differences in mean mood scores. Perform *post hoc comparisons* to determine which levels differ significantly, comparing the mean mood scores for $0 vs. $10, $0 vs. $20, and $10 vs. $20. The probability of a Type I error in the study—the *experiment-wise error rate*—equals α.

For Practice

1. A study involving one independent variable is a _____ design.
2. Perform the _____ when a study involves independent samples; perform the _____ when it involves related samples.
3. An independent variable is also called a _____, and a condition is also called a _____, or _____.
4. The _____ will indicate whether any of the conditions differ, and then the _____ will indicate which specific conditions differ.
5. The probability of a Type I error in the study is called the _____.

Answers

1. one-way
2. between-subjects ANOVA; within-subjects ANOVA
3. factor; level; treatment
4. F_{obt}; post hoc comparisons
5. experiment-wise error rate

UNDERSTANDING THE ANOVA

The logic and components of all versions of the ANOVA are very similar. In each case, the analysis of variance does exactly that—it "analyzes variance." This is the same concept of variance that we've talked about since Chapter 5. But we do not *call* it variance.

Instead, ANOVA has its own terminology. We begin with the formula for the estimated population variance:

$$S_X^2 = \frac{\Sigma(X - \overline{X})^2}{N - 1} = \frac{sum\ of\ squares}{degrees\ of\ freedom} = \frac{SS}{df} = mean\ square = MS$$

In the numerator we find the sum of the squared deviations between the mean and each score. In ANOVA, the "sum of the squared deviations" is shortened to **sum of squares**, which is abbreviated as *SS*. In the denominator we divide by $N - 1$, which is our *degrees of freedom* or *df*. Recall that dividing the sum of the squared deviations by *df* produces something like the average or the *mean of the squared deviations*. In ANOVA, this is shortened to **mean square**, which is symbolized as *MS*.

Thus, when we compute *MS* we *are* computing an estimate of the population variance. In the ANOVA we compute variance from two perspectives, called the *mean square within groups* and the *mean square between groups*.

The Mean Square within Groups

The **mean square within groups** describes the variability of scores *within* the conditions of an experiment. Its symbol is MS_{wn}. You can conceptualize the computation of MS_{wn} as shown in Table 13.2: First, we find the variance in level 1 (finding how much the scores in level 1 differ from $\overline{X}_1$), then we find the variance of scores in level 2 around $\overline{X}_2$, and then we find the variance of scores in level 3 around $\overline{X}_3$. Although each sample provides an estimate of the population variability, we obtain a better estimate by averaging or "pooling" them together, like we did in the independent-samples *t*-test. Thus, the MS_{wn} is the "average" variability of the scores in each condition.

We compute MS_{wn} by looking at the scores *within* one condition at a time, so the differences among those scores are not influenced by our independent variable. Therefore, the value of MS_{wn} stays the same, regardless of whether H_0 is true or false. Either way, MS_{wn} is an estimate of the variance of scores in the population (σ_X^2). If H_0 is true and we are representing one population, then MS_{wn} estimates the σ_X^2 of that population. If H_a is correct and we are representing more than one population, because of homogeneity of variance, MS_{wn} estimates the one value of σ_X^2 that would occur in each population.

> *REMEMBER* The MS_{wn} is an estimate of the variability of individual scores in the population.

The Mean Square between Groups

The other variance computed in ANOVA is the mean square between groups. The **mean square between groups** describes the variability *between* the means of our

TABLE 13.2

How to Conceptualize the Computation of MS_{wn}

Here, we find the difference between each score in a condition and the mean of that condition.

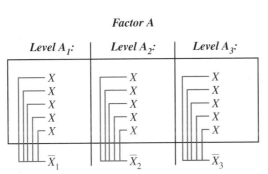

levels. It is symbolized by MS_{bn}. You can conceptualize the computation of MS_{bn} as shown in Table 13.3.

We compute variability as the differences between a set of scores and their mean, so here, we treat the level means as scores, finding the "average" amount they deviate from their mean, which is the overall mean of the experiment. In the same way that the deviations of raw scores around the mean describe how different the scores are from each other, the deviations of the sample means around the overall mean describe how different the sample means are from each other.

Thus, MS_{bn} is our way of measuring how much the means of our levels differ from each other. This serves as an estimate of the differences between sample means that would be found in *one* population. That is, we are testing the H_0 that our data all come from the same, one population. If so, sample means from that population will not necessarily equal μ or each other every time, because of sampling error. Therefore, when H_0 is true, the differences between our means as measured by MS_{bn} will not be zero. Instead, MS_{bn} is an estimate of the "average" amount that sample means from that one population differ from each other due to chance, sampling error.

> *REMEMBER* The MS_{bn} describes the differences between our means as an estimate of the differences between means found in the population.

As we'll see, performing the ANOVA involves first using our data to compute the MS_{wn} and MS_{bn}. The final step is to then compare them by computing F_{obt}.

Comparing the Mean Squares: The Logic of the *F*-Ratio

The test of H_0 is based on the fact that statisticians have shown that when samples of scores are selected from one population, the size of the differences among the sample means will *equal* the size of the differences among individual scores. This makes sense because how much the sample means differ depends on how much the individual scores differ. Say that the variability in the population is small so that all scores are very close to each other. When we select samples of such scores, we will have little variety in scores to choose from, so each sample will contain close to the same scores as the next and their means also will be close to each other. However, if the variability is very large, we have many different scores available. When we select samples of these scores, we will often encounter a very different batch each time, so the means also will be very different each time.

> *REMEMBER* In one population, the variability of sample means will equal the variability of individual scores.

TABLE 13.3

How to Conceptualize the Computation of MS_{bn}

Here, we find the difference between the mean of each condition and the overall mean of the study.

	Factor A	
Level A₁:	*Level A₂:*	*Level A₃:*
X	X	X
X	X	X
X	X	X
X	X	X
X	X	X
$\overline{X}_1$	$\overline{X}_2$	$\overline{X}_3$ Overall $\overline{X}$

Here is the key: the MS_{bn} estimates the variability of sample means in the population and the MS_{wn} estimates the variability of individual scores in the population. We've just seen that when we are dealing with only one population, sample means and individual scores will differ to the same degree. Therefore, when we are dealing with only one population, the MS_{bn} should equal MS_{wn}: the answer we compute for MS_{bn} should be the same answer as for MS_{wn}. *Our H_0 always says that we are dealing with only one population, so if H_0 is true for our study, then our MS_{bn} should equal our MS_{wn}.*

An easy way to determine if two numbers are equal is to make a fraction out of them, which is what we do when computing F_{obt}.

The formula for F_{obt} is

$$F_{obt} = \frac{MS_{bn}}{MS_{wn}}$$

This fraction is referred to as the *F-ratio*. The **F-ratio** equals the MS_{bn} divided by the MS_{wn}. (The MS_{bn} is always on top!)

If we place the same number in the numerator as in the denominator, the ratio will equal 1. Thus, when H_0 is true and we are representing one population, the MS_{bn} should equal the MS_{wn}, so F_{obt} should equal 1. Or, conversely, when F_{obt} equals 1, it tells us that H_0 is true.

Of course F_{obt} may not equal 1 *exactly* when H_0 is true, because we may have sampling error in either MS_{bn} or MS_{wn}. That is, either the differences among our individual scores and/or among our level means may be "off" in representing the corresponding differences in the population. Therefore, realistically, we expect that, if H_0 is true, F_{obt} will equal 1 or at least will be close to 1. In fact, if F_{obt} is less than 1, mathematically it can only be that H_0 is true and we have sampling error in representing this. (Each MS is a variance, in which we square differences, so F_{obt} cannot be negative.)

It gets interesting, however, as F_{obt} becomes *larger* than 1. No matter what our data show, H_0 implies that F_{obt} is "trying" to equal 1, and if it does not, it's because of sampling error. Let's think about that. If $F_{obt} = 2$, it is twice what H_0 says it should be, although according to H_0, we should conclude "No big deal—a little sampling error." Or, say that $F_{obt} = 4$, so the MS_{bn} is *four times* the size of MS_{wn} (and F_{obt} is four times what it should be). Yet, H_0 says that MS_{bn} would have equaled MS_{wn} but by chance we happened to get a few unrepresentative scores. If F_{obt} is, say, 10, then it, and the MS_{bn}, are *ten times* what H_0 says they should be! Still, H_0 says this is because we had a little bad luck in representing the population.

As this illustrates, the larger the F_{obt}, the more difficult it is to believe that our data are poorly representing the situation where H_0 is true. Of course, if sampling error won't explain so large an F_{obt}, then we need something else that will. The answer is our independent variable. When H_a is true so that changing our conditions produces different populations of scores, MS_{bn} will not equal MS_{wn}, and F_{obt} will not equal 1. Further, the more that changing the levels of our factor changes scores, the larger will be the differences between our level means, and so the larger will be MS_{bn}. However, recall that the value of MS_{wn} stays the same regardless of whether H_0 is true. Thus, greater differences produced by our factor will produce only a larger MS_{bn}, which produces a larger F_{obt}. Turning this around, the larger the F_{obt}, the more it appears that H_a is true. Putting this all together:

> **The larger the F_{obt}, the less likely it is that H_0 is true and the more likely it is that H_a is true.**

If our F_{obt} is large enough to be beyond F_{crit}, we will conclude that H_0 is so unlikely to be true that we will reject H_0 and accept H_a.

> **REMEMBER** If H_0 is true, F_{obt} should equal 1 or be close to 1. The larger the F_{obt}, the less likely that H_0 is true and the more likely that H_a is true.

Before moving on to the computations, we will briefly discuss the underlying components that F_{obt} represents in the population.

The Theoretical Components of the *F*-ratio

To fully understand the *F*-ratio, we need to understand what MS_{bn} and MS_{wn} represent in the population. We saw that MS_{wn} estimates the variance of individual scores in the population (σ_X^2). Statisticians also call this variance the **error *variance*,** symbolized by σ_{error}^2. Thus, the MS_{wn} is an estimate of σ_{error}^2. (The MS_{wn} is also referred to as the "error term" in the *F*-ratio.)

When H_0 is true and we have only one population, the MS_{bn} also estimates σ_{error}^2. We saw that with one population, the variability of sample means depends on the variability of individual scores. Thus, although MS_{bn} is computed using sample means, it ultimately reflects the variability among the scores, which is σ_{error}^2. Therefore, when H_0 is true, the reason that MS_{bn} should equal MS_{wn} is because both reflect the error variance in that one population. In symbols then, here is what the *F*-ratio represents in the population when H_0 is true.

$$\textit{Sample} \qquad \textit{Estimates} \qquad \textit{Population}$$
$$F_{obt} = \frac{MS_{bn}}{MS_{wn}} \qquad \begin{array}{c}\rightarrow\\\rightarrow\end{array} \qquad \frac{\sigma_{error}^2}{\sigma_{error}^2} = 1$$

Both mean squares are merely estimates of the one value of σ_{error}^2, so they should be equal, and so their ratio equals 1.

On the other hand, if H_0 is false and H_a is true, then more than one population is involved. By measuring the differences between the means of our conditions, MS_{bn} will measure this treatment effect. Statisticians refer to the differences between the populations produced by a factor as the **treatment variance,** which is symbolized as σ_{treat}^2. Thus, MS_{bn} is an estimate of σ_{treat}^2.

However, even if a factor does produce different populations, our samples will not perfectly represent them because of sampling error. Therefore, to some extent, the differences between our means, as measured by the MS_{bn}, will still reflect the variability in scores, which we call error variance. Thus, MS_{bn} estimates both treatment variance *plus* error variance. Altogether, here is what the *F*-ratio represents in the population when H_0 is false and H_a is true.

$$\textit{Sample} \qquad \textit{Estimates} \qquad \textit{Population}$$
$$F_{obt} = \frac{MS_{bn}}{MS_{wn}} \qquad \begin{array}{c}\rightarrow\\\rightarrow\end{array} \qquad \frac{\sigma_{error}^2 + \sigma_{treat}^2}{\sigma_{error}^2} = F > 1$$

In the denominator, the MS_{wn} is still the same estimate of σ_{error}^2. In the numerator, however, the larger the differences between the conditions, the larger is the σ_{treat}^2

component, and so the larger will be MS_{bn}. A larger numerator produces an F_{obt} that is greater than 1. Thus, when H_a is true, regardless of whether we have a positive, negative, or curvilinear relationship, MS_{bn} simply will be larger than MS_{wn}, so that F_{obt} is greater than 1. This is why we test only two-tailed hypotheses. (Technically, this formula is always used to describe what F_{obt} represents, even when H_0 is true. The MS_{bn} always estimates the amount of σ^2_{treat} that is present, but when H_0 is true, this amount is zero. This produces the equivalent of the previous formula where $F_{obt} = 1$.)

> *REMEMBER*   An F_{obt} equal to (or less than) 1 indicates that H_0 is true. An F_{obt} greater than 1 may result from sampling error, or it may indicate a treatment effect in the population. A significant F_{obt} indicates that the means from the conditions are likely to represent two or more μs.

PERFORMING THE ANOVA

Now we can discuss the computations involved in performing the ANOVA. Recall from the beginning of this chapter that we changed the formula for the variance into computing a *mean square* by dividing the *sum of squares* by the *degrees of freedom*. In symbols, this is

$$MS = \frac{SS}{df}$$

Adding subscripts, we will compute the mean square between groups (MS_{bn}) by computing the sum of squares between groups (SS_{bn}) and dividing by the degrees of freedom between groups (df_{bn}). Likewise, we will compute the mean square within groups (MS_{wn}) by computing the sum of squares within groups (SS_{wn}) and dividing by the degrees of freedom within groups (df_{wn}). Once we have MS_{bn} and MS_{wn}, we can compute F_{obt}.

If all this strikes you as the most confusing thing ever devised, you'll find an *ANOVA summary table* very helpful. Here is its general format:

Summary Table of One-Way ANOVA

Source	*Sum of Squares*	df	*Mean Square*	F
Between	SS_{bn}	df_{bn}	MS_{bn}	F_{obt}
Within	SS_{wn}	df_{wn}	MS_{wn}	
Total	SS_{tot}	df_{tot}		

The source column identifies each source of variation, either *between, within,* or *total.* In the following sections, we'll compute the components for the other columns.

Computing the F_{obt}

Say that we performed the perceived difficulty study discussed earlier, telling participants that some math problems are easy, of medium difficulty, or difficult and then measured the number of problems they solved. The data are presented in Table 13.4. As shown in the following sections, there are four parts to computing F_{obt}, finding (1) the sum of squares, (2) the degrees of freedom, (3) the mean squares, and (4) F_{obt}. So that you don't get lost, fill in the ANOVA summary table as you complete each step. (There *will* be a test later.)

Computing the Sums of Squares   The first task is to compute the sum of squares. Do this in four steps.

Step 1 *Compute the sums and means.* As in Table 13.4, compute ΣX, ΣX^2, and $\overline{X}$ for each level (each column). Then add together the ΣXs from all levels to get the total, which is ΣX_{tot}. Also, add together the ΣX^2s from all levels to get the total, which is ΣX^2_{tot}.

TABLE 13.4

Data from Perceived Difficulty Experiment

Factor A: Perceived Difficulty

Level A$_1$: *Easy*	*Level A$_2$:* *Medium*	*Level A$_3$:* *Difficult*	
9	4	1	
12	6	3	
4	8	4	
8	2	5	
7	10	2	*Totals*
$\Sigma X = 40$	$\Sigma X = 30$	$\Sigma X = 15$	$\Sigma X_{tot} = 85$
$\Sigma X^2 = 354$	$\Sigma X^2 = 220$	$\Sigma X^2 = 55$	$\Sigma X^2_{tot} = 629$
$n_1 = 5$	$n_2 = 5$	$n_3 = 5$	$N = 15$
$\overline{X}_1 = 8$	$\overline{X}_2 = 6$	$\overline{X}_3 = 3$	$k = 3$

Step 2 *Compute the total sum of squares* (SS_{tot}).

> **The formula for the total sum of squares is**
>
> $$SS_{tot} = \Sigma X_{tot}^2 - \left(\frac{(\Sigma X_{tot})^2}{N} \right)$$

The ΣX_{tot} is the sum of all Xs, and ΣX_{tot}^2 is the sum of all squared Xs. N is the total N in the study.

Using the data from Table 13.4, $\Sigma X_{tot}^2 = 629$, $\Sigma X_{tot} = 85$, and $N = 15$, so

$$SS_{tot} = 629 - \frac{(85)^2}{15}$$

so

$$SS_{tot} = 629 - \frac{7225}{15}$$

and

$$SS_{tot} = 629 - 481.67 = 147.33$$

Step 3 *Compute the sum of squares between groups* (SS_{bn}).

> **The formula for the sum of squares between groups is**
>
> $$SS_{bn} = \Sigma \left(\frac{(\text{Sum of scores in the column})^2}{n \text{ of scores in the column}} \right) - \left(\frac{(\Sigma X_{tot})^2}{N} \right)$$

Back in Table 13.4, each column represents a level of the factor. Thus, find the ΣX for a level, square the ΣX, and then divide by the n in that level. After doing this for all levels, add the results together and subtract the quantity $(\Sigma X_{tot})^2/N$. Thus, we have

$$SS_{bn} = \left(\frac{(40)^2}{5} + \frac{(30)^2}{5} + \frac{(15)^2}{5} \right) - \left(\frac{(85)^2}{15} \right)$$

so

$$SS_{bn} = (320 + 180 + 45) - 481.67$$

and

$$SS_{bn} = 545 - 481.67 = 63.33$$

Step 4 *Compute the sum of squares within groups* (SS_{wn}). We use a shortcut to compute SS_{wn}. Mathematically, SS_{tot} equals SS_{bn} plus SS_{wn}. Therefore, the total minus the between leaves the within.

> **The formula for the sum of squares within groups is**
>
> $$SS_{wn} = SS_{tot} - SS_{bn}$$

In the example, SS_{tot} is 147.33 and SS_{bn} is 63.33, so

$$SS_{wn} = 147.33–63.33 = 84.00$$

Filling in the first column of the ANOVA summary table, we have

Source	Sum of Squares	df	Mean Square	F
Between	63.33	df_{bn}	MS_{bn}	F_{obt}
Within	84.00	df_{wn}	MS_{wn}	
Total	147.33	df_{tot}		

As a double check, make sure that the total equals the between plus the within. Here, $63.33 + 84.00 = 147.33$.

Now compute the degrees of freedom.

Computing the Degrees of Freedom We compute df_{bn}, df_{wn}, and df_{tot}, so there are three steps.

1. *The degrees of freedom between groups equals* k − *1*, where k is the number of levels in the factor. This is because when computing MS_{bn} we essentially have a sample containing the means from our levels that we determine the variability of. For *df* we reduce the number in the sample by 1, so out of k means, $df = k − 1$. In the example, with three levels of perceived difficulty, $k = 3$. Thus, $df_{bn} = 2$.
2. *The degrees of freedom within groups equals* N − k, where N is the total N in the experiment and k is the number of levels in the factor. This is because when computing MS_{wn} we essentially find the estimated variance in each condition and pool them. In each condition we compute the variance using $n − 1$, so out of the entire experiment (N), we subtract one per condition, subtracting a total of k. In the example, N is 15 and k is 3, so $df_{wn} = 15–3 = 12$.
3. *The degrees of freedom total equals* N − *1*, where N is the total N in the experiment. This is because when computing the total SS we treat the experiment as one sample. Then we have N scores, so the df is $N − 1$. In the example, N is 15, so $df_{tot} = 15 − 1 = 14$.

The $df_{tot} = df_{bn} + df_{wn}$. Thus, to check our work, $df_{bn} + df_{wn} = 2 + 12$, which equals 14, the df_{tot}.

Adding the *df* to the summary table, it looks like this:

Source	Sum of Squares	df	Mean Square	F
Between	63.33	2	MS_{bn}	F_{obt}
Within	84.00	12	MS_{wn}	
Total	147.33	14		

Now find each mean square.

Computing the Mean Squares You can work directly from the summary table to compute the mean squares. Any mean square equals the appropriate sum of squares divided by the corresponding *df*. Thus,

> **The formula for the mean square between groups is**
>
> $$MS_{bn} = \frac{SS_{bn}}{df_{bn}}$$

From the summary table, we see that

$$MS_{bn} = \frac{63.33}{2} = 31.67$$

> **The formula for the mean square within groups is**
>
> $$MS_{wn} = \frac{SS_{wn}}{df_{wn}}$$

For the example,

$$MS_{wn} = \frac{84}{12} = 7.00$$

Do *not* compute the mean square for SS_{tot} because it has no use. Now in the summary table we have

Source	Sum of Squares	df	Mean Square	F
Between	63.33	2	31.67	F_{obt}
Within	84.00	12	7.00	
Total	147.33	14		

Computing the F   Finally, compute F_{obt}.

The formula for F_{obt} is

$$F_{obt} = \frac{MS_{bn}}{MS_{wn}}$$

Notice that in the formula, MS_{bn} is "over" MS_{wn}, and this is how they are positioned in the summary table. In our example, MS_{bn} is 31.67 and MS_{wn} is 7.00, so

$$F_{obt} = \frac{MS_{bn}}{MS_{wn}} = \frac{31.67}{7.00} = 4.52$$

Now the completed ANOVA summary table is

Source	Sum of Squares	df	Mean Square	F
Between	63.33	2	31.67	4.52
Within	84.00	12	7.00	
Total	147.33	14		

The F_{obt} is always placed in the row labeled "Between" because the F_{obt} is testing for significant differences *between* our conditions. Also, in the source column you may see (1) the name of the factor at "Between," and (2) the word "Error" in place of "Within."

Interpreting F_{obt}

We interpret F_{obt} by comparing it to F_{crit}, and for that we examine the F-distribution. The **F-distribution** is the sampling distribution showing the various values of F that occur when H_0 is true and all conditions represent one population. To create it, it is as if, using our ns and k, we select the scores for all of our conditions from one raw score population (like H_0 says we did in our experiment) and compute MS_{wn}, MS_{bn} and then F_{obt}. We do this an infinite number of times, and plotting the Fs, produce the sampling distribution, as shown in Figure 13.1.

The F-distribution is skewed because there is no limit to how large F_{obt} can be, but it cannot be less than zero. The mean of the distribution is 1 because, most often when H_0 is true, MS_{bn} will equal MS_{wn}, so F will equal 1. The right-hand tail shows that sometimes, by chance, F is greater than 1. Because our F_{obt} can reflect a relationship in the population only when it is greater than 1, the entire region of rejection is in this upper tail of the F-distribution. (That's right, ANOVA involves two-tailed hypotheses, but they are tested using only the upper tail of the sampling distribution.)

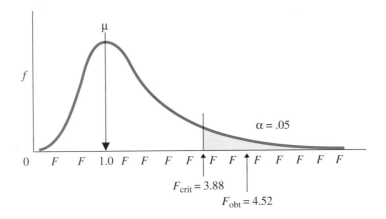

FIGURE 13.1

Sampling distribution of F when H_0 is true for $df_{bn} = 2$ and $df_{wn} = 12$

There are actually many F-distributions, each having a slightly different shape, depending on our degrees of freedom. However, *two* values of df determine the shape of an F-distribution: the df used when computing the mean square between groups (df_{bn}) and the df used when computing the mean square within groups (df_{wn}). Therefore, to obtain F_{crit}, turn to Table 5 in Appendix C, entitled "Critical Values of F." Across the top of the tables, the columns are labeled "Degrees of Freedom Between Groups." Along the left-hand side, the rows are labeled "Degrees of Freedom Within Groups." Locate the appropriate column and row, using the dfs from your study. The critical values in dark type are for $\alpha = .05$, and those in light type are for $\alpha = .01$. For our example, $df_{bn} = 2$ and $df_{wn} = 12$. For $\alpha = .05$, the F_{crit} is 3.88. (If your df_{wn} are not in the table, then for df from 30 to 50, your F_{crit} is the average of the two critical values shown for the bracketing df that are given. For df above 50, compare F_{obt} to the two critical values for the df in the table that bracket your df, using the same strategy we discussed for t-tests in Chapter 11.)

Thus, in our perceived difficulty study, F_{obt} is 4.52 and F_{crit} is 3.88 as above in Figure 13.1. Our H_0 says that F_{obt} is greater than 1 because of sampling error and that actually we are poorly representing no relationship in the population. However, our F_{obt} is beyond F_{crit}, so we reject H_0: Our F_{obt} is so unlikely to occur if our samples were representing no difference in the population that we reject that this is what they represent. Therefore, we conclude that the F_{obt} is significant and that the factor of perceived difficulty produces a significant difference in mean performance scores.

Of course, had F_{obt} been less than F_{crit}, then the corresponding differences between our means would *not* be too unlikely to occur when H_0 is true, so we would not reject H_0. Then, as usual, we'd draw no conclusion about our independent variable, one way or the other. We would also consider if we had sufficient *power* to prevent a Type II error (missing the relationship). We increase the power of an ANOVA using the same strategies discussed in previous chapters: maximize the differences in scores *between* conditions, minimize variability *within* conditions, and maximize n. (These strategies also increase the power of post hoc comparisons.)

A QUICK REVIEW

- To compute F_{obt}, compute SS_{tot}, SS_{bn}, and SS_{wn} and df_{tot}, df_{bn}, and df_{wn}. Dividing SS_{bn} by df_{bn} gives MS_{bn}; dividing SS_{wn} by df_{wn} gives MS_{wn}. Dividing MS_{bn} by MS_{wn} gives F_{obt}. Compare F_{obt} to F_{crit}.

MORE EXAMPLES

We test participants under conditions A_1 and A_2.

A_1	A_2
4	6
5	8
3	9
5	8

$\overline{X}_1 = 4.25$ $\overline{X} = 7.75$

$\Sigma X = 17$ $\Sigma X = 31$ $\Sigma X_{tot} = 48$

$\Sigma X^2 = 75$ $\Sigma X^2 = 245$ $\Sigma X_{tot}^2 = 320$

$n_1 = 4$ $n_2 = 4$ $N = 8$

1. *Compute the sums of squares:*

$$SS_{tot} = \Sigma X_{tot}^2 - \frac{(\Sigma X_{tot})^2}{N} = 320 - \left(\frac{48^2}{8}\right) = 32$$

$$SS_{bn} = \Sigma\left(\frac{(\Sigma X \text{ in column})^2}{n \text{ in column}}\right) - \frac{(\Sigma X_{tot})^2}{N}$$

$$= \left(\frac{17^2}{4} + \frac{31^2}{4}\right) - \left(\frac{48^2}{8}\right) = 24.5$$

$$SS_{wn} = SS_{tot} - SS_{bn} = 32 - 24.5 = 7.5$$

2. *Compute the degrees of freedom:*

$$df_{bn} = k - 1 = 2 - 1 = 1$$

$$df_{wn} = N - k = 8 - 2 = 6$$

$$df_{tot} = N - 1 = 8 - 1 = 7$$

3. *Compute the mean squares:*

$$MS_{bn} = SS_{bn}/df_{bn} = 24.5/1 = 24.5$$

$$MS_{wn} = SS_{wn}/df_{wn} = 7.5/6 = 1.25$$

4. *Compute F_{obt}:*

$$F_{obt} = MS_{bn}/MS_{wn} = 24.5/1.25 = 19.60$$

5. *Compare F_{obt} to F_{crit}:* With $\alpha = .05$, $df_{bn} = 1$, and $df_{wn} = 6$, $F_{crit} = 5.99$. The F_{obt} is beyond F_{crit}. Therefore, the means of the conditions differ significantly.

For Practice

1. What two components are needed to compute any mean square?

2. For between groups, to compute _____ we divide _____ by _____. For within groups, to compute _____ we divide _____ by _____.

3. Finally, F_{obt} equals _____ divided by _____.

Answers

1. The sums of squares and the *df*
2. MS_{bn}, SS_{bn}, df_{bn}; MS_{wn}, SS_{wn}, df_{wn}
3. MS_{bn}, MS_{wn}

PERFORMING POST HOC COMPARISONS

Because we rejected H_0 and accepted H_a, we return to the means from the levels of our factor:

Perceived Difficulty

Easy	Medium	Difficult
$\overline{X}_1 = 8$	$\overline{X}_2 = 6$	$\overline{X}_3 = 3$

We are confident that these means represent a relationship in the population, in which increasing perceived difficulty is associated with fewer problems solved. However, we do not know whether *every* increase in difficulty produces a significant drop in performance. Therefore, we must determine which specific means differ significantly, and to do that, we perform post hoc comparisons.

Statisticians have developed a variety of post hoc procedures that differ in how likely they are to produce Type I or Type II errors. We'll discuss two procedures that have acceptably low error rates. Depending on whether or not your ns are equal, perform either *Fisher's protected t-test* or *Tukey's HSD test.*

Fisher's Protected *t*-Test

Perform **Fisher's protected *t*-test** when the ns in all levels are not equal.

The formula for Fisher's protected *t*-test is

$$t_{obt} = \frac{\overline{X}_1 - \overline{X}_2}{\sqrt{MS_{wn}\left(\dfrac{1}{n_1} + \dfrac{1}{n_2}\right)}}$$

This is a variation of the independent-samples t-test. We are testing H_0: $\mu_1 - \mu_2 = 0$, where $\overline{X}_1$ and $\overline{X}_2$ are the means for any two levels of the factor and n_1 and n_2 are the corresponding ns in those levels. The MS_{wn} is from our ANOVA.

It is not incorrect to perform the protected t-test even when all ns are equal. For example, we can compare the mean from our easy level (8) to the mean from our difficult level (3). Each n is 5, and our MS_{wn} is 7. Filling in the formula gives

$$t_{obt} = \frac{8 - 3}{\sqrt{7\left(\dfrac{1}{5} + \dfrac{1}{5}\right)}}$$

Then

$$t_{obt} = \frac{+5}{\sqrt{7(.4)}} = \frac{+5}{\sqrt{2.80}} = \frac{+5}{1.673} = +2.99$$

Next, we compare t_{obt} to t_{crit}, which we obtain from the t-tables (Table 2 in Appendix C). *The df here equals the df_{wn} from the ANOVA.* For the example, with $\alpha = .05$ and $df_{wn} = 12$, t_{crit} is ± 2.179. Because the t_{obt} of $+2.99$ is beyond this t_{crit}, the means from the easy and difficult levels differ significantly.

To complete these comparisons, perform the protected t-test on all possible pairs of means in the factor. Thus, we would also test the means from easy and medium, and the means from medium and difficult.

If a factor contains many levels, the protected t-test becomes very tedious. If you think there *must* be an easier way, you're right.

Tukey's *HSD* Multiple Comparisons Test

Perform the **Tukey *HSD* multiple comparisons test** when the ns in all levels are equal. The *HSD* is a rearrangement of the t-test that computes the minimum difference between two means that is required for the means to differ significantly (*HSD* stands for the honestly significant difference). There are four steps to performing the *HSD* test.

Step 1 *Find* q_k. Using the appropriate value of q_k in the computations is what protects the experiment-wise error for the number of means being compared. Find the value of q_k in Table 6 in Appendix C, entitled "Values of Studentized Range Statistic." In the table, locate the column labeled with the k corresponding to the number of means in your factor. Next, find the row labeled with the df_{wn} used to compute your F_{obt}. Then find the value of q_k for the appropriate α.

For our study above, $k = 3$, $df_{wn} = 12$, and $\alpha = .05$, so $q_k = 3.77$.

Step 2 *Compute the* HSD.

The formula for Tukey's *HSD* test is

$$HSD = (q_k)\left(\sqrt{\frac{MS_{wn}}{n}}\right)$$

MS_{wn} is from the ANOVA, and n is the number of scores in each level of the factor. In the example, MS_{wn} was 7 and n was 5, so

$$HSD = (3.77)\left(\sqrt{\frac{7}{5}}\right) = 4.46$$

Step 3 *Determine the differences between each pair of means.* Subtract each mean from every other mean. Ignore whether differences are positive or negative (for each pair, this is a two-tailed test of H_0: $\mu_1 - \mu_2 = 0$).

The differences for the perceived difficulty study can be diagramed as shown below:

Perceived Difficulty

On the line connecting any two levels is the absolute difference between their means.

Step 4 *Compare each difference to the* HSD. If the absolute difference between two means is *greater than* the *HSD*, then these means differ significantly. (It's as if you performed a *t*-test on these means and t_{obt} was significant.) If the absolute difference between two means is less than or equal to the *HSD*, then it is *not* a significant difference (and would not produce a significant t_{obt}).

Above, the *HSD* was 4.46. The means from the easy level (8) and the difficult level (3) differ by more than 4.46, so they differ significantly. The mean from the medium level (6), however, differs from the other means by less than 4.46, so it does not differ significantly from them.

Thus, our final conclusion about this study is that we demonstrated a relationship between scores and perceived difficulty, but only for the easy and difficult conditions. If these two conditions were given to the population, we would expect to find one population for easy with a μ around 8 and another population for difficult with a μ around 3. We cannot say anything about the medium level, however, because it did not produce a significant difference. Finally, as usual, we would now interpret the results in terms of the behaviors being studied, explaining why this manipulation worked as it did.

A QUICK REVIEW

- Perform post hoc comparisons when F_{obt} is significant to determine which levels differ significantly.

- Perform Tukey's *HSD* test when all *n*s are equal and Fisther's *t*-test when *n*s are unequal.

MORE EXAMPLES

An F_{obt} is significant, with $\overline{X}_1 = 4.0$, $\overline{X}_2 = 1.5$, and $\overline{X}_3 = 6.8$ and $n = 11$, $MS_{wn} = 20.61$, and $df_{wn} = 30$. To compute Fisher's *t*-test on $\overline{X}_1$ and $\overline{X}_3$,

$$t_{obt} = \frac{\overline{X}_1 - \overline{X}_2}{\sqrt{MS_{wn}\left(\dfrac{1}{n_1} + \dfrac{1}{n_2}\right)}} = \frac{4.0 - 6.8}{\sqrt{20.61\left(\dfrac{1}{11} + \dfrac{1}{11}\right)}}$$

$$= \frac{-2.8}{\sqrt{3.75}}$$

$t_{obt} = -1.446$ and $t_{crit} = \pm2.042$. These means do not differ significantly.

To compute Tukey's *HSD*, find q_k. For $k = 3$ and $df_{wn} = 30$, $q_k = 3.49$. Then:

$$HSD = (q_k)\left(\sqrt{\dfrac{MS_{wn}}{n}}\right) = (3.49)\left(\sqrt{\dfrac{20.61}{11}}\right) = 4.78$$

The differences are $\overline{X}_1 - \overline{X}_2 = 4.0 - 1.5 = 2.5$; $\overline{X}_2 - \overline{X}_3 = 1.5 - 6.8 = -5.3$; $\overline{X}_1 - \overline{X}_3 = 4.0 - 6.8 = -2.8$. Comparing each difference to $HSD = 4.78$, only $\overline{X}_2$ and $\overline{X}_3$ differ significantly.

For Practice

We have $\overline{X}_1 = 16.50$, $\overline{X}_2 = 11.50$, and $\overline{X}_3 = 8.92$, with $n = 21$ in each condition, $MS_{wn} = 63.44$, and $df_{wn} = 60$.

1. Which post hoc test should we perform?
2. What is q_k here?
3. What is the *HSD*?
4. Which means differ significantly?

Answers

1. Tukey's *HSD*
2. For $k = 3$ and $df_{wn} = 60$, $q_k = 3.40$.
3. $HSD = (3.40)(\sqrt{63.44/21}) = 5.91$
4. Only $\overline{X}_1$ and $\overline{X}_3$ differ significantly.

SUMMARY OF STEPS IN PERFORMING A ONE-WAY ANOVA

It's been a long haul, but here is everything we do when performing a one-way, between-subjects ANOVA.

1. *Create the hypotheses.* The null hypothesis is H_0: $\mu_1 = \mu_2 = \ldots \mu_k$, and the alternative hypothesis is H_a: not all μs are equal.

2. *Compute F_{obt}.* Compute the sums of squares and the degrees of freedom. Then compute MS_{bn} and MS_{wn}. Then compute F_{obt}.

3. *Compare F_{obt} to F_{crit}.* Find F_{crit} using df_{bn} and df_{wn}. If F_{obt} is larger than F_{crit}, then F_{obt} is significant, indicating that the means in at least two conditions differ significantly.

4. *Perform post hoc tests.* If F_{obt} is significant and there are more than two levels of the factor, determine which levels differ significantly by performing post hoc comparisons. Perform Fisher's *t*-test if the *n*s are not equal or perform Tukey's *HSD* test if all *n*s are equal.

If you followed all of that, then congratulations, you're getting *good* at this stuff. Of course, all of this merely determines whether there is a relationship. Now you must describe that relationship.

ADDITIONAL PROCEDURES IN THE ONE-WAY ANOVA

The following presents procedures for describing a significant relationship that we have discussed in previous chapters, except that here they have been altered to accommodate the computations in ANOVA.

The Confidence Interval for Each Population μ

As usual, we can compute a confidence interval for the μ represented by the mean of any condition. This is the same confidence interval for μ that was discussed in Chapter 11, but the formula is slightly different.

The formula for the confidence interval for a single μ is

$$\left(\sqrt{\frac{MS_{wn}}{n}}\right)(-t_{crit}) + \overline{X} \le \mu \le \left(\sqrt{\frac{MS_{wn}}{n}}\right)(+t_{crit}) + \overline{X}$$

The t_{crit} is the two-tailed value found in the *t*-tables using the appropriate α and using df_{wn} from the ANOVA as the *df*. MS_{wn} is also from the ANOVA. The $\overline{X}$ and *n* are from the level we are describing.

For example, in the easy condition, $\overline{X} = 8.0$, $MS_{wn} = 7.0$, $df_{wn} = 12$, and $n = 5$. The two-tailed t_{crit} (at $df = 12$ and $\alpha = .05$) is ±2.179. Thus,

$$\left(\sqrt{\frac{7.0}{5}}\right)(-2.179) + 8.0 \le \mu \le \left(\sqrt{\frac{7.0}{5}}\right)(+2.179) + 8.0$$

This becomes

$$(-2.578) + 8.0 \le \mu \le (+2.578) + 8.0$$

and finally,

$$5.42 \le \mu \le 10.58$$

Thus, if we were to test the entire population under our easy condition, we are 95% confident that the population mean would fall between 5.42 and 10.58.

Follow the same procedure to describe the μ from any other *significant* level of the factor.

Graphing the Results in ANOVA

As usual, graph your results by placing the dependent variable on the Y axis and the independent variable on the X axis. Then plot the mean for each condition. Figure 13.2 shows the line graph for the perceived difficulty study. Note that we include the medium level of difficulty, even though it did not produce significant differences.

As usual, the line graph summarizes the relationship that is present, and here it indicates a largely negative linear relationship.

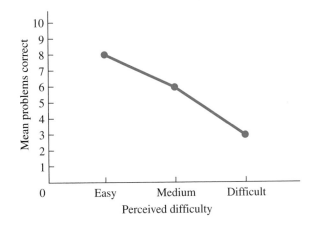

Describing Effect Size in the ANOVA

Recall that in an experiment we describe the *effect size,* which tells us how large of an impact the independent variable had on dependent scores. The way to do this is to compute the *proportion of variance accounted for,* which tells us the proportional improvement in predicting participants' scores that we achieve by predicting the mean of their condition. In ANOVA, we compute this by computing a new correlation coefficient, called *eta* (pronounced "ay-tah"), and then squaring it: **eta squared** indicates the proportion of variance in the dependent variable that is accounted for by changing the levels of a factor in an ANOVA. The symbol for eta squared is η^2.

> ### The formula for eta squared is
>
> $$\eta^2 = \frac{SS_{bn}}{SS_{tot}}$$

Both SS_{bn} and SS_{tot} are computed in the ANOVA. The SS_{bn} reflects the differences between the conditions. The SS_{tot} reflects the total differences between all scores in the experiment. Thus, η^2 reflects the proportion of all differences in scores that are associated with the different conditions.

For example, for our study, SS_{bn} was 63.33 and SS_{tot} was 147.33. So,

$$\eta^2 = \frac{SS_{bn}}{SS_{tot}} = \frac{63.33}{147.33} = .43$$

This is interpreted in the same way that we previously interpreted r_{pb}^2. The larger the η^2, the more consistently the factor "caused" participants to have a particular score in a particular condition, and thus the more scientifically important the factor is for explaining and predicting differences in the underlying behavior. Thus, our η^2 of .43 indicates that we are 43% more accurate at predicting participants' scores when we predict for them the mean of the difficulty level they were tested under. In other words, 43% of all differences in these scores are accounted for ("caused") by changing the levels of perceived difficulty. Because 43% is a very substantial amount, this factor is important in determining participants' performance, so it is important for scientific study.

Recall that our other measure of effect size is Cohen's d, which describes the magnitude of the differences between our means. However, with three or more levels, the procedure is extremely complicated. Therefore, instead, η^2 is the preferred measure of effect size in ANOVA.

STATISTICS IN PUBLISHED RESEARCH: REPORTING ANOVA

In research reports, the results of an ANOVA are reported in the same ways as with previous procedures. However, now we are getting to more complicated designs, so there is an order and logic to the report. Typically, we report the means and standard deviation from each condition first. (Rather than use an incredibly long sentence for this, often a table is presented.) Then we describe the characteristics of the primary ANOVA and report the results. Then we report any secondary procedures. Thus, for our perceived difficulty study, you might see:

> **"A one-way, between-subjects ANOVA was performed on the scores from the three levels of perceived difficulty. The results were significant, F(212) = 4.52, p < .05 A Tukey _HSD_ test revealed that only the means for the easy and difficult conditions differed significantly (p < .05). This manipulation accounted for .43 of the variance in scores (using η^2). The 95% confidence interval for the easy condition is . . ."**

Notice that for F_{obt}, we report df_{bn} and then df_{wn}. We also indicate that the Tukey test was performed, although usually the _HSD_ value is not reported. The alpha level we used is reported (as p < .05), as is a summary of the levels that differ significantly. Then we report other secondary analyses, such as η^2 and confidence intervals.

PUTTING IT ALL TOGETHER

When all is said and done, the F-ratio is a convoluted way of measuring the differences between the means of our conditions and then fitting those differences to a sampling distribution. The larger the F_{obt}, the less likely that the means are representing the same μ. A significant F_{obt} indicates that the means are unlikely to represent one population mean. Then we determine which sample means actually differ significantly and describe the relationship they form. That's all there is to it.

There is, however, one other type of procedure that you should be aware of. All of the research designs in this book involve _one dependent_ variable, and the statistics we perform are called **univariate statistics.** We can, however, measure participants on two or more _dependent_ variables in one experiment. Statistics for multiple dependent variables are called **multivariate statistics.** These include the multivariate t-test and the multivariate analysis of variance (MANOVA). Even though these are very complex procedures, the basic logic still holds: The larger the t_{obt} or F_{obt}, the less likely it is that the samples represent no relationship in the population. To discuss multivariates further would require another book.

Using the SPSS Appendix As discussed in Appendix B.7, SPSS will perform the one-way between-subjects ANOVA. This includes reporting the significance level of

F_{obt}, performing the *HSD* test, and graphing the means. The program also computes, the $\overline{X}$, s_X, and 95% confidence interval for μ for each level. It does not compute η^2. (The "partial eta squared" is not what we've discussed.

Appendix B.11 describes using SPSS to perform the one-way, within-subjects ANOVA (discussed in Appendix A.3). The program provides the same information as above, except that it does not perform the *HSD* test.

CHAPTER SUMMARY

1. The general terms used previously and their corresponding ANOVA terms are shown in this table:

General Term	=	*ANOVA Term*
independent variable	=	factor
condition	=	level of treatment
sum of squared deviations	=	sum of squares (*SS*)
variance (s_X^2)	=	mean square (*MS*)
effect of independent variable	=	treatment effect

2. A *one-way analysis of variance* tests for significant differences between the means from two or more levels of a factor. In a *between-subjects factor,* each condition involves an independent sample. In a *within-subjects factor,* the conditions involve related samples.

3. The *experiment-wise error rate* is the probability that a Type I error will occur in an experiment. ANOVA keeps the experiment-wise error rate equal to α.

4. ANOVA tests the H_0 that all μs being represented are equal; H_a is that not all μs are equal.

5. The *mean square within groups (MS$_{wn}$),* measures the differences among the scores within the conditions. The *mean square between groups* (MS_{bn}), measures the differences among the level means.

6. When H_0 is true, MS_{wn} and MS_{bn} estimate the *error variance.* When H_0 is false, MS_{bn} also reflects added *treatment variance.*

7. F_{obt} is computed using the *F-ratio,* which equals the mean square between groups divided by the mean square within groups.

8. F_{obt} may be greater than 1 because either (a) there is no treatment effect, but the sample data are not perfectly representative of this, or (b) two or more sample means represent different population means.

9. If F_{obt} is significant with more than two levels, perform *post hoc comparisons* to determine which means differ significantly. When the *ns* are *not* equal, perform *Fisher's protected t-test* on each pair of means. If all *ns* are equal, perform *Tukey's* HSD *test.*

10. *Eta squared* (η^2) describes the *effect size*—the *proportion of variance* in dependent scores accounted for by the levels of the independent variable.

KEY TERMS

k F_{obt} F_{crit} MS_{wn} σ^2_{error} MS_{bn}
σ^2_{treat} df_{bn} df_{wn} η^2 SS_{bn} SS_{wn}
HSD

analysis of variance *292*
ANOVA *290*
between-subjects ANOVA *291*
between-subjects factor *291*
error variance *298*
eta squared *311*
experiment-wise error rate *293*
factor *290*
F-distribution *304*
Fisher's protected *t*-test *307*
F-ratio *297*
level *290*

mean square *295*
mean square between groups *295*
mean square within groups *295*
multivariate statistics *312*
one-way ANOVA *291*
post hoc comparisons *294*
sum of squares *295*
treatment *290*
treatment effect *290*
treatment variance *298*
Tukey's *HSD* multiple comparisons
   test *307*
univariate statistics *312*
within-subjects ANOVA *291*
within-subjects factor *291*

REVIEW QUESTIONS

(Answers for odd-numbered questions are in Appendix D.)

1. What does each of the following terms mean? (a) ANOVA (b) one-way design (c) factor (d) level (e) treatment (f) between subjects (g) within subjects
2. (a) What is the difference between n and N? (b) What does k stand for?
3. What are two reasons for conducting a study with more than two levels of a factor?
4. (a) What are error variance and treatment variance? (b) What are the two types of mean squares, and what does each estimate?
5. (a) What is the experiment-wise error rate? (b) Why does ANOVA solve the problem with the experiment-wise error rate created by multiple *t*-tests?
6. Summarize the steps involved in analyzing an experiment when $k > 2$.
7. (a) When is it necessary to perform post hoc comparisons? Why? (b) When is it unnecessary to perform post hoc comparisons?
8. When do you use each of the two post hoc tests discussed in this chapter?
9. What does η^2 indicate?
10. (a) Why should F_{obt} equal 1 if the data represent the H_0 situation? (b) Why is F_{obt} greater than 1 when the data represent the H_a situation? (c) What does a significant F_{obt} indicate about differences between the levels of a factor?
11. A research article reports the results of a "multivariate" analysis. What does this term communicate about the study?

APPLICATION QUESTIONS

12. A researcher conducts an experiment with three levels of the independent variable. (a) Which two versions of a parametric procedure are available to her? (b) How does she choose between them?
13. (a) In a study comparing four conditions, what is H_0 for the ANOVA? (b) What is H_a in the same study? (c) Describe in words what H_0 and H_a say for this study.

14. Foofy obtained a significant F_{obt} from an experiment with five levels. She therefore concludes that changing each condition of the independent variable results in a significant change in the dependent variable. (a) Is she correct? Why or why not? (b) What must she now do?

15. (a) Poindexter computes an F_{obt} of .63. How should this be interpreted? (b) He computes another F_{obt} of -1.7. How should this be interpreted?

16. A report says that the between-subjects factor of participants' salary produced significant differences in self-esteem. (a) What does this tell you about the design? (b) What does it tell you about the results?

17. A report says that a new diet led to a significant decrease in weight for a group of participants. (a) What does this tell you about the design? (b) What do we call this design?

18. A researcher investigated the number of viral infections people contract as a function of the amount of stress they experienced during a 6-month period. She obtained the following data:

Amount of Stress

Negligible Stress	Minimal Stress	Moderate Stress	Severe Stress
2	4	6	5
1	3	5	7
4	2	7	8
1	3	5	4

(a) What are H_0 and H_a? (b) Compute F_{obt} and complete the ANOVA summary table. (c) With $\alpha = .05$, what is F_{crit}? (d) Report your statistical results. (e) Perform the appropriate post hoc comparisons. (f) What do you conclude about this study? (g) Describe the effect size and interpret it. (h) Estimate the value of μ that is likely to be found in the severe stress condition.

19. Here are data from an experiment studying the effect of age on creativity scores:

Age 4	Age 6	Age 8	Age 10
3	9	9	7
5	11	12	7
7	14	9	6
4	10	8	4
3	10	9	5

(a) Compute F_{obt} and create an ANOVA summary table. (b) With $\alpha = .05$, what do you conclude about F_{obt}? (c) Perform the appropriate post hoc comparisons. What should you conclude about this relationship? (d) Statistically, how important is the relationship in this study? (e) Describe how you would graph these results.

20. In a study in which $k = 3$, $n = 21$, $\overline{X}_1 = 45.3$, $\overline{X}_2 = 16.9$, and $\overline{X}_3 = 8.2$, you compute the following sums of squares:

Source	Sum of Squares	df	Mean Square	F
Between	147.32	___	___	___
Within	862.99	___	___	
Total	1010.31	___		

(a) Complete the ANOVA summary table. (b) With $\alpha = .05$, what do you conclude about F_{obt}? (c) Report your results in the correct format. (d) Perform the appropriate post hoc comparisons. What do you conclude about this relationship? (e) What is the effect size in this study, and what does this tell you about the influence of the independent variable?

21. Performing ANOVA protects our experimentwise error rate. (a) Name and explain the error that this avoids. (b) Why is this an important error to avoid?

22. A researcher investigated the effect of volume of background noise on participants' accuracy rates while performing a boring task. He tested three groups of randomly selected students and obtained the following means and sums of squares:

	Low Volume	Moderate Volume	High Volume
$\overline{X}$	61.5	65.5	48.25
n	4	5	7

Source	Sum of Squares	df	Mean Square	F
Between groups	652.16	_____	_____	_____
Within groups	612.75	_____	_____	
Total	1264.92	_____		

(a) Complete the ANOVA (b) At $\alpha = .05$, what is F_{crit}? (c) Report the statistical results in the proper format. (d) Perform the appropriate post hoc tests. (e) What do you conclude about this study? (f) Compute the effect size and interpret it.

INTEGRATION QUESTIONS

23. (a) How do we create related samples? (b) In part (a) what two inferential procedures are appropriate? (c) How do we create independent samples? (d) In part (c) what two inferential procedures are appropriate? (Chs. 12, 13)

24. (a) In this chapter we tested the relationship between performance scores and perceived difficulty. Describe this relationship using "as a function of."
(b) An experimenter computes the mean anxiety level for samples of freshmen, sophomores, juniors, and seniors. To graph these results, how would you label the axes? (c) Would this be a bar or line graph and why? (Chs. 2, 4)

25. (a) How is η^2 similar to r_{pb}^2? (b) How do we interpret this proportion? (Chs. 12, 13)

26. (a) Name and explain the error that power prevents. (b) Why is it important to maximize the power of any experiment? (c) How is this done in a design using ANOVA? (d) How does influencing the differences between and within conditions influence the size of F_{obt} and its likelihood of being significant? (Chs. 10, 11, 13)

27. (a) How do you identify which variable in a study is the factor? (b) How do you identify the levels of a factor? (c) How do you identify the variable that is the dependent variable? (Chs. 2, 13)

28. For the following, identify the inferential procedure to perform and the key information for answering the research question. *Note*: If no inferential procedure is appropriate, indicate why. (a) Doing well in statistics should reduce students'

math phobia. We measure their math phobia after selecting groups who received either an A, B, C, or D in statistics. (b) To determine if recall is better or worse than recognition, participants study a list of words, and then half of them recall the words and the other half performs a recognition test. (c) We repeatedly test the aggressiveness of rats after 1, 3, 5 and 7 weeks, to see if they become more aggressive as they grow older. (d) Using the grades of 100 students, we want to use students' score on the first exam to predict their final exam grade. (e) We test for gender differences in voting history by counting the number of males and females who voted in the last election. (f) We ask if pilots are quicker than navigators, comparing the reaction time of a group of pilots to that of their navigators. (Chs. 7, 8, 10, 12, 13)

29. In question 28, identify the levels of the factor and the dependent variable in experiments, and the predictor/criterion variables in correlational studies. (Chs. 8, 12, 13)

■■■ SUMMARY OF FORMULAS

The format for the summary table for a one-way ANOVA is

Summary Table of One-Way ANOVA

Source	*Sum of Squares*	**df**	*Mean Square*	**F**
Between	SS_{bn}	df_{bn}	MS_{bn}	F_{obt}
Within	SS_{wn}	df_{wn}	MS_{wn}	
Total	SS_{tot}	df_{tot}		

1. To perform the one-way, between-subjects ANOVA

a. Compute the sum of squares,

$$SS_{tot} = \Sigma X_{tot}^2 - \left(\frac{(\Sigma X_{tot})^2}{N}\right)$$

$$SS_{bn} = \Sigma\left(\frac{(\text{Sum of scores in the column})^2}{n \text{ of scores in the column}}\right) - \left(\frac{(\Sigma X_{tot})^2}{N}\right)$$

$$SS_{wn} = SS_{tot} - SS_{bn}$$

b. Compute the mean squares,

$$MS_{bn} = \frac{SS_{bn}}{df_{bn}} \qquad df_{bn} = k - 1$$

$$MS_{wn} = \frac{SS_{wn}}{df_{wn}} \qquad df_{wn} = N - k$$

c. Compute the *F*-ratio

$$F_{obt} = \frac{MS_{bn}}{MS_{wn}}$$

2. The formula for the protected *t*-test is

$$t_{obt} = \frac{\overline{X}_1 - \overline{X}_2}{\sqrt{MS_{wn}\left(\frac{1}{n_1} + \frac{1}{n_2}\right)}}$$

Use the two-tailed t_{crit} for $df = df_{wn}$.

3. The formula for the HSD is

$$HSD = (q_k)\left(\sqrt{\frac{MS_{wn}}{n}}\right)$$

Values of q_k are found in Appendix C, Table 6.

4. The formula for the confidence interval for μ is

$$\left(\sqrt{\frac{MS_{wn}}{n}}\right)(-t_{crit}) + \overline{X} \leq \mu \leq \left(\sqrt{\frac{MS_{wn}}{n}}\right)(+t_{crit}) + \overline{X}$$

Use the two-tailed t_{crit} for $df = df_{wn}$.

5. The formula for eta squared is

$$\eta^2 = \frac{SS_{bn}}{SS_{tot}}$$

14

The Two-Way Analysis of Variance

GETTING STARTED

To understand this chapter, recall the following:

- From Chapter 13, the terms *factor* and *level,* how to calculate *F,* what a significant *F* indicates, when to perform post hoc tests, and what η^2 indicates.

Your goals in this chapter are to learn

- What a *two-way factorial* ANOVA is.
- How to *collapse* across a factor to find main effect means.
- How to calculate *main effect means* and *cell means.*
- How to compute the *F*s in a two-way ANOVA
- What a significant *main effect* indicates.
- What a significant *interaction* indicates.
- How to perform *post hoc tests.*
- How to interpret the results of a two-way experiment.

In the previous chapter, you saw that ANOVA tests the means from one factor. In this chapter, we'll expand the experiment to involve *two* factors. Then the analysis is similar to the previous ANOVA, except that here we compute several values of F_{obt}. Therefore, be forewarned that the computations are rather involved (although they are more tedious than difficult). Don't try to memorize the formulas, because nowadays we usually analyze such experiments using a computer. However, you still need to understand the basic logic, terminology, and purpose of the calculations. Therefore, the following sections present (1) the general layout of a two-factor experiment, (2) what the ANOVA indicates, (3) how to compute the ANOVA, and (4) how to interpret a completed study.

NEW STATISTICAL NOTATION

When a study involves two factors, it is called a *two-way* design. The **two-way ANOVA** is the parametric inferential procedure that is applied to designs that involve two independent variables. However, we have different versions of this depending on whether we have *independent* or *related* samples. When both factors involve independent samples, we perform the **two-way, between-subjects ANOVA.** When both factors involve related samples (either because of *matching* or *repeated measures*), we perform the **two-way, within-subjects ANOVA.** If one factor is tested using independent samples and the other factor involves related samples, we perform the **two-way, mixed-design ANOVA.** These ANOVAs are identical except for slight differences buried in their formulas. In this chapter, we discuss the between-subjects version.

Each of our factors may contain any number of levels, so we have a code for describing a specific design. The generic format is to identify one independent variable as factor A and the other independent variable as factor B. To describe a particular ANOVA, we use the number of levels in each factor. If, for example, factor A has two levels and factor B has two levels, we have a two-by-two ANOVA, which is written as 2×2. Or if one factor has four levels and the other factor has three levels, we have a 4×3 or a 3×4 ANOVA, and so on.

WHY IS IT IMPORTANT TO KNOW ABOUT THE TWO-WAY ANOVA?

It is important for you to understand the two-way ANOVA because you, and other researchers, will often study two factors in one experiment. This is because, first, a two-factor design tells us everything about the influence of each factor that we would learn if it were the only independent variable. But we can also study something that we'd otherwise miss—the *interaction effect*. For now, think of an interaction effect as the influence of combining the two factors. Interactions are important because, in nature, many variables that influence a behavior are often simultaneously present. By manipulating more than one factor in an experiment, we can examine the influence of such combined variables. Thus, the primary reason for conducting a study with two (or more) factors is to observe the interaction between them.

A second reason for multifactor studies is that once you've created a design for studying one independent variable, often only a minimum of additional effort is required to study additional factors. Multifactor studies are an efficient and cost-effective way of determining the effects of—and interactions among—several independent variables. Thus, you'll often encounter two-way ANOVAs in behavioral research. And by understanding them, you'll be prepared to understand the even more complex ANOVAs that will occur.

UNDERSTANDING THE TWO-WAY DESIGN

The key to understanding the two-way ANOVA is to understand how to envision it. As an example, say that we are again interested in the effects of a "smart pill" on a person's IQ. We'll manipulate the number of smart pills given to participants, calling this factor A, and test two levels (one or two pills). Our dependent variable is a participant's IQ score. We want to show the *effect* of factor A, showing how IQ scores change as we increase dosage.

Say that we're also interested in studying the influence of a person's age on IQ. We'll call age factor B and test two levels (10- or 20-year-olds). Here, we want to show the effect of factor B, showing how IQ scores change with increasing age.

To create a two-way design, we would simultaneously manipulate both the participants' age and the number of pills they receive. The way to envision this is to use the matrix in Table 14.1. We always place participants' dependent scores inside the matrix, so here each X represents a participant's IQ score. Understand the following about this matrix:

1. Each *column* represents a level of one independent variable, which here is our pill factor. (For our formulas, we will always call the column factor "factor A.") Thus, for example, any score in column A_1 is from someone tested with one pill.

TABLE 14.1

Two-Way Design for Studying the Factors of Number of Smart Pills and Participant's Age

Each column contains scores for one level of number of pills; each row contains scores for one level of age.

Factor A: Number of Pills

	Level A₁: 1 Pill	Level A₂: 2 Pills	

Factor B: Age

Level B₁: 10-Year-Olds — X X X $\overline{X}_{A_1B_1}$ | X X X $\overline{X}_{A_2B_1}$ — Scores

Level B₂: 20-Year-Olds — X X X $\overline{X}_{A_1B_2}$ | X X X $\overline{X}_{A_2B_2}$

↑ One of the four cells

2. Each *row* represents one level of the other independent variable, which here is the age factor. (The row factor is always called "factor B.") Thus, for example, any score in row B₁ is from a 10-year-old.

3. Each small square produced by combining a level of factor A with a level of factor B is called a **cell.** Here we have four cells, each containing a sample of three participants, who are all a particular age and given the same dose of pills. For example, the highlighted cell contains scores from 20-year-olds given two pills.

4. Because we have two levels of each factor, we have a 2×2 design (it produces a 2×2 matrix).

5. We can identify each cell using the levels of the two factors. For example, the cell formed by combining level 1 of factor A and level 1 of factor B is cell A₁B₁. We can identify the mean and *n* from each cell in the same way, so, for example, in cell A₁B₁ we will compute $\overline{X}_{A_1B_1}$.

6. The *n* in each cell is 3, so $N = 12$.

7. We have combined all of our levels of one factor with all levels of the other factor, so we have a **complete factorial design.** On the other hand, in an **incomplete factorial design,** not all levels of the two factors are combined. For example, if we had not collected scores from 20-year-olds given one pill, we would have an incomplete factorial design. Incomplete designs require procedures not discussed here.

OVERVIEW OF THE TWO-WAY, BETWEEN-SUBJECTS ANOVA

Now that you understand a two-way design, we can perform the ANOVA. But, enough about smart pills. Here's a semi-fascinating idea for a new study. Television commercials are often much louder than the programs themselves because advertisers believe that increased volume makes the commercial more persuasive. To test this, we will play a recording of an advertising message to participants at one of three volumes. Volume is measured in decibels, but to simplify things we'll call the three volumes soft, medium, and loud. Say that we're also interested in the differences between how males and females are persuaded, so our other factor is the gender of the listener. Therefore, we have a two-factor experiment involving three levels of volume and two levels of gender, so we have a 3 × 2 design. The dependent variable indicates how persuasive the message is, on a scale of 0 (not at all) to 25 (totally convincing).

This study is diagramed in Table 14.2 The numbers inside the cells are the persuasiveness scores. Each column represents a level of the volume factor (factor A), and each row represents a level of the gender factor (factor B). For simplicity we have the unpowerful N of 18: Nine men and nine women were tested, with three men and three women hearing the message at each volume, so we have three persuasiveness scores per cell.

But now what? As usual in any experiment, we want to conclude that if we tested the entire population under our different conditions of volume or gender, we'd find different populations of scores located at different μs. But, there is the usual problem: Our sample data may reflect sampling error, so we might actually find the same population and μ for all conditions. Therefore, once again we must eliminate the idea of sampling error, and to do this, we perform ANOVA. As usual, first we set alpha (usually at .05) and then check the assumptions.

With a complete factorial design, the assumptions of the two-way, between-subjects ANOVA are

1. Each *cell* contains an independent sample.

2. The dependent variable measures normally distributed interval or ratio scores.

3. The populations have homogeneous variance.

Each cell in the persuasiveness study contains an independent sample, so we will perform a 3 × 2 between-subjects ANOVA. We want to determine the effect on persuasiveness when we change (1) the levels of the volume, (2) the levels of gender, and (3) the interaction of volume and gender. To do so, we will examine one effect at a time, as if we had only a one-way ANOVA involving that effect. You already understand a one-way ANOVA, so the following is simply a guide for computing the various Fs. Any two-way ANOVA involves examining three things: the two *main effects* and the *interaction effect*.

The Main Effect of Factor A

The **main effect** of a factor is the effect that changing the levels of that factor has on dependent scores, while we ignore all other factors in the study. In the persuasiveness study, to find the main effect of factor A (volume), we will ignore the levels of factor B

TABLE 14.2

A 3 × 2 Design for the Factors of Volume and Gender

Each column represents a level of the volume factor; each row represents a level of the gender factor.

		Factor A: Volume		
		Level A₁: *Soft*	*Level A₂:* *Medium*	*Level A₃:* *Loud*
Factor B: *Gender*	*Level B₁:* *Male*	9 4 11	8 12 13	18 17 15
	Level B₂: *Female*	2 6 4	9 10 17	6 8 4

$N = 18$

(gender). Literally erase the horizontal line that separates the rows of males and females back in Table 14.2, and treat the experiment as if it were this:

Factor A: Volume

Level A_1: Soft	Level A_2: Medium	Level A_3: Loud	
9	8	18	
4	12	17	
11	13	15	
2	9	6	$k_A = 3$
6	10	8	
4	17	4	
$\overline{X}_{A_1} = 6$	$\overline{X}_{A_2} = 11.5$	$\overline{X}_{A_3} = 11.33$	
$n_{A_1} = 6$	$n_{A_2} = 6$	$n_{A_3} = 6$	

We ignore whether there are males or females in each condition, so we simply have *people*. Therefore, for example, we started with three males and three females who heard the soft message, so ignoring gender, we have six people in that level. Thus, when we look at the main effect of A, our entire experiment consists of one factor, with three levels of volume. The number of levels in factor A is called k_A, so $k_A = 3$. With six scores in each column, $n_A = 6$.

Notice that the means in the three columns are 6, 11.5, and 11.33, respectively. So, for example, the mean for people (male and female) tested under the soft condition is 6. By averaging together the scores in a column, we produce the main effect means for the column factor. A **main effect mean** is the overall mean of one level of a factor while ignoring the influence of the other factor. Here, we have the "main effect means for volume."

In statistical terminology, we produce the main effect means for volume when we *collapse* across the gender factor. **Collapsing** across a factor means averaging together all scores from all levels of that factor. (We averaged together the scores from males and females.) Thus, whenever we collapse across one factor, we have the main effect means for the remaining factor.

To see the *main effect* of volume, look at the overall pattern in the three main effect means to see how persuasiveness scores change as volume increases: Scores go up from around 6 (at soft) to around 11.5 (at medium), but then scores drop slightly to around 11.3 (at high). To determine if these are significant differences—if there is a *significant main effect of the volume factor*—we essentially perform a one-way ANOVA that compares these three main effect means.

> **REMEMBER** When we examine the *main effect* of factor A, we look at the overall mean (the *main effect mean*) of each level of A, examining the column means.

The H_0 says that no difference exists between the levels of factor A in the population, so

$$H_0: \mu_{A_1} = \mu_{A_2} = \mu_{A_3}$$

In our study, this says that changing volume has no effect, so the three levels of volume represent the same population of persuasiveness scores. If we reject H_0, then we will accept the alternative hypothesis, which is

$$H_a: \text{not all } \mu_A \text{ are equal}$$

For our study, this says that at least two levels of volume represent different populations of scores, having different μs.

We test H_0 by computing an F_{obt} called F_A. If F_A is significant, it indicates that at least two main effect means from factor A differ significantly. Then we describe this relationship by graphing the main effect means, performing post hoc comparisons to determine which means differ significantly, and determining the proportion of variance that is accounted for by this factor.

The Main Effect of Factor B

After analyzing the main effect of factor A, we move on to the main effect of factor B. Therefore, we collapse across factor A (volume), so erase the vertical lines separating the levels of volume back in Table 14.2. Then we have

	Level B₁: *Male*	9 8 18 4 12 17 11 13 15	$\overline{X}_{B_1} = 11.89$ $n_{B_1} = 9$	
Factor B: *Gender*				$k_B = 2$
	Level B₂: *Female*	2 9 6 6 10 8 4 17 4	$\overline{X}_{B_2} = 7.33$ $n_{B_2} = 9$	

Now we simply have the persuasiveness scores of some males and some females, ignoring the fact that some of each heard the message at different volumes. Thus, when we look at the main effect of B, now our entire experiment consists of one factor with *two* levels. Notice that this changes things: With only two levels, k_B is 2. And each n has changed! For example, we started with three males in soft, three in medium, and three in loud. So, ignoring volume, we have a total of nine males (and nine females). With nine scores per row, $n_B = 9$.

Averaging the scores in each row yields the mean persuasiveness score for each gender, which are the *main effect means for factor B*. To see the main effect of this factor, again look at the pattern of the means: Apparently, changing from males to females leads to a drop in scores from around 11.89 to around 7.33.

> REMEMBER   When we examine the *main effect of factor B*, we look at the overall mean (the main effect mean) for each level of B, examining the row means.

To determine if this is a significant difference—if there is a significant *main effect of gender*—we essentially perform another one-way ANOVA that compares these means. Our H_0 says that no difference exists in the population, so

$$H_0: \mu_{B_1} = \mu_{B_2}$$

In our study, this says that our males and females represent the same population. The alternative hypothesis is

$$H_a: \text{not all } \mu_B \text{ are equal}$$

In our study, this says that our males and females represent different populations.

To test H_0 for factor B, we compute a separate F_{obt}, called F_B. If F_B is significant, then at least two of the main effect means for factor B differ significantly. Then we graph these means, perform post hoc comparisons, and compute the proportion of variance accounted for.

A QUICK REVIEW

■ Collapsing (averaging together) the scores from the levels of factor B produces the main effect means for factor A. Differences among these means reflect the main effect of A. Collapsing the levels of A produces the main effect means for factor B. Differences among these means reflect the main effect of B.

MORE EXAMPLES

Let's say that our previous "smart pill" and age study produced the following IQ scores:

Factor A: Dose

	One Pill	*Two Pills*	
10 years	100 105 110	140 145 150	$\overline{X} = 125$
20 years	110 115 120	110 115 120	$\overline{X} = 115$
	$\overline{X} = 110$	$\overline{X} = 130$	

Factor B: Age

The column means are the main effect means for dose: The main effect is that mean IQ increases from 110 to 130 as dosage increases. The row means are the main effect means for age: The main effect is that mean IQ decreases from 125 to 115 as age increases.

For Practice

In this study,

	A_1	A_2
B_1	2 2 2	5 4 3
B_2	11 10 9	7 6 5

1. Each n_A equals _____, and each n_B equals _____.
2. The means produced by collapsing across factor B equal _____ and _____. They are called the _____ means for factor _____.
3. What is the main effect of A?
4. The means produced by collapsing across factor A are _____ and _____. They are called the _____ means for factor _____.
5. What is the main effect of B?

Answers

1. 6; 6
2. $\overline{X}_{A_1} = 6; \overline{X}_{A_2} = 5$; main effect; A
3. Changing from A_1 to A_2 produces a decrease in scores.
4. $\overline{X}_{B_1} = 3; \overline{X}_{B_2} = 8$; main effect; B
5. Changing from B_1 to B_2 produces an increase in scores.

Interaction Effects

After examining the main effects, we examine the effect of the interaction. The interaction of two factors is called a two-way interaction, and results from combining the levels of factor A with the levels of factor B. In our example, the interaction is produced by combining volume with gender. An interaction is identified as $A \times B$. Here, factor A has three levels, and factor B has two levels, so it is a 3×2 interaction (but say "3 *by* 2").

Because an interaction is the influence of combining the levels of both factors, we do not collapse across, or ignore, either factor. Instead, we treat each *cell* in the study as a level of the interaction and compare the *cell means*.

REMEMBER   For the interaction effect we compare the cell means.

In our study, we start with the original six cells back in Table 14.2. Using the three scores per cell, we compute the mean in each cell, obtaining the interaction means shown in Table 14.3.

TABLE 14.3

Cell Means for the
Volume by Gender
Interaction

Factor A: Volume

		Soft	*Medium*	*Loud*
Factor B: **Gender**	*Male*	$\overline{X} = 8$	$\overline{X} = 11$	$\overline{X} = 16.67$
	Female	$\overline{X} = 4$	$\overline{X} = 12$	$\overline{X} = 6$

$k = 6$
$n = 3$

Notice that things have changed again. For k, now we have six cells, so $k_{A \times B} = 6$. For n, we are looking at the scores in only one cell at a time, so our "cell size" is 3, so $n_{A \times B} = 3$.

Thus, now our experiment is somewhat like one "factor" with six levels, with three scores per level. We will determine if the mean in the male–soft cell is different from in the male–medium cell or from in the female–soft cell, and so on. However, examining an interaction is not as simple as saying that the cell means are significantly different. Interpreting an interaction is difficult because both independent variables are changing, as well as the dependent scores. To simplify the process, look at the influence of changing the levels of factor A under *one* level of factor B. Then see if this effect—this pattern—for factor A is *different* when you look at the other level of factor B. For example, here is the first row from Table 14.3, showing the relationship between volume and scores for the males. What happens? As volume increases, mean persuasiveness scores also increase, in an apparently positive, linear relationship.

Factor A: Volume

	Soft	*Medium*	*Loud*
B_1: *male*	$\overline{X} = 8$	$\overline{X} = 11$	$\overline{X} = 16.67$

But now look at the relationship between volume and persuasiveness scores for the females.

Factor A: Volume

	Soft	*Medium*	*Loud*
B_2: *female*	$\overline{X} = 4$	$\overline{X} = 12$	$\overline{X} = 6$

Here, as volume increases, mean persuasiveness scores first increase but then decrease, producing a nonlinear relationship.

Thus, there is a different relationship between volume and persuasiveness scores for each gender level. A **two-way interaction effect** is present when the relationship between one factor and the dependent scores changes with, or *depends on*, the level of the other factor that is present. Thus, whether increasing volume always increases scores *depends* on whether we're talking about males or females. In other words, an interaction effect occurs when the influence of changing one factor is not the same for each level of the other factor. Here we have an interaction effect because increasing the volume does not have the same effect for males as it does for females.

You can also see the interaction by looking at the difference between males and females at each volume back in Table 14.3. Who scores higher, males or females? It *depends* on which level of volume we're talking about.

Conversely, an interaction effect would not be present if the cell means formed the *same* pattern for males and females. For example, say that the cell means had been as follows:

Factor A: Volume

		Soft	Medium	Loud
Factor B: **Gender**	**Male**	$\overline{X} = 5$	$\overline{X} = 10$	$\overline{X} = 15$
	Female	$\overline{X} = 20$	$\overline{X} = 25$	$\overline{X} = 30$

Here, each increase in volume increases scores by about 5 points, *regardless* of whether it's for males or females. (Or, females always score higher, regardless of volume.) Thus, an interaction effect is not present when the influence of changing the levels of one factor does not depend on which level of the other variable is present. Or, in other words, there's no interaction when there is the same relationship between the scores and one factor for each level of the other factor.

> **REMEMBER** *A two-way interaction effect* indicates that the influence that one factor has on scores depends on which level of the other factor is present.

To determine if there is a *significant* interaction effect in our data, we essentially perform another one-way ANOVA that compares the cell means. To write the H_0 and H_a in symbols is complicated, but in words, H_0 is that the cell means do not represent an interaction effect in the population, and H_a is that at least some of the cell means do represent an interaction effect in the population.

To test H_0, we compute another F_{obt}, called $F_{A \times B}$. If $F_{A \times B}$ is significant, it indicates that at least two of the cell means differ significantly in a way that produces an interaction effect. Then, as always, we graph the interaction, perform post hoc comparisons to determine which cell means differ significantly, and compute the proportion of variance accounted for by the interaction.

A QUICK REVIEW

- We examine the interaction effect by looking at the cell means. An effect is present if the relationship between one factor and the dependent scores changes as the levels of the other factor change.

MORE EXAMPLES

Here are the data again when factor A is dose of the smart pill and factor B is age of participants.

Factor A: Dose

		One Pill	Two Pills
	10 years	100 105 110 $\overline{X} = 105$	140 145 150 $\overline{X} = 145$
Factor B: **Age**			
	20 years	110 115 120 $\overline{X} = 115$	110 115 120 $\overline{X} = 115$

(continued)

Look at the cell means in one *row* at a time: We see an interaction effect because the influence of increasing dose depends on participants' age. Dosage increases mean IQ for 10-year-olds from 105 to 145, but it does not change mean IQ for 20-year-olds (always at 115). Or, looking at each *column,* the influence of increasing age depends on dose. With 1 pill, 20-year-olds score higher (115) than 10-year-olds (105), but with 2 pills 10-year-olds score higher (145) than 20-year-olds (115).

For Practice

A study produces these data:

	A_1	A_2
B_1	2	5
	2	4
	2	3
B_2	11	7
	10	6
	9	5

1. The means to examine for the interaction are called the _____ means.
2. When we change from A_1 to A_2 for B_1, the cell means are _____ and _____.
3. When we change from A_1 to A_2 for B_2, the cell means are _____ and _____.
4. How does the influence of changing from A_1 to A_2 depend on the level of *B* that is present?
5. Is an interaction effect present?

Answers

1. cell
2. 2, 4
3. 10, 6
4. Under B_1 the means increase, under B_2 they decrease.
5. Yes

COMPUTING THE TWO-WAY ANOVA

We've seen that a two-way ANOVA involves three *F*s: one for the main effect of factor A, one for the main effect of factor B, and one for the interaction effect of $A \times B$. The logic for each of these is the same as in the one-way ANOVA discussed in Chapter 13: F_{obt} should equal 1 if H_0 is true. The larger the F_{obt}, however, the less likely that H_0 is true. If H_0 is larger than F_{crit}, we will reject H_0.

Previously when we computed F_{obt} we saw two important formulas to keep in mind:

$$F_{obt} = \frac{MS_{bn}}{MS_{wn}} \qquad MS = \frac{SS}{df}$$

The MS_{wn} describes the variability within groups and is our estimate of σ^2_{error}. It is computed as the "average" variability in the *cells.* The MS_{wn} is our *one* estimate of the error variance used in all three *F* ratios. We compute MS_{wn} by computing SS_{wn} and dividing by df_{wn}.

The MS_{bn} indicates the differences between our sample means, as an estimate of σ^2_{treat}. However, because we have two main effects and the interaction, we have three sources of between-groups variance, so we compute three mean squares. Thus, for the main effect of factor A, we compute the sum of squares between groups for A (SS_A), divide by the degrees of freedom for A (df_A), and have the mean square between groups for A (MS_A). For the main effect of B, we compute the sum of squares between groups for B (SS_B), divide by the degrees of freedom for B (df_B), and have the mean square between groups for B (MS_B). For the interaction, we compute the sum of squares between groups $(SS_{A \times B})$, divide by the degrees of freedom $(df_{A \times B})$, and have the mean square between groups $(MS_{A \times B})$.

The summary table in Table 14.4 shows the preceding components. (Notice that if a component is not labeled "within" or "total," then it is one of our many between-groups components.) To complete the ANOVA, for factor A, divide MS_A by MS_{wn} to produce F_A. For factor B, divide MS_B by MS_{wn} to produce F_B. For the interaction, divide $MS_{A \times B}$ by MS_{wn} to produce $F_{A \times B}$.

The following sections show how to compute the above components.

Computing the Sums and Means

Your first step is to organize the data in each cell. Table 14.5 shows the persuasiveness data, as well as the various components to compute.

Step 1 *Compute ΣX and ΣX^2 in each cell.* For example, in the male–soft cell, $\Sigma X = 4 + 9 + 11 = 24$; $\Sigma X^2 = 4^2 + 9^2 + 11^2 = 218$. Also note the n of the cell (here $n = 3$.) Then compute the mean for each cell (for the male–soft cell, $\overline{X} = 8$). These are the interaction means.

Step 2 *Compute ΣX vertically in each column.* Sum the ΣXs from the cells in a column (for example, for soft, $\Sigma X = 24 + 12$). Note the n in each column (here $n = 6$) and compute the mean for the scores in each column (for example, $\overline{X}_{soft} = 6$). (Or average the cell means in the column.) These are the main effect means for factor A.

Step 3 *Compute ΣX horizontally in each row.* Sum the ΣXs from the cells in a row (for males, $\Sigma X = 24 + 33 + 50 = 107$). Note the n in each row (here $n = 9$.) Compute the mean of the scores in each row (for example, $\overline{X}_{male} = 11.89$). (Or average the cell means in the row.) These are the main effect means for factor B.

Step 4 *Compute ΣX_{tot}.* Sum the ΣX from the levels (columns) of factor A, so $\Sigma X_{tot} = 36 + 69 + 68 = 173$. (Or add the ΣX from the levels of factor B.)

Step 5 *Compute ΣX_{tot}^2.* Sum the ΣX^2 from each cell, so $\Sigma X_{tot}^2 = 218 + 377 + 838 + 56 + 470 + 116 = 2075$. Note the N (here $N = 18$.)

We will use the components from the previous five steps to compute the sums of squares, then the degrees of freedom, then the mean squares, and finally the Fs. To keep track of your computations and prevent brain strain, fill in the ANOVA summary table as you go along.

TABLE 14.4

Summary Table of Two-Way ANOVA

Source	Sum of Squares	/	df	=	Mean Square	F
Between						
Factor A (volume)	SS_A		df_A		MS_A	F_A
Factor B (gender)	SS_B		df_B		MS_B	F_B
Interaction (vol × gen)	$SS_{A \times B}$		$df_{A \times B}$		$MS_{A \times B}$	$F_{A \times B}$
Within	SS_{wn}		df_{wn}		MS_{wn}	
Total	SS_{tot}		df_{tot}			

TABLE 14.5

Summary of Data for
3×2 ANOVA

		Factor A: Volume		
		A_1: Soft	A_2: Medium	A_3: Loud	
	B_1: Male	4 9 11 $\overline{X} = 8$ $\Sigma X = 24$ $\Sigma X^2 = 218$ $n = 3$	8 12 13 $\overline{X} = 11$ $\Sigma X = 33$ $\Sigma X^2 = 377$ $n = 3$	18 17 15 $\overline{X} = 16.67$ $\Sigma X = 50$ $\Sigma X^2 = 838$ $n = 3$	$\overline{X}_{male} = 11.89$ $\Sigma X = 107$ $n = 9$
Factor B: Gender	B_2: Female	2 6 4 $\overline{X} = 4$ $\Sigma X = 12$ $\Sigma X^2 = 56$ $n = 3$	9 10 17 $\overline{X} = 12$ $\Sigma X = 36$ $\Sigma X^2 = 470$ $n = 3$	6 8 4 $\overline{X} = 6$ $\Sigma X = 18$ $\Sigma X^2 = 116$ $n = 3$	$\overline{X}_{fem} = 7.33$ $\Sigma X = 66$ $n = 9$
		$\overline{X}_{soft} = 6$ $\Sigma X = 36$ $n = 6$	$\overline{X}_{med} = 11.5$ $\Sigma X = 69$ $n = 6$	$\overline{X}_{loud} = 11.33$ $\Sigma X = 68$ $n = 6$	$\Sigma X_{tot} = 173$ $\Sigma X_{tot}^2 = 2075$ $N = 18$

Computing the Sums of Squares

First, compute the sums of squares.

Step 1 *Compute the total sum of squares.*

> **The formula for the total sum of squares is**
>
> $$SS_{tot} = \Sigma X_{tot}^2 - \left(\frac{(\Sigma X_{tot})^2}{N} \right)$$

This says to divide $(\Sigma X_{tot})^2$ by N and then subtract the answer from ΣX_{tot}^2.
From Table 14.5, $\Sigma X_{tot} = 173$, $\Sigma X_{tot}^2 = 2075$, and $N = 18$. Filling in the formula
gives

$$SS_{tot} = 2075 - \left(\frac{(173)^2}{18} \right)$$

$$SS_{tot} = 2075 - 1662.72$$

$$SS_{tot} = 412.28$$

Note: The quantity $(\Sigma X_{tot})^2/N$ is also used when computing other sums of squares.
We call it the *correction* (here the correction equals 1662.72).

Step 2 *Compute the sum of squares for factor A.* As in the diagrams here, always have
factor A form your *columns*.

> **The formula for the sum of squares between groups for factor A is**
>
> $$SS_A = \Sigma\left(\frac{(\text{Sum of scores in the column})^2}{n \text{ of scores in the column}}\right) - \left(\frac{(\Sigma X_{tot})^2}{N}\right)$$

This says to square the ΣX in each column of factor A and divide by the n in the column. Then add the answers together and subtract the correction.

From Table 14.5, the three columns produced sums of 36, 69, and 68, and n was 6. Filling in the above formula gives

$$SS_A = \left(\frac{(36)^2}{6} + \frac{(69)^2}{6} + \frac{(68)^2}{6}\right) - \left(\frac{(173)^2}{18}\right)$$

$$SS_A = (216 + 793.5 + 770.67) - 1662.72$$

$$SS_A = 1780.17 - 1662.72$$

$$SS_A = 117.45$$

Step 3 *Compute the sum of squares for factor B.* In your diagram, the levels of factor B should form the *rows*.

> **The formula for the sum of squares between groups for factor B is**
>
> $$SS_B = \Sigma\left(\frac{(\text{Sum of scores in the row})^2}{n \text{ of scores in the row}}\right) - \left(\frac{(\Sigma X_{tot})^2}{N}\right)$$

This says to square the ΣX for each level of factor B and divide by the n in the level. Then add the answers and subtract the correction.

In Table 14.5, the two rows produced sums of 107 and 66, and n was 9. Filling in the formula gives

$$SS_B = \left(\frac{(107^2)}{9} + \frac{(66^2)}{9}\right) - 1662.72$$

$$SS_B = 1756.11 - 1662.72$$

$$SS_B = 93.39$$

Step 4 *Compute the sum of squares between groups for the interaction.* This requires two substeps. First, compute something called the total sum of squares between groups, identified as SS_{bn}.

> **The formula for the total sum of squares between groups is**
>
> $$SS_{bn} = \Sigma \left(\frac{(\text{Sum of scores in the cell})^2}{n \text{ of scores in the cell}} \right) - \left(\frac{(\Sigma X_{tot})^2}{N} \right)$$

Find $(\Sigma X)^2$ for each cell and divide by the n of the cell. Then add the answers together and subtract the correction.

From Table 14.5,

$$SS_{bn} = \left(\frac{(24)^2}{3} + \frac{(33)^2}{3} + \frac{(50)^2}{3} + \frac{(12)^2}{3} + \frac{(36)^2}{3} + \frac{(18)^2}{3} \right) - 1662.72$$

$$SS_{bn} = 1976.33 - 1662.72$$

$$SS_{bn} = 313.61$$

The SS_{bn} equals the sum of squares for factor A plus the sum of squares for factor B plus the sum of squares for the interaction. Therefore, we find $SS_{A \times B}$ by subtracting the SS_A and SS_B found above (in Steps 2 and 3) from the total SS_{bn}. Thus,

> **The formula for the sum of squares between groups for the interaction is**
>
> $$SS_{A \times B} = SS_{bn} - SS_A - SS_B$$

Above, $SS_{bn} = 313.61$, $SS_A = 117.45$, and $SS_B = 93.39$, so

$$SS_{A \times B} = 313.61 - 117.45 - 93.39$$

$$SS_{A \times B} = 102.77$$

Step 5 *Compute the sum of squares within groups.* The sum of squares within groups plus the total sum of squares between groups equals the total sum of squares. Therefore, subtract the total SS_{bn} in Step 4 from the SS_{tot} in Step 1 to obtain the SS_{wn}.

> **The formula for the sum of squares within groups is**
>
> $$SS_{wn} = SS_{tot} - SS_{bn}$$

Above, $SS_{tot} = 412.28$ and $SS_{bn} = 313.61$, so

$$SS_{wn} = 412.28 - 313.61$$

$$SS_{wn} = 98.67$$

The previous sums of squares are shown in Table 14.6.
Now determine the *df*.

TABLE 14.6

Summary Table of Two-Way ANOVA showing the Sums of Squares

Source	Sum of Squares	df	Mean Square	F
Between				
Factor A (volume)	117.45	df_A	MS_A	F_A
Factor B (gender)	93.39	df_B	MS_B	F_B
Interaction (vol × gen)	102.77	$df_{A\times B}$	$MS_{A\times B}$	$F_{A\times B}$
Within	98.67	df_{wn}	MS_{wn}	
Total	412.28	df_{tot}		

Computing the Degrees of Freedom

1. *The degrees of freedom between groups for factor A is* $k_A - 1$, *where* k_A *is the number of levels in factor A.* (In our example, k_A is the three levels of volume, so $dk_A = 2$.)

2. *The degrees of freedom between groups for factor B is* $k_B - 1$, *where* k_B *is the number of levels in factor B.* (In our example, k_B is the two levels of gender, so $df_B = 1$.)

3. *The degrees of freedom between groups for the interaction is the* df *for factor A multiplied times the* df *for factor B.* (In our example, $df_A = 2$ and $df_B = 1$, so $df_{A\times B} = 2$.)

4. *The degrees of freedom within groups equals* $N - k_{A\times B}$, *where* N *is the total* N *of the study and* $k_{A\times B}$ *is the number of cells in the study.* (In our example, N is 18 and we have six cells, so $df_{wn} = 18 - 6 = 12$.)

5. *The degrees of freedom total equals* $N - 1$. Use this to check your previous calculations, because the sum of the above *df*s should equal df_{tot}. (In our example $df_{tot} = 17$.)

Place each *df* in the ANOVA summary table as in Table 14.7. Perform the remainder of the computations by working directly from this table. Next, we compute the mean squares.

Computing the Mean Squares

Any mean square equals the appropriate sum of squares divided by the appropriate *df*. Therefore, for factor A,

> **The formula for the mean square between groups for factor A is**
> $$MS_A = \frac{SS_A}{df_A}$$

From Table 14.7 we find

$$MS_A = \frac{117.45}{2} = 58.73$$

TABLE 14.7

Summary Table of Two-Way ANOVA with *df* and Sums of Squares

Source	*Sum of Squares*	df	*Mean Square*	F
Between				
Factor A (volume)	117.45	2	MS_A	F_A
Factor B (gender)	93.39	1	MS_B	F_B
Interaction (vol × gen)	102.77	2	$MS_{A \times B}$	$F_{A \times B}$
Within	98.67	12	MS_{wn}	
Total	412.28	17		

> **The formula for the mean square between groups for factor B is**
>
> $$MS_B = \frac{SS_B}{df_B}$$

From Table 14.7

$$MS_B = \frac{93.39}{1} = 93.39$$

> **The formula for the mean square between groups for the interaction is**
>
> $$MS_{A \times B} = \frac{SS_{A \times B}}{df_{A \times B}}$$

Using Table 14.7

$$MS_{A \times B} = \frac{102.77}{2} = 51.39$$

> **The formula for the mean square within groups is**
>
> $$MS_{wn} = \frac{SS_{wn}}{df_{wn}}$$

Table 14.7 gives

$$MS_{wn} = \frac{98.67}{12} = 8.22$$

Putting the above values into the summary table gives Table 14.8.

Now, finally, compute the Fs.

Computing F

Any F equals the MS_{bn} divided by the MS_{wn}. Therefore,

The formula for F_A for the main effect of factor A is

$$F_A = \frac{MS_A}{MS_{wn}}$$

In our example, from Table 14.8 we have

$$F_A = \frac{58.73}{8.22} = 7.14$$

The formula for F_B for the main effect of factor B is

$$F_B = \frac{MS_B}{MS_{wn}}$$

Thus, we have

$$F_B = \frac{93.39}{8.22} = 11.36$$

The formula for $F_{A \times B}$ for the interaction effect is

$$F_{A \times B} = \frac{MS_{A \times B}}{MS_{wn}}$$

TABLE 14.8

Summary Table of Two-Way ANOVA Showing the Mean Squares, *df*, and Sums of Squares

Source	Sum of Squares	df	Mean Square	F
Between				
Factor A (volume)	117.45	2	58.73	F_A
Factor B (gender)	93.39	1	93.39	F_B
Interaction (vol × gen)	102.77	2	51.39	$F_{A \times B}$
Within	98.67	12	8.22	
Total	412.28	17		

So

$$F_{A \times B} = \frac{51.39}{8.22} = 6.25$$

And now the finished summary table is in Table 14.9.

Interpreting Each *F*

Each F_{obt} is tested in the same way as in the previous chapter: The F_{obt} may be larger than 1 because (1) H_0 is true but we have sampling error or (2) H_0 is false and at least two means represent a relationship in the population. The larger an F_{obt}, the less likely that H_0 is true. If F_{obt} is larger than F_{crit}, then we reject H_0.

To find the F_{crit} for a particular F_{obt}, in the *F*-tables (Table 5 in Appendix *C*), use the df_{bn} that you used in computing that F_{obt} and your df_{wn}. Thus,

1. To find F_{crit} for testing F_A, use df_A as the *df* between groups and df_{wn}. In our example, $df_A = 2$ and $df_{wn} = 12$. So, for $\alpha = .05$, the F_{crit} is 3.88.
2. To find F_{crit} for testing F_B, use df_B as the *df* between groups and df_{wn}. In our example, $df_B = 1$ and $df_{wn} = 12$. So, at $\alpha = .05$, the F_{crit} is 4.75.
3. To find F_{crit} for the interaction, use $df_{A \times B}$ as the *df* between groups and df_{wn}. In our example, $df_{A \times B} = 2$ and $df_{wn} = 12$. Thus, at $\alpha = .05$, the F_{crit} is 3.88.

Notice that because factors A and B have different *df* between groups, they have different critical values.

Thus, we end up comparing the following:

	F_{obt}	F_{crit}
Main effect of volume (A)	7.14	3.88
Main effect of gender (B)	11.36	4.75
Interaction (A × B)	6.25	3.88

By now you can do this with your eyes closed: Imagine a sampling distribution with a region of rejection and F_{crit} in the positive tail. (If you can't imagine this, look back in Chapter 13 at Figure 13.1.) First, our F_A of 7.14 is larger than the F_{crit}, so it lies in the region of rejection. Therefore, we conclude that changing the volume of a message produced significant differences in persuasiveness scores.

Likewise, the F_B of 11.36 is significant, so we conclude that the males and females in this study represent different populations of scores.

TABLE 14.9

Completed Summary Table of Two-Way ANOVA

Source	Sum of Squares	df	Mean Square	F
Between				
Factor A (volume)	117.45	2	58.73	7.14
Factor B (gender)	93.39	1	93.39	11.36
Interaction (vol × gen)	102.77	2	51.39	6.25
Within	98.67	12	8.22	
Total	412.28	17		

Finally, the $F_{A \times B}$ of 6.25 is significant, so we conclude that the effect that changing volume has in the population *depends* on whether it is a population of males or a population of females. Or we can say that the difference between the male and female populations we'd see *depends* on whether a message is played at soft, medium, or loud volume.

Note: It is just a coincidence of your particular data which *F*s will be significant: any combination of the main effects and/or the interaction may or may not be significant.

A QUICK REVIEW

- A two-way ANOVA produces an F_{obt} for each main effect and for the interaction.
- Each *MS* equals the corresponding *SS* divided by the *df*.
- Each F_{obt} equals the appropriate *MS* between groups divided by the MS_{wn}.

MORE EXAMPLES

In a new study, our factor A tests the effect of three doses of a smart pill (one, two, or three pills), and factor B tests two ages (10- or 20-year olds). We obtain the *SS*s and *df*s shown here.

Source	SS	df	MS	F
A(dose)	48	2	24	6.00
B (age)	8	1	8	2.00
Interaction	38	2	19	4.75
Within	72	18	4	
Total	166			

Then, MS_A is $48/2 = 24$; MS_B is $8/1 = 8$; $MS_{A \times B}$ is $38/2 = 19$; MS_{wn} is $72/18 = 4$. Then, F_A is $24/4 = 6.00$; F_B is $8/4 = 2.00$; $F_{A \times B}$ is $19/4 = 4.75$. The F_{crit} for A is 3.55, so F_A is significant: Changing

dose influenced IQ. The F_{crit} for B is 4.41, so F_B is not significant. The F_{crit} for $F_{A \times B}$ is 3.55, so it is significant: The influence that changing dose had on IQ depends on participants' age.

For Practice

1. For this 4×3 study, complete the summary table:

Source	SS	df	MS	F
A	108	3	____	____
B	104	2	____	____
Interaction	252	6	____	____
Within	672	48	____	____
Total	536	59	____	

2. The F_{crit} for F_A is ____, the F_{crit} for F_B is ____, and the F_{crit} for $F_{A \times B}$ is ____.

3. Which effects are significant?

Answers

1. $MS_A = 36$, $MS_B = 52$, $MS_{A \times B} = 42$, and $MS_{wn} = 14$; $F_A = 2.57$, $F_B = 3.71$, and $F_{A \times B} = 3.00$
2. 2.80; 3.19; 2.30
3. The main effect of *B* and the interaction effect

INTERPRETING THE TWO-WAY EXPERIMENT

To understand and interpret the results of a two-way ANOVA, you should examine the means from each significant main effect and interaction by graphing them and performing post hoc comparisons.

Graphing and Post Hoc Comparisons with Main Effects

We graph each main effect separately, plotting all main effect means, even those that the post hoc tests may indicate are not significant. As usual, label the *Y* axis as the mean of the dependent scores and place the levels of the factor on the *X* axis. For our volume

FIGURE 14.1

Graphs showing main effects of volume and gender

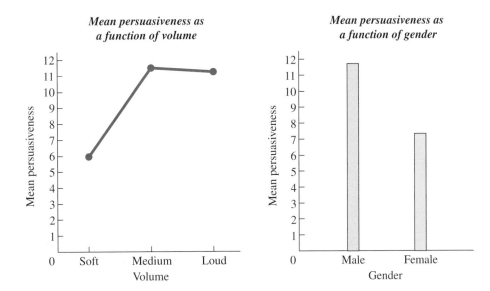

factor, the main effect means were 6.00 for low, 11.50 for medium, and 11.33 for loud. For gender, the means were 11.89 for males and 7.33 for females. Figure 14.1. shows the resulting graphs. Volume is a continuous, ratio variable, so we create a line graph. Gender is a discrete, nominal variable, so we create a bar graph.

To determine which main effect means differ significantly, we usually perform Tukey's *HSD* procedure because we usually have equal *n*s. (With unequal *n*s, perform Fisher's *t*-test, as in Chapter 13.) Recall that the formula for the *HSD* is

$$HSD = (q_k)\left(\sqrt{\frac{MS_{wn}}{n}}\right)$$

where MS_{wn} is from the ANOVA, q_k is from Table 6 in Appendix C for df_{wn} and k (where k is the number of levels in the factor), and n is the number of scores in a level. But be careful here: Recall that the k and n were different for each main effect. In particular, *the n is always the number of scores used to compute each mean you are comparing right now!* Also, because q_k depends on k, when factors have a different k, they have different values of q_k. Therefore, you will have a different *HSD* for each main effect when k or n is different.

Our volume factor has three means, and each n is 6. The $MS_{wn} = 8.22$. With $\alpha = .05$, $k = 3$, and $df_{wn} = 12$, the $q_k = 3.77$. Thus,

$$HSD = (q_k)\left(\sqrt{\frac{MS_{wn}}{n}}\right) = (3.77)\left(\sqrt{\frac{8.22}{6}}\right) = 4.41$$

The *HSD* for factor A is 4.41. The mean for soft (6.00) differs from the means for medium (11.50) and loud (11.33) by more than 4.41. Thus, soft produces a significant difference from the other volumes: increasing volume from soft to medium and from soft to loud produced significant differences (increases) in scores. However, the means for medium and loud differ by *less* than 4.41, so these conditions do *not* differ significantly: Increasing volume from medium to loud did not produce a significant change in scores.

After comparing the main effect means of factor A, we move on the main effect means of factor B. When a factor contains only two levels (like our gender factor), we do not need to perform post hoc comparisons (it must be that the mean for males differs significantly from the mean for females). If, however, a significant factor B had more than two levels, you would compute the *HSD* using the *n* and *k* in that factor and compare the differences between these main effect means as we did above.

Graphing the Interaction Effect

An interaction can be a beast to interpret, so always graph it! As usual, place the dependent variable along the *Y* axis. To produce the simplest graph, place the factor with the most levels on the *X* axis. You'll show the other factor by drawing a separate line on the graph for each of its levels.

Thus, we'll label the *X* axis with our three volumes. Then we plot the *cell* means. The resulting graph is shown in Figure 14.2. As in any graph, you're showing the relationship between the *X* and *Y* variables, but here you're showing the relationship between volume and persuasiveness, first for males and then for females. Thus, approach this in the same way that we examined the means back in Table 14.3. There, we first looked at the relationship between volume and persuasiveness scores for males: Their cell means are $\overline{X}_{soft} = 8$, $\overline{X}_{medium} = 11$, and $\overline{X}_{loud} = 16.67$. Plot these three means and connect the adjacent data points with straight lines. Then we looked at the relationship between volume and scores for *females:* Their cell means are $\overline{X}_{soft} = 4$, $\overline{X}_{medium} = 12$, and $\overline{X}_{loud} = 6$. Plot these means and connect their adjacent data points with straight lines. (*Note:* Always provide a key to identify each line.)

The way to read the graph is to look at one line at a time. For males (the dashed line), as volume increases, mean persuasiveness scores increase. However, for females (the solid line), as volume increases, persuasiveness scores first increase but then decrease. Thus, we see a linear relationship for males and a different, nonlinear relationship for females. Therefore, the graph shows an interaction effect by showing that the effect of increasing volume depends on whether the participants are male or female.

REMEMBER Graph the interaction by drawing a separate line that shows the relationship between the factor on the *X* axis and the dependent (*Y*) scores for each level of the other factor.

FIGURE 14.2

Graph of cell means, showing the interaction of volume and gender

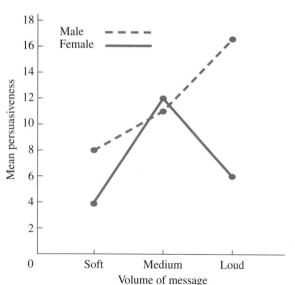

Note one final aspect of an interaction. An interaction effect can produce an infinite variety of different graphs, but *it always produces lines that are not parallel.* Each line summarizes a relationship, and a line that is shaped or oriented differently from another line indicates a *different* relationship. Therefore, when the lines are not parallel they indicate that the relationship between *X* and *Y* changes depending on the level of the second factor, so an interaction effect is present. Conversely, when an interaction effect is not present, the lines will be essentially parallel, with each line depicting essentially the same relationship. To see this distinction, say that our data had produced one of the two graphs in Figure 14.3. On the left, as the levels of A change, the mean scores either increase or decrease depending on the level of B, so an interaction is present. However, on the right the lines are parallel, so as the levels of A change, the scores increase, regardless

FIGURE 14.3

Two graphs showing when an interaction is and is not present

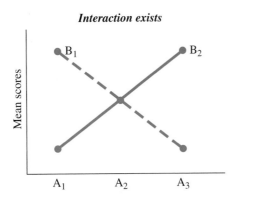

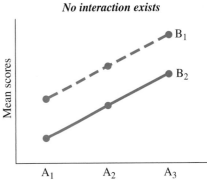

of the level of B. Therefore, this graph does not depict an interaction effect. (The fact that, *overall,* the scores are higher in B_1 than in B_2 is the main effect of—difference due to—factor B.)

Think of significance testing of the interaction $F_{A \times B}$ as testing whether the lines are significantly different from parallel. When an interaction is not significant, the lines may represent parallel lines that would be found for the population. When an interaction is significant, the lines we'd find in the population probably would not be parallel, so there would be an interaction effect in the population.

> *REMEMBER* An *interaction effect* is present when its graph produces lines that are *not* parallel.

Performing Post Hoc Comparisons on the Interaction Effect

To determine which *cell* means in a interaction effect differ significantly, we usually perform Tukey's *HSD* procedure because we usually have equal cell *n*s. (Otherwise, perform Fisher's *t*-test.) However, we do *not* compare every cell mean to every other cell mean. Look at Table 14.10. We would not, for example, compare the mean for males at loud volume to the mean for females at soft volume. This is because we would not know what caused the difference: The two cells differ in terms of both gender *and* volume. Therefore, we would have a confused, or confounded, comparison. A **confounded comparison** occurs when two cells differ along more than one factor. When performing post hoc comparisons on an interaction, we perform only **unconfounded comparisons,** in which two cells differ along only one factor. Therefore, compare only cell means within the same column because these differences result from factor B. Compare means within the same row because these differences

TABLE 14.10

Summary of Interaction Means for Persuasiveness Study

Horizontal and vertical lines between two cells show unconfounded comparisons; diagonal lines show confounded comparisons.

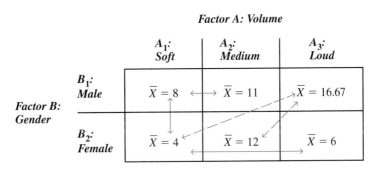

TABLE 14.11

Values of Adjusted k

Design of Study	Number of Cell Means in Study	Adjusted Value of k
2×2	4	3
2×3	6	5
2×4	8	6
3×3	9	7
3×4	12	8
4×4	16	10
4×5	20	12

result from factor A. Do not, however, make any diagonal comparisons because these are confounded comparisons.

We have one other difference when performing the *HSD* on an interaction, and that involves how we obtain q_k. Previously, we found q_k in Table 6 (Appendix C) using k, the number of means being compared. Each q_k in the table is appropriate for making *all* possible comparisons between k means, as in a main effect. However, in an interaction we make fewer comparisons, because we only make unconfounded comparisons. Therefore, we compensate for fewer comparisons by using an *adjusted k*. Obtain the *adjusted k* from Table 14.11 (or at the beginning of Table 6 of Appendix C). In the left-hand column, locate the design of your study. Our study is a 3×2 or a 2×3 design. As a double-check, confirm that the middle column contains the number of cell means in the interaction: we have 6. In the right-hand column is the *adjusted k*: for our study it is 5.

The *adjusted k* is the value of k to use to obtain q_k. Thus, for our study, in Table 6 (Appendix C), we look in the column labeled for k equal to 5. With $\alpha = .05$ and $df_{wn} = 12$, the $q_k = 4.51$. This q_k is appropriate for the number of unconfounded comparisons that we'll actually make.

Now compute the *HSD* using the usual formula. Our MS_{wn} is 8.22, and the n in each mean that we're comparing right now is 3. So

$$HSD = (q_k)\left(\sqrt{\frac{MS_{wn}}{n}}\right) = (4.51)\left(\sqrt{\frac{8.22}{3}}\right) = 7.47$$

Thus, the *HSD* for the interaction is 7.47.

Now determine the differences between all cell means vertically within each column and horizontally within each row. To see these differences, arrange them as in Table 14.12. On the line connecting two cells is the absolute difference between their means. Only three differences are larger than the *HSD* and thus significant: (1) between the mean for females at the soft volume and the mean for females at the medium volume, (2) between the mean for males at the soft volume and the mean for males at the loud volume, and (3) between the mean for males at the loud volume and the mean for females at the loud volume.

TABLE 14.12

Table of the Interaction Cells Showing the Differences Between Unconfounded Means

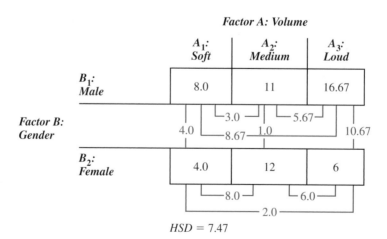

$HSD = 7.47$

A QUICK REVIEW

- The graph of an interaction shows the relationship between one factor on X and dependent scores on Y for each level of the other factor.

- When performing Tukey's *HSD* test on an interaction effect, determine the adjusted value of k and make only unconfounded comparisons.

MORE EXAMPLES

We obtain the cell means on the left. To produce the graph of the interaction on the right, we plot data points at 2 and 6 for B_1 and connect them with the solid line. We plot data points at 10 and 4 for B_2 and connect them with the dashed line.

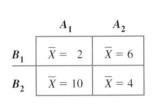

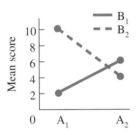

Say that $df_{wn} = 16$, $MS_{wn} = 5.19$ and the n per cell is 5. For the *HSD*, from Table 14.11, the adjusted k is 3. In Table 6 (Appendix C), at $\alpha = .05$, the q_k is 3.65. Then

$$HSD = (q_k)\left(\sqrt{\frac{MS_{wn}}{n}}\right) = (3.65)\left(\sqrt{\frac{5.19}{5}}\right) = 3.72$$

The unconfounded comparisons involve subtracting the means in each column and each row. All differences are significant except when comparing 6 versus 4.

For Practice

We obtain the following data:

	A_1	A_2
B_1	$\bar{X} = 13$	$\bar{X} = 14$
B_2	$\bar{X} = 12$	$\bar{X} = 22$

The $df_{wn} = 12$, $MS_{wn} = 4.89$, and $n = 4$.

1. The adjusted k is ___.
2. The q_k is ___.
3. The *HSD* is ___.
4. Which cell means differ significantly?

Answers

1. 3
2. 3.77
3. 4.17
4. Only 12 versus 22 and 14 versus 22

Interpreting the Overall Results of the Experiment

We report each F_{obt} using the same format as in the one-way ANOVA. Typically we report the F, the means and their significant differences, and the effect size for the main effect of A, then for the main effect of B, and then for the interaction. (Include nonsignificant effects as well.) Then, to interpret the results of the study, focus on the means that differ significantly in each main effect and interaction. All of these differences found in the persuasiveness study are summarized in Table 14.13. Each line connecting two means indicates that they differ significantly. (*Note*: You may also compute the confidence interval for the μ represented by any significant main effect mean or cell mean. Use the formula in Chapter 13.)

Usually, the interpretation of a two-way study rests with the interaction, even when main effects are significant. This is because the conclusions about main effects are contradicted by the interaction. For example, our main effect means for gender suggest that males score higher than females. However, in the interaction, we see that gender

TABLE 14.13

Summary of Significant Differences in the Persuasiveness Study

Each line connects two means that differ significantly.

Factor A: Volume

Factor B: Gender		Level A_1: Soft	Level A_2: Medium	Level A_3: Loud	
	Level B_1: Male	8.0	11	16.67	$\overline{X} = 11.89$
	Level B_2: Female	4.0	12	6	$\overline{X} = 7.33$

$\overline{X}_{soft} = 6$ $\qquad$ $\overline{X}_{med} = 11.5$ $\qquad$ $\overline{X}_{loud} = 11.33$

differences *depend* on volume: Only in the loud condition is there a significant difference between males and females. Therefore, because the interaction contradicts the pattern suggested by the main effect, we *cannot* make an overall, general conclusion about differences between males and females.

Likewise, the main effect of volume showed that increasing volume from soft to medium and from soft to loud produced significant differences. However, the interaction indicates that increasing the volume from soft to medium actually produced a significant difference *only* for females, while increasing the volume from soft to loud produced a significant difference *only* for males.

Thus, as above, usually you cannot draw clear conclusions about significant main effects when the interaction is significant. After all, the interaction indicates that the influence of one factor *depends* on the levels of the other factor and vice versa, so you should not turn around and act like either factor has a consistent effect by itself. When the interaction is not significant, then focus on any significant main effects. (For completeness, however, always perform the entire ANOVA for all main effects and the interaction.)

> REMEMBER The primary interpretation of a two-way ANOVA rests on the interaction when it is significant.

Thus, we conclude that increasing the volume of a message beyond soft tends to increase persuasiveness scores in the population, but this increase occurs for females with medium volume and for males with loud volume. Further, we conclude that differences in persuasiveness scores occur between males and females in the population but only if the volume of the message is loud. (And, after all of the above shenanigans, for all of these conclusions together, the probability of a Type I error in the study—the experiment-wise error rate—is still $p < .05$.)

Describing the Effect Size: Eta Squared

Finally, we again compute eta squared (η^2) to describe effect size—the proportion of variance in dependent scores that is accounted for by a variable. Compute a separate eta squared for each *significant* main and interaction effect. The formula for eta squared is

$$\eta^2 = \frac{\text{Sum of squares between groups for the effect}}{SS_{tot}}$$

This says to divide the SS_{tot} into the sum of squares for the factor, either SS_A, SS_B, or $SS_{A \times B}$. Thus, for our factor A (volume), SS_A was 117.45 and SS_{tot} was 412.28. Therefore,

$$\eta_A^2 = \frac{SS_A}{SS_{tot}} = \frac{117.45}{412.28} = .28$$

Thus, if we predict participants' scores using the main effect mean of the volume condition they were tested under, we can account for 28% of the total variance in persuasiveness scores. Likewise, for the gender factor, SS_B is 93.39, so η_B^2 is .23: Predicting the mean of their condition for male and female participants will account for 23% of the variance in scores. Finally, for the interaction, $SS_{A \times B}$ is 102.77, so $\eta_{A \times B}^2$ is .25: By using the mean of the cell to predict a participant's score, we can account for 25% of the variance.

Recall that the greater the effect size, the more important the effect is in determining participants' scores. Because each of the above has about the same size, they are all of equal importance in understanding differences in persuasiveness scores in this experiment. However, suppose that one effect accounted for only 1% of the total variance. Such a small η^2 indicates that this relationship is very inconsistent, so it is not useful or informative. Therefore, we should emphasize the other, larger significant effects. In essence, if eta squared indicates that an effect was not a big deal in the experiment, then we should not make a big deal out of it when interpreting the experiment.

The effect size is especially important when dealing with interactions. The one exception to the rule of always focusing on the significant interaction is when it has a very small effect size. If the interaction's effect is small (say, only .02), then although the interaction contradicts the main effect, it is only slightly and inconsistently contradictory. In such cases, you may focus your interpretation on any significant main effects that had a more substantial effect size.

SUMMARY OF THE STEPS IN PERFORMING A TWO-WAY ANOVA

The following summarizes the steps in a two-way ANOVA:

1. *Compute the Fs*: Compute *SS, df,* and *MS* for each main effect, the interaction, and for within groups. Dividing each mean square between groups by the mean square within groups produces each F_{obt}.

2. *Find F$_{crit}$*: For each factor or interaction, if F_{obt} is larger than F_{crit}, then there is a significant difference between two or more means from the factor or interaction.

3. *For each significant main effect*: Perform post hoc tests when the factor has more than two levels. Graph each main effect.

4. *For a significant interaction effect*: Perform post hoc tests by making only unconfounded comparisons. For the *HSD,* determine the adjusted *k*. Graph the effect by labeling the *X* axis with one factor and using a separate line to connect the cell means from each level of the other factor.

5. *Compute eta squared*: Describe the proportion of variance in dependent scores accounted for by each significant main effect or interaction.

6. *Compute the confidence interval*: This can be done for the μ represented by the mean in any relevant level or cell.

7. *Interpret the experiment*: Based on the significant main and/or interaction effects and their values of η^2, develop an overall conclusion regarding the relationships formed by the specific means from the cells and levels that differ significantly.

Congratulations, you are getting *very* good at this stuff.

PUTTING IT ALL TOGETHER

Recognize that, although there is no limit to the number of factors we can have in an ANOVA, there is a practical limit to how many factors we can *interpret*. Say that we added a third factor to the persuasiveness study—the sex of the speaker of the message. This would produce a three-way ($3 \times 2 \times 2$) ANOVA in which we compute an F_{obt} for three main effects (*A*, *B*, and *C*), for three two-way interactions (A $\times$ B, A $\times$ C, and B $\times$ C), and for a three-way interaction (A $\times$ B $\times$ C)! (If significant, it indicates that the two-way interaction between volume and participant gender changes, depending on the sex of the speaker.) If this sounds very complicated, it's because it *is* very complicated. Therefore, unless you have a very good reason for including many factors in one study, it is best to limit yourself to two or, at most, three factors. You may not learn about many variables at once, but what you do learn you will understand.

Using the SPSS Appendix As shown in Appendix B.8, SPSS will perform the complete two-way between-subjects ANOVA. It also performs the *HSD* test, but only for main effects. SPSS also computes the $\overline{X}$ and s_X^2 for the levels of all main effects and for the cells of the interaction, as well as computing the 95% confidence interval for each mean. And, it graphs the means for the main effects and interaction.

CHAPTER SUMMARY

1. A *two-way, between-subjects ANOVA* involves two independent variables, and all conditions of both factors contain independent samples. A *two-way, within-subjects* ANOVA is performed when both factors involve related samples. A *two-way, mixed-design* ANOVA is performed when one factor has independent samples and one factor has related samples.

2. In a *complete factorial design,* all levels of one factor are combined with all levels of the other factor. Each *cell* is formed by a particular combination of a level from each factor.

3. In a two-way ANOVA, we compute an F_{obt} for the main effect of A, for the main effect of B, and for the interaction of A $\times$ B.

4. The *main effect means* for a factor are obtained by *collapsing* across (combining the scores from) the levels of the other factor. Collapsing across factor B produces the main effect means for factor A. Collapsing across factor A produces the main effect means for factor B.

5. A significant *main effect* indicates significant differences between the main effect means, indicating a relationship is produced when we manipulate one independent variable by itself.

6. A significant *two-way interaction effect* indicates that the *cell means* differ significantly such that the relationship between one factor and the dependent scores depends on the level of the other factor that is present. When graphed, an interaction produces nonparallel lines.

7. Perform *post hoc comparisons* on each significant effect having more than two levels to determine which specific means differ significantly.

8. Post hoc comparisons on the interaction are performed for *unconfounded* comparisons only. The means from two cells are unconfounded if the cells differ along only one factor. Two means are *confounded* if the cells differ along more than one factor.

9. An interaction is graphed by plotting cell means on Y and the levels of one factor on X. Then a separate line connects the data points for the cell means from each level of the other factor.

10. Conclusions from a two-way ANOVA are based on the significant main and interaction effects and upon which level or cell means differ significantly. Usually, conclusions about the main effects are contradicted when the interaction is significant.

11. *Eta squared* describes the effect size of each significant main effect and interaction.

KEY TERMS

F_A F_B $F_{A \times B}$ SS_A df_A MS_A
SS_B df_B MS_B $SS_{A \times B}$ $df_{A \times B}$
$MS_{A \times B}$
cell *320*
collapsing *322*
complete factorial design *320*
confounded comparison *339*
incomplete factorial design *320*
main effect *321*

main effect mean *322*
two-way ANOVA *318*
two-way, between-subjects
  ANOVA *318*
two-way interaction effect *325*
two-way, mixed-design ANOVA *318*
two-way, within-subjects ANOVA *318*
unconfounded comparison *339*

REVIEW QUESTIONS

(Answers for odd-numbered questions are in Appendix D.)

1. What are the two reasons for conducting a two-factor experiment?
2. Identify the following terms: (a) two-way design, (b) complete factorial, and (c) cell.
3. What is the difference between a main effect mean and a cell mean?
4. Which type of ANOVA is used in a two-way design when (a) both factors are tested using independent samples? (b) One factor involves independent samples and one factor involves related samples? (c) Both factors involve related samples?
5. (a) What is a confounded comparison, and when does it occur? (b) What is an unconfounded comparison, and when does it occur? (c) Why don't we perform post hoc tests on confounded comparisons?
6. What does it mean to collapse across a factor?
7. For a 2×2 ANOVA, describe the following in words: (a) the statistical hypotheses for factor A, (b) the statistical hypotheses for factor B, and (c) the statistical hypotheses for A $\times$ B.
8. (a) What does a significant main effect indicate? (b) What does a significant interaction effect indicate? (c) Why do we usually base the interpretation of a two-way design on the interaction effect when it is significant?

APPLICATION QUESTIONS

9. One more time, using a factorial design, we study the effect of changing the dose for one, two, three, or four smart pills and test participants who are 10-, 15-, and 20-years old. We test ten participants in each cell. (a) Using two numbers, describe this design. (b) When computing the main effect means for the factor of dose, what will be the n in each group? (c) When computing the main effect means for the factor of age, what will be the n in each group? (d) When performing Tukey's *HSD* on the interaction, what will be the n?

10. (a) When is it appropriate to compute the effect size in a two-way ANOVA? (b) For each effect, what does the effect size tell you?

11. Below are the cell means of three experiments. For each experiment, compute the main effect means and decide whether there appears to be an effect of A, B, and/or A $\times$ B.

	Study 1	
	A_1	A_2
B_1	2	4
B_2	12	14

	Study 2	
	A_1	A_2
B_1	10	5
B_2	5	10

	Study 3	
	A_1	A_2
B_1	8	14
B_2	8	2

12. In question 11, if you label the X axis with factor A and graph the cell means, what pattern will we see for each interaction?

13. After performing a 3 $\times$ 4 ANOVA with equal ns, you find that all Fs are significant. What other procedures should you perform?

14. A 2 $\times$ 2 design studies participants' frustration levels when solving problems as a function of the difficulty of the problem and whether they are math or logic problems. The results are that logic problems produce significantly more frustration than math problems, greater difficulty leads to significantly greater frustration, and difficult math problems produce significantly greater frustration than difficult logic problems, but the reverse is true for easy problems. In the ANOVA in this study, what effects are significant?

15. In question 14, say instead that the researcher found no difference between math and logic problems, frustration significantly increases with greater difficulty, and this is true for both math and logic problems. In the ANOVA in this study, what effects are significant?

16. In an experiment, you measure the popularity of two brands of soft drinks (factor A), and for each brand you test males and females (factor B). The following table shows the main effect and cell means from the study:

		Factor A	
		Level A$_1$: *Brand X*	*Level A$_2$:* *Brand Y*
	Level B$_1$: *Males*	14	23
Factor B			
	Level B$_2$: *Females*	25	12

(a) Describe the graph of the interaction when factor A is on the X axis.
(b) Does there appear to be an interaction effect? Why? (c) What are the main effect means and thus the main effect of changing brands? (d) What are the main effect means and thus the main effect of changing gender? (e) Why will a significant interaction prohibit you from making conclusions based on the main effects?

17. A researcher examines performance on an eye–hand coordination task as a function of three levels of reward and three levels of practice, obtaining the following cell means:

		Reward		
		Low	*Medium*	*High*
Practice	*Low*	4	10	7
	Medium	5	5	14
	High	15	15	15

(a) What are the main effect means for reward, and what do they appear to indicate about this factor? (b) What are the main effect means for practice, and what do they appear to indicate? (c) Does it appear that there is an interaction effect? (d) How would you perform unconfounded post hoc comparisons of the cell means?

18. (a) In question 17, why does the interaction contradict your conclusions about the effect of reward? (b) Why does the interaction contradict your conclusions about practice?

19. A study compared the performance of males and females tested by either a male or a female experimenter. Here are the data:

		Factor A: Participants	
		Level A$_1$: Males	*Level A$_2$:* Females
Factor B: Experimenter	*Level B$_1$:* Male Experimenter	6 11 9 10 9	8 14 17 16 19
	Level B$_2$: Female Experimenter	8 10 9 7 10	4 6 5 5 7

(a) Using $\alpha = .05$, perform an ANOVA and complete the summary table.
(b) Compute the main effect means and interaction means. (c) Perform the appropriate post hoc comparisons. (d) What do you conclude about the relationships this study demonstrates? (e) Compute the effect size where appropriate.

20. You conduct an experiment involving two levels of self-confidence (A_1 is low and A_2 is high) and examine participants' anxiety scores after they speak to one of four groups of differing sizes (B_1 through B_4 represent speaking to a small, medium, large, or extremely large group, respectively). You compute the following sums of squares ($n = 4$ and $N = 32$):

Source	Sum of Squares	df	Mean Square	F
Between				
Factor A	8.42	_____	_____	_____
Factor B	76.79	_____	_____	_____
Interaction	23.71	_____	_____	_____
Within	110.72	_____	_____	
Total	219.64	_____		

(a) Complete the ANOVA summary table. (b) With $\alpha = .05$, what do you conclude about each F_{obt}? (c) Compute the appropriate values of HSD. (d) For the levels of factor B, the means are $\overline{X}_1 = 18.36$, $\overline{X}_2 = 20.02$, $\overline{X}_3 = 24.6$, and $\overline{X}_4 = 28.3$. What should you conclude about the main effect of B? (e) How important is the size of the audience in determining a person's anxiety score? How important is the person's self-confidence?

21. You measure the dependent variable of participants' relaxation level as a function of whether they meditate before being tested, and whether they were shown a film containing a low, medium, or high amount of fantasy. Perform all appropriate statistical analyses, and determine what you should conclude about this study.

Amount of Fantasy

	Low	Medium	High
Meditation	5	7	9
	6	5	8
	2	6	10
	2	9	10
	5	5	10
No Meditation	10	2	5
	10	5	6
	9	4	5
	10	3	7
	10	2	6

INTEGRATION QUESTIONS

22. What does a 95% confidence interval for μ indicate? (Ch.11)

23. What does "significant" indicate about the results of any study? (Ch. 10)

24. (a) How do you recognize the independent variable in a study? (b) How do you recognize the levels in a study? (c) How do you recognize the dependent variable in a study? (Chs. 2, 12, 13, 14)

25. To select a statistical procedure for an experiment, what must you ask about how participants are selected? (Chs. 12, 13, 14)

26. (a) How do you recognize a two-sample *t*-test design? (b) What must be true about the dependent variable? (c) Which versions of the *t*-test are available? (Chs. 2, 10, 12)

27. (a) How do you recognize a correlational design? (b) To select a correlation coefficient, what must you ask about the variables? (c) What is the parametric correlation coefficient? (d) What is the nonparametric correlation coefficient? (Chs. 2, 7, 10, 11)

28. (a) How do you recognize a design that requires a one-way ANOVA? (b) What must be true about the dependent variable? (c) Which versions of ANOVA are available? (Chs. 10, 12 13)

29. (a) How do you recognize a design that fits a two-way ANOVA? (b) What must be true about the dependent variable? (c) Which versions of ANOVA are available? (Chs. 10, 12, 14)

30. For the following, identify the factor(s), the primary inferential procedure to perform and the key findings we'd look for. If a correlational design, indicate the predictor and criterion. If no parametric procedure can be used, indicate why. (Chs. 7, 8, 10, 11, 12, 13, 14)
(a) We measure babies' irritability when their mother is present and when she is absent. (b) We test the driving ability of participants who are either high, medium, or low in the personality trait of "thrill seeker." For each type, we test some participants who have had either 0, 1, or 2 oz. of alcohol. (c) Parents with alcoholism may produce adult children who are more prone to alcoholism. We compare the degree of alcoholism in participants with alcoholic parents to those with nonalcoholic parents. (d) To study dark adaptation, participants were asked to identify stimuli after sitting in a dim room for 1 minute, again after 15 minutes, and again after 30 minutes. (e) We study whether people who smoke cigarettes are more prone to be drug abusers. We identify participants who are smokers or non-smokers, and for each, count the number who are high or low drug abusers. (f) To test if creativity scores change with age, we test groups of 5-, 10-, or 15-year-olds. We also identify them as Caucasian or non-Caucasian to determine if age-related changes in creativity depend on race. (g) As in part (f) we measure the age scores and creativity scores of all students in the school district, again separating Caucasian and non-Caucasian. (Chs. 8, 12, 13, 14)

■ ■ ■ SUMMARY OF FORMULAS

The general format for the summary table for a two-way, between-subjects ANOVA is

Summary Table of Two-Way ANOVA

Source	Sum of Squares	df	Mean Square	F
Between				
Factor A	SS_A	df_A	MS_A	F_A
Factor B	SS_B	df_B	MS_B	F_B
Interaction	$SS_{A \times B}$	$df_{A \times B}$	$MS_{A \times B}$	$F_{A \times B}$
Within	SS_{wn}	df_{wn}	MS_{wn}	
Total	SS_{tot}	df_{tot}		

1. To perform the two-way, between-subjects ANOVA

a. Compute the sum of squares,

$$SS_{tot} = \Sigma X_{tot}^2 - \left(\frac{(\Sigma X_{tot})^2}{N}\right)$$

$$SS_A = \Sigma\left(\frac{(\text{Sum of scores in the column})^2}{n \text{ of scores in the column}}\right) - \left(\frac{(\Sigma X_{tot})^2}{N}\right)$$

$$SS_B = \Sigma\left(\frac{(\text{Sum of scores in the row})^2}{n \text{ of scores in the row}}\right) - \left(\frac{(\Sigma X_{tot})^2}{N}\right)$$

$$SS_{bn} = \Sigma\left(\frac{(\text{Sum of scores in the row})^2}{n \text{ of scores in the row}}\right) - \left(\frac{(\Sigma X_{tot})^2}{N}\right)$$

$$SS_{A \times B} = SS_{bn} - SS_A - SS_B$$
$$SS_{wn} = SS_{tot} - SS_{bn}$$

b. Compute the degrees of freedom,

  i. The *df* between groups for factor A (df_A) equals $k_A - 1$, where k_A is the number of levels in factor A.

  ii. The *df* between groups for factor B (df_B) equals $k_B - 1$, where k_B is the number of levels in factor B.

  iii. The *df* between groups for the interaction $(df_{A \times B})$ equals df_A multiplied times df_B.

  iv. The *df* within groups (df_{wn}) equals $N - k_{A \times B}$, where N is the total N of the study and $k_{A \times B}$ is the total number of cells in the study.

c. Compute the mean square,

$$MS_A = \frac{SS_A}{df_A}$$

$$MS_B = \frac{SS_B}{df_B}$$

$$MS_{A \times B} = \frac{SS_{A \times B}}{df_{A \times B}}$$

$$MS_{wn} = \frac{SS_{wn}}{df_{wn}}$$

d. Compute F_{obt},

$$F_A = \frac{MS_A}{MS_{wn}}$$

$$F_B = \frac{MS_B}{MS_{wn}}$$

$$F_{A \times B} = \frac{MS_{A \times B}}{MS_{wn}}$$

Critical values of F are found in Table 5 of Appendix C.

2. The formula for Tukey's *HSD* post hoc comparisons is

$$HSD = (q_k)\left(\sqrt{\frac{MS_{wn}}{n}}\right)$$

a. *For a significant main effect:* Find q_k in Table 6 in Appendix C for k equal to the number of levels in the factor. In the formula, n is the number of scores used to compute each main effect mean in the factor.

b. *For a significant interaction effect:* Determine the *adjusted k* using the small table at the top of Table 6 in Appendix C. Find q_k in Table 6 using the *adjusted k*. In the formula for *HSD*, n is the number of scores in each cell.

3. The formula for eta squared is

$$\eta^2 = \frac{\text{Sum of squares between groups for the factor}}{SS_{tot}}$$

15

Chi Square and Other Nonparametric Procedures

GETTING STARTED

To understand this chapter, recall the following:

- From Chapter 2, the four types of measurement scales (nominal, ordinal, interval, and ratio).
- From Chapter 12, the types of designs that call for either the independent-samples *t*-test or the related-samples *t*-test.
- From Chapter 13, the one-way ANOVA, post hoc tests, and eta squared.
- From Chapter 14, what a two-way interaction indicates.

Your goals in this chapter are to learn

- When to use *nonparametric* statistics.
- The logic and use of the *one-way chi square.*
- The logic and use of the *two-way chi square.*
- The *nonparametric procedures* corresponding to the independent-samples and related-samples *t*-test and to the between-subjects and within-subjects ANOVA.

Previous chapters have discussed the category of inferential statistics called parametric procedures. Now we'll turn to the other category, called *nonparametric procedures.* Nonparametric procedures are still inferential statistics for deciding whether the differences between samples accurately represent differences in the populations, so the logic here is the same as in past procedures. What's different is the type of dependent variable involved. In this chapter, we will discuss (1) two common procedures used with nominal scores called the one-way and two-way chi square and (2) review several less common procedures used with ordinal scores.

WHY IS IT IMPORTANT TO KNOW ABOUT NONPARAMETRIC PROCEDURES?

Previous parametric procedures have required that dependent scores reflect an interval or ratio scale, that the scores are normally distributed, and that the population variances are homogeneous. It is better to design a study that allows you to use parametric procedures because they are more *powerful* than nonparametric procedures. However, sometimes researchers don't obtain data that fit parametric procedures. Some dependent variables are nominal variables (for example, whether someone is male or female). Sometimes we can measure a dependent variable only by assigning ordinal scores (for example, judging this participant as showing the most of the variable, this one second-most, and so on). And sometimes a variable involves an interval or ratio scale, but the populations are severely skewed and/or do not have homogeneous variance (for example, previously we saw that yearly income forms a positively skewed distribution).

Parametric procedures will tolerate *some* violation of their assumptions. But if the data severely violate the rules, then the result is to *increase* the probability of a Type I error so that it is much *larger* than the alpha level we think we have.

Therefore, when data do not fit a parametric procedure, we turn to **nonparametric statistics**. They do not assume a normal distribution or homogeneous variance, and the scores may be nominal or ordinal. By using these procedures, we keep the probability of a Type I error equal to the alpha level that we've selected. Therefore, it is important to know about nonparametric procedures because you may use them in your own research, and you will definitely encounter them when reading the research of others.

> *REMEMBER* Use *nonparametric statistics* when dependent scores form very nonnormal distributions, when the population variance is not homogeneous, or when scores are measured using ordinal or nominal scales.

CHI SQUARE PROCEDURES

Chi square procedures are used when participants are measured using a *nominal* variable. With nominal variables, we do not measure an amount, but rather we *categorize* participants. Thus, we have nominal variables when counting how many individuals answer yes, no, or maybe to a question; how many claim to vote Republican, Democratic, or Socialist; how many say that they were or were not abused as children; and so on. In each case, we count the number, or *frequency,* of participants in each category.

The next step is to determine what the data represent. For example, we might find that out of 100 people, 40 say yes to a question and 60 say no. These numbers indicate how the *frequencies are distributed* across the categories of yes/no. As usual, we want to draw inferences about the population: Can we infer that if we asked the entire population this question, 40% would say yes and 60% would say no? Or would the frequencies be distributed in a different manner? To make inferences about the frequencies in the population, we perform chi square (pronounced "kigh square"). The **chi square procedure** is the nonparametric inferential procedure for testing whether the frequencies in each category in sample data represent specified frequencies in the population. The symbol for the chi square statistic is χ^2.

> *REMEMBER* Use the *chi square procedure* (χ^2) when you count the number of participants falling into different categories.

Theoretically, there is no limit to the number of categories—levels—you may have in a variable and no limit to the number of variables you may have. Therefore, we describe a chi square design in the same way we described ANOVAs: When a study has only one variable, perform the *one-way chi square;* when a study has two variables, perform the *two-way chi square;* and so on.

ONE-WAY CHI SQUARE

The **one-way chi square** is used when data consist of the frequencies with which participants belong to the different categories of *one* variable. Here we examine the relationship between the different categories and the frequency with which participants fall into each. We ask, "As the categories change, do the frequencies in the categories also change?"

Here is an example. Being right-handed or left-handed is related to brain organization, and many of history's great geniuses were left-handed. Therefore, using an IQ test, we select a sample of 50 geniuses. Then we ask them whether they are left- or right-handed (ambidextrous is not an option). The total numbers of left- and right-handers are the frequencies in the two categories. The results are shown here:

Handedness

Left-Handers	*Right-Handers*
$f_o = 10$	$f_o = 40$

$$k = 2$$
$$N = \text{total } f_o = 50$$

Each column contains the frequency in that category. We call this the **observed frequency,** symbolized by f_o. The sum of the f_os from all categories equals N, the total number of participants. Notice that k stands for the number of categories, or levels, and here $k = 2$.

Above, 10 of the 50 geniuses (20%) are left-handers, and 40 of them (80%) are right-handers. Therefore, we might argue that the same distribution of 20% left-handers and 80% right-handers would occur in the population of geniuses. But, there is the usual problem: sampling error. Maybe, by luck, the people in our sample are unrepresentative, so in the population of geniuses, we would not find this distribution of right- and left-handers. Maybe our results poorly represent some *other* distribution.

What is that "other distribution" of frequencies that the sample poorly represents? To answer this, we create a *model* of the distribution of the frequencies we expect to find in the population when H_0 is true. The H_0 model describes the distribution of frequencies in the population if there is *not* the predicted relationship. It is because we test this model that the one-way chi square procedure is also called a goodness-of-fit test. Essentially, we test how "good" the "fit" is between our data and the H_0 model. Thus, the **goodness-of-fit test** is another way of asking whether sample data are likely to represent the distribution of frequencies in the population as described by H_0.

Hypotheses and Assumptions of the One-Way Chi Square

The one-way χ^2 tests *only* two-tailed hypotheses. Usually, researchers test the H_0 that there is no difference among the frequencies in the categories in the population, meaning that there is no relationship in the population. For the handedness study, for the moment we'll ignore that there are more right-handers than left-handers in the world. Therefore, if there is no relationship in the population, then our H_0 is that the frequencies of left- and right-handed geniuses are equal in the population. There is no conventional way to write this in symbols, so simply write H_0: *all frequencies in the population are equal.* This implies that, if the observed frequencies in the sample are not equal, it's because of sampling error.

The alternative hypothesis always implies that the study did demonstrate the predicted relationship, so we have H_a: *not all frequencies in the population are equal.* For our handedness study, H_a implies that the observed frequencies represent different frequencies of left- and right-handers in the population of geniuses.

The one-way χ^2 has five assumptions:

1. Participants are categorized along one variable having two or more categories, and we count the frequency in each category.

2. Each participant can be in only one category (that is, you cannot have repeated measures).

3. Category membership is independent: The fact that an individual is in a category does not influence the probability that another participant will be in any category.

4. We include the responses of all participants in the study (that is, you would not count only the number of right-handers, or in a different study, you would count both those who do and those who do not agree with a statement).

5. For theoretical reasons, each "expected frequency" discussed below must be at least 5.

Computing the One-Way Chi Square

The first step in computing χ^2 is to translate H_0 into the expected frequency for each category. The **expected frequency** is the frequency we expect in a category if the sample data perfectly represent the distribution in the population described by the null hypothesis. The symbol for an expected frequency is f_e. Our H_0 is that the frequencies of left- and right-handedness are equal. If the sample perfectly represents this, then out of our 50 participants, 25 should be right-handed and 25 should be left-handed. Thus, the expected frequency in each category is $f_e = 25$.

Notice that, whenever we are testing the H_0 of no difference among the categories, the f_e will be the same for all categories, and it will always equal N/k. Thus, above, our $f_e = 50/2 = 25$. Also notice that sometimes f_e may contain a decimal. For example, if we included a third category, ambidextrous, then $k = 3$, and each f_e would be 16.67.

After computing each f_e, the next step is to compute χ^2, which we call χ^2_{obt}.

The formula for chi square is

$$\chi^2_{obt} = \Sigma \left(\frac{(f_o - f_e)^2}{f_e} \right)$$

This says to find the difference between f_o and f_e in each category, square that difference, and then divide it by the f_e for that category. After doing this for all categories, sum the quantities, and the answer is χ^2_{obt}.

For the handedness study, we have these frequencies:

Handedness

Left-Handers	Right-Handers
$f_o = 10$ $f_e = 25$	$f_o = 40$ $f_e = 25$

Filling in the formula gives

$$\chi^2_{obt} = \Sigma\left(\frac{(f_o - f_e)^2}{f_e}\right) = \left(\frac{(10 - 25)^2}{25}\right) + \left(\frac{(40 - 25)^2}{25}\right)$$

In the numerator of each fraction, first subtract the f_e from the f_o, giving

$$\chi^2_{obt} = \left(\frac{(-15)^2}{25}\right) + \left(\frac{(15)^2}{25}\right)$$

Squaring each numerator gives

$$\chi^2_{obt} = \left(\frac{225}{25}\right) + \left(\frac{225}{25}\right)$$

After dividing,

$$\chi^2_{obt} = 9 + 9$$

So our $\chi^2_{obt} = 18.0$

Interpreting the One-Way Chi Square

If the sample perfectly represents the situation where H_0 is true, then each f_o "should" equal its corresponding f_e. Then the difference between f_e and f_o should equal zero, and so χ^2_{obt} should equal zero. If not, H_0 says this is due to sampling error. However, the larger the differences between f_e and f_o (and the larger the χ^2_{obt}), the harder it is for us to accept that this is simply due to sampling error. At the same time, larger differences between f_e and f_o are produced because of a larger observed frequency in one category and a smaller one in another, so the more it looks like we are really representing a relationship. Therefore, the larger the χ^2_{obt}, the less likely it is that H_0 is true and the more likely it is that H_a is true.

To determine if our χ^2_{obt} is large enough, we examine the χ^2-distribution. Like previous sampling distributions, it is as if we have infinitely selected samples from the situation where H_0 is true. The **χ^2-distribution** is the sampling distribution containing all possible values of χ^2 when H_0 is true. Thus, for the handedness study, the χ^2-distribution is the distribution of all possible values of χ^2 when there are two categories and the frequencies in the two categories in the population are equal. You can envision the χ^2-distribution as shown in Figure 15.1.

Even though the χ^2-distribution is not at all normal, it is used in the same way as previous sampling distributions. Most often the data perfectly represent the H_0 situation so that each f_o equals its f_e, and then χ^2 is zero. However, sometimes by chance, the observed frequencies differ from the expected frequencies, producing a χ^2 greater than zero. The larger the χ^2, the larger are the differences between the observed and the expected and then the less likely they are to occur when H_0 is true.

Because χ^2_{obt} can only get larger, we again have two-tailed hypotheses but one region of rejection. To determine if χ^2_{obt} is significant, we

FIGURE 15.1

Sampling distribution of χ^2 when H_0 is true

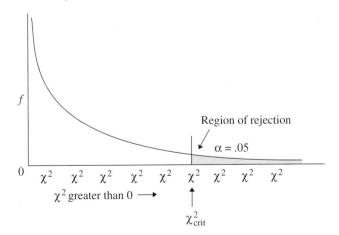

compare it to the critical value, symbolized by χ^2_{crit}. As with previous statistics, the χ^2-distribution changes shape as the degrees of freedom change, so we must first determine the degrees of freedom. Here, $df = k - 1$, where k is the number of categories.

> *REMEMBER* In a one-way chi square, $df = k - 1$.

Find the critical value of χ^2 in Table 7 in Appendix C, entitled "Critical Values of Chi Square." For the handedness study, $k = 2$, so $df = 1$, and with $\alpha = .05$, the $\chi^2_{crit} = 3.84$. Our χ^2_{obt} of 18.0 is larger than this χ^2_{crit}, so the results are significant: This χ^2_{obt} indicates that the differences between our observed and expected frequencies are very unlikely to occur if our data represent no relationship in the population. Therefore we reject that this is what the data represent. We then accept the H_a that the sample represents frequencies in the population that are not equal. In fact, we would expect to find about 20% left-handers and 80% right-handers in the population of geniuses. We conclude that we have evidence of a relationship between handedness and genius. Then, as usual, we interpret the relationship in terms of the behaviors and variables involved.

If χ^2_{obt} had not been significant, we would have no evidence—one way or the other—regarding how handedness is distributed among geniuses.

Note: If a study involves three categories, a significant chi square is not followed by post hoc comparisons. We simply assume that the observed frequency in each category represents frequencies that would be found in the population. Also, there is no measure of effect size for this design.

Testing Other Hypotheses with the One-Way Chi Square

The χ^2 procedure can also be used to test an H_0 other than that there is no difference among the categories. For example, only about 10% of the general population is actually left-handed, so we should test whether handedness in geniuses is distributed differently than this. Our H_0 is that geniuses are like the general population, being 10% left-handed and 90% right-handed. Our H_a is that our data represent a population of geniuses that does not have this distribution (or for simplicity, we can write H_a as "not H_0").

Each f_e is again based on our H_0. Say that we test our previous 50 geniuses. Our H_0 says that left-handed geniuses should occur 10% of the time: 10% of 50 is 5, so $f_e = 5$. Right-handed geniuses should occur 90% of the time: 90% of 50 is 45, so $f_e = 45$. We found $f_o = 10$ for left-handers and $f_o = 40$ for right-handers. We compute χ^2_{obt} using the previous formula, comparing the f_e to f_o for left-handers and the f_e to f_o for right-handers.

$$\chi^2_{obt} = \Sigma\left(\frac{(f_o - f_e)^2}{f_e}\right) = \left(\frac{(10 - 5)^2}{5}\right) + \left(\frac{(40 - 45)^2}{45}\right) = 5.56$$

With $\alpha = .05$ and $k = 2$, the critical value of χ^2 is again 3.84. Because the χ^2_{obt} of 5.56 is larger than χ^2_{crit}, we reject H_0 and conclude that the observed frequencies are significantly different from what we would expect if handedness in the population of geniuses was distributed as it is in the general population. Instead, we estimate that in the population of geniuses, 20% are left-handers and 80% are right-handers.

- The one-way χ^2 is used when counting the frequency of category membership on one variable.

MORE EXAMPLES

Below are the number of acts of graffiti that occur on walls painted white, painted blue, or covered with chalkboard. H_0 is that there are no differences in the population. With $N = 30$, $f_e = N/k = 30/3 = 10$ for each category.

White	Blue	Chalk
$f_o = 8$ $f_e = 10$	$f_o = 5$ $f_e = 10$	$f_o = 17$ $f_e = 10$

$$\chi^2_{obt} = \Sigma\left(\frac{(f_e - f_o)^2}{f_e}\right)$$

$$\chi^2_{obt} = \frac{(10 - 8)^2}{10} + \frac{(10 - 5)^2}{10} + \frac{(10 - 17)^2}{10}$$

$$\chi^2_{obt} = \frac{4}{10} + \frac{25}{10} + \frac{49}{10} = 7.80$$

With $df = k - 1 = 2$ and $\alpha = .05$, $\chi^2_{crit} = 5.99$, the wall coverings produce a significant difference in the frequency of graffiti acts. In the population, we expect 27% of graffiti on white walls, 17% on blue walls, and 57% on chalkboard walls.

For Practice

1. The one-way chi square is used when we count the _____ with which participants fall into different _____.
2. We find $f_o = 21$ in category A and $f_o = 39$ in category B. H_0 is that the frequencies are equal. The f_e for A is _____, and the f_e for B is _____.
3. Compute χ^2_{obt}.
4. The df is _____, so at $\alpha = .05$, χ^2_{crit} is _____.
5. The χ^2_{obt} is _____, so in the population we expect membership is around _____% in A and around _____% in B.

Answers

1. frequency; categories
2. Each $f_e = 60/2 = 30$
3. $\chi^2_{obt} = \frac{(30 - 21)^2}{30} + \frac{(30 - 39)^2}{30} = 5.40$
4. 1; 3.84
5. significant; 35%; 65%

THE TWO-WAY CHI SQUARE

TABLE 15.1

A Two-Way Chi Square Design Comparing Participants' Personality Type and Health

The **two-way chi square** procedure is used when you count the frequency of category membership along *two* variables. (The assumptions of the two-way chi square are the same as for the one-way chi square.) Here is an example. At one time, psychologists claimed that someone with a Type A personality tends to be a very pressured, hostile individual who never seems to have enough time. The Type B personality tends not to be so time pressured and is more relaxed and mellow. A controversy developed over whether Type A people are less healthy, especially when it comes to the "big one"—having heart attacks. Therefore, say that we select a sample of 80 people and determine how many are Type A and how many Type B. We then count the frequency of heart attacks in each type. We must also count how many in each type have *not* had heart attacks (see our assumption 4). Therefore, we have two categorical variables: personality type (A or B) and health (heart attack or no heart attack). Table 15.1 shows the layout of this study. Notice, this is a 2×2 matrix, so it is called a 2×2 design. Depending on the number of categories in each variable, a study might be a 2×3, a 3×4, and so on.

		Personality Type	
		Type A	Type B
Health	Heart Attack	f_o	f_o
	No Heart Attack	f_o	f_o

Although Table 15.1 looks like a two-way ANOVA, it is not analyzed like one. Instead of testing for main effects, *the two-way χ^2 procedure tests only what is essentially the interaction.* Recall that with an interaction, the influence of one variable *depends* on the other. The two-way χ^2 is also called the **test of independence** because it tests whether the frequency that participants fall into the categories of one variable depends on the frequency of falling into the categories on the other variable. Thus, our study will test whether the frequencies of having or not having a heart attack are independent of the frequencies of being Type A or Type B.

To understand *independence,* the left-hand matrix in Table 15.2 shows an example of data we might get if category membership on our variables was perfectly independent. Here, the frequency of having or not having a heart attack does not depend on the frequency of being Type A or Type B. Another way to view the two-way χ^2 is as a test of whether a *correlation* exists between the two variables. When variables are independent, there is no correlation, and using the categories from one variable is no help in predicting the frequencies for the other variable. Here, knowing if people are Type A or Type B does not help to predict if they do or do not have heart attacks (and the health categories do not help in predicting personality type).

On the other hand, the right-hand matrix in Table 15.2 shows an example of when category membership on the two variables is totally dependent. Here, the frequency of a heart attack or no heart attack *depends* on personality type. Likewise, a perfect correlation exists because whether people are Type A or Type B is a perfect predictor of whether or not they have had a heart attack (and vice versa).

But, say that the actual observed frequencies from our participants are those shown in Table 15.3. There is a *degree* of dependence here because heart attacks tend to be more frequent for Type A personalities, while no heart attack is more frequent for Type B personalities. Therefore, some degree of correlation exists between the variables. On the one hand, we'd like to conclude that this relationship exists in the population. On the other hand, perhaps there really is no correlation in the population, but by chance we obtained frequencies that poorly represent this. The above translate into our null and alternative hypotheses. In the two-way χ^2, H_0 is that category membership on one variable is independent of (not correlated with) category membership on the other variable. If the sample data look correlated, this is due to sampling error. The H_a is that category membership on the two variables in the population is dependent (correlated).

TABLE 15.2

Examples of Independence and Dependence

On the left, personality type and heart attacks are perfectly independent. On the right, personality type and heart attacks are perfectly dependent.

		Personality Type					Personality Type	
		Type A	**Type B**				**Type A**	**Type B**
Health	**Heart Attack**	$f_o = 20$	$f_o = 20$		**Health**	**Heart Attack**	$f_o = 40$	$f_o = 0$
	No Heart Attack	$f_o = 20$	$f_o = 20$			**No Heart Attack**	$f_o = 0$	$f_o = 40$

TABLE 15.3

Observed Frequencies as a Function of Personality Type and Health

Personality Type

		Type A	*Type B*
Health	*Heart Attack*	$f_o = 25$	$f_o = 10$
	No Heart Attack	$f_o = 5$	$f_o = 40$

Computing the Two-Way Chi Square

Again the first step in computing χ^2 is to compute the expected frequencies. To do so, as shown in Table 15.4, first compute the total of the observed frequencies in each column and the total of the observed frequencies in each row. Also, note N, the total of all observed frequencies.

Each f_e is based on the probability of a participant falling into a *cell* if the two variables are independent. For example, for the cell of Type A and heart attack, we determine the probability of someone in our study being Type A and the probability of someone in our study reporting a heart attack, when these variables are independent. The expected frequency in this cell then equals this probability multiplied times N. Luckily, the steps involved in this can be combined to produce this formula.

> **The formula for computing the expected frequency in a cell of a two-way chi square is**
>
> $$f_e = \frac{(\text{Cell's row total } f_o)(\text{Cell's column total } f_o)}{N}$$

For each cell we multiply the total observed frequencies for the row containing the cell times the total observed frequencies for the column containing the cell. Then divide by the N of the study.

Table 15.4 shows the completed computations of f_e for our study. To check your work, confirm that the sum of the f_e in each column or row equals the column or row total.

TABLE 15.4

Diagram Containing f_o and f_e for Each Cell

Each f_e *equals the row total times the column total, divided by* N.

Personality Type

		Type A	*Type B*	
Health	*Heart Attack*	$f_o = 25$ $f_e = 13.125$ (35)(30)/80	$f_o = 10$ $f_e = 21.875$ (35)(50)/80	row total = 35
	No Heart Attack	$f_o = 5$ $f_e = 16.875$ (45)(30)/80	$f_o = 40$ $f_e = 28.125$ (45)(50)/80	row total = 45
		column total = 30	column total = 50	total = 80 $N = 80$

Compute the two-way χ^2_{obt} using the same formula used in the one-way design, which is

$$\chi^2_{obt} = \Sigma\left(\frac{(f_o - f_e)^2}{f_e}\right)$$

With the data in Table 15.4 we have

$$\chi^2_{obt} = \left(\frac{(25 - 13.125)^2}{13.125}\right) + \left(\frac{(10 - 21.875)^2}{21.875}\right) + \left(\frac{(5 - 16.875)^2}{16.875}\right)$$
$$+ \left(\frac{(40 - 28.125)^2}{28.125}\right)$$

As you did previously with this formula, in the numerator of each fraction subtract f_e from f_o. Then square that difference and then divide by the f_e in the denominator. Then

$$\chi^2_{obt} = 10.74 + 6.45 + 8.36 + 5.01$$

so

$$\chi^2_{obt} = 30.56$$

To evaluate χ^2_{obt}, compare it to the appropriate χ^2_{crit}. First, determine the degrees of freedom by looking at the number of rows and columns in the diagram of your study.

In a two-way chi square, df = (Number of rows − 1)(Number of columns − 1).

For our 2 × 2 design, df is $(2 - 1)(2 - 1) = 1$. Find the critical value of χ^2 in Table 7 in Appendix C. At $\alpha = .05$ and $df = 1$, the χ^2_{crit} is 3.84.

Our χ^2_{obt} of 30.56 is larger than χ^2_{crit}, so it is significant. Therefore, envision the sampling distribution back in Figure 15.1, with χ^2_{obt} in the region of rejection. This indicates that the differences between our observed and expected frequencies are so unlikely to occur if our data represent variables that are independent in the population, that we reject that this is what the data represent. If the variables are not independent, then they must be dependent. Therefore, we accept the H_a that the frequency of participants falling into each category on one of our variables depends on the category they fall into on the other variable. In other words, we conclude that there is a significant correlation such that the frequency of having or not having a heart attack depends on the frequency of being Type A or Type B (and vice versa). If χ^2_{obt} is not larger than the critical value, do not reject H_0. Then, we cannot say whether these variables are independent or not.

REMEMBER A significant *two-way* χ^2 indicates that the sample data are likely to represent two variables that are dependent (or correlated) in the population.

A QUICK REVIEW

- The two-way χ^2 is used when counting the frequency of category membership on two variables.
- The H_0 is that category membership for one variable is independent of category membership for the other variable.

MORE EXAMPLES

We count the participants who indicate (1) whether they like or dislike statistics and (2) their gender. The H_0 is that liking/disliking is independent of gender. The results are

(continued)

	Like	*Dislike*	
Male	$f_o = 20$ $f_e = 15$	$f_o = 10$ $f_e = 15$	Total $f_o = 30$
Female	$f_o = 5$ $f_e = 10$	$f_o = 15$ $f_e = 10$	Total $f_o = 20$
	Total $f_o = 25$	Total $f_o = 25$	$N = 50$

Compute each f_e:

For male-like: $f_e = (30)(25)/50 = 15$
For male-dislike: $f_e = (30)(25)/50 = 15$
For female-like: $f_e = (20)(25)/50 = 10$
For female-dislike: $f_e = (20)(25)/50 = 10$

$$\chi^2_{obt} = \frac{(20-15)^2}{15} + \frac{(10-15)^2}{15} + \frac{(5-10)^2}{10}$$
$$+ \frac{(15-10)^2}{10}$$

$\chi^2_{obt} = 25/15 + 25/15 + 25/10 + 25/10$

$\chi^2_{obt} = 8.334$

$df = (2-1)(2-1) = 1$

With $\alpha = .05$, $\chi^2_{crit} = 3.84$, so χ^2_{obt} is significant: The frequency of liking/disliking statistics *depends on*—is correlated with—whether participants are male or female.

For Practice

1. The two-way χ^2 is used when counting the _____ with which participants fall into the _____ of two variables.

2. The H_0 is that the frequencies in the categories of one variable are _____ of those of other variable.

3. Below are the frequencies for people who are satisfied/dissatisfied with their job and who do/don't work overtime. What is the f_e in each cell?

	Overtime	*No Overtime*
Satisfied	$f_o = 11$	$f_o = 3$
Dissatisfied	$f_o = 8$	$f_o = 12$

4. Compute χ^2_{obt}.
5. The $df = $ _____ and $\chi^2_{crit} = $ _____.
6. What do you conclude about these variables?

Answers

1. frequency; categories
2. independent
3. For satisfied–overtime, $f_e = 7.824$; for satisfied–no overtime, $f_e = 6.176$; for dissatisfied–overtime, $f_e = 11.176$; for dissatisfied–no overtime, $f_e = 8.824$.
4. $\chi^2_{obt} = \frac{(11-7.824)^2}{7.824} + \frac{(3-6.176)^2}{6.176} + \frac{(8-11.176)^2}{11.176} + \frac{(12-8.824)^2}{8.824} = 4.968$
5. 1; 3.84
6. χ^2_{obt} is significant: The frequency of job satisfaction/dissatisfaction depends on the frequency of overtime/no overtime.

Describing the Relationship in a Two-Way Chi Square

A significant two-way chi square indicates a significant correlation between the variables. To determine the size of this correlation, we have two new correlation coefficients: We compute either the *phi coefficient* or the *contingency coefficient*.

If you have performed a 2 × 2 chi square and it is significant, compute the **phi coefficient.** Its symbol is ɸ, and its value can be between 0 and +1. Think of phi as comparing your data to the ideal situations shown back in Table 15.2. A coefficient of 0 indicates that the variables are perfectly independent. The larger the coefficient, the closer the variables are to forming a pattern that is perfectly dependent.

> **The formula for the phi coefficient is**
>
> $$\phi = \sqrt{\frac{\chi_{obt}^2}{N}}$$

N equals the total number of participants in the study.

For the heart attack study, χ_{obt}^2 was 30.56 and N was 80, so

$$\phi = \sqrt{\frac{\chi_{obt}^2}{N}} = \sqrt{\frac{30.56}{80}} = \sqrt{.382} = .62$$

Thus, on a scale of 0 to $+1$, where $+1$ indicates perfect dependence, the correlation is .62 between the frequency of heart attacks and the frequency of personality types.

Remember that another way to describe a relationship is to *square* the correlation coefficient, computing the proportion of variance accounted for. If you didn't take the square root in the above formula, you would have ϕ^2 (phi squared). This is analogous to r^2 or η^2, indicating how much more accurately we can predict scores by using the relationship. Above, $\phi^2 = .38$, so we are 38% more accurate in predicting the frequency of heart attacks/no heart attacks when we know personality type (or vice versa).

The other correlation coefficient is the **contingency coefficient,** symbolized by C. This is used to describe a significant two-way chi square that is *not* a 2×2 design (it's a 2×3, a 3×3, and so on).

> **The formula for the contingency coefficient is**
>
> $$C = \sqrt{\frac{\chi_{obt}^2}{N + \chi_{obt}^2}}$$

N is the number of participants in the study. Interpret C in the same way you interpret ϕ. Likewise, C^2 is analogous to ϕ^2.

STATISTICS IN PUBLISHED RESEARCH: REPORTING CHI SQUARE

The results of a one-way or two-way χ^2 are reported like previous results, except that in addition to the df, we also include the N. For example, in our handedness study, N was 50, df was 1, and the significant χ_{obt}^2 was 18. We report this as $\chi^2 (1, N = 50) = 18.00, p < .05$.

To graph a one-way design, label the Y axis with frequency and the X axis with the categories, and then plot the f_o in each category. With a nominal X variable, create a *bar graph*. Thus, the graph on the left in Figure 15.2 shows the results of our handedness study. The graph on the right shows the results of our heart attack study. For a two-way design, place frequency on the Y axis and one of the nominal variables on the X axis. The levels of the other variable are indicated in the body of the graph. (This is similar to the way a two-way interaction was plotted in the previous chapter, except that here we create bar graphs.)

FIGURE 15.2

Frequencies of
(a) left- and right-
handed geniuses and
(b) heart attacks and
personality type

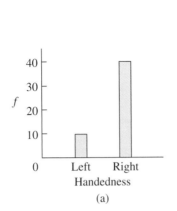

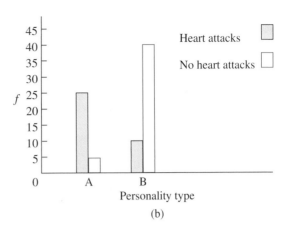

NONPARAMETRIC PROCEDURES FOR RANKED DATA

The χ^2 is the only common inferential statistic for nominal scores. The only other type of nonparametric procedure is for when the dependent variable involves rank-ordered (ordinal) scores. You obtain ranked scores for one of two reasons. First, sometimes you'll directly measure participants using ranked scores (directly assigning participants a score of 1st, 2nd, and so on). Second, sometimes you'll initially measure interval or ratio scores, but they violate the assumptions of parametric procedures by not being normally distributed or not having homogeneous variance. Then you transform these scores to ranks (the highest raw score is ranked 1, the next highest score is ranked 2, and so on). Either way, you then compute one of the following nonparametric inferential statistics to determine whether there are significant differences between the conditions of your independent variable.

The Logic of Nonparametric Procedures for Ranked Data

Instead of computing the mean of each condition in the experiment, with nonparametric procedures we summarize the individual ranks in a condition by computing the *sum of ranks*. The symbol for a sum of ranks is ΣR. In each procedure, we compare the *observed* sum of ranks to an *expected* sum of ranks. To see the logic of this, say we have the following scores:

Condition 1	*Condition 2*
1	2
4	3
5	6
8	7

$\Sigma R = 18$ $\Sigma R = 18$

Here, the conditions do not differ, with each containing both high and low ranks. When the ranks are distributed equally between two groups, the sums of ranks are also equal (here, ΣR is 18 in each). This is true even for two populations. Our H_0 is always that the populations are equal, so with ranked data, H_0 is that the sums of ranks for each population are equal. Therefore, when H_0 is true, we *expect* the sums of ranks from the samples to be equal. Thus, the $\Sigma R = 18$ observed above is exactly what we would expect if H_0 is true, so such an outcome supports H_0.

But say the data had turned out differently, as here:

Condition 1	*Condition 2*
1	5
2	6
3	7
4	8
$\Sigma R = 10$	$\Sigma R = 26$

Condition 1 contains all of the low ranks, and Condition 2 contains all of the high ranks. Because these samples are different, they may represent two different populations. With ranked data H_a says that one population contains predominantly low ranks and the other contains predominantly high ranks. When our data are consistent with H_a, the *observed* sum of ranks in each sample is different from the *expected* sum of ranks produced when H_0 is true: Here, each ΣR does not equal 18.

Thus, the observed sum of ranks in each condition should equal the expected sum if H_0 is true, but the observed sum will *not* equal the expected sum if H_a is true. Of course, it may be that H_0 is true, but we have sampling error in representing this, in which case, the observed sum will not equal the expected sum. However, the larger the difference between the expected and observed sum of ranks, the less likely it is that this difference is due to sampling error, and the more likely it is that each sample represents a different population.

In each of the following procedures, we compute a statistic that measures the difference between the expected and the observed sum of ranks. If we can then reject H_0 and accept H_a, we are confident that the reason the observed sum is different from the expected sum is that the samples represent different populations. And, if the ranks reflect underlying interval or ratio scores, a significant difference in ranks indicates that the raw scores also differ significantly.

Resolving Tied Ranks

Each of the following procedures assumes you have resolved any tied ranks, in which two participants receive the same rank on the same variable. Say that two people are ranked 2nd. If not tied, one of them would have been 2nd and the other 3rd. Therefore, resolve ties by assigning the mean of the ranks that would have been used had there not been a tie. The mean of 2 and 3 is 2.5, so each tied participant would be given the new rank of 2.5. Now, in a sense, you've used 2 and 3, so the next participant (originally 3rd) is assigned the new rank of 4, the next is given 5, and so on. (If three participants had tied at 2nd, we'd assign them the mean of 2, 3, and 4, and assign the next person the rank of 5. And so on.) Doing this makes your sum of ranks work out correctly.

Choosing a Nonparametric Procedure

Each of the major parametric procedures found in previous chapters has a corresponding nonparametric procedure for ranked data. Your first task is to know which nonparametric procedure to choose for your type of research design. Table 15.5 shows the name of the nonparametric version of each parametric procedure we have discussed.

TABLE 15.5

Parametric Procedures and Their Nonparametric Counterparts

Type of Design	Parametric Test	Nonparametric Test
Two independent samples	Independent-samples t-test	Mann–Whitney U or rank sums test
Two related samples	Related-samples t-test	Wilcoxon T test
Three or more independent samples	Between-subjects ANOVA (Post hoc test: protected t-test)	Kruskal–Wallis H test (Post hoc test: rank sums test)
Three or more repeated-measures samples	Within-subjects ANOVA (Post hoc test: Tukey's HSD)	Friedman χ^2 test (Post hoc test: Nemenyi's test)
Correlational study	Pearson r	Spearman r_S

Note: The table includes from Chapter 7 the Pearson r, the parametric correlation coefficient, and the Spearman r_S, the nonparametric correlation coefficient for ranks. My advice is to memorize this table!

The steps in calculating each new nonparametric procedure are described in the following sections.

Tests for Two Independent Samples: The Mann–Whitney U Test and the Rank Sums Test

Two nonparametric procedures are analogous to the t-test for two independent samples: the Mann–Whitney U test and the rank sums test. Which test we use depends on the n in each condition.

The Mann–Whitney U Test   Perform the **Mann–Whitney U test** when the n in each condition is equal to or less than 20 and there are two independent samples of ranks. For example, say that we measure the reaction times of people to different visual symbols that are printed in either black or red ink. Reaction times tend to be highly positively skewed, so we cannot perform the t-test. Therefore, we convert the reaction time scores to ranks. Say that each n is 5 (but unequal ns are acceptable). Table 15.6 gives the reaction times (measured in milliseconds) and their corresponding ranks.

TABLE 15.6

Ranked Data from Two Independent Samples

Red Symbols		Black Symbols	
Reaction Time	*Ranked Score*	*Reaction Time*	*Ranked Score*
540	2	760	7
480	1	890	8
600	5	1105	10
590	3	595	4
605	6	940	9
$\Sigma R = 17$		$\Sigma R = 38$	
$n = 5$		$n = 5$	

To perform the Mann–Whitney U test,

1. *Assign ranks to all scores in the experiment.* Assign the rank of 1 to the lowest score in the experiment, regardless of which group it is in. Assign the rank of 2 to the second-lowest score, and so on.

2. *Compute the sum of the ranks for each group.* Compute ΣR for each group.

3. *Compute two versions of the Mann–Whitney* U. First, compute U_1 for Group 1, using the formula

$$U_1 = (n_1)(n_2) + \frac{n_1(n_1 + 1)}{2} - \Sigma R_1$$

where n_1 is the n of Group 1, n_2 is the n of Group 2, and ΣR_1 is the sum of ranks from Group 1. Let's identify the red symbol condition as Group 1, so from Table 15.6, we have

$$U_1 = (5)(5) + \frac{5(5 + 1)}{2} - 17 = 40 - 17 = 23.0$$

Second, compute U_2 for Group 2, using the formula

$$U_2 = (n_1)(n_2) + \frac{n_2(n_2 + 1)}{2} - \Sigma R_2$$

Our black symbol condition is Group 2, so

$$U_2 = (5)(5) + \frac{5(5 + 1)}{2} - 38 = 40 - 38 = 2.0$$

4. *Determine the Mann–Whitney* U_{obt}. In a two-tailed test, the value of U_{obt} equals the smaller of U_1 or U_2. In the example, $U_1 = 23.0$ and $U_2 = 2.0$, so $U_{obt} = 2.0$. In a one-tailed test, we predict that one of the groups has the *larger* sum of ranks. The corresponding value of U_1 or U_2 from that group becomes U_{obt}.

5. *Find the critical value of* U *in Table 8 of Appendix C entitled "Critical Values of the Mann–Whitney* U." Choose the table for either a two-tailed or a one-tailed test. Then locate U_{crit} using n_1 and n_2. For our example, with a two-tailed test and $n_1 = 5$ and $n_2 = 5$, the U_{crit} is 2.0.

6. *Compare* U$_{obt}$ *to* U$_{crit}$. WATCH OUT! Unlike any statistic we've discussed, the U_{obt} is significant if it is *equal to or less than* U$_{crit}$. (This is because the *smaller* the U_{obt}, the more likely that H_0 is false. In the example, $U_{obt} = 2.0$ and $U_{crit} = 2.0$, so the samples differ significantly, representing different populations of ranks. Because the ranks reflect reaction time scores, the samples of reaction times also differ significantly and represent different populations ($p < .05$).

7. *To describe the effect size, compute eta squared.* If U_{obt} is significant, then ignore the rule about the ns and reanalyze the data using the following rank sums test to get to η^2.

The Rank Sums Test   Perform the **rank sums test** when you have two independent samples of ranks and *either n* is greater than 20. To illustrate the calculations, we'll violate this rule and use the data from the previous reaction time study.

To perform the rank sums test,

1. *Assign ranks to the scores in the experiment.* As we did back in Table 15.6, rank-order all scores in the experiment.

2. *Choose one group and compute the sum of the ranks.* Compute ΣR for one group.

3. *Compute the expected sum of ranks (ΣR_{exp}) for the chosen group.* Use the formula

$$\Sigma R_{exp} = \frac{n(N + 1)}{2}$$

where n is the n of the chosen group and N is the total N of the study. We'll compute ΣR_{exp} for the red symbol group, which had $n = 5$ (N is 10). Thus,

$$\Sigma R_{exp} = \frac{n_1(N + 1)}{2} = \frac{5(10 + 1)}{2} = \frac{55}{2} = 27.5$$

4. *Compute the rank sums statistic, symbolized by z_{obt}.* Use the formula

$$z_{obt} = \frac{\Sigma R - \Sigma R_{exp}}{\sqrt{\dfrac{(n_1)(n_2)(N + 1)}{12}}}$$

where ΣR is the sum of the ranks for the chosen group, ΣR_{exp} is the expected sum of ranks for the chosen group, n_1 and n_2 are the ns of the two groups, and N is the total N of the study.

For our example, $\Sigma R = 17$ so

$$z_{obt} = \frac{\Sigma R - \Sigma R_{exp}}{\sqrt{\dfrac{(n_1)(n_2)(N + 1)}{12}}} = \frac{17 - 27.5}{\sqrt{\dfrac{(5)(5)(10 + 1)}{12}}}$$

$$z_{obt} = \frac{-10.5}{\sqrt{22.92}} = \frac{-10.5}{4.79} = -2.19$$

5. *Find the critical value of z in the z-tables (Table 1 in Appendix C).* At $\alpha = .05$, the two-tailed $z_{crit} = \pm 1.96$. (If we had predicted whether the sum of ranks of the chosen group would be greater than or less than the expected sum of ranks, then we would use the one-tailed value of either $+1.645$ or -1.645.)

6. *Compare z_{obt} to z_{crit}.* If the absolute value of z_{obt} is *larger* than z_{crit}, then the samples differ significantly. In our example, $z_{obt} = -2.19$ and $z_{crit} = \pm 1.96$. Therefore, we conclude that the samples of ranked scores—as well as the underlying samples of reaction times—differ significantly ($p < .05$).

7. *Describe a significant relationship using eta squared.* Here eta squared is analogous to r_{pb}^2. Use the formula

$$\eta^2 = \frac{(z_{obt})^2}{N - 1}$$

where z_{obt} is computed in the above rank sums test and N is the total number of participants.

In the example, z_{obt} is -2.19 and N is 10, so we have $(2.19)^2/9$, or $.53$. Thus, the color of the symbols accounts for $.53$ of the variance in the ranks. Because the ranks reflect reaction time scores, *approximately* 53% of the differences in reaction time scores are associated with the color of the symbol.

The Wilcoxon *T* Test for Two Related Samples

Perform the **Wilcoxon *T* test** when you have *related samples* of ranked data. Recall that related samples occur when you *match* samples or have *repeated measures*. For example, say that we perform a study similar to the previous reaction time study, but this time we measure the reaction times of the *same* participants to both the red and black symbols. Table 15.7 gives the data we might obtain.

To compute the Wilcoxon T_{obt}

1. *Determine the difference score for each pair of scores.* Subtract the score in one condition from the score in the other for each pair. It makes no difference which score is subtracted from which, but subtract the scores the same way for all pairs.

2. *Determine the N of the nonzero difference scores.* Ignore any differences equal to zero and count the number of the other difference scores. In our study, one difference equals zero, so $N = 9$.

3. *Assign ranks to the nonzero difference scores.* Ignore the sign (+ or −) of each difference. Assign the rank of 1 to the smallest difference, the rank of 2 to the second-smallest difference, and so on. Record the ranked scores in a column.

4. *Separate the ranks, using the sign of the difference scores.* Create two columns of ranks, labeled "$R-$" and "$R+$." The $R-$ column contains the ranks assigned to negative differences in step 3. The $R+$ column contains the ranks assigned to positive differences.

5. *Compute the sums of ranks for the positive and negative difference scores.* Compute ΣR for the column labeled "$R+$" and for the column labeled "$R-$."

6. *Determine the Wilcoxon* T_{obt}. In the two-tailed test, the Wilcoxon T_{obt} equals the *smallest* ΣR. In the example, the smallest ΣR equals 3, so $T_{obt} = 3$. In the one-tailed test, we predict whether most differences are positive or negative, depending on our experimental hypotheses. Thus, we predict whether $R+$ or $R-$ contains the smaller ΣR, and the one we predict is T_{obt}. (If we predicted that red symbols would produce the largest reaction time scores, given the way we subtracted, we would predict that ΣR for the $R-$ column would be smaller, so T_{obt} would be 42.)

TABLE 15.7

Data for the Wilcoxon *T* Test for Two Related Samples

Participant	Reaction Time to Red Symbols	Reaction Time to Black Symbols	Difference	Ranked Scores	$R-$	$R+$
1	540	760	− 220	6	6	
2	580	710	− 130	4	4	
3	600	1105	− 505	9	9	
4	680	880	− 200	5	5	
5	430	500	− 70	3	3	
6	740	990	− 250	7	7	
7	600	1050	− 450	8	8	
8	690	640	+ 50	2		2
9	605	595	+ 10	1		1
10	520	520	0			
			$N = 9$		$\Sigma R = 42$	$\Sigma R = 3$

7. *Find the critical value of* T *in Table 9 of Appendix C, entitled "Critical Values of the Wilcoxon* T*."* Find T_{crit} for the appropriate α and N, the number of nonzero difference scores. In our study, $N = 9$ and $\alpha = .05$, so T_{crit} is 5.0.

8. *Compare* T_{obt} *to* T_{crit}. Again, watch out: T_{obt} is significant if it is *equal to or less than* T_{crit}. (The critical value is the largest value that our smallest ΣR can be and still reflect a significant difference.)

In the example, the T_{obt} is less than the T_{crit}, so we have a significant difference. Therefore, we conclude that each sample represents a different distribution of ranks and thus a different population of reaction time scores ($p < .05$).

There is no way to compute η^2 for this procedure.

The Kruskal–Wallis *H* Test

The **Kruskal–Wallis *H* test** is analogous to a one-way, between-subjects ANOVA for ranks. It assumes that the study involves one factor involving at least *three* conditions, and each is tested using *independent samples,* with at least five participants in each sample. The null hypothesis is that all conditions represent the same population of ranks.

As an example, say that we examine the independent variable of a golfer's height and the dependent variable of the distance he or she hits the ball. We test golfers classified as either short, medium, or tall, measuring the distance each drives the ball in meters. However, say that we cannot assume that the distance scores have homogeneous variance, so we cannot perform ANOVA. Instead, we perform the Kruskal–Wallis *H* test. The data are shown in Table 15.8.

To compute the Kruskal–Wallis *H* test,

1. *Assign ranks, using all scores in the experiment.* Assign a rank of 1 to the lowest score in the experiment, a 2 to the second-lowest score, and so on.

2. *Compute the sum of the ranks in each condition.* Compute the ΣR in each column.

3. *Compute the sum of squares between groups* (SS_{bn}). Use the formula

$$SS_{bn} = \frac{(\Sigma R_1)^2}{n_1} + \frac{(\Sigma R_2)^2}{n_2} + \cdots + \frac{(\Sigma R_k)^2}{n_k}$$

For our example,

$$SS_{bn} = \frac{(21)^2}{5} + \frac{(35)^2}{5} + \frac{(64)^2}{5} = 88.2 + 245 + 819.2 = 1152.4$$

TABLE 15.8

Data for the Kruskal–Wallis *H* Test

	Height				
Short		**Medium**		**Tall**	
Score	*Rank*	*Score*	*Rank*	*Score*	*Rank*
10	2	24	3	68	14
28	6	27	5	71	15
26	4	35	7	57	10
39	8	44	9	60	12
6	1	58	11	62	13
$\Sigma R_1 = 21$		$\Sigma R_2 = 35$		$\Sigma R_2 = 64$	
$n_1 = 5$		$n_2 = 5$		$n_2 = 5$	$N = 15$

4. *Compute* H_{obt}. Use the formula

$$H_{obt} = \left(\frac{12}{N(N + 1)} \right)(SS_{bn}) - 3(N + 1)$$

where N is the total N of the study, and SS_{bn} is computed as above.
  In the example

$$H_{obt} = \left(\frac{12}{15(15 + 1)} \right)(1152.4) - 3(15 + 1) = (.05)(1152.4) - 48$$

$$H_{obt} = 57.62 - 48 = 9.62$$

5. *Find the critical value of* H *in the* χ^2 *tables* (Table 7 in Appendix C). Values of H have the same sampling distribution as χ^2. The degrees of freedom are

   $$df = k - 1$$

   where k is the number of levels in the factor.
     In the example, k is 3, so for $\alpha = .05$ and $df = 2$, χ^2_{crit} is 5.99.

6. *Compare the obtained value of* H *to the critical value of* χ^2. The H_{obt} is significant if it is *larger* than the critical value. Above, the H_{obt} of 9.62 is larger than the χ^2_{crit} of 5.99, so it is significant. This means that at least two samples represent different populations of ranks. Because the distance participants hit the ball underlies each rank, we conclude that at least two of the populations of distances for short, medium, and tall golfers are not the same ($p < .05$).

7. *Perform post hoc comparisons using the rank sums test.* When H_{obt} is significant, determine which specific conditions differ by performing the rank sums test on every pair of conditions. This is analogous to Fisher's protected t-test (discussed in Chapter 13) and is used regardless of the n in each group. For each pair, treat the two conditions being compared as if they comprised the entire study: re-rank the scores using only the two conditions being compared, and then perform the previous rank sums test.

   In the example, comparing short to medium golfers produces a z_{obt} of 1.36, comparing short to tall golfers produces a z_{obt} of 2.62, and comparing medium to tall golfers produces a z_{obt} of 2.40. With $\alpha = .05$, from the z-tables z_{crit} is ± 1.96. Therefore, the scores of short and medium participants are not significantly different, but they both differ significantly from those in the tall condition. We conclude that tall golfers produce one population of distances that is different from the population for short and medium golfers.

8. *If* H_{obt} *is significant, compute eta squared.* Use the formula

   $$\eta^2 = \frac{H_{obt}}{N - 1}$$

   where H_{obt} is computed in the Kruskal–Wallis test and N is the total number of participants. Above, $H_{obt} = 9.62$ and $N = 15$, so $\eta^2 = 9.62/14$, or .69. Therefore, the variable of a player's height accounts for approximately 69% of the variance in the distance scores.

The Friedman χ^2 Test

The **Friedman χ^2 test** is analogous to a one-way, within-subjects ANOVA for ranks. It assumes that the study involves one factor having at least *three* levels and that the samples in each are *related* (because of either *matching* or *repeated measures*). With only three levels, we must have at least ten participants in the study. With only four levels, we must have at least five participants.

As an example, say that we consider the three teaching styles of Dr. Highman, Dr. Shyman, and Dr. Whyman. A sample of students who have taken courses from all three instructors is repeatedly measured. Table 15.9 shows the data.

To perform the Friedman χ^2 test,

1. *Assign ranks within the scores of each participant.* If the scores are not already ranks, assign the rank of 1 to the lowest score for participant 1, assign the rank of 2 to the second-lowest score for participant 1, and so on. Repeat the process for each participant.

2. *Compute the sum of the ranks in each condition.* Find ΣR in each column.

3. *Compute the sum of squares between groups* (SS_{bn}). Use the formula

$$SS_{bn} = (\Sigma R_1)^2 + (\Sigma R_2)^2 + \cdots + (\Sigma R_k)^2$$

In the example,

$$SS_{bn} = (12)^2 + (23)^2 + (25)^2 = 1298$$

4. *Compute the Friedman χ^2 statistic.* Use the formula

$$\chi^2_{obt} = \left(\frac{12}{(k)(N)(k+1)} \right)(SS_{bn}) - 3(N)(k+1)$$

where N is the number of participants and k is the number of levels of the factor.
In the example

$$\chi^2_{obt} = \left(\frac{12}{(3)(10)(3+1)} \right)(1298) - 3(10)(3+1)$$

$$\chi^2_{obt} = (.10)(1298) - 120 = 129.8 - 120 = 9.80$$

TABLE 15.9

Data for the Friedman Test

Rankings for Three Instructors

Participant	*Dr. Highman*	*Dr. Shyman*	*Dr. Whyman*
1	1	2	3
2	1	3	2
3	1	2	3
4	1	3	2
5	2	1	3
6	1	3	2
7	1	2	3
8	1	3	2
9	1	3	2
10	2	1	3
$N = 10$	$\Sigma R_1 = 12$	$\Sigma R_2 = 23$	$\Sigma R_3 = 25$

5. *Find the critical value of χ^2 in the χ^2-tables* (Table 7 in Appendix C). The degrees of freedom are

$$df = k - 1$$

where k is the number of levels in the factor.
   For the example, $k = 3$, so for $df = 2$ and $\alpha = .05$, the critical value is 5.99.

6. *Compare χ^2_{obt} to the critical value of χ^2. If χ^2_{obt} is larger than χ^2_{crit}, the results are significant.* Our χ^2_{obt} of 9.80 is larger than the χ^2_{crit} of 5.99, so at least two of the samples represent different populations ($p < .05$).

7. *When the χ^2_{obt} is significant, perform post hoc comparisons using **Nemenyi's Procedure**.* This procedure is analogous to Tukey's *HSD* procedure. To perform Nemenyi's procedure,

   a. *Compute the critical difference.* Use the formula

$$\text{Critical difference} = \sqrt{\left(\frac{k(k + 1)}{6(N)}\right)(\chi^2_{crit})}$$

   where k is the number of levels of the factor, N is the number of participants (or rows in the study's diagram), and χ^2_{crit} is the critical value used to test the Friedman χ^2.
      In the example, $\chi^2_{crit} = 5.99$, $k = 3$, and $N = 10$, so

$$\text{Critical difference} = \sqrt{\left(\frac{k(k + 1)}{6(N)}\right)(\chi^2_{crit})} = \sqrt{\left(\frac{3(3 + 1)}{6(10)}\right)(5.99)}$$

$$\text{Critical difference} = \sqrt{(.2)(5.99)} = \sqrt{1.198} = 1.09$$

   b. *Compute the mean rank for each condition.* For each condition, divide the sum of ranks (ΣR) by the number of participants. In the example, the mean ranks are 1.2, 2.3, and 2.5 for Highman, Shyman, and Whyman, respectively.

   c. *Compute the differences between all pairs of mean ranks.* Subtract each mean rank from the other mean ranks. Any absolute difference between two means that is greater than the critical difference indicates that the two conditions differ significantly. In the example, the differences between Dr. Highman and the other two instructors are 1.10 and 1.30, respectively, and the difference between Shyman and Whyman is .20. The critical difference is 1.09, so only Dr. Highman's ranking is significantly different from those of the other two instructors. Thus, we conclude that Dr. Highman would be ranked superior to the other two instructors by the population.

8. *Describe a significant relationship, using eta squared.* Use the formula

$$\eta^2 = \frac{\chi^2_{obt}}{(N)(k) - 1}$$

where χ^2_{obt} is from the Friedman χ^2 test, N is the number of participants, and k is the number of levels of the factor. For the example,

$$\eta^2 = \frac{\chi^2_{obt}}{(N)(k) - 1} = \frac{9.80}{(10)(3) - 1} = \frac{9.80}{30 - 1} = .34$$

Thus, the instructor variable accounts for 34% of the variability in rankings.

PUTTING IT
ALL TOGETHER

Congratulations! You have read an entire statistics book, and that's an accomplishment! You should be proud of the sophisticated level of your knowledge because you are now familiar with the vast majority of statistical procedures used in psychology and other behavioral sciences. Even if you someday go to graduate school, you'll find that there is little in the way of basics for you to learn.

Using the SPSS Appendix As shown in Appendix B.9, SPSS will perform the one-way or the two-way χ^2 procedure. This includes reporting its significance level and, in the two-way design, computing ϕ or C. Appendix B.10 describes how to use SPSS to compute the nonparametric procedures for ranked scores discussed in this chapter, except for post hoc comparisons and η^2.

CHAPTER SUMMARY

1. *Nonparametric procedures* are used when data do not meet the assumptions of parametric procedures. Nonparametric procedures are less powerful than parametric procedures.

2. *Chi square* (χ^2) is used with one or more nominal (categorical) variables, and the data are the frequencies with which participants fall into each category.

3. The *one-way* χ^2 compares the the frequency of category membership along one variable. A significant χ^2_{obt} indicates that the observed frequencies are unlikely to represent the distribution of frequencies in the population described by H_0.

4. The *two-way* χ^2 tests whether category membership for one variable is independent of category membership for the other variable. A significant χ^2 indicates that the data represent dependent or correlated variables in the population.

5. With a significant two-way χ^2, describe the strength of the relationship with (a) the *phi coefficient* (Φ) if the design is a 2×2, or (b) the *contingency coefficient* (C) if the design is not a 2×2. Squaring ϕ or C gives the proportion of variance accounted for, which indicates how much more accurately the frequencies of category membership on one variable can be predicted by knowing category membership on the other variable.

6. The two nonparametric versions of the independent-samples *t*-test for ranks are the *Mann–Whitney* U *test*, performed when both *n*s are less than 20, and the *rank sums test*, performed when either *n* is greater than 20.

7. The *Wilcoxon* T *test* is the nonparametric, related-samples *t*-test for ranks.

8. The *Kruskal–Wallis* H *test* is the nonparametric, one-way, between-subjects ANOVA for ranks. The rank sums test is the post hoc test for identifying the specific conditions that differ.

9. The *Friedman* χ^2 *test* is the nonparametric, one-way, within-subjects ANOVA for ranks. *Nemenyi's test* is the post hoc test for identifying the specific conditions that differ.

10. *Eta squared* describes the relationship found in experiments involving ranked data.

KEY TERMS

f_o f_e χ^2_{obt} χ^2_{crit} ϕ ϕ^2 C C^2

χ^2-distribution *355*

chi square procedure *352*

contingency coefficient *362*

expected frequency *354*

Friedman χ^2 test *371*

goodness-of-fit test *353*

Kruskal–Wallis H test *369*

Mann–Whitney U test *365*

Nemenyi's procedure *372*

nonparametric statistics *352*

observed frequency *353*

one-way chi square *352*

phi coefficient *361*

rank sums test *366*

test of independence *358*

two-way chi square *357*

Wilcoxon T test *368*

REVIEW QUESTIONS

(Answers for odd-numbered questions are in Appendix D.)

1. What do all nonparametric inferential procedures have in common with all parametric procedures?

2. (a) Which variable in an experiment determines whether to use parametric or nonparametric procedures? (b) In terms of the dependent variable, what are the two categories into which all nonparametric procedures can be grouped?

3. (a) Which two scales of measurement always require nonparametric procedures? (b) What two things can be "wrong" with interval/ratio scores that lead to nonparametric procedures (c) What must you do to the interval/ratio scores first?

4. (a) Why, if possible, should we design a study that meets the assumptions of a parametric procedure? (b) Why shouldn't you use parametric procedures for data that clearly violate their assumptions?

5. (a) When do you use the one-way chi square? (b) When do you use the two-way chi square?

6. (a) What is the symbol for observed frequency? What does it mean? (b) What is the symbol for expected frequency? What does it mean?

7. (a) What does a significant one-way chi square indicate? (b) What does a significant two-way chi square indicate?

8. What is the logic of H_0 and H_a in all procedures for ranked data?

9. (a) What is the *phi coefficient,* and when is it used? (b) What does the squared phi coefficient indicate? (c) What is the *contingency coefficient,* and when is it used? (d) What does the *squared contingency coefficient* indicate?

10. What is the nonparametric version of each of the following? (a) A one-way, between-subjects ANOVA (b) An independent-samples t-test ($n < 20$) (c) A related-samples t-test (d) An independent-samples t-test ($n > 20$) (e) A one-way, within-subjects ANOVA (f) Fisher's protected t-test (g) Tukey's *HSD* test

APPLICATION QUESTIONS

11. In the population, political party affiliation is 30% Republican, 55% Democratic, and 15% other. To determine whether this distribution is also found among the elderly, in a sample of 100 senior citizens, we find 18 Republicans, 64 Democrats, and 18 other. (a) What procedure should we perform? (b) What are H_0 and H_a?

(c) What is f_e for each group? (d) Compute χ^2_{obt}. (e) With $\alpha = .05$, what do you conclude about party affiliation in the population of senior citizens?

12. A survey finds that, given the choice, 34 females prefer males much taller than themselves, and 55 females prefer males only slightly taller than themselves. (a) What procedure should we perform? (b) What are H_0 and H_a? (c) With $\alpha = .05$, what do you conclude about the preference of females in the population? (d) Describe how you would graph these results.

13. Foofy counts the students who like Professor Demented and those who like Professor Randomsampler. She then performs a one-way χ^2 to determine if there is a significant difference between the frequency with which students like each professor. (a) Why is this approach incorrect? (*Hint:* Check the assumptions of χ^2.) (b) How should she analyze the data?

14. The following data reflect the frequency with which people voted in the last election and were satisfied with the officials elected:

		Satisfied	
		Yes	*No*
Voted	*Yes*	48	35
	No	33	52

(a) What procedure should we perform? (b) What are H_0 and H_a? (c) What is f_e in each cell? (d) Compute χ^2_{obt}. (e) With $\alpha = .05$, what do you conclude about the correlation between these variables? (f) How consistent is this relationship?

15. A study determines the frequency of the different political party affiliations for male and female senior citizens. The following data are obtained:

		Affiliation		
		Republican	*Democrat*	*Other*
Gender	*Male*	18	43	14
	Female	39	23	18

(a) What procedure should we perform? (b) What are H_0 and H_a? (c) What is f_e in each cell? (d) Compute χ^2_{obt}. (e) With $\alpha = .05$, what do you conclude about gender and party affiliation in the population of senior citizens? (f) How consistent is this relationship?

16. Select the noparametric procedure to use when we study: (a) The effect of a pain reliever on rankings of the emotional content of words describing pain. One group is tested before and after taking the drug. (b) The effect of eight colors of spaghetti sauce on its tastiness. A different sample tastes each color of sauce, and tastiness scores are ranked. (c) The (skewed) reaction time scores after one group of participants consumed 1, 3, and then 5 alcoholic drinks. (d) Whether two levels of family income influence the percentage of income spent on clothing last year. Percentages are then ranked.

17. After testing 40 participants, a significant χ^2_{obt} of 13.31 was obtained. With $\alpha = .05$ and $df = 2$, how would this result be reported in a publication?

18. We compare the attitude scores of people tested in the morning to their scores when tested in the afternoon. We obtain the following interval data but it does not have homogeneous variance. With $\alpha = .05$, is there a significant difference in scores as a function of testing times?

Morning	Afternoon
14	36
18	31
20	19
28	48
3	10
34	49
20	20
24	29

19. We measure the maturity level of students who have completed statistics and students who have not. Maturity scores tend to be skewed. For the following interval scores,

Nonstatistics	Statistics
43	51
52	58
65	72
23	81
31	92
36	64

(a) Do the groups differ significantly ($\alpha = .05$)? (b) What do you conclude about maturity scores in the population?

20. A therapist evaluates a sample in a new treatment program after 1 month, after 2 months, and again after 3 months. Such data do not have homogeneous variance. (a) What procedure should be used? Why? (b) What must the therapist do to the data first? (c) If the results are significant, what procedure should be performed? (d) Ultimately, what will the therapist be able to identify?

21. An investigator evaluated the effectiveness of a therapy on three types of patients. She collected the following improvement ratings, but these data form skewed distributions.

Depressed	Manic	Schizophrenic
16	7	13
11	9	6
12	6	10
20	4	15
21	8	9

(a) Which procedure should be used? Why? (b) What should the investigator do to the data first? (c) If the results are significant, what should she do next? (d) Ultimately, what will we learn from this study?

22. A report indicates that the Friedman χ^2 test was significant. (a) What does this test indicate about the design of the study? (b) What does it indicate about the raw scores? (c) What two procedures do you also expect to be reported? (d) What will you learn about the relationship?

23. A report indicates that the Wilcoxon T test was significant. (a) What does this test indicate about the design of the study? (b) What does it indicate about the raw scores? (c) What will you learn about the relationship here?

24. You show participants a picture of a person either smiling, frowning, or smirking. For each, they indicate whether the person was happy or sad. (a) What are the factor(s) and level(s) in this design, and (b) What is the dependent variable? (c) How will you analyze the results? (d) What potential flaw is built into the study in terms of the statistical errors we may make? (c) When would this flaw be a concern? (d) How can you eliminate the flaw?

INTEGRATION QUESTIONS

25. Thinking back on the previous few chapters, what three aspects of your independent variable(s) and one aspect of your dependent variable determine the specific inferential procedure to perform in a particular experiment? (Ch's. 10, 11, 12, 13, 14, 15)

26. For the following, what inferential statistical procedures should be performed and what is the key information for answering the research question? Unless described otherwise, assume scores are parametric. (a) To test the influence of dietary fat levels on visual accuracy, rats are placed on one of four different diets. Three weeks later, all are tested for visual accuracy. (b) We measure the preferences of three groups of students for their classmates after either 1, 5, or 10 weeks into the semester. To be friendly, everyone's scores are *skewed* toward being positive. (c) To examine if body image is related to personality traits, we measure the self-confidence and degree of obesity in a sample of adults. (d) We find that among all Americans without marital problems, people nod their head 32 times during a conversation with a spouse. In a sample of people having problems, the mean number of nods is only 22, with $s_X = 5.6$. (e) We select 30 left-handed and 30 right-handed men and measure their manual dexterity scores. Ten of each type is tested under either low, medium, or high illumination levels. (f) Last semester a teacher gave 25 As, 35 Bs, 20 Cs, 10 Ds, and 10 Fs. According to college policy, each grade should occur 20% of the time. Is the teacher different from the college's model? (g) We compare the introversion of two groups of autistic children, one after receiving a new improved therapy and the other after traditional therapy. (h) To determine if handedness is related to aphasia, in a sample we count the number of left- and right-handed individuals and in each group count those who do and do not exhibit aphasia. (i) We create two groups of 15 participants, who either slept 4 or 8 hours the night before. Then we rank their driving ability. (j) To determine how colors influence consumer behavior, we compare the mean attractiveness scores from a group of participants who viewed a product's label when printed in red, in green, and in blue. (k) Does chewing gum while taking an exam lead to better performance? We obtain exam grades for students who do or do not chew gum during an exam, when the exam is either multiple choice or essay. (Chs. 7, 8, 11, 12, 13, 14, 15)

■ ■ ■ SUMMARY OF FORMULAS

1. The formula for chi square is

$$\chi^2_{obt} = \Sigma\left(\frac{(f_o - f_e)^2}{f_e}\right)$$

In a one-way chi square:

$$df = k - 1$$

In a two-way chi square:

$$f_e = \frac{(\text{Cell's row total } f_o)(\text{Cell's column total } f_o)}{N}$$

$$df = (\text{Number of rows} - 1)(\text{Number of columns} - 1)$$

2. The formula for the phi coefficient is

$$\phi = \sqrt{\frac{\chi^2_{obt}}{N}}$$

3. The formula for the contingency coefficient (C) is

$$C = \sqrt{\frac{\chi^2_{obt}}{N + \chi^2_{obt}}}$$

Additional Statistical Formulas

A.1   Creating Grouped Frequency Distributions

A.2   Performing Linear Interpolation

A.3   The One-Way, Within-Subjects Analysis of Variance

A.1 CREATING GROUPED FREQUENCY DISTRIBUTIONS

In a grouped distribution, different scores are grouped together, and then the total f, *rel. f,* or cf of each group is reported. For example, say that we measured the level of anxiety exhibited by 25 participants, obtaining the following scores:

$$3 \quad 4 \quad 4 \quad 18 \quad 4 \quad 28 \quad 26 \quad 41 \quad 5 \quad 40 \quad 4 \quad 6 \quad 5$$
$$18 \quad 22 \quad 3 \quad 17 \quad 12 \quad 26 \quad 4 \quad 20 \quad 8 \quad 15 \quad 38 \quad 36$$

First, determine the number of scores the data span. The number spanned between any two scores is

$$\text{Number of scores} = (\text{High score} - \text{Low score}) + 1$$

Thus, there is a span of 39 values between 41 and 3.

Next, decide how many scores to put into each group, with the same range of scores in each. You can operate as if the sample contained a wider range of scores than is actually in the data. For example, we'll operate as if these scores are from 0 to 44, spanning 45 scores. This allows nine groups, each spanning 5 scores, resulting in the grouped distribution shown in Table A.1.

TABLE A.1

Grouped Distribution Showing *f, rel. f,* and *cf* for Each Group of Anxiety Scores

The column on the left identifies the lowest and highest score in each class interval.

Anxiety Scores	f	rel. f	cf
40–44	2	.08	25
35–39	2	.08	23
30–34	0	.00	21
25–29	3	.12	21
20–24	2	.08	18
15–19	4	.16	16
10–14	1	.04	12
5– 9	4	.16	11
0– 4	7	.28	7
Total: 25		1.00	

The group labeled 0–4 contains the scores 0, 1, 2, 3, and 4, the group 5–9 contains 5 through 9, and so on. Each group is called a **class interval,** and the number of scores spanned by an interval is called the **interval size.** Here, the interval size is 5, so each group includes five scores. Choose an interval size that is easy to work with (such as 2, 5, 10, or 20). Also, an interval size that is an odd number is preferable because later we'll use the middle score of the interval.

Notice several things about the score column in Table A.1. First, each interval is labeled with the low score on the left. Second, the low score in each interval is a whole-number multiple of the interval size of 5. Third, every class interval is the same size. (Even though the highest score in the data is only 41, we have the complete interval of 40–44.) Finally, the intervals are arranged so that higher scores are located toward the top of the column.

To complete the table, find the f for each class interval by summing the individual frequencies of all scores in the group. In the example, there are no scores of 0, 1, or 2, but there are two 3s and five 4s. Thus, the 0–4 interval has a total f of 7. For the 5–9 interval, there are two 5s, one 6, no 7s, one 8, and no 9s, so f is 4. And so on.

Compute the relative frequency for each interval by dividing the *f* for the interval by *N*. Remember, *N* is the total number of raw scores (here, 25), not the number of class intervals. Thus, for the 0–4 interval, *rel. f* equals 7/25, or .28.

Compute the cumulative frequency for each interval by counting the number of scores that are at or below the *highest* score in the interval. Begin with the lowest interval. There are 7 scores at 4 or below, so the *cf* for interval 0–4 is 7. Next, *f* is 4 for the scores between 5 and 9, and adding the 7 scores below the interval produces a *cf* of 11 for the interval 5–9. And so on.

Real versus Apparent Limits

What if one of the scores in the above example were 4.6? This score seems too large for the 0–4 interval, but too small for the 5–9 interval. To allow for such scores, we consider the "real limits" of each interval. These are different from the upper and lower numbers of each interval seen in the frequency table, which are called the *apparent upper limit* and the *apparent lower limit,* respectively. As in Table A.2, the apparent limits for each interval imply corresponding real limits. Thus, for example, the interval having the apparent limits of 40–44 actually contains any score between the real limits of 39.5 and 44.5.

Note that (1) each real limit is halfway between the lower apparent limit of one interval and the upper apparent limit of the interval below it, and (2) the lower real limit of one interval is always the same number as the upper real limit of the interval below it. Thus, 4.5 is halfway between 4 and 5, so 4.5 is the lower real limit of the 5–9 interval and the upper real limit of the 0–4 interval. Also, the difference between the lower real limit and the upper real limit equals the interval size (9.5 − 4.5 = 5).

Real limits eliminate the gaps between intervals, so now a score such as 4.6 falls into the interval 5–9 because it falls between 4.5 and 9.5. If scores equal a real limit (such as two scores of 4.5), put half in the lower interval and half in the upper interval. If one score is left over, just pick an interval.

The principle of real limits also applies to ungrouped data. Implicitly, each individual score is a class interval with an interval size of 1. Thus, when a score in an ungrouped distribution is labeled 6, this is both the upper and the lower *apparent* limits. However, the lower *real* limit for this interval is 5.5, and the upper *real* limit is 6.5.

Graphing Grouped Distributions

TABLE A.2

Real and Apparent Limits

The apparent limits in the column on the left imply the real limits in the column on the right.

Apparent Limits (Lower–Upper)	Imply	Real Limits (Lower–Upper)
40–44	→	39.5–44.5
35–39	→	34.5–39.5
30–34	→	29.5–34.5
25–29	→	24.5–29.5
20–24	→	19.5–24.5
15–19	→	14.5–19.5
10–14	→	9.5–14.5
5– 9	→	4.5– 9.5
0– 4	→	−0.5– 4.5

Grouped distributions are graphed in the same way as ungrouped distributions, *except* that the *X* axis is labeled differently. To graph simple frequency or relative frequency, label the *X* axis using the *midpoint* of each class interval. To find the midpoint, multiply .5 times the interval size and add the result to the lower real limit. Above, the interval size is 5, which multiplied times .5 is 2.5. For the 0–4 interval, the lower real limit is −.5. Adding 2.5 to −.5 yields 2. Thus, the score of 2 on the *X* axis identifies the class interval of 0–4. Similarly, for the 5–9 interval, 2.5 plus 4.5 is 7, so this interval is identified using 7.

As usual, for nominal or ordinal scores create a bar graph, and for interval or ratio scores, create a histogram or polygon. Figure A.1 presents a histogram and polygon for the grouped distribution from Table A.1. The height of each data point or bar corresponds to the total simple frequency of all scores in the class interval. Plot a relative frequency distribution in the same way, except that the *Y* axis is labeled in increments between 0 and 1.

FIGURE A.1

Grouped frequency polygon and histogram

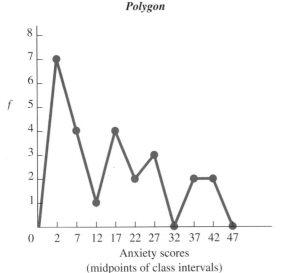

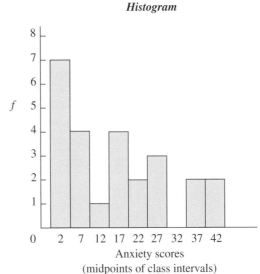

Application Questions

(Answers for odd-numbered questions are in Appendix D.)

1. Organize the scores below into an ungrouped distribution showing simple frequency, cumulative frequency, and relative frequency.

   49   52   47   52   52   47   49   47   50
   51   50   49   50   50   50   53   51   49

2. Using an interval size of 5, group these scores and construct a table that shows simple, relative, and cumulative frequency. The highest apparent limit is 95.

   76   66   80   82   76   80   84   86   80   86
   85   87   74   90   92   87   91   94   94   91
   94   93   57   82   76   76   82   90   87   91
   66   80   57   66   74   76   80   84   94   66

3. Using an interval size of 4, group these scores and construct a table showing simple, relative, and cumulative frequency. The lowest apparent limit is 100.

   122   117   116   114   110   109   107
   105   103   102   129   126   123   123
   122   122   119   118   117   112   108
   117   117   126   123   118   113   112

A.2 PERFORMING LINEAR INTERPOLATION

This section presents the procedures for linear interpolation of z-scores as discussed in Chapter 6 and of values of t_{crit} as discussed in Chapter 11.

Interpolating from the z-Tables

You interpolate to find an exact proportion not shown in the z-table or when dealing with a z-score that has three decimal places. Carry all computations to four decimal places.

Finding an Unknown z-Score

Say that we seek a z-score that corresponds to exactly .45 (.4500) of the curve between the mean and z. First, from the z-tables, identify the two bracketing proportions that are above and below the target proportion. Note their corresponding z-scores. For .4500, the bracketing proportions are .4505 at $z = 1.6500$ and .4495 at $z = 1.6400$. Arrange the values this way:

	Known *Proportion under Curve*	*Unknown* *z-score*
Upper bracket	.4505	1.6500
Target	.4500	?
Lower bracket	.4495	1.6400

Because the "known" target proportion is bracketed by .4505 and .4495, the "unknown" target z-score falls between 1.6500 and 1.6400.

First, deal with the known side. The target of .4500 is halfway between .4495 and .4505. That is, the difference between the lower known proportion and the target proportion is one-half of the difference between the two known proportions. We assume that the z-score corresponding to .4500 is also halfway between the two bracketing z-scores of 1.6400 and 1.6500. The difference between these z-scores is .010, and one-half of that is .005. To go to halfway between 1.6400 and 1.6500, we add .005 to 1.6400. Thus, a z-score of 1.6450 corresponds to .4500 of the curve between the mean and z.

The answer will not always be as obvious as in this example, so use the following steps.

Step 1 Determine the difference between the upper and lower known brackets. In the example .4505 − .4495 = .0010. This is the total distance between the two proportions.

Step 2 Determine the difference between the known target and the lower known bracket. Above, .4500 − .4495 = .0005.

Step 3 Form a fraction with the answer from Step 2 as the numerator and the answer from Step 1 as the denominator. Above, the fraction is .0005/.0010 = .5. Thus, .4500 is one-half of the distance from .4495 to .4505.

Step 4 Find the difference between the two brackets in the unknown column. Above, 1.6500 − 1.6400 = .010. This is the total distance between the two z-scores that bracket the unknown target z-score.

Step 5 Multiply the answer in Step 3 by the answer in Step 4. Above, (.5)(.010) = .005. The unknown target z-score is .005 larger than the lower bracketing z-score.

Step 6 Add the answer in Step 5 to the lower bracketing z-score. Above, .005 + 1.640 = 1.6451.640 = 1.645. Thus, .4500 of the normal curve lies between the mean and $z = 1.645$.

Finding an Unknown Proportion

Apply the above steps to find an unknown proportion for a known three-decimal z-score. For example, say that we seek the proportion between the mean and a z of 1.382. From the z-tables, the upper and lower brackets around this z are 1.390 and 1.380. Arrange the z-scores and corresponding proportions as shown below:

	Known z-score	*Unknown Proportion under Curve*
Upper bracket	1.390	.4177
Target	1.382	?
Lower bracket	1.380	.4162

To find the target proportion, use the preceding steps.

Step 1 $1.390 - 1.380 = .010$
This is the total difference between the known bracketing z-scores.

Step 2 $1.382 - 1.380 = .002$
This is the distance between the lower known bracketing z-score and the target z-score.

Step 3 $\dfrac{.002}{.010} = .20$
This is the proportion of the distance that the target z-score lies from the lower bracket. A z of 1.382 is .20 of the distance between 1.380 and 1.390.

Step 4 $.4177 - .4162 = .0015$
The total distance between the brackets of .4177 and .4162 in the unknown column is .0015.

Step 5 $(0.20)(0.0015) = .0003$
Thus, .20 of the distance separating the bracketing proportions in the unknown column is .0003.

Step 6 $.4162 + .0003 = .4165$
Increasing the lower proportion in the unknown column by .0003 takes us to the point corresponding to .20 of the distance between the bracketing proportions. This point is .4165, which is the proportion that corresponds to $z = 1.382$.

Interpolating Critical Values

Sometimes you must interpolate between the critical values in a table. Apply the same steps described above, except now use degrees of freedom and critical values.

For example, say that we seek the t_{crit} corresponding to 35 df (with $\alpha = .05$, two-tailed test). The t-tables have values only for 30 df and 40 df, giving the following:

	Known df	*Unknown Critical Value*
Upper bracket	30	2.042
Target	35	?
Lower bracket	40	2.021

Because 35 *df* is halfway between 30 *df* and 40 *df*, the corresponding critical value is halfway between 2.042 and 2.021. Following the steps described for *z*-scores, we have

Step 1 $40 - 30 = 10$
This is the total distance between the known bracketing *df*s.

Step 2 $35 - 30 = 5$
Notice a change here: This is the distance between the *upper* bracketing *df* and the target *df*.

Step 3 $\dfrac{5}{10} = .50$
This is the proportion of the distance that the target *df* lies from the upper known bracket. Thus, the *df* of 35 is .50 of the distance from 30 to 40.

Step 4 $2.042 - 2.021 = .021$
The total distance between the bracketing critical values of 2.042 and 2.021 in the unknown column is .021.
   The *df* of 35 is .50 of the distance between the bracketing *df*s, so the target critical value is .50 of the distance between 2.042 and 2.021, or .50 of .021.

Step 5 $(.50)(.021) = .0105$
Thus, .50 of the distance between the bracketing critical values is .0105. Because critical values *decrease* as *df* increases, we are going from 30 *df* to 35 *df*, so subtract .0105 from the larger value, 2.042.

Step 6 $2.042 - .0105 = 2.0315$
Thus, $t = 2.0315$ is the critical value for 35 *df* at $\alpha = .05$ for a two-tailed test.
   The same logic can be applied to find critical values for any other statistic.

Application Questions

(Answers for odd-numbered questions are in Appendix D.)

1. What is the *z*-score you must score above to be in the top 25% of scores?
2. Foofy obtains a *z*-score of 1.909. What proportion of scores are between her score and the mean?
3. For $\alpha = .05$, what is the two-tailed t_{crit} for $df = 50$?
4. For $\alpha = .05$, what is the two-tailed t_{crit} for $df = 55$?

A.3 THE ONE-WAY, WITHIN-SUBJECTS ANALYSIS OF VARIANCE

This section contains formulas for the one-way, within-subjects ANOVA discussed in Chapter 13. This ANOVA is similar to the two-way ANOVA discussed in Chapter 14, so read that chapter first.

Assumptions of the Within-Subjects ANOVA

In a within-subjects ANOVA, either the same participants are measured repeatedly or different participants are matched under all levels of one factor. (Statistical terminology still uses the old fashioned term *subjects* instead of the more modern *participants*.) The other assumptions here are (1) the dependent variable is a ratio or interval variable, (2) the populations are normally distributed, and (3) the population variances are homogeneous.

Logic of the One-Way, Within-Subjects ANOVA

As an example, say that we're interested in whether a person's form of dress influences how comfortable he or she feels in a social setting. On three consecutive days, we ask each participant to act as a "greeter" for other people participating in a different experiment. On the first day, participants dress casually; on the second day, they dress semiformally; on the third day, they dress formally. We test the very unpowerful N of 5. At the end of each day, participants complete a questionnaire measuring the dependent variable of their comfort level while greeting people. Labeling the independent variable of type of dress as factor A, the layout of the study is shown in Table A.3.

To describe the relationship that is present, we'll find the mean of each level (column) under factor A. As usual, we test whether the means from the levels represent different μs. Therefore, the hypotheses are the same as in a between-subjects design:

H_0: $\mu_1 = \mu_2 = \mu_3$
H_a: Not all μs are equal

Elements of the Within-Subjects ANOVA

Notice that this one-way ANOVA can be viewed as a two-way ANOVA: Factor A (the columns) is one factor, and the different participants or subjects (the rows) are a second factor, here with five levels. The interaction is between subjects and type of dress.

In Chapters 13 and 14, we computed the F-ratio by dividing by the mean square within groups (MS_{wn}). This estimates the error variance (σ^2_{error}), the variability among scores in the population. We computed MS_{wn} using the differences between the scores in each *cell* and the mean of the cell. However, in Table A.3, each cell contains only one score. Therefore, the mean of each cell *is* the score in the cell, and the differences within a cell are always zero. Obviously, we cannot compute MS_{wn} in the usual way.

Instead, the mean square for the interaction between factor A and subjects (abbreviated $MS_{A \times subs}$) reflects the inherent variability of scores. Recall that an interaction indicates that the effect of one factor changes as the levels of the other factor change. It is because of the inherent variability among people that the effect of type of dress will change as we change the "levels" of which participant we test. Therefore, $MS_{A \times subs}$ is our estimate of the error variance, and it is used as the denominator of the F-ratio. (If the study involved matching, each triplet of matched participants would provide the

TABLE A.3

One-Way Repeated-Measures Study of the Factor of Type of Dress

Each X represents a participant's score on the dependent variable of comfort level.

Factor A: Type of Dress

		Level A_1: Casual	Level A_2: Semiformal	Level A_3: Formal
	1	X	X	X
	2	X	X	X
Subjects Factor	3	X	X	X
	4	X	X	X
	5	$\overline{X}$	X	X
		$\overline{X}_{A_1}$	$\overline{X}_{A_2}$	$\overline{X}_{A_3}$

scores in each row, and the $MS_{A \times subs}$ here would still show the variability among their scores.)

As usual, MS_A describes the difference between the means in factor A, and it estimates the variability due to error plus the variability due to treatment. Thus, the F-ratio here is

Sample		*Estimates*	*Population*
$F_{obt} = \dfrac{MS_A}{MS_{A \times subs}}$		$\rightarrow$ $\rightarrow$	$\dfrac{\sigma^2_{error} + \sigma^2_{treat}}{\sigma^2_{error}}$

If H_0 is true and all μs are equal, then both the numerator and the denominator will contain only σ^2_{error}, so F_{obt} will equal 1. However, the larger the F_{obt}, the less likely it is that the means for the levels of factor A represent one population μ. If F_{obt} is significant, then at least two of the means represent different μs.

Computing the One-Way, Within-Subjects ANOVA

Say that we obtained these data:

Factor A: Type of Dress

Subjects	Level A_1: Casual	Level A_2: Semiformal	Level A_3: Formal	
1	4	9	1	$\Sigma X_{sub} = 14$
2	6	12	3	$\Sigma X_{sub} = 21$
3	8	4	4	$\Sigma X_{sub} = 16$
4	2	8	5	$\Sigma X_{sub} = 15$
5	10	7	2	$\Sigma X_{sub} = 19$

			Total:
$\Sigma X = 30$	$\Sigma X = 40$	$\Sigma X = 15$	$\Sigma X_{tot} = 30 + 40 + 15 = 85$
$\Sigma X^2 = 220$	$\Sigma X^2 = 354$	$\Sigma X^2 = 55$	$\Sigma X^2_{tot} = 220 + 354 + 55 = 629$
$n_1 = 5$	$n_2 = 5$	$n_3 = 5$	$N = 15$
$\overline{X}_1 = 6$	$\overline{X}_2 = 8$	$\overline{X}_3 = 3$	$k = 3$

Step 1 Compute the ΣX, the $\overline{X}$, and the ΣX^2 for each level of factor A (each column). Then compute ΣX_{tot} and ΣX^2_{tot}. Also, compute ΣX_{sub}, which is the ΣX for each participant's scores (each row). Notice that the *ns* and *N* are based on the number of *scores*, not the number of participants.

Then follow these steps.

Step 2 Compute the total sum of squares.

The formula for the total sums of squares is

$$SS_{tot} = \Sigma X^2_{tot} - \left(\frac{(\Sigma X_{tot})^2}{N} \right)$$

From the example, we have

$$SS_{tot} = 629 - \left(\frac{85^2}{15}\right)$$

$$SS_{tot} = 629 - 481.67 = 147.33$$

Note that the quantity $(\Sigma X_{tot})^2/N$ is the *correction* in the following computations. (Here, the correction is 481.67.)

Step 3 Compute the sum of squares for the column factor, factor A.

> **The formula for the sum of squares between groups for factor A is**
>
> $$SS_A = \Sigma\left(\frac{(\text{Sum of scores in the column})^2}{n \text{ of scores in the column}}\right) - \left(\frac{(\Sigma X_{tot})^2}{N}\right)$$

Find ΣX in each level (column) of factor A, square the sum and divide by the n of the level. After doing this for all levels, add the results together and subtract the correction. In the example

$$SS_A = \left(\frac{(30)^2}{5} + \frac{(40)^2}{5} + \frac{(15)^2}{5}\right) - 481.67$$

$$SS_A = 545 - 481.67 = 63.33$$

Step 4 Find the sum of squares for the row factor, for subjects.

> **The formula for the sum of squares for subjects is**
>
> $$SS_{subs} = \frac{(\Sigma X_{sub1})^2 + (\Sigma X_{sub2})^2 + \cdots + (\Sigma X_n)^2}{k} - \frac{(\Sigma X_{tot})^2}{N}$$

Square the sum for each subject (ΣX_{sub}). Then add the squared sums together. Next, divide by k, the number of levels of factor A. Finally, subtract the correction. In the example,

$$SS_{subs} = \frac{(14)^2 + (21)^2 + (16)^2 + (15)^2 + (19)^2}{3} - 481.67$$

$$SS_{subs} = 493 - 481.67 = 11.33$$

Step 5 Find the sum of squares for the interaction. To do this, subtract the sums of squares for the other factors from the total.

> **The formula for the interaction of factor a by subjects is**
>
> $$SS_{A \times subs} = SS_{tot} - SS_A - SS_{subs}$$

In the example

$$SS_{A \times subs} = 147.33 - 63.33 - 11.33 = 72.67$$

Step 6 Determine the degrees of freedom.

> ### The degrees of freedom between groups for factor A is
>
> $$df_A = k_A - 1$$

k_A is the number of levels of factor A. (In the example, $k_A = 3$, so df_A is 2.)

> ### The degrees of freedom for the interaction is
>
> $$df_{A \times subs} = (k_A - 1)(k_{subs} - 1)$$

k_A is the number of levels of factor A, and k_{subs} is the number of participants. In the example with three levels of factor A and five subjects, $df_{A \times subs} = (2)(4) = 8$.

Compute df_{subs} and df_{tot} to check the above df. The $df_{subs} = k_{subs} - 1$. The $df_{tot} = N - 1$, where N is the total number of *scores* in the experiment. The df_{tot} is also equal to the sum of all other *dfs*.

Step 7 Place the sum of squares and the *dfs* in the summary table. For the example,

Summary Table of One-Way, Within-Subjects ANOVA

Source	*Sum of Squares*	*df*	*Mean Square*	**F**
Subjects	11.33	4		
Factor A (dress)	63.33	2	MS_A	F_A
Interaction				
(A × subjects)	72.67	8	$MS_{A \times subs}$	
Total	147.33	14		

Because there is only one factor of interest here (type of dress), we will find the F_{obt} only for factor A.

Step 8 Find the mean square for factor A and the interaction.

> ### The mean square for factor A is
>
> $$MS_A = \frac{SS_A}{df_A}$$

In our example,

$$MS_A = \frac{SS_A}{df_A} = \frac{63.33}{2} = 31.67$$

The mean square for the interaction between factor A and subjects is

$$MS_{A \times subs} = \frac{SS_{A \times subs}}{df_{A \times subs}}$$

In the example,

$$MS_{A \times subs} = \frac{SS_{A \times subs}}{df_{A \times subs}} = \frac{72.67}{8} = 9.08$$

Step 9 Find F_{obt}.

The within-subjects F-ratio is

$$F_{obt} = \frac{MS_A}{MS_{A \times subs}}$$

In the example,

$$F_{obt} = \frac{MS_A}{MS_{A \times subs}} = \frac{31.67}{9.08} = 3.49$$

The finished summary table is

Source	Sum of Squares	df	Mean Square	F
Subjects	11.33	4		
Factor A (dress)	63.33	2	31.67	3.49
Interaction				
(A × subjects)	72.67	8	9.08	
Total	147.33	14		

Step 10 Find the critical value of F in Table 5 of Appendix C. Use df_A as the degrees of freedom between groups and $df_{A \times subs}$ as the degrees of freedom within groups. In the example for $\alpha = .05$, $df_A = 2$, and $df_{A \times subs} = 8$, the F_{crit} is 4.46.

Interpreting the Within-Subjects *F*

Interpret the above F_{obt} the same way that you would a between-subjects F_{obt}. Because F_{obt} in the above example is *not* larger than F_{crit}, it is not significant. Thus, we do not have evidence that the means from at least two levels of type of dress represent different populations of scores. Had F_{obt} been significant, it would indicate that at least two of the level means differ significantly. Then, for post hoc comparisons, graphing, eta squared, and confidence intervals, follow the procedures discussed in Chapter 13. However, in any of those formulas, in place of the term MS_{wn} use $MS_{A \times subs}$.

Note: The related-samples *t*-test in Chapter 12 is more powerful than the independent-samples *t*-test because the variability in the scores is less. For the same reason, a within-subjects ANOVA is more powerful than a between-subjects ANOVA for the same data.

Application Questions

(Answers for odd-numbered questions are in Appendix D.)

1. You read in a research report that the repeated-measures factor for a person's weight gain led to a decrease in his or her mood. (a) What does this tell you about the design? (b) What does it tell you about the results?

2. Which of these relationships suggest using a repeated-measures design? (a) Examining the improvement in language ability as children grow older (b) Measuring participants' reaction when the experimenter surprises them by unexpectedly shouting, under three levels of volume of shouting (c) Comparing the dating strategies of males and females (d) Comparing memory ability under the conditions of participants' consuming different amounts of alcoholic beverages.

3. We study the influence of practice on performing a task requiring eye–hand coordination. We test people with no practice, after 1 hour of practice, and after 2 hours of practice. In the following data, higher scores indicate better performance. (a) What are H_0 and H_a? (b) Complete the ANOVA summary table. (c) With $\alpha = .05$, what do you conclude about F_{obt}? (d) Perform the appropriate post hoc comparisons. (e) What is the effect size in this study? (f) What should you conclude about this relationship?

Amount of Practice

Subjects	None	1 Hour	2 Hours
1	4	3	6
2	3	5	5
3	1	4	3
4	3	4	6
5	1	5	6
6	2	6	7
7	2	4	5
8	1	3	8

4. You measure 21 students' degree of positive attitude toward statistics at four equally spaced intervals during the semester. The mean score for each level is time 1, 62.50; time 2, 64.68; time 3, 69.32; time 4, 72.00. You obtain the following sums of squares:

Source	Sum of Squares	df	Mean Square	F
Subjects	402.79			
Factor A	189.30			
A × subjects	688.32			
Total	1280.41			

(a) What are H_0 and H_a? (b) Complete the ANOVA summary table (c) With $\alpha = .05$, what do you conclude about F_{obt}? (d) Perform the appropriate post hoc comparisons. (e) What is the effect size in this study? (f) What should you conclude about this relationship?

B

Using SPSS

B.1   **Entering Data**

B.2   **Frequency Distributions and Percentile**

B.3   **Central Tendency, Variability, and *z*-Scores**

B.4   **Correlation Coefficients and the Linear Regression Equation**

B.5   **The One-Sample *t*-Test and Significance Testing of Correlation Coefficients**

B.6   **Two-Sample *t*-Tests**

B.7   **The One-Way, Between-Subjects ANOVA**

B.8   **The Two-Way, Between-Subjects ANOVA**

B.9   **Chi Square Procedures**

B.10   **Nonparametric Tests for Ranked Scores**

B.11   **The One-Way, Within-Subjects ANOVA**

This appendix describes how to use the computer program called *SPSS* (or *PASW*) to compute the statistics described in this textbook. These instructions are appropriate for most recent versions of SPSS, although copies of "screens" were made from Student Version 17.0 and different versions may have minor differences.

It is best if you perform the following sections in order. For each, type in the example data and perform the other steps as they are described. We'll refer to example problems from earlier chapters so that you can compare your answer when using a formula to the answer when using the computer. (Slight differences may arise because of rounding.)

B.1 ENTERING DATA

Install and start SPSS using the instructions included with the program. The opening window is shown in Screen B.1.

The window in the foreground gives several options. Usually we will input data, so using the left mouse button click at *Type in data* and then click the *OK* button. The foreground window disappears, leaving the window in the background, containing the grid of rectangles. This is the "Data Editor" and in it we enter our data. Use the horizontal scroll bar at the bottom of the Data Editor to see columns to the right.

Across the top of the Data Editor is the typical "Menu Bar," with buttons for *File, Edit,* and so on. Some useful information is available by clicking *Help:* Click *Topics* for an index, *Tutorial* for instructions and examples, and *Statistics Coach* for help in selecting a procedure.

The first step in any data analysis is to name the variables and enter the scores.

SCREEN B.1
Opening Windows of SPSS

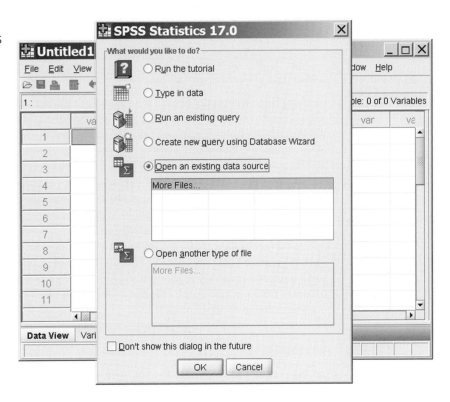

Naming Variables

The program's default will name your variables var00001, var00002, and so on, but give them a more meaningful name. For example, say that we want to input the following scores, which measure creativity and intelligence:

Participant	Creativity	Intelligence
1	50	110
2	47	116
3	65	125
4	66	127
5	48	100

Begin with a blank Data Editor: To clear old data, on the Menu Bar click *File,* point to *New,* and click *Data.*

Name the variable(s): At the bottom left of the Data Editor, click on *Variable View.* In the left column under "Name" click on the first rectangle and type the variable's name. (For example, Creativity.) Press *Enter* on the keyboard.

The information that appears next to a variable's name are the SPSS defaults, which will work for our statistics, assuming that you want scores to have no more than two decimal places. (Type in a three decimal number, and it will be rounded.) Click any rectangle in a row to change the default value. In the "Label" column, you can add information about the variable that you want to remember.

In the "Name" column, click the second rectangle and enter the second variable's name (Intelligence).

Entering Data

Click on *Data View* at the bottom left of the Data Editor. Input the scores under the corresponding variable. When participants' are measured on both variables, each row in the Data Editor holds the scores from the same participant. Thus, participant 1 scored 50 on creativity, so in the Creativity column, click the rectangle next to "1" and type 50. Next you have three choices: (1) Press the *Tab* key on the keyboard, and the rectangle to the right of 50 is highlighted, ready for participant 1's intelligence score; (2) press the *Enter* key and the rectangle below 50 is highlighted, ready for participant 2's creativity score; (3) click any rectangle in either column to highlight it and type in the score that belongs there. Enter the remaining scores.

You can enter data from several variables and then only analyze some of it at one time. (We can analyze only creativity, only intelligence, or both.)

To correct a mistake: Click on the score or rectangle and retype. Always check that the Data Editor contains the correct scores. (Here the columns should look like our original columns of data above.)

Saving Data

To save data: The first time that you save, on the Menu Bar click *File,* click *Save as,* and name the file. For saves after that, click *File* and *Save.*

To open a saved file: Open a file to add more data or to analyze it. Below the Menu Bar, click the file folder icon. Click your file's name and click *Open.*

Saving or Printing the Output

Later we'll see that SPSS shows the results of calculations by opening a new window called the "Output." Although you'll have many options with this window, the most useful are to select the appropriate icon to save or print the results.

Close *SPSS* by clicking the X in the upper right-hand corner.

For Practice

Enter the following scores. Your Data Editor should look like that shown in Screen B.2. Then save and then retrieve the file.

SCREEN B.2

Practice Data Editor

Participant	Test 1	Test 2
1	19	11
2	17	7
3	15	5
4	16	7
5	17	11
6	19	12
7	19	3.431

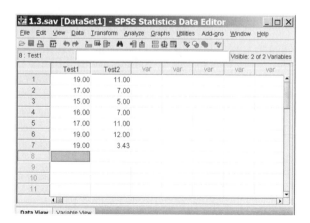

B.2 FREQUENCY DISTRIBUTIONS AND PERCENTILE

SPSS creates frequency tables and graphs, and it computes percentiles as discussed in Chapter 3. (Graphs can also be exported into a report that you are writing.) For example, say that we have these 15 scores:

6   4   5   6   7   9   6   3   4   6   5   8   5   7   7

Creating a Frequency Table with Percentiles

Enter the data: Name one variable (for example, Scores) and enter the data as above.

   Select a frequency table: We'll often do the type of move shown in Screen B.3. On the Menu Bar, click *Analyze,* move the cursor to *Descriptive Statistics,* and then click *Frequencies.* A "box" appears that is labeled "Frequencies," as shown in Screen B.4

SCREEN B.3
Creating Frequency
Tables

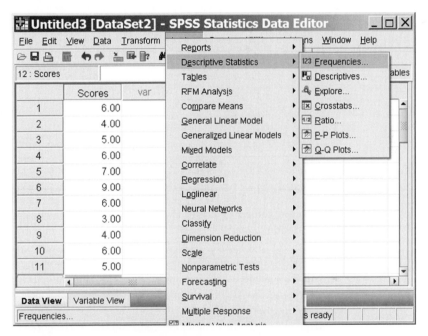

SCREEN B.4
Frequencies Box

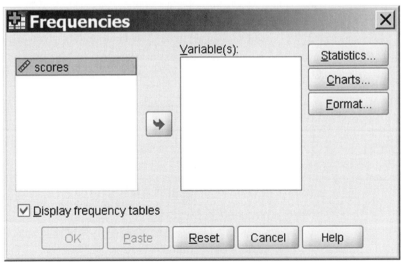

Select the variable(s): Click [▶] and the highlighted variable will move to under "Variable(s)." Or, you may "drag and drop" the highlighted variable into the space under "Variable(s)."

To remove a variable: Highlight the variable under "Variable(s)" and click the reversed-arrow button.

The output: Click *OK*. The frequency table in Screen B.5 appears. In the left-hand column are the original scores, but arranged in *increasing* order. In the next column is each score's simple frequency. Under "Percent" is the percent each score's *f* is of *N* (its *rel. f* times 100). Under "Cumulative Percent" is percentile. (Ignore "Valid Percent.")

SCREEN B.5

Frequency Table Output

scores

		Frequency	Percent	Valid Percent	Cumulative Percent
Valid	3.00	1	6.7	6.7	6.7
	4.00	2	13.3	13.3	20.0
	5.00	3	20.0	20.0	40.0
	6.00	4	26.7	26.7	66.7
	7.00	3	20.0	20.0	86.7
	8.00	1	6.7	6.7	93.3
	9.00	1	6.7	6.7	100.0
	Total	15	100.0	100.0	

Determining the Score at a Particular Percentile

Select a frequency table: On the Menu Bar, again click *Analyze, Descriptive Statistics,* and *Frequencies.* Then move each variable to "Variables(s)."

Select a percentile: Click *Statistics.* A "Frequencies: Statistics" box appears. Click to place a check mark at *Percentile(s)* and in the space next to it, type the percentile that you seek. Say that we seek the score at the 10th percentile, so type 10. Click *Add* and 10 appears in the larger white box. Add other percentiles in the same way. Or click *Quartiles* to produce the 25th, 50th, and 75th percentile. Or, use *Cut points* for the score at the 10th percentile, the 20th, and so on. (To remove a percentile from the white box, click it and click *Remove.*) Click *Continue.*

The output: Back at the "Frequencies" box, click *OK*. The percentile(s) will be listed in the "Statistics" output table shown below. (The score at the 10th percentile is 3.6.)

Statistics

scores

N	Valid	15
	Missing	0
Percentiles	10	3.6000

Plotting Bar Graphs and Histograms

Select a frequency table: Repeat the original steps for a frequency table: On the Menu Bar, click *Analyze, Descriptive Statistics,* and *Frequencies.* In the "Frequencies" box, move your variable to "Variables(s)."

Select Charts: Click *Charts.* In the "Frequencies: Charts" box, click *Bar charts* or *Histograms.* Click *Continue,* and click *OK.* (You may need to scroll down in the output window to see the graph.)

Plotting Frequency Polygons

On the Menu Bar, click *Graphs* and click *Chart Builder:* Click *OK* on the Chart Builder box that appears. Under *Choose from,* click *Histogram.* Place the cursor over the *Frequency Polygon* icon and drag it into the *Chart preview* area. Close the *Element Properties* box. Drag your variable from the *Variables* column to the area under the *X* axis. Click *OK.* The polygon will not show zero frequency when a score did not occur—the line will connect the frequencies of the scores above and below it. (For older versions of *SPSS,* on the *Menu Bar* click *Graphs,* then *Line* then *Simple.* Click *Define* and move the variable to *Category Axis.* Click *N* of *cases* and click *OK.*)

For Practice

1. Create a frequency table for the data in Application Question 21 in Chapter 3.
2. For the above data, create a bar graph and a frequency polygon.

Answers

1. Your output should give the same answers as in Appendix D.
2.

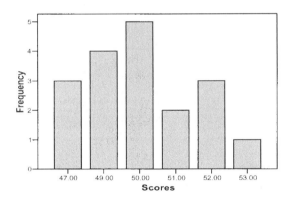 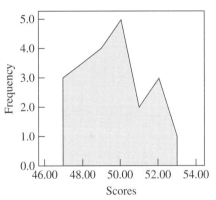

B.3 CENTRAL TENDENCY, VARIABILITY, AND *z*-SCORES

This section describes how to (1) compute measures of central tendency (as in Chapter 4) and variability (as in Chapter 5) for a sample of scores and (2) transform a sample of raw scores into *z*-scores (as in Chapter 6).

Summarizing a Sample of Scores

SPSS often computes the mean and standard deviation of scores as part of other procedures, including computing them for each condition when analyzing an experiment. The following section describes how, for one sample of scores, we can also simultaneously compute the mean, median, and/or mode, as well as the standard deviation, variance, and/or range. For example, say we have these ten test scores:

$$40 \quad 33 \quad 36 \quad 38 \quad 49 \quad 39 \quad 40 \quad 38 \quad 36 \quad 37$$

Enter the data: Name the variable and enter the scores into the Data Editor as usual.

Select a frequency table: Repeat the previous steps for a frequency table: On the Menu Bar, select *Analyze, Descriptive Statistics,* and *Frequencies.* Then move each variable to "Variables(s)."

Select statistics: Click *Statistics.* The "Frequencies: Statistics" box appears. Under *Central Tendency,* click to place a checkmark at each measure that you seek. *Dispersion* means variability, so check these measures that you seek.

Note: The standard (std.) deviation and variance given here are the estimated population versions, what we called s_X and s_X^2, in which the final division involves $N - 1$. Also, the "S.E. mean" is the standard error of the mean introduced in Chapter 6.

The output: Click *Continue* and click *OK.* You'll see Screen B.6.

Transforming a Sample to *z*-Scores

Enter the data: Enter the data as above. (Let's use the previous scores.)

Select Descriptives: On the Menu Bar click *Analyze, Descriptive Statistics,* and *Descriptives.* The "Descriptives" box appears.

Select the variable(s): Move each variable to "Variable(s)."

Select *z*-scores: Check *Save standardized values as variables.*

The output: Click *OK.* The output includes the mean and standard deviation of the raw scores. But back in your Data Editor, a new variable (a new column) will appear containing the *z*-score for each raw score.

For Practice

Using the data in questions 14 and 15 in Chapter 5, determine the mean, median, mode, estimated standard deviation and variance, and range.

SCREEN B.6

Output for Central Tendency and Variability

Statistics

Test

N	Valid	10
	Missing	0
Mean		38.6000
Median		38.0000
Mode		36.00[a]
Std. Deviation		4.22164
Variance		17.822
Range		16.00

a. Multiple modes exist. The smallest value is shown.

Answer

Statistics

scores

N	Valid	17
	Missing	0
Mean		76.2941
Median		76.0000
Mode		76.00[a]
Std. Deviation		4.10434
Variance		16.846
Range		17.00

a. Multiple modes exist. The smallest value is shown.

B.4 CORRELATION COEFFICIENTS AND THE LINEAR REGRESSION EQUATION

This section describes how to compute the Pearson r and the Spearman r_S (as in Chapter 7) and the linear regression equation (as in Chapter 8).

Computing the Pearson Correlation Coefficient

SPSS computes r, as well as the mean and estimated standard deviation of the Xs and Ys. (You may have more than two variables and correlate all pairs of variables.) As an example, say that we wonder if a person's extroversion level is correlated with his or her aggression level. We obtain these scores:

Participant	Extroversion	Aggression
1	14	23
2	12	10
3	10	11
4	13	18
5	19	20
6	20	18
7	9	10
8	4	9

Enter the data: Name the two variables and enter the scores in the two columns. (Save the file for use later with regression.)

Select the correlation: On the Menu Bar, select *Analyze, Correlate,* and *Bivariate.* The "Bivariate Correlations" box appears. Be sure *Pearson* is checked. (Anything referring to "significant" will make sense after you've read Chapter 11.)

Select the variable(s): Move the two variables you want to correlate to "Variables."

Select Descriptives: Click *Options* and then check *Means and standard deviations.* Click *Continue.*

The output: Click *OK.* In the "Descriptive Statistics" table are the $\overline{X}$ and s_X for each variable. In the "Correlations" table is the "correlation matrix," shown in Screen B.7.

SCREEN B.7

Output for the Pearson *r*

Correlations

		Extroversion	Aggression
Extroversion	Pearson Correlation	1	.747*
	Sig. (2-tailed)	.	.033
	N	8	8
Aggression	Pearson Correlation	.747*	1
	Sig. (2-tailed)	.033	.
	N	8	8

* Correlation is significant at the 0.05 level (2-tailed).

The matrix contains the correlation coefficients produced when the variable in every row is correlated with the variable in every column. Thus, in the first row, the (meaningless) r_{obt} between extroversion scores and extroversion scores is $+1$. However, the r_{obt} that we seek between extroversion and aggression is $+.747$. (If it were negative, a minus sign would appear.) If we had a third variable, we'd also see the r_{obt} between it and extroversion and between it and aggression.

Computing the Spearman Correlation Coefficient

Say that we rank-ordered ten participants in terms of their attractiveness and their apparent honesty and we wish to correlate these ranks.

Participant	*Attractive*	*Honest*
1	8	7
2	6	8
3	5	4
4	7	6
5	3	3
6	1	2
7	10	9
8	2	2
9	4	1
10	9	5

Enter the data: Enter the scores as we did for *r*. SPSS automatically resolves tied ranks as discussed in Chapter 15. Also, you may enter interval or ratio scores, and SPSS will transform them to ranks.

Select the correlation: On the Menu Bar, select *Analyze, Correlate,* and *Bivariate.* The "Bivariate Correlations" box appears.

Select the variable(s): Move the variables you want to correlate to "Variables."

Select Spearman: Check *Spearman;* uncheck *Pearson.*

The output: Click *OK.* A correlation matrix again appears. Notice that r_S is called "rho" here. Our r_S between attractiveness and honesty is $+.815$.

Computing the Linear Regression Equation

SPSS computes the components of the linear regression equation. For example, in the Pearson r above, we correlated extroversion and aggression. To perform linear regression on these data, retrieve the file (or re-enter the scores).

Identify X and Y: First decide which variable is the predictor variable (your X) and which is the criterion variable (your Y). Say that we want to use extroversion scores to predict aggression scores.

Select Regression: On the Menu Bar, select *Analyze, Regression,* and *Linear.* The "Linear Regression" box appears.

Select the variable(s): Move your predictor or X variable (here "Extroversion") under "Independent(s)." Move your criterion or Y variable (here "Aggression") under "Dependent(s)."

Select Descriptives: Click *Statistics* and, in the box check *Descriptives* to compute the $\overline{X}$ and s_X for each variable and the r between them. Click *Continue.*

The output: Click *OK.* Considerable information is provided, but the basic material is shown in Screen B.8.

In the "Model Summary" table is r_{obt} (called R), r^2 (called R Square), and the standard error of the estimate. (This is computed differently than our $S_{Y'}$, because ours describes the standard error for our sample, but in SPSS it is an estimate of the standard error that would be found in the population. You'll get the same answer if, in our defining formula for $S_{Y'}$ you divide by $N - 2$ instead of N.) The components of the regression equation are in the "Coefficients" table. Locate the column under B. In the row at "Constant" is the Y intercept (our a). Here, a is 5.030. In the row at our predictor variable's name (here, "Extroversion") is the slope (our b). Here, b is .780. Thus, the linear equation here is $Y' = .780X + 5.030$.

For Practice

1. Compute the Pearson r in Application Question 19 in Chapter 7. Compare your output to the answers in Appendix D.

2. As in Application Question 21 in Chapter 8, compute the linear regression equation when using Burnout to predict Absences. Compare your output to the answers in Appendix D.

3. Compute the Spearman r_S Application Question 21 in Chapter 7. Compare your output to the answers in Appendix D.

SCREEN B.8

Output for Linear Regression

Model Summary

Model	R	R Square	Adjusted R Square	Std. Error of the Estimate
1	.747[a]	.559	.485	3.91996

a. Predictors: (Constant), Extroversion

Coefficients[a]

Model		Unstandardized Coefficients		Standardized Coefficients	t	Sig.
		B	Std. Error	Beta		
1	(Constant)	5.030	3.832		1.313	.237
	Extroversion	.780	.283	.747	2.756	.033

a. Dependent Variable: Aggression

B.5 THE ONE-SAMPLE *t*-TEST AND SIGNIFICANCE TESTING OF CORRELATION COEFFICIENTS

As discussed in Chapter 11, this section describes how to (1) perform the one-sample *t*-test and compute the confidence interval for μ, and (2) perform significance testing of the Pearson and Spearman coefficients.

The One-Sample *t*-Test

SPSS simultaneously computes $\overline{X}$ and s_X for the sample, performs the *t*-test, and computes the confidence interval. For example, we want to test if poor readers score differently on a grammar test than the national population of readers (where $\mu = 89$; so H_0: $\mu = 89$). Our dependent (grammar) scores are

72 67 59 76 93 90 75 81 71 93

Enter the data: Name the variable and enter the scores as usual.

Select the *t*-Test: On the Menu Bar, select *Analyze, Compare Means,* and *One-sample T Test.* The "One-Sample T Test" box appears.

Select the variable: Move your dependent variable to "Test Variable(s)."

Enter μ: Click the space at *Test Value* and enter your value of μ. (For the example, enter 89.)

The output: Click *OK*. The "One-Sample Statistics" table shows the $\overline{X}$ (77.70) and s_X (11.461) for our data. The "Std. Error Mean" is $s_{\overline{X}}$. The *t*-test results are in Screen B.9.

Deciding if a Result Is Significant

In Screen B.9, the t_{obt} is -3.424 with $df = 9$. For a two-tailed test you do not need to look up t_{crit}. Under "Sig. (2-tailed)" is the smallest possible value of α that is needed for the results to be significant. This must be *less than* or *equal to* your α for the results to be significant. In the example is .012, indicating that t_{obt} is significant if the region of rejection is the extreme .012 of the sampling distribution. Thus, t_{obt} is in a region that is smaller than our usual .05 of the curve. Therefore, this t_{obt} would be beyond our t_{crit}, so it is significant. If the "Sig. (2-tailed)" value is .000, it means that α—and *p*—is less than .001.

On the other hand, if "Sig. (2-tailed)" had been, say, .06, then this t_{obt} is beyond t_{crit} only when the region of rejection is .06 of the curve. This is larger than our .05 region, so t_{obt} is not beyond our t_{crit}, and is not significant.

SCREEN B.9

Output for the One Sample *t*-Test

One-Sample Test

| | \multicolumn{6}{c|}{Test Value = 89} |
| | | | | | 95% Confidence Interval of the Difference | |
	t	df	Sig. (2-tailed)	Mean Difference	Lower	Upper
Grammar	-3.118	9	.012	-11.30000	-19.4984	-3.1016

The Confidence Interval for μ

Also in Screen B.9 is a slightly different version of a confidence interval. It indicates the minimum and maximum difference that is likely between the μ in H_0 and the μ represented by our sample. To convert this to our confidence interval, add the values shown under "Lower" and "Upper" to the μ in your H_0. Here we happen to be adding negative numbers, so adding -19.4984 to 89 gives 69.5016; adding -3.1016 to 89 gives 85.8984. Thus, after rounding, the 95% confidence interval for the μ of poor readers is $69.50 \leq \mu \leq 85.90$.

Significance Testing of the Pearson Correlation Coefficient

SPSS performs significance testing when it calculates *r*. For example, previously in Section B.4, we correlated extroversion and aggression. Retrieve that file and on the Menu Bar again select *Analyze, Correlate,* and *Bivariate*. In the "Bivariate Correlations" box, be sure *Flag significant correlations* is checked. Also check whether you want a one-tailed or a two-tailed test. Click *OK*. The relevant output is the "Correlations" table in Screen B.10.

The $r_{obt} = .747$. The second row shows "Sig. (2-tailed)" and to its right is .033. This is interpreted as discussed previously in the one-sample *t*-test: .033 is less than .05, so the r_{obt} is significant in a two-tailed test. In fact, notice the * at .747 and the footnote under the table. For a one-tailed test, we interpret "Sig. (1-tailed)" as above.

Significance Testing of the Spearman Correlation Coefficient

Interpret the output for a Spearman r_S like the Pearson *r*. For example, in Section B.4, we correlated rankings that reflected attractiveness and honesty, producing Screen B.11.

The $r_S = +.815$. The "Sig. (2-tailed)" is .004. Because .004 is less than .05, this r_S is significant.

For Practice

1. Perform the *t*-test in Application Question 11 in Chapter 11.

2. For these SPSS results, determine whether each is significant (α = .05).
   (a) $r_S = +.42$, $N = 28$, Sig. (2-tailed) = .041; (b) $r_{obt} = -.531$, $N = 180$, Sig. (2-tailed) = .000; (c) $t_{obt} = -2.21$, $N = 8$, Sig. (2-tailed) = .20

Answers

1. Compare your output to the answers in Appendix D.

2. (a) Significant; (b) significant; (c) not significant.

SCREEN B.10

Significance Test of *r*

Correlations

		Extroversion	Aggression
Extroversion	Pearson Correlation	1	.747*
	Sig. (2-tailed)	.	.033
	N	8	8
Aggression	Pearson Correlation	.747*	1
	Sig. (2-tailed)	.033	.
	N	8	8

*Correlation is significant at the 0.05 level (2-tailed).

Correlations

			Honest	Attractive
Spearman's rho	Honest	Correlation Coefficient	1.000	.815**
		Sig. (2-tailed)	.	.004
		N	10	10
	Attractive	Correlation Coefficient	.815**	1.000
		Sig. (2-tailed)	.004	.
		N	10	10

** Correlation is significant at the 0.01 level (2-tailed).

B.6 TWO-SAMPLE *t*-TESTS

This section describes how to compute the independent-samples and related-samples *t*-test (as in Chapter 12) along with the corresponding confidence intervals and descriptive statistics.

Identifying the Conditions of the Independent Variable

First, we must tell SPSS the condition in which each participants' dependent score belongs. For example, say that we test the influence of the independent variable of the color of a product's label (Blue or Green) on the dependent variable of how desirable it is, obtaining these scores:

Independent Variable: Color

Condition 1: Blue	Condition 2: Green
10	20
12	24
14	28
17	19
16	21

Name the variables: In the Data Editor, name one variable using the independent variable (Color) and one using the dependent variable (Desire.)

Label the output: We will use a number to tell SPSS in which condition each score belongs. So first, arbitrarily choose a number to identify each condition. Lets use "1" for Blue and "2" for Green. However, it is *very* helpful to have output in which the conditions are labeled with words and not 1s and 2s. Therefore, while in *variable view* in the Data Editor, in the row for the independent variable, click on the rectangle under "Values" and then in it click the gray square with the three dots. In the "Values Label" box, at "Value," enter the number for one condition (e.g., 1). At "Value Label," enter the name of the condition for the output (e.g., Blue). Click *Add.* Likewise, in the same "Values Label" box, enter "2" and "Green," and click *Add.* Click *OK.*

Enter the data: Return to *data view.* To enter each dependent score, first identify the condition by entering the condition's number under "color." Then, while in the same row, enter the score under "desire." Thus, for the first participant in Blue who scored 10, in the first row of the Data Editor, enter a 1 under "color" and a 10 under "desire." Next enter 1 and 12 and so on. In the sixth row, enter 2 (for Green) under "color," with 20 under "desire," and so on. The completed Data Editor is in Screen B.12.

SCREEN B.12

Data Editor for the Independent-Samples *t*-Test

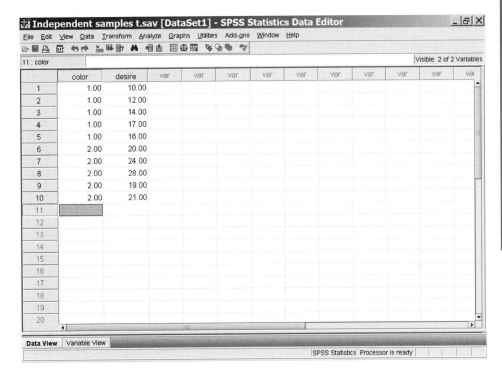

The Independent-Samples *t*-Test

Enter the data: Input the data as described above. Save this file. (The following assumes that you have equal *n*s.)

Select the *t*-test: On the Menu Bar, select *Analyze, Compare Means,* and *Independent-Samples T-test.* The "Independent-Samples T-test" box appears.

Select the variables: Move your dependent variable (Desire) to "Test variable(s)." Move your independent variable to "Grouping Variable."

Identify the conditions: Click *Define groups.* In the "Define Groups" box, enter the number used to identify each condition. (We used 1 and 2.) Click *Continue.*

The output: Click *OK.* In the "Group Statistics" box is the $\overline{X}$ and s_X for each condition. The *t*-test is in the "Independent Samples Test" box, as in Screen B.13.

Read the row for "Equal variances assumed": The t_{obt} is −4.147, with $df = 8$. The "Sig. (2-tailed)" is .003, so t_{obt} is significant. The "Mean Difference" is the difference between the means of the two conditions. The "Std. Error Difference" is our $s_{\overline{X}_1 - \overline{X}_2}$. The confidence interval is for the difference between the μs, so with rounding, $-13.38 \leq \mu_1 - \mu_2 \leq -3.82$. You must compute the effect size using our formula for r^2_{pb}.

SCREEN B.13

Output for the Independent-Samples *t*-Test

Independent Samples Test

		Levene's Test for Equality of Variances		t-test for Equality of Means						
									95% Confidence Interval of the Difference	
		F	Sig.	t	df	Sig. (2-tailed)	Mean Difference	Std. Error Difference	Lower	Upper
desire	Equal variances assumed	.421	.534	-4.147	8	.003	-8.60000	2.07364	-13.38183	-3.81817
	Equal variances not assumed			-4.147	7.574	.004	-8.60000	2.07364	-13.42907	-3.77093

The Related-Samples *t*-Test

For a related-samples *t*-test, we enter the data differently than above. For example, say that we study the total errors made in estimating distance by the same people when using one or both eyes. We obtain these data:

One Eye	Two Eyes
10	2
12	4
9	2
6	1
8	3

Enter the data: In the Data Editor, create two variables, each the name of a condition of the independent variable (for example, One and Two). Then in each row of the Data Editor, enter the two dependent scores from the same participant; for example, in row 1, enter 10 under One and 2 under Two. (In a matched-samples design, each row contains the two scores from a matched pair.)

Select the *t*-test: On the Menu Bar, select *Analyze, Compare Means,* and *Paired-Samples T-test.* The "Paired-Samples T-test" box appears.

Select the variables: In the area under "Paired Variables," drag and drop each of your variables into the highlighted row labeled "1." Drop one variable under "Variable 1" and the other under "Variable 2."

The output: Click *OK.* The output for the *t*-test is in the "Paired Samples Test" table as in Screen B.14.

The "Mean" (6.6) and "Std. Deviation" (1.52) are our $\overline{D}$ and s_D. The "Std. Error Mean" is our $s_{\overline{D}}$. The confidence interval is for μ_D, so with rounding, $4.72 \leq \mu_D \leq 8.48$. The t_{obt} is +9.731, with $df = 4$. The "Sig. (2-tailed)" is .001, so the results are significant ($p = .001$).

The output also includes the "Paired Samples Statistics" table, containing the $\overline{X}$ and s_X in each condition. In the "Paired Samples Correlations" table is the Pearson r between the scores in the two conditions. This is not effect size! For that, you must use our formula for r_{pb}^2.

For Practice

1. Perform the independent-samples *t*-test in Application Question 21 in Chapter 12. Your answers should match those in Appendix D.

SCREEN B.14
Output for the Related-Samples *t*-Test

Paired Samples Test

		Paired Differences						
					95% Confidence Interval of the Difference				
		Mean	Std. Deviation	Std. Error Mean	Lower	Upper	t	df	Sig. (2-tailed)
Pair 1	One - Two	6.60000	1.51658	.67823	4.71692	8.48308	9.731	4	.001

2. Perform the related-samples *t*-test in Application Question 15 in Chapter 12. Compare your answers to those in Appendix D.

B.7 THE ONE-WAY, BETWEEN-SUBJECTS ANOVA

This section describes how to perform the one-way, between-subjects ANOVA, as described in Chapter 13. This includes Tukey's *HSD* comparisons, descriptive statistics for each level, and graphs. (These instructions assume equal *n*s.)

Say that we expand the study discussed with the independent-samples *t*-test, testing the influence of three colors on a product's desirability. We have these data:

Condition 1: Blue	Condition 2: Green	Condition 3: Yellow
10	20	24
12	24	25
14	28	26
17	19	21
16	21	23

Enter the data: Enter the data as we did in the independent-samples *t*-test: Name one variable for the independent variable (for example, Color) and one for the dependent variable (Desire). Again identify a participant's condition by entering the condition's number in the Color column (either a 1, 2, or 3). In the same row, enter that participant's dependent score in the Desire column.

Label the output: Use words to label each level, as we did in the independent-samples *t*-test.

Select the ANOVA: On the Menu Bar, select *Analyze, Compare Means,* and *One-way ANOVA.* The "One-Way ANOVA" box appears.

Select the Variables: Move your dependent variable (Desire) to "Dependent list." Move your independent variable (Color) to "Factor."

Select the post hoc test: Click *Post Hoc* and, in the box that appears, checkmark the box with only the word *Tukey.* Click *Continue.*

Select Descriptive: Click *Options* and, in the "Options" box, checkmark *Descriptive* to get the $\overline{X}$ and s_X of each level.

SCREEN B.15

One-Way ANOVA Summary Table

ANOVA

desire

	Sum of Squares	df	Mean Square	F	Sig.
Between Groups	293.200	2	146.600	17.452	.000
Within Groups	100.800	12	8.400		
Total	394.000	14			

Select the graph: While at the "Options" box, checkmark *Means plot* to produce a line graph. Click *Continue.*

The output: Click *OK.* In the "Descriptives" table, the first three rows give the $\overline{X}$, s_X and confidence interval for μ in each level.

The ANOVA summary table is shown in Screen B.15 above. Our F_{obt} is 17.452. Under "Sig." is the minimum α needed for F_{obt} to be significant, so interpret this as in previous procedures. Here the results are significant, with $p < .001$.

In Screen B.16 is the "Multiple Comparisons" table, showing the Tukey *HSD* procedure. Under "(I) color" is first Blue, and in the rows here are the comparisons between Blue and the other conditions. Thus, the first row compares the mean of Blue to the mean of Green and the difference is -8.6. The asterisk indicates that this difference is larger than the *HSD* value when $\alpha = .05$, and under "Sig." we see that $\alpha = .001$. The confidence interval is for the difference between the μs represented by these two level means. Under "(I) color" at Green are the comparisons involving the mean of Green, including again comparing it with Blue. And so on.

Note in your output the line graph of the means, which may be exported to a report you are writing. Compute effect size using our formula for η^2.

SCREEN B.16

Output for Tukey's *HSD* Test

Multiple Comparisons

Dependent Variable: desire

Tukey HSD

(I) color	(J) color	Mean Difference (I-J)	Std. Error	Sig.	95% Confidence Interval	
					Lower Bound	Upper Bound
Blue	Green	-8.60000*	1.83303	.001	-13.4903	-3.7097
	Yellow	-10.00000*	1.83303	.000	-14.8903	-5.1097
Green	Blue	8.60000*	1.83303	.001	3.7097	13.4903
	Yellow	-1.40000	1.83303	.731	-6.2903	3.4903
Yellow	Blue	10.00000*	1.83303	.000	5.1097	14.8903
	Green	1.40000	1.83303	.731	-3.4903	6.2903

*. The mean difference is significant at the .05 level.

For Practice

Perform the ANOVA in Application Question 19 in Chapter 13. Confirm that your answers match those in Appendix D.

B.8 THE TWO-WAY, BETWEEN-SUBJECTS ANOVA

This section describes how to perform the two-way, between-subjects ANOVA, as described in Chapter 14. This includes Tukey's *HSD* comparisons for main effects only, descriptive statistics for the cells and main effects, and graphing. (These instructions assume equal *n*s in all cells.) As an example, in Chapter 14, we tested the factors of volume of a message and gender on the dependent variable of persuasiveness. We had the data shown in Table B.1.

Name the variables: In the Data Editor, name three variables: one for factor A (Volume), one for factor B (Gender), and one for the dependent variable (Persuasion). Arbitrarily assign a number to each level in each factor. Let's use 1, 2, and 3 for soft, medium, and loud, and 1 and 2 for male and female, respectively.

Label the output: Enter word labels for each factor as described in the independent-samples *t*-test (B.6). Otherwise, the output is mind-boggling!

Enter the data: We must identify the cell in which each score belongs. In the Data Editor, enter a participant's level of A in the Volume column and, in the same row, enter that participant's level of B in the Gender column. While still in the same row, enter that participant's dependent score in the Persuasion column. Thus, in the male-soft cell is the score of 9, so we enter 1 under Volume, 1 under Gender, and 9 under Persuasion. See Screen B.17. Enter all scores from a cell together, so in the second row enter 1, 1, and 4. In the third row, enter 1, 1, and 11. For the next cell, change only one factor. Let's change to medium volume. In row 4 of the Data Editor, for the first male-medium score, enter 2 under Volume, 1 under Gender, and 8 under Persuasion, and so on. In row 10, go back to 1 for Volume, but enter 2 for Gender (females), and so on.

Select the ANOVA: On the Menu Bar, select *Analyze, General Linear Model,* and *Univariate.* The "Univariate" box appears.

Select the variables: Move your dependent variable (Persuasion) to "Dependent variable." Move both factors to "Fixed Factors."

Select the post hoc test: Click *Post Hoc* and, in the box that appears, move each factor, to "Post Hoc Test for." Checkmark *Tukey.* Click *Continue.*

Select Descriptives: Click *Options* and, in the box that appears, checkmark *Descriptive Statistics.* This will give the $\overline{X}$ and s_X for all levels and cells. Click *Continue.*

TABLE B.1

A 3 × 2 Design for the Factors of Volume and Gender

Each column represents a level of the volume factor; each row represents a level of the gender factor.

		Factor A: Volume		
		Level A_1: Soft	Level A_2: Medium	Level A_3: Loud
Factor B: Gender	Level B_1: Male	9 4 11	8 12 13	18 17 15
	Level B_2: Female	2 6 4	9 10 17	6 8 4

$N = 18$

SCREEN B.17

Data Editor for the
Two-Way ANOVA

	Volume	Gender	Persuasion	var	var	var	var	var	var	var	var
1	1.00	1.00	9.00								
2	1.00	1.00	4.00								
3	1.00	1.00	11.00								
4	2.00	1.00	8.00								
5	2.00	1.00	12.00								
6	2.00	1.00	13.00								
7	3.00	1.00	18.00								
8	3.00	1.00	17.00								
9	3.00	1.00	15.00								
10	1.00	2.00	2.00								
11	1.00	2.00	6.00								
12	1.00	2.00	4.00								
13	2.00	2.00	9.00								
14	2.00	2.00	10.00								
15	2.00	2.00	17.00								
16	3.00	2.00	6.00								
17	3.00	2.00	8.00								
18	3.00	2.00	4.00								
19											
20											

Select graphs: Click *Plots.* In the box that appears, to plot the main effect means, move a factor to "Horizontal Axis" and click *Add.* Do the same with the other factor. To plot the interaction, click the factor with the most levels (Volume) and move it to "Horizontal Axis." Click the other factor (Gender) and move it to "Separate Lines." Click *Add* and then *Continue.*

The output: Click *OK.* The ANOVA summary table is in Screen B.18.

In the row at the name of each factor is the relevant information. For Volume, the F_{obt} is 7.142, which is significant ($p = .009$). For Gender, F_{obt} is 11.358, which is significant ($p = .006$). SPSS describes the interaction here as "Volume * Gender." The interaction

SCREEN B.18

Two-Way ANOVA Summary Table

Tests of Between-Subjects Effects

Dependent Variable: Persuasion

Source	Type III Sum of Squares	df	Mean Square	F	Sig.
Corrected Model	313.611[a]	5	62.722	7.628	.002
Intercept	1662.722	1	1662.722	202.223	.000
Volume	117.444	2	58.722	7.142	.009
Gender	93.389	1	93.389	11.358	.006
Volume * Gender	102.778	2	51.389	6.250	.014
Error	98.667	12	8.222		
Total	2075.000	18			
Corrected Total	412.278	17			

a. R Squared = .761 (Adjusted R Squared = .661)

SCREEN B. 19

Descriptive Statistics
Output for the
Two-Way ANOVA

Descriptive Statistics

Dependent Variable: Persuasion

Volume	Gender	Mean	Std. Deviation	N
Soft	Male	8.0000	3.60555	3
	Female	4.0000	2.00000	3
	Total	6.0000	3.40588	6
Medium	Male	11.0000	2.64575	3
	Female	12.0000	4.35890	3
	Total	11.5000	3.27109	6
Loud	Male	16.6667	1.52753	3
	Female	6.0000	2.00000	3
	Total	11.3333	6.05530	6
Total	Male	11.8889	4.48454	9
	Female	7.3333	4.44410	9
	Total	9.6111	4.92459	18

USING SPSS

F_{obt} is 6.250, and it is significant ($p = .014$). The MS_{wn} is in the row labeled "Error" (8.222) and $df_{wn} = 12$. Use these numbers when computing the HSD for the interaction. Use the "Corrected Total" sum of squares when computing η squared.

Screen B.19 above shows the "Descriptive Statistics" table. First are the cells under soft volume. The first row is labeled "Male," so for the cell of male-soft, $\overline{X} = 8.0000$ and $s_X = 3.60555$. The second row, "Female," describes the female-soft cell. The row labeled "Total" has the $\overline{X}$ and s_X for the main effect of soft volume, after collapsing across gender. In the next group of rows labeled "Medium" are the male-medium cell, the female-medium cell, and the main effect for medium volume, and so on. In the bottom rows labeled "Total" are the main effect means for gender.

Your output will also include the "Multiple Comparisons" table which shows Tukey's HSD test for the main effect of Volume. Interpret this as we did for the one-way ANOVA. If the Gender factor had involved more than two levels, a separate "Multiple Comparisons" table for it would appear.

Remember that SPSS does not compute Tukey's HSD for the interaction effect, so use the procedure described in Chapter 14. Likewise, compute effect size—using our formula for η^2—for each significant effect.

For Practice

Perform the two-way ANOVAs in Application Questions 19 and 21 in Chapter 14. Confirm that your answers match those in Appendix D.

B.9 CHI SQUARE PROCEDURES

This section describes how to perform the one-way and two-way chi square procedure, as described in Chapter 15.

The One-Way Chi Square

In Chapter 15, we discussed a study involving the frequency of left- or right-handed geniuses. There we had a total 10 left-handers and 40 right-handers. However, we must let SPSS count the participants in each category. For example, let's look at a small portion of the original handedness data before the totals. We used a 1 to indicate left-handed and a 2 to indicate right-handed.

Participant	Handedness
1	1
2	2
3	2
4	2
5	2
6	2
7	2
8	2
9	2
10	2
11	1
12	2

Enter the data: In the Data Editor, name one variable (for example, Handedness). (Label the output as in previous procedures.) Then enter the scores.

Select Chi Square: On the Menu Bar, select *Analyze, Nonparametric Tests,* and *Chi-Square.* The "Chi-Square Test" box appears.

Select the variables: Move your variable to "Test Variable List." (If your H_0 is *not* that all frequencies are equal, under "Expected values" check *Values.* Then type in the expected frequency for the lowest labeling score: We'd enter the f_e for the number of 1s. Click *Add.* Enter the f_e for the second labeling score (2s), click *Add,* and so on.)

The output: Click *OK.* As in Screen B.20, the "Handedness" table shows the f_o and f_e for each category. The "Test Statistics" table shows that $\chi^2_{obt} = 5.33$ with $df = 1$. The "Asymp. Sig." is .021, which is interpreted as in previous procedures: This χ^2_{obt} is significant, with $p = .021$.

The Two-Way Chi Square

Here SPSS must count the participants in each category of two variables. For example, let's examine the study comparing Type A or B personalities and the incidence of heart attacks from Chapter 15. We'll violate the assumptions and look at a very small N.

SCREEN B.20

Output for the One-Way Chi Square

Handedness

	Observed N	Expected N	Residual
left	2	6.0	-4.0
right	10	6.0	4.0
Total	12		

Test Statistics

	Handedness
Chi-Square	5.333
df	1
Asymp. Sig.	.021

Participant	Type	Attack
1	1	1
2	2	2
3	2	2
4	1	2
5	1	1
6	2	1
7	1	1
8	2	2
9	1	1
10	2	2
11	1	1
12	2	2

Enter the data: Name the two variables (for example, Type and Attack). For each, determine the scores that will indicate the categories. Above, a 1 is Type A and a 2 is Type B. Also, a 1 indicates a heart attack, and a 2 indicates no attack. (Label the output as we've done for previous procedures.) Then, for each participant, enter in the same row the scores that indicate category membership on each variable: For participant 1, enter 1 under Type and 1 under Attack. For participant 2, enter 2 and 2, and so on.

Select the Chi Square: On the Menu Bar, select *Analyze, Descriptive Statistics,* and *Crosstabs.* The "Crosstabs" box appears.

Select the variables: This 2 × 2 design forms a 2 × 2 matrix. Move the variable you want to form the rows of the matrix to "Rows." Move the variable that forms the columns to "Columns."

Select a bar chart: Checkmark *Display clustered bar charts.*

Select the statistics: Click *Statistics.* In the box that appears, checkmark *Chi-square.* Below that, checkmark *Contingency coefficient* or *Phi and . . .* (in this example, its the latter). Click *Continue.*

Select the expected frequencies: To see the f_e in each cell, click *Cells.* In the box that appears, under "Counts" checkmark *Expected.* Click *Continue.*

The output: Click *OK.* In your output is the table labeled "HeartAttack*Type Crosstabulation." In it is the 2 × 2 matrix containing the observed and expected frequencies (count). In the "Chi-square Tests" table, at "Pearson Chi-Square," is the χ^2_{obt} of 5.333, with $df = 1$. The "Asymp. Sig." is .021, which is interpreted as in previous procedures, so χ^2_{obt} is significant ($p = .021$). In the "Symmetric Measures" table, the phi coefficient is .667.

For Practice

1. In a survey, Foofy finds that three people prefer country music, nine prefer hip-hop, and two prefer classical. Compute the χ^2 for these data. (a) What is χ^2_{obt} and *df*? (b) Are the results significant? (c) What is p?

2. In another survey, Foofy asks if people like (1) or dislike (2) country music and if they like (1) or dislike (2) classical music. Compute the χ^2 on the following data:

Participant	Country	Classical
1	2	1
2	1	2
3	2	2
4	1	2
5	2	1
6	2	1
7	1	2
8	2	2
9	1	1
10	2	2
11	2	2
12	2	2

(a) What is χ^2_{obt} and df? (b) Are the results significant? (c) What is p? (d) Does the bar chart show an interaction?

Answers

1. (a) $\chi^2_{obt} = 6.143$, $df = 2$; (b) Significant; (c) $p = .046$

2. (a) $\chi^2_{obt} = .188$, $df = 1$; (b) Not significant; (c) $p = .665$; (d) No: Connecting the tops of the "Like" (blue) bars with one line, and the tops of the "Dislike" (green) bars with another produces close to parallel lines.

B.10 NONPARAMETRIC TESTS FOR RANKED SCORES

The following describes how to compute the Mann–Whitney U, the Wilcoxon T, the Kruskal–Wallis H, and the Friedman χ^2 tests, which are discussed in Chapter 15. However, use our formulas for computing effect size and post hoc tests. For each, the dependent scores can be ranks, or they can be interval or ratio scores that SPSS will automatically convert to ranks. You can compute descriptive statistics for these raw interval/ratio scores by selecting *Options* and then checking *Descriptive.*

The Mann–Whitney *U* Test

Enter the data: Create the Data Editor as in the independent-samples *t*-test (B.6).

 Select the nonparametric test: On the Menu Bar, select *Analyze, Nonparametric Tests,* and *2 Independent Samples.* The "Two Independent Samples" box appears. Under "Test Type" check *Mann-Whitney U.*

 Select the variables: Move your dependent variable to "Test Variable List." Move your independent variable to "Grouping Variable."

 Define the groups: Click *Define groups* and, in the box that appears, enter the number that you used to identify each condition. Click *Continue.*

 The output: Click *OK.* In the "Test Statistics" table, U_{obt} is at "Mann–Whitney U." The minimum α needed to be significant is at "Asymp. Sig. (2-tailed)."

The Wilcoxon *T* Test

Enter the data: Create the Data Editor as in the related-samples *t*-test (B.6).

Select the nonparametric test: On the Menu Bar, select *Analyze, Nonparametric Tests,* and *2 Related Samples.* The "Two Related Samples" box appears. Under "Test Type" check *Wilcoxon.*

Select the variables: In the area under "Test Pairs," drag and drop each of your variables into the highlighted row labeled "1." Drop one variable under "Variable 1" and the other under "Variable 2."

The output: Click *OK.* In the "Ranks" table are the sums of the positive and negative ranks. You may use the smaller sum as T_{obt} as described in Chapter 15 and compare it to T_{crit}. Or, in the output's "Test Statistics" table, the smaller sum is transformed to a *z*-score. The minimum α for this *z* to be significant is at "Asymp. Sig. (2-tailed)."

The Kruskal–Wallis *H* Test

Enter the data: Create the Data Editor as we did for the one-way, between-subjects ANOVA (B.7).

Select the nonparametric test: On the Menu Bar, select *Analyze, Nonparametric Tests,* and *K Independent Samples.* The "Test for Several Independent Samples" box appears. Under "Test Type" check *Kruskal-Wallis H.*

Select the variables: Move your dependent variable to "Test Variable List." Move your independent variable to "Grouping Variable."

Define the groups: Click *Define Range.* Enter the lowest score that you used to identify a condition at "Minimum." Enter the highest score at "Maximum." Click *Continue.*

The output: Click *OK.* In the "Ranks" table are the mean of the ranks in each condition. In the "Test Statistics" table, the H_{obt} is at "Chi-Square," under which is the *df.* The minimum α for H_{obt} to be significant is at "Asymp. Sig. (2-tailed)."

The Friedman χ^2 Test

Enter the data: Create the Data Editor as in the Wilcoxon *T* above, but name a variable for each level. For example, if you have three conditions then name three variables. Then in each row, put the three scores from the same participant in the appropriate columns.

Select the nonparametric test: On the Menu Bar, select *Analyze, Nonparametric Tests,* and *K Related Samples.* The "Test for Several Related Samples" box appears. Under "Test Type" check *Friedman.*

Select the variables: Move all variables to "Test Variables."

The output: Click *OK.* In the "Ranks" table is the mean rank for each condition. In the "Test Statistics" table, *N* is the number of participants, and at "Chi-Square" is the χ^2_{obt}. At "Asymp. Sig. (2-tailed)" is the minimum a needed for χ^2_{obt} to be significant.

For Practice

Perform each of the above procedures using the data from the example study that we discussed when that statistic was introduced in Chapter 15. Compare the answers in your output to those we obtained using the formulas.

B.11 THE ONE-WAY, WITHIN-SUBJECTS ANOVA

This section describes the one-way, within-subjects ANOVA that is discussed in Appendix A.3, including descriptive statistics and graphs. The student version of SPSS will not perform the Tukey *HSD* test here, so follow the instructions in Appendix A.3.

As an example, in Appendix A.3 we discussed a repeated-measures design that measured participants' comfort level when wearing either casual, semiformal, or formal clothes. The data are shown again in Table B.2.

Enter the data: As in Appendix A.3, we treat this as a two-way design, so create the Data Editor as we did in the two-way ANOVA: Name *three* variables. Name the first Participants, the next for factor A (Dress), and the third for the dependent variable (Comfort). Assign a number to identify each level in factor A and *label the output!* Then enter the participants' number (1, 2, or 3, . . .) under "Participants," the level of A under "Dress," and the dependent score under "Comfort." In this study, three rows in the Data Editor will belong to the same participant, each containing the score from a level of A. Thus, for participant 1: In the first row, enter 1, 1, and 4; in the next row, enter 1, 2, and 9; in the third row, enter 1, 3, and 1, and so on. The completed Data Editor is in Screen B.21.

Select the ANOVA: On the Menu Bar, select *Analyze, General Linear Model,* and *Univariate.* The "Univariate" box appears.

Select the variables: Move your dependent variable (Comfort) to "Dependent Variable." Move your independent variable (Dress) to "Fixed Factors." Move "participants" to "Random Factor(s)."

Select Plots: Click *Plots* but, to simplify matters, move only the factor (Dress) to "Horizontal Axis." Click *Add* and *Continue.*

Select Descriptive: Click *Options* and checkmark "Descriptive Statistics." Click *Continue.*

The output: Click *OK.* The ANOVA summary table is in Screen B.22.

We are interested in only the following: In the row labeled "Dress Hypothesis" is the F_{obt} for factor A (here, 3.486). In the column labeled "Sig." is .082, so this F_{obt} is not significant. In the row labeled "Dress * Participant Hypothesis," under Mean Square is our MS_{wn}. Use this and its *df* when computing the *HSD*. When computing η^2, add the *SS* in the above two rows to that in the "Subject Hypothesis" row to obtain SS_{tot}; use the *SS* at "Dress Hypothesis" as SS_{bn}.

TABLE B.2

Data from One-Way, Within-Subjects Study

	Factor A: Type of Dress		
Participants	*Level A₁:* *Casual*	*Level A₂:* *Semiformal*	*Level A₃:* *Formal*
1	4	9	1
2	6	12	3
3	8	4	4
4	2	8	5
5	10	7	2

SCREEN B.21

Data Editor for the Within-Subjects ANOVA

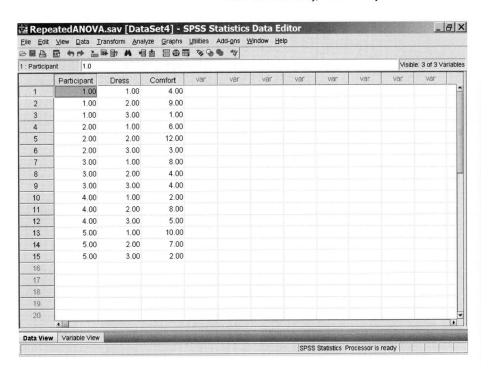

Also in your output is the "Descriptive Statistics" box: In the rows under "Casual," look at the row at "Total." This contains the $\overline{X}$ and s_X for the casual level. Likewise, find the $\overline{X}$ and s_X for the other levels.

In your line graph, ignore that it says "estimated marginal means. . . ." They are our real means.

For Practice

Perform the one-way, within-subjects ANOVA in Application Question 3 in Appendix A.3. Confirm that your answers match those in Appendix D.

SCREEN B.22

Within-Subjects ANOVA Summary Table

Tests of Between-Subjects Effects

Dependent Variable: Comfort

Source		Type III Sum of Squares	df	Mean Square	F	Sig.
Intercept	Hypothesis	481.667	1	481.667	170.000	.000
	Error	11.333	4	2.833[a]		
Dress	Hypothesis	63.333	2	31.667	3.486	.082
	Error	72.667	8	9.083[b]		
Participant	Hypothesis	11.333	4	2.833	.312	.862
	Error	72.667	8	9.083[b]		
Dress * Participant	Hypothesis	72.667	8	9.083	.	.
	Error	.000	0	.[c]		

C

Statistical Tables

Table C.1   Proportions of Area under the Standard Normal Curve: The z-Tables

Table C.2   Critical Values of t: The t-Tables

Table C.3   Critical Values of the Pearson Correlation Coefficient: The r-Tables

Table C.4   Critical Values of the Spearman Rank-Order Correlation Coefficient: The r_S-Tables

Table C.5   Critical Values of F: The F-Tables

Table C.6   Values of Studentized Range Statistic, q_k

Table C.7   Critical Values of Chi Square: The χ^2-Tables

Table C.8   Critical Values of the Mann–Whitney U

Table C.9   Critical Values of the Wilcoxon T

TABLE C.1

Proportions of Area under the Standard Normal Curve: The z-Tables

Column A lists z-score values. Column B lists the proportion of the area between the mean and the z-score value. Column C lists the proportion of the area beyond the z-score in the tail of the distribution. (*Note:* Because the normal distribution is symmetrical, areas for negative z-scores are the same as those for positive z-scores.)

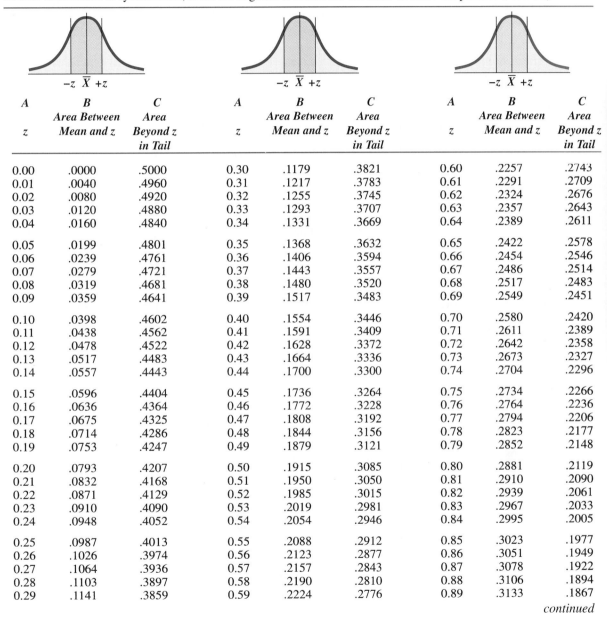

A	B	C	A	B	C	A	B	C
	Area Between	Area		Area Between	Area		Area Between	Area
z	Mean and z	Beyond z in Tail	z	Mean and z	Beyond z in Tail	z	Mean and z	Beyond z in Tail
0.00	.0000	.5000	0.30	.1179	.3821	0.60	.2257	.2743
0.01	.0040	.4960	0.31	.1217	.3783	0.61	.2291	.2709
0.02	.0080	.4920	0.32	.1255	.3745	0.62	.2324	.2676
0.03	.0120	.4880	0.33	.1293	.3707	0.63	.2357	.2643
0.04	.0160	.4840	0.34	.1331	.3669	0.64	.2389	.2611
0.05	.0199	.4801	0.35	.1368	.3632	0.65	.2422	.2578
0.06	.0239	.4761	0.36	.1406	.3594	0.66	.2454	.2546
0.07	.0279	.4721	0.37	.1443	.3557	0.67	.2486	.2514
0.08	.0319	.4681	0.38	.1480	.3520	0.68	.2517	.2483
0.09	.0359	.4641	0.39	.1517	.3483	0.69	.2549	.2451
0.10	.0398	.4602	0.40	.1554	.3446	0.70	.2580	.2420
0.11	.0438	.4562	0.41	.1591	.3409	0.71	.2611	.2389
0.12	.0478	.4522	0.42	.1628	.3372	0.72	.2642	.2358
0.13	.0517	.4483	0.43	.1664	.3336	0.73	.2673	.2327
0.14	.0557	.4443	0.44	.1700	.3300	0.74	.2704	.2296
0.15	.0596	.4404	0.45	.1736	.3264	0.75	.2734	.2266
0.16	.0636	.4364	0.46	.1772	.3228	0.76	.2764	.2236
0.17	.0675	.4325	0.47	.1808	.3192	0.77	.2794	.2206
0.18	.0714	.4286	0.48	.1844	.3156	0.78	.2823	.2177
0.19	.0753	.4247	0.49	.1879	.3121	0.79	.2852	.2148
0.20	.0793	.4207	0.50	.1915	.3085	0.80	.2881	.2119
0.21	.0832	.4168	0.51	.1950	.3050	0.81	.2910	.2090
0.22	.0871	.4129	0.52	.1985	.3015	0.82	.2939	.2061
0.23	.0910	.4090	0.53	.2019	.2981	0.83	.2967	.2033
0.24	.0948	.4052	0.54	.2054	.2946	0.84	.2995	.2005
0.25	.0987	.4013	0.55	.2088	.2912	0.85	.3023	.1977
0.26	.1026	.3974	0.56	.2123	.2877	0.86	.3051	.1949
0.27	.1064	.3936	0.57	.2157	.2843	0.87	.3078	.1922
0.28	.1103	.3897	0.58	.2190	.2810	0.88	.3106	.1894
0.29	.1141	.3859	0.59	.2224	.2776	0.89	.3133	.1867

continued

STATISTICAL TABLES

TABLE C.1 (CONT.)
Proportions of Area under the Standard Normal Curve: The z-Tables

A	B	C	A	B	C	A	B	C
z	Area Between Mean and z	Area Beyond z in Tail	z	Area Between Mean and z	Area Beyond z in Tail	z	Area Between Mean and z	Area Beyond z in Tail
0.90	.3159	.1841	1.25	.3944	.1056	1.60	.4452	.0548
0.91	.3186	.1814	1.26	.3962	.1038	1.61	.4463	.0537
0.92	.3212	.1788	1.27	.3980	.1020	1.62	.4474	.0526
0.93	.3238	.1762	1.28	.3997	.1003	1.63	.4484	.0516
0.94	.3264	.1736	1.29	.4015	.0985	1.64	.4495	.0505
0.95	.3289	.1711	1.30	.4032	.0968	1.65	.4505	.0495
0.96	.3315	.1685	1.31	.4049	.0951	1.66	.4515	.0485
0.97	.3340	.1660	1.32	.4066	.0934	1.67	.4525	.0475
0.98	.3365	.1635	1.33	.4082	.0918	1.68	.4535	.0465
0.99	.3389	.1611	1.34	.4099	.0901	1.69	.4545	.0455
1.00	.3413	.1587	1.35	.4115	.0885	1.70	.4554	.0446
1.01	.3438	.1562	1.36	.4131	.0869	1.71	.4564	.0436
1.02	.3461	.1539	1.37	.4147	.0853	1.72	.4573	.0427
1.03	.3485	.1515	1.38	.4162	.0838	1.73	.4582	.0418
1.04	.3508	.1492	1.39	.4177	.0823	1.74	.4591	.0409
1.05	.3531	.1469	1.40	.4192	.0808	1.75	.4599	.0401
1.06	.3554	.1446	1.41	.4207	.0793	1.76	.4608	.0392
1.07	.3577	.1423	1.42	.4222	.0778	1.77	.4616	.0384
1.08	.3599	.1401	1.43	.4236	.0764	1.78	.4625	.0375
1.09	.3621	.1379	1.44	.4251	.0749	1.79	.4633	.0367
1.10	.3643	.1357	1.45	.4265	.0735	1.80	.4641	.0359
1.11	.3665	.1335	1.46	.4279	.0721	1.81	.4649	.0351
1.12	.3686	.1314	1.47	.4292	.0708	1.82	.4656	.0344
1.13	.3708	.1292	1.48	.4306	.0694	1.83	.4664	.0336
1.14	.3729	.1271	1.49	.4319	.0681	1.84	.4671	.0329
1.15	.3749	.1251	1.50	.4332	.0668	1.85	.4678	.0322
1.16	.3770	.1230	1.51	.4345	.0655	1.86	.4686	.0314
1.17	.3790	.1210	1.52	.4357	.0643	1.87	.4693	.0307
1.18	.3810	.1190	1.53	.4370	.0630	1.88	.4699	.0301
1.19	.3830	.1170	1.54	.4382	.0618	1.89	.4706	.0294
1.20	.3849	.1151	1.55	.4394	.0606	1.90	.4713	.0287
1.21	.3869	.1131	1.56	.4406	.0594	1.91	.4719	.0281
1.22	.3888	.1112	1.57	.4418	.0582	1.92	.4726	.0274
1.23	.3907	.1093	1.58	.4429	.0571	1.93	.4732	.0268
1.24	.3925	.1075	1.59	.4441	.0559	1.94	.4738	.0262

continued

TABLE C.1 (CONT.)

Proportions of Area under the Standard Normal Curve: The z-Tables

A z	B Area Between Mean and z	C Area Beyond z in Tail	A z	B Area Between Mean and z	C Area Beyond z in Tail	A z	B Area Between Mean and z	C Area Beyond z in Tail
1.95	.4744	.0256	2.30	.4893	.0107	2.65	.4960	.0040
1.96	.4750	.0250	2.31	.4896	.0104	2.66	.4961	.0039
1.97	.4756	.0244	2.32	.4898	.0102	2.67	.4962	.0038
1.98	.4761	.0239	2.33	.4901	.0099	2.68	.4963	.0037
1.99	.4767	.0233	2.34	.4904	.0096	2.69	.4964	.0036
2.00	.4772	.0228	2.35	.4906	.0094	2.70	.4965	.0035
2.01	.4778	.0222	2.36	.4909	.0091	2.71	.4966	.0034
2.02	.4783	.0217	2.37	.4911	.0089	2.72	.4967	.0033
2.03	.4788	.0212	2.38	.4913	.0087	2.73	.4968	.0032
2.04	.4793	.0207	2.39	.4916	.0084	2.74	.4969	.0031
2.05	.4798	.0202	2.40	.4918	.0082	2.75	.4970	.0030
2.06	.4803	.0197	2.41	.4920	.0080	2.76	.4971	.0029
2.07	.4808	.0192	2.42	.4922	.0078	2.77	.4972	.0028
2.08	.4812	.0188	2.43	.4925	.0075	2.78	.4973	.0027
2.09	.4817	.0183	2.44	.4927	.0073	2.79	.4974	.0026
2.10	.4821	.0179	2.45	.4929	.0071	2.80	.4974	.0026
2.11	.4826	.0174	2.46	.4931	.0069	2.81	.4975	.0025
2.12	.4830	.0170	2.47	.4932	.0068	2.82	.4976	.0024
2.13	.4834	.0166	2.48	.4934	.0066	2.83	.4977	.0023
2.14	.4838	.0162	2.49	.4936	.0064	2.84	.4977	.0023
2.15	.4842	.0158	2.50	.4938	.0062	2.85	.4978	.0022
2.16	.4846	.0154	2.51	.4940	.0060	2.86	.4979	.0021
2.17	.4850	.0150	2.52	.4941	.0059	2.87	.4979	.0021
2.18	.4854	.0146	2.53	.4943	.0057	2.88	.4980	.0020
2.19	.4857	.0143	2.54	.4945	.0055	2.89	.4981	.0019
2.20	.4861	.0139	2.55	.4946	.0054	2.90	.4981	.0019
2.21	.4864	.0136	2.56	.4948	.0052	2.91	.4982	.0018
2.22	.4868	.0132	2.57	.4949	.0051	2.92	.4982	.0018
2.23	.4871	.0129	2.58	.4951	.0049	2.93	.4983	.0017
2.24	.4875	.0125	2.59	.4952	.0048	2.94	.4984	.0016
2.25	.4878	.0122	2.60	.4953	.0047	2.95	.4984	.0016
2.26	.4881	.0119	2.61	.4955	.0045	2.96	.4985	.0015
2.27	.4884	.0116	2.62	.4956	.0044	2.97	.4985	.0015
2.28	.4887	.0113	2.63	.4957	.0043	2.98	.4986	.0014
2.29	.4890	.0110	2.64	.4959	.0041	2.99	.4986	.0014

continued

TABLE C.1 (CONT.)
Proportions of Area under the Standard Normal Curve: The *z*-Tables

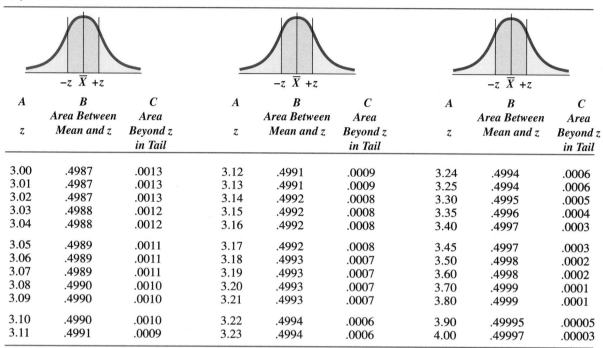

A	B	C	A	B	C	A	B	C
z	Area Between Mean and *z*	Area Beyond *z* in Tail	*z*	Area Between Mean and *z*	Area Beyond *z* in Tail	*z*	Area Between Mean and *z*	Area Beyond *z* in Tail
3.00	.4987	.0013	3.12	.4991	.0009	3.24	.4994	.0006
3.01	.4987	.0013	3.13	.4991	.0009	3.25	.4994	.0006
3.02	.4987	.0013	3.14	.4992	.0008	3.30	.4995	.0005
3.03	.4988	.0012	3.15	.4992	.0008	3.35	.4996	.0004
3.04	.4988	.0012	3.16	.4992	.0008	3.40	.4997	.0003
3.05	.4989	.0011	3.17	.4992	.0008	3.45	.4997	.0003
3.06	.4989	.0011	3.18	.4993	.0007	3.50	.4998	.0002
3.07	.4989	.0011	3.19	.4993	.0007	3.60	.4998	.0002
3.08	.4990	.0010	3.20	.4993	.0007	3.70	.4999	.0001
3.09	.4990	.0010	3.21	.4993	.0007	3.80	.4999	.0001
3.10	.4990	.0010	3.22	.4994	.0006	3.90	.49995	.00005
3.11	.4991	.0009	3.23	.4994	.0006	4.00	.49997	.00003

TABLE C.2
Critical Values of *t*: The *t*-Tables

(*Note:* Values of $-t_{crit}$ = values of $+t_{crit}$.)

	Two-Tailed Test				One-Tailed Test	

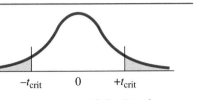
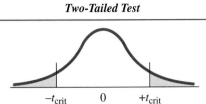

	Alpha Level				Alpha Level	
df	$\alpha = .05$	$\alpha = .01$		df	$\alpha = .05$	$\alpha = .01$
1	12.706	63.657		1	6.314	31.821
2	4.303	9.925		2	2.920	6.965
3	3.182	5.841		3	2.353	4.541
4	2.776	4.604		4	2.132	3.747
5	2.571	4.032		5	2.015	3.365
6	2.447	3.707		6	1.943	3.143
7	2.365	3.499		7	1.895	2.998
8	2.306	3.355		8	1.860	2.896
9	2.262	3.250		9	1.833	2.821
10	2.228	3.169		10	1.812	2.764
11	2.201	3.106		11	1.796	2.718
12	2.179	3.055		12	1.782	2.681
13	2.160	3.012		13	1.771	2.650
14	2.145	2.977		14	1.761	2.624
15	2.131	2.947		15	1.753	2.602
16	2.120	2.921		16	1.746	2.583
17	2.110	2.898		17	1.740	2.567
18	2.101	2.878		18	1.734	2.552
19	2.093	2.861		19	1.729	2.539
20	2.086	2.845		20	1.725	2.528
21	2.080	2.831		21	1.721	2.518
22	2.074	2.819		22	1.717	2.508
23	2.069	2.807		23	1.714	2.500
24	2.064	2.797		24	1.711	2.492
25	2.060	2.787		25	1.708	2.485
26	2.056	2.779		26	1.706	2.479
27	2.052	2.771		27	1.703	2.473
28	2.048	2.763		28	1.701	2.467
29	2.045	2.756		29	1.699	2.462
30	2.042	2.750		30	1.697	2.457
40	2.021	2.704		40	1.684	2.423
60	2.000	2.660		60	1.671	2.390
120	1.980	2.617		120	1.658	2.358
∞	1.960	2.576		∞	1.645	2.326

From Table 12 of E. Pearson and H. Hartley, *Biometrika Tables for Statisticians,* Vol. 1, 3rd ed. (Cambridge: Cambridge University Press, 1966.) Reprinted with the permission of the Biometrika Trustees.

STATISTICAL TABLES

TABLE C.3

Critical Values of the Pearson Correlation Coefficient: The *r*-Tables

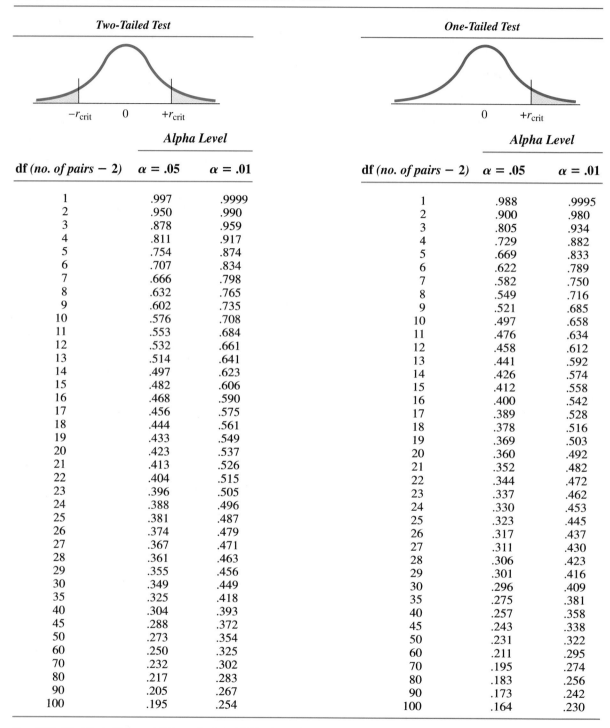

Two-Tailed Test			**One-Tailed Test**		
	Alpha Level			**Alpha Level**	
df *(no. of pairs − 2)*	$\alpha = .05$	$\alpha = .01$	df *(no. of pairs − 2)*	$\alpha = .05$	$\alpha = .01$
1	.997	.9999	1	.988	.9995
2	.950	.990	2	.900	.980
3	.878	.959	3	.805	.934
4	.811	.917	4	.729	.882
5	.754	.874	5	.669	.833
6	.707	.834	6	.622	.789
7	.666	.798	7	.582	.750
8	.632	.765	8	.549	.716
9	.602	.735	9	.521	.685
10	.576	.708	10	.497	.658
11	.553	.684	11	.476	.634
12	.532	.661	12	.458	.612
13	.514	.641	13	.441	.592
14	.497	.623	14	.426	.574
15	.482	.606	15	.412	.558
16	.468	.590	16	.400	.542
17	.456	.575	17	.389	.528
18	.444	.561	18	.378	.516
19	.433	.549	19	.369	.503
20	.423	.537	20	.360	.492
21	.413	.526	21	.352	.482
22	.404	.515	22	.344	.472
23	.396	.505	23	.337	.462
24	.388	.496	24	.330	.453
25	.381	.487	25	.323	.445
26	.374	.479	26	.317	.437
27	.367	.471	27	.311	.430
28	.361	.463	28	.306	.423
29	.355	.456	29	.301	.416
30	.349	.449	30	.296	.409
35	.325	.418	35	.275	.381
40	.304	.393	40	.257	.358
45	.288	.372	45	.243	.338
50	.273	.354	50	.231	.322
60	.250	.325	60	.211	.295
70	.232	.302	70	.195	.274
80	.217	.283	80	.183	.256
90	.205	.267	90	.173	.242
100	.195	.254	100	.164	.230

From Table IV of R. A. Fisher and F. Yates, *Statistical Tables for Biological, Agricultural and Medical Research,* 6th ed. (London: Longman Group Ltd., 1974). Reprinted by permission of Pearson Education Limited.

TABLE C.4
Critical Values of the Spearman Rank-Order Correlation Coefficient: The r_s-Tables

To interpolate the critical value for an N not given, find the critical values for the N above and below your N, add them together, and then divide the sum by 2. When N is greater than 30, transform r_S to a z-score using the formula $z_{obt} = (r_S)(\sqrt{N-1})$. For $\alpha = .05$, the two-tailed $z_{crit} = \pm 1.96$ and the one-tailed $z_{crit} = 1.645$.

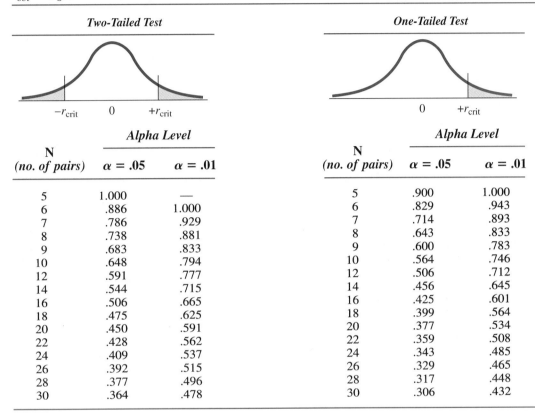

Two-Tailed Test			*One-Tailed Test*		
	Alpha Level			*Alpha Level*	
N *(no. of pairs)*	$\alpha = .05$	$\alpha = .01$	**N** *(no. of pairs)*	$\alpha = .05$	$\alpha = .01$
5	1.000	—	5	.900	1.000
6	.886	1.000	6	.829	.943
7	.786	.929	7	.714	.893
8	.738	.881	8	.643	.833
9	.683	.833	9	.600	.783
10	.648	.794	10	.564	.746
12	.591	.777	12	.506	.712
14	.544	.715	14	.456	.645
16	.506	.665	16	.425	.601
18	.475	.625	18	.399	.564
20	.450	.591	20	.377	.534
22	.428	.562	22	.359	.508
24	.409	.537	24	.343	.485
26	.392	.515	26	.329	.465
28	.377	.496	28	.317	.448
30	.364	.478	30	.306	.432

From E. G. Olds (1949), "The 5 Percent Significance Levels of Sums of Squares of Rank Differences and a Correction," *Annals of Math Statistics*, 20, pp. 117–118; and E. G. Olds (1938), "Distribution of Sums of Squares of Rank Differences for Small Numbers of Individuals," *Annals of Math Statistics*, 9, pp. 133–148. Reprinted with permission of the Institute of Mathematical Statistics.

TABLE C.5
Critical Values of *F:* The *F*-Tables

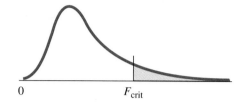

Critical values for α = .05 are in **dark numbers.**
Critical values for α = .01 are in light numbers.

Degrees of Freedom Within Groups (degrees of freedom in denominator of F-ratio)	α	\multicolumn{15}{c}{Degrees of Freedom between Groups (degrees of freedom in numerator of F-ratio)}														
		1	**2**	**3**	**4**	**5**	**6**	**7**	**8**	**9**	**10**	**11**	**12**	**14**	**16**	**20**
1	.05	**161**	**200**	**216**	**225**	**230**	**234**	**237**	**239**	**241**	**242**	**243**	**244**	**245**	**246**	**248**
	.01	4,052	4,999	5,403	5,625	5,764	5,859	5,928	5,981	6,022	6,056	6,082	6,106	6,142	6,169	6,208
2	.05	**18.51**	**19.00**	**19.16**	**19.25**	**19.30**	**19.33**	**19.36**	**19.37**	**19.38**	**19.39**	**19.40**	**19.41**	**19.42**	**19.43**	**19.44**
	.01	98.49	99.00	99.17	99.25	99.30	99.33	99.34	99.36	99.38	99.40	99.41	99.42	99.43	99.44	99.45
3	.05	**10.13**	**9.55**	**9.28**	**9.12**	**9.01**	**8.94**	**8.88**	**8.84**	**8.81**	**8.78**	**8.76**	**8.74**	**8.71**	**8.69**	**8.66**
	.01	34.12	30.82	29.46	28.71	28.24	27.91	27.67	27.49	27.34	27.23	27.13	27.05	26.92	26.83	26.69
4	.05	**7.71**	**6.94**	**6.59**	**6.39**	**6.26**	**6.16**	**6.09**	**6.04**	**6.00**	**5.96**	**5.93**	**5.91**	**5.87**	**5.84**	**5.80**
	.01	21.20	18.00	16.69	15.98	15.52	15.21	14.98	14.80	14.66	14.54	14.45	14.37	14.24	14.15	14.02
5	.05	**6.61**	**5.79**	**5.41**	**5.19**	**5.05**	**4.95**	**4.88**	**4.82**	**4.78**	**4.74**	**4.70**	**4.68**	**4.64**	**4.60**	**4.56**
	.01	16.26	13.27	12.06	11.39	10.97	10.67	10.45	10.27	10.15	10.05	9.96	9.89	9.77	9.68	9.55
6	.05	**5.99**	**5.14**	**4.76**	**4.53**	**4.39**	**4.28**	**4.21**	**4.15**	**4.10**	**4.06**	**4.03**	**4.00**	**3.96**	**3.92**	**3.87**
	.01	13.74	10.92	9.78	9.15	8.75	8.47	8.26	8.10	7.98	7.87	7.79	7.72	7.60	7.52	7.39
7	.05	**5.59**	**4.47**	**4.35**	**4.12**	**3.97**	**3.87**	**3.79**	**3.73**	**3.68**	**3.63**	**3.60**	**3.57**	**3.52**	**3.49**	**3.44**
	.01	12.25	9.55	8.45	7.85	7.46	7.19	7.00	6.84	6.71	6.62	6.54	6.47	6.35	6.27	6.15
8	.05	**5.32**	**4.46**	**4.07**	**3.84**	**3.69**	**3.58**	**3.50**	**3.44**	**3.39**	**3.34**	**3.31**	**3.28**	**3.23**	**3.20**	**3.15**
	.01	11.26	8.65	7.59	7.01	6.63	6.37	6.19	6.03	5.91	5.82	5.74	5.67	5.56	5.48	5.36
9	.05	**5.12**	**4.26**	**3.86**	**3.63**	**3.48**	**3.37**	**3.29**	**3.23**	**3.18**	**3.13**	**3.10**	**3.07**	**3.02**	**2.98**	**2.93**
	.01	10.56	8.02	6.99	6.42	6.06	5.80	5.62	5.47	5.35	5.26	5.18	5.11	5.00	4.92	4.80
10	.05	**4.96**	**4.10**	**3.71**	**3.48**	**3.33**	**3.22**	**3.14**	**3.07**	**3.02**	**2.97**	**2.94**	**2.91**	**2.86**	**2.82**	**2.77**
	.01	10.04	7.56	6.55	5.99	5.64	5.39	5.21	5.06	4.95	4.85	4.78	4.71	4.60	4.52	4.41
11	.05	**4.84**	**3.98**	**3.59**	**3.36**	**3.20**	**3.09**	**3.01**	**2.95**	**2.90**	**2.86**	**2.82**	**2.79**	**2.74**	**2.70**	**2.65**
	.01	9.65	7.20	6.22	5.67	5.32	5.07	4.88	4.74	4.63	4.54	4.46	4.40	4.29	4.21	4.10
12	.05	**4.75**	**3.88**	**3.49**	**3.26**	**3.11**	**3.00**	**2.92**	**2.85**	**2.80**	**2.76**	**2.72**	**2.69**	**2.64**	**2.60**	**2.54**
	.01	9.33	6.93	5.95	5.41	5.06	4.82	4.65	4.50	4.39	4.30	4.22	4.16	4.05	3.98	3.86
13	.05	**4.67**	**3.80**	**3.41**	**3.18**	**3.02**	**2.92**	**2.84**	**2.77**	**2.72**	**2.67**	**2.63**	**2.60**	**2.55**	**2.51**	**2.46**
	.01	9.07	6.70	5.74	5.20	4.86	4.62	4.44	4.30	4.19	4.10	4.02	3.96	3.85	3.78	3.67
14	.05	**4.60**	**3.74**	**3.34**	**3.11**	**2.96**	**2.85**	**2.77**	**2.70**	**2.65**	**2.60**	**2.56**	**2.53**	**2.48**	**2.44**	**2.39**
	.01	8.86	6.51	5.56	5.03	4.69	4.46	4.28	4.14	4.03	3.94	3.86	3.80	3.70	3.62	3.51
15	.05	**4.54**	**3.68**	**3.29**	**3.06**	**2.90**	**2.79**	**2.70**	**2.64**	**2.59**	**2.55**	**2.51**	**2.48**	**2.43**	**2.39**	**2.33**
	.01	8.68	6.36	5.42	4.89	4.56	4.32	4.14	4.00	3.89	3.80	3.73	3.67	3.56	3.48	3.36
16	.05	**4.49**	**3.63**	**3.24**	**3.01**	**2.85**	**2.74**	**2.66**	**2.59**	**2.54**	**2.49**	**2.45**	**2.42**	**2.37**	**2.33**	**2.28**
	.01	8.53	6.23	5.29	4.77	4.44	4.20	4.03	3.89	3.78	3.69	3.61	3.55	3.45	3.37	3.25

continued

TABLE C.5 (CONT.)
Critical Values of *F:* The *F*-Tables

Degrees of Freedom Within Groups (degrees of freedom in denominator of F-ratio)	α	Degrees of Freedom between Groups (degrees of freedom in numerator of F-ratio)														
		1	2	3	4	5	6	7	8	9	10	11	12	14	16	20
17	.05	4.45	3.59	3.20	2.96	2.81	2.70	2.62	2.55	2.50	2.45	2.41	2.38	2.33	2.29	2.23
	.01	8.40	6.11	5.18	4.67	4.34	4.10	3.93	3.79	3.68	3.59	3.52	3.45	3.35	3.27	3.16
18	.05	4.41	3.55	3.16	2.93	2.77	2.66	2.58	2.51	2.46	2.41	2.37	2.34	2.29	2.25	2.19
	.01	8.28	6.01	5.09	4.58	4.25	4.01	3.85	3.71	3.60	3.51	3.44	3.37	3.27	3.19	3.07
19	.05	4.38	3.52	3.13	2.90	2.74	2.63	2.55	2.48	2.43	2.38	2.34	2.31	2.26	2.21	2.15
	.01	8.18	5.93	5.01	4.50	4.17	3.94	3.77	3.63	3.52	3.43	3.36	3.30	3.19	3.12	3.00
20	.05	4.35	3.49	3.10	2.87	2.71	2.60	2.52	2.45	2.40	2.35	2.31	2.28	2.23	2.18	2.12
	.01	8.10	5.85	4.94	4.43	4.10	3.87	3.71	3.56	3.45	3.37	3.30	3.23	3.13	3.05	2.94
21	.05	4.32	3.47	3.07	2.84	2.68	2.57	2.49	2.42	2.37	2.32	2.28	2.25	2.20	2.15	2.09
	.01	8.02	5.78	4.87	4.37	4.04	3.81	3.65	3.51	3.40	3.31	3.24	3.17	3.07	2.99	2.88
22	.05	4.30	3.44	3.05	2.82	2.66	2.55	2.47	2.40	2.35	2.30	2.26	2.23	2.18	2.13	2.07
	.01	7.94	5.72	4.82	4.31	3.99	3.76	3.59	3.45	3.35	3.26	3.18	3.12	3.02	2.94	2.83
23	.05	4.28	3.42	3.03	2.80	2.64	2.53	2.45	2.38	2.32	2.28	2.24	2.20	2.14	2.10	2.04
	.01	7.88	5.66	4.76	4.26	3.94	3.71	3.54	3.41	3.30	3.21	3.14	3.07	2.97	2.89	2.78
24	.05	4.26	3.40	3.01	2.78	2.62	2.51	2.43	2.36	2.30	2.26	2.22	2.18	2.13	2.09	2.02
	.01	7.82	5.61	4.72	4.22	3.90	3.67	3.50	3.36	3.25	3.17	3.09	3.03	2.93	2.85	2.74
25	.05	4.24	3.38	2.99	2.76	2.60	2.49	2.41	2.34	2.28	2.24	2.20	2.16	2.11	2.06	2.00
	.01	7.77	5.57	4.68	4.18	3.86	3.63	3.46	3.32	3.21	3.13	3.05	2.99	2.89	2.81	2.70
26	.05	4.22	3.37	2.98	2.74	2.59	2.47	2.39	2.32	2.27	2.22	2.18	2.15	2.10	2.05	1.99
	.01	7.72	5.53	4.64	4.14	3.82	3.59	3.42	3.29	3.17	3.09	3.02	2.96	2.86	2.77	2.66
27	.05	4.21	3.35	2.96	2.73	2.57	2.46	2.37	2.30	2.25	2.20	2.16	2.13	2.08	2.03	1.97
	.01	7.68	5.49	4.60	4.11	3.79	3.56	3.39	3.26	3.14	3.06	2.98	2.93	2.83	2.74	2.63
28	.05	4.20	3.34	2.95	2.71	2.56	2.44	2.36	2.29	2.24	2.19	2.15	2.12	2.06	2.02	1.96
	.01	7.64	5.45	4.57	4.07	3.76	3.53	3.36	3.23	3.11	3.03	2.95	2.90	2.80	2.71	2.60
29	.05	4.18	3.33	2.93	2.70	2.54	2.43	2.35	2.28	2.22	2.18	2.14	2.10	2.05	2.00	1.94
	.01	7.60	5.42	4.54	4.04	3.73	3.50	3.33	3.20	3.08	3.00	2.92	2.87	2.77	2.68	2.57
30	.05	4.17	3.32	2.92	2.69	2.53	2.42	2.34	2.27	2.21	2.16	2.12	2.09	2.04	1.99	1.93
	.01	7.56	5.39	4.51	4.02	3.70	3.47	3.30	3.17	3.06	2.98	2.90	2.84	2.74	2.66	2.55
32	.05	4.15	3.30	2.90	2.67	2.51	2.40	2.32	2.25	2.19	2.14	2.10	2.07	2.02	1.97	1.91
	.01	7.50	5.34	4.46	3.97	3.66	3.42	3.25	3.12	3.01	2.94	2.86	2.80	2.70	2.62	2.51
34	.05	4.13	3.28	2.88	2.65	2.49	2.38	2.30	2.23	2.17	2.12	2.08	2.05	2.00	1.95	1.89
	.01	7.44	5.29	4.42	3.93	3.61	3.38	3.21	3.08	2.97	2.89	2.82	2.76	2.66	2.58	2.47
36	.05	4.11	3.26	2.86	2.63	2.48	2.36	2.28	2.21	2.15	2.10	2.06	2.03	1.98	1.93	1.87
	.01	7.39	5.25	4.38	3.89	3.58	3.35	3.18	3.04	2.94	2.86	2.78	2.72	2.62	2.54	2.43
38	.05	4.10	3.25	2.85	2.62	2.46	2.35	2.26	2.19	2.14	2.09	2.05	2.02	1.96	1.92	1.85
	.01	7.35	5.21	4.34	3.86	3.54	3.32	3.15	3.02	2.91	2.82	2.75	2.69	2.59	2.51	2.40
40	.05	4.08	3.23	2.84	2.61	2.45	2.34	2.25	2.18	2.12	2.07	2.04	2.00	1.95	1.90	1.84
	.01	7.31	5.18	4.31	3.83	3.51	3.29	3.12	2.99	2.88	2.80	2.73	2.66	2.56	2.49	2.37
42	.05	4.07	3.22	2.83	2.59	2.44	2.32	2.24	2.17	2.11	2.06	2.02	1.99	1.94	1.89	1.82
	.01	7.27	5.15	4.29	3.80	3.49	3.26	3.10	2.96	2.86	2.77	2.70	2.64	2.54	2.46	2.35

continued

STATISTICAL TABLES

TABLE C.5 (CONT.)

Critical Values of *F:* The *F*-Tables

Degrees of Freedom Within Groups (degrees of freedom in denominator of F-ratio)	α	1	2	3	4	5	6	7	8	9	10	11	12	14	16	20
44	.05	4.06	3.21	2.82	2.58	2.43	2.31	2.23	2.16	2.10	2.05	2.01	1.98	1.92	1.88	1.81
	.01	7.24	5.12	4.26	3.78	3.46	3.24	3.07	2.94	2.84	2.75	2.68	2.62	2.52	2.44	2.32
46	.05	4.05	3.20	2.81	2.57	2.42	2.30	2.22	2.14	2.09	2.04	2.00	1.97	1.91	1.87	1.80
	.01	7.21	5.10	4.24	3.76	3.44	3.22	3.05	2.92	2.82	2.73	2.66	2.60	2.50	2.42	2.30
48	.05	4.04	3.19	2.80	2.56	2.41	2.30	2.21	2.14	2.08	2.03	1.99	1.96	1.90	1.86	1.79
	.01	7.19	5.08	4.22	3.74	3.42	3.20	3.04	2.90	2.80	2.71	2.64	2.58	2.48	2.40	2.28
50	.05	4.03	3.18	2.79	2.56	2.40	2.29	2.20	2.13	2.07	2.02	1.98	1.95	1.90	1.85	1.78
	.01	7.17	5.06	4.20	3.72	3.41	3.18	3.02	2.88	2.78	2.70	2.62	2.56	2.46	2.39	2.26
55	.05	4.02	3.17	2.78	2.54	2.38	2.27	2.18	2.11	2.05	2.00	1.97	1.93	1.88	1.83	1.76
	.01	7.12	5.01	4.16	3.68	3.37	3.15	2.98	2.85	2.75	2.66	2.59	2.53	2.43	2.35	2.23
60	.05	4.00	3.15	2.76	2.52	2.37	2.25	2.17	2.10	2.04	1.99	1.95	1.92	1.86	1.81	1.75
	.01	7.08	4.98	4.13	3.65	3.34	3.12	2.95	2.82	2.72	2.63	2.56	2.50	2.40	2.32	2.20
65	.05	3.99	3.14	2.75	2.51	2.36	2.24	2.15	2.08	2.02	1.98	1.94	1.90	1.85	1.80	1.73
	.01	7.04	4.95	4.10	3.62	3.31	3.09	2.93	2.79	2.70	2.61	2.54	2.47	2.37	2.30	2.18
70	.05	3.98	3.13	2.74	2.50	2.35	2.23	2.14	2.07	2.01	1.97	1.93	1.89	1.84	1.79	1.72
	.01	7.01	4.92	4.08	3.60	3.29	3.07	2.91	2.77	2.67	2.59	2.51	2.45	2.35	2.28	2.15
80	.05	3.96	3.11	2.72	2.48	2.33	2.21	2.12	2.05	1.99	1.95	1.91	1.88	1.82	1.77	1.70
	.01	6.96	4.88	4.04	3.56	3.25	3.04	2.87	2.74	2.64	2.55	2.48	2.41	2.32	2.24	2.11
100	.05	3.94	3.09	2.70	2.46	2.30	2.19	2.10	2.03	1.97	1.92	1.88	1.85	1.79	1.75	1.68
	.01	6.90	4.82	3.98	3.51	3.20	2.99	2.82	2.69	2.59	2.51	2.43	2.36	2.26	2.19	2.06
125	.05	3.92	3.07	2.68	2.44	2.29	2.17	2.08	2.01	1.95	1.90	1.86	1.83	1.77	1.72	1.65
	.01	6.84	4.78	3.94	3.47	3.17	2.95	2.79	2.65	2.56	2.47	2.40	2.33	2.23	2.15	2.03
150	.05	3.91	3.06	2.67	2.43	2.27	2.16	2.07	2.00	1.94	1.89	1.85	1.82	1.76	1.71	1.64
	.01	6.81	4.75	3.91	3.44	3.14	2.92	2.76	2.62	2.53	2.44	2.37	2.30	2.20	2.12	2.00
200	.05	3.89	3.04	2.65	2.41	2.26	2.14	2.05	1.98	1.92	1.87	1.83	1.80	1.74	1.69	1.62
	.01	6.76	4.71	3.88	3.41	3.11	2.90	2.73	2.60	2.50	2.41	2.34	2.28	2.17	2.09	1.97
400	.05	3.86	3.02	2.62	2.39	2.23	2.12	2.03	1.96	1.90	1.85	1.81	1.78	1.72	1.67	1.60
	.01	6.70	4.66	3.83	3.36	3.06	2.85	2.69	2.55	2.46	2.37	2.29	2.23	2.12	2.04	1.92
1000	.05	3.85	3.00	2.61	2.38	2.22	2.10	2.02	1.95	1.89	1.84	1.80	1.76	1.70	1.65	1.58
	.01	6.66	4.62	3.80	3.34	3.04	2.82	2.66	2.53	2.43	2.34	2.26	2.20	2.09	2.01	1.89
∞	.05	3.84	2.99	2.60	2.37	2.21	2.09	2.01	1.94	1.88	1.83	1.79	1.75	1.69	1.64	1.57
	.01	6.64	4.60	3.78	3.32	3.02	2.80	2.64	2.51	2.41	2.32	2.24	2.18	2.07	1.99	1.87

Reprinted by permission from *Statistical Methods,* 8th edition by G. Snedecor and W. Cochran. © 1989 by The Iowa State University Press, Ames, Iowa.

TABLE C.6
Values of Studentized Range Statistic, q_k

For a one-way ANOVA, or a comparison of the means from a main effect, the value of k is the number of means in the factor.

To compare the means from an interaction, find the appropriate design (or number of cell means) in the table below and obtain the adjusted value of k. Then use adjusted k as k to find the value of q_k.

Values of Adjusted k

Design of Study	Number of Cell Means in Study	Adjusted Value of k
2×2	4	3
2×3	6	5
2×4	8	6
3×3	9	7
3×4	12	8
4×4	16	10
4×5	20	12

Values of q_k for $\alpha = .05$ are **dark numbers** and for $\alpha = .01$ are light numbers.

Degrees of Freedom Within Groups (degrees of freedom in denominator of F-ratio)	α	k = Number of Means Being Compared										
		2	3	4	5	6	7	8	9	10	11	12
1	.05	**18.00**	**27.00**	**32.80**	**37.10**	**40.40**	**43.10**	**45.40**	**47.40**	**49.10**	**50.60**	**52.00**
	.01	90.00	135.00	164.00	186.00	202.00	216.00	227.00	237.00	246.00	253.00	260.00
2	.05	**6.09**	**8.30**	**9.80**	**10.90**	**11.70**	**12.40**	**13.00**	**13.50**	**14.00**	**14.40**	**14.70**
	.01	14.00	19.00	22.30	24.70	26.60	28.20	29.50	30.70	31.70	32.60	33.40
3	.05	**4.50**	**5.91**	**6.82**	**7.50**	**8.04**	**8.48**	**8.85**	**9.18**	**9.46**	**9.72**	**9.95**
	.01	8.26	10.60	12.20	13.30	14.20	15.00	15.60	16.20	16.70	17.10	17.50
4	.05	**3.93**	**5.04**	**5.76**	**6.29**	**6.71**	**7.05**	**7.35**	**7.60**	**7.83**	**8.03**	**8.21**
	.01	6.51	8.12	9.17	9.96	10.60	11.10	11.50	11.90	12.30	12.60	12.80
5	.05	**3.64**	**4.60**	**5.22**	**5.67**	**6.03**	**6.33**	**6.58**	**6.80**	**6.99**	**7.17**	**7.32**
	.01	5.70	6.97	7.80	8.42	8.91	9.32	9.67	9.97	10.20	10.50	10.70
6	.05	**3.46**	**4.34**	**4.90**	**5.31**	**5.63**	**5.89**	**6.12**	**6.32**	**6.49**	**6.65**	**6.79**
	.01	5.24	6.33	7.03	7.56	7.97	8.32	8.61	8.87	9.10	9.30	9.49
7	.05	**3.34**	**4.16**	**4.69**	**5.06**	**5.36**	**5.61**	**5.82**	**6.00**	**6.16**	**6.30**	**6.43**
	.01	4.95	5.92	6.54	7.01	7.37	7.68	7.94	8.17	8.37	8.55	8.71
8	.05	**3.26**	**4.04**	**4.53**	**4.89**	**5.17**	**5.40**	**5.60**	**5.77**	**5.92**	**6.05**	**6.18**
	.01	4.74	5.63	6.20	6.63	6.96	7.24	7.47	7.68	7.87	8.03	8.18
9	.05	**3.20**	**3.95**	**4.42**	**4.76**	**5.02**	**5.24**	**5.43**	**5.60**	**5.74**	**5.87**	**5.98**
	.01	4.60	5.43	5.96	6.35	6.66	6.91	7.13	7.32	7.49	7.65	7.78

continued

STATISTICAL TABLES

TABLE C.6 (CONT.)

Values of Studentized Range Statistic, q_k

Degrees of Freedom Within Groups (degrees of freedom in denominator of F-ratio)	α	2	3	4	5	6	7	8	9	10	11	12
						k = *Number of Means Being Compared*						
10	.05	3.15	3.88	4.33	4.65	4.91	5.12	5.30	5.46	5.60	5.72	5.83
	.01	4.48	5.27	5.77	6.14	6.43	6.67	6.87	7.05	7.21	7.36	7.48
11	.05	3.11	3.82	4.26	4.57	4.82	5.03	5.20	5.35	5.49	5.61	5.71
	.01	4.39	5.14	5.62	5.97	6.25	6.48	6.67	6.84	6.99	7.13	7.26
12	.05	3.08	3.77	4.20	4.51	4.75	4.95	5.12	5.27	5.40	5.51	5.62
	.01	4.32	5.04	5.50	5.84	6.10	6.32	6.51	6.67	6.81	6.94	7.06
13	.05	3.06	3.73	4.15	4.45	4.69	4.88	5.05	5.19	5.32	5.43	5.53
	.01	4.26	4.96	5.40	5.73	5.98	6.19	6.37	6.53	6.67	6.79	6.90
14	.05	3.03	3.70	4.11	4.41	4.64	4.83	4.99	5.13	5.25	5.36	5.46
	.01	4.21	4.89	5.32	5.63	5.88	6.08	6.26	6.41	6.54	6.66	6.77
16	.05	3.00	3.65	4.05	4.33	4.56	4.74	4.90	5.03	5.15	5.26	5.35
	.01	4.13	4.78	5.19	5.49	5.72	5.92	6.08	6.22	6.35	6.46	6.56
18	.05	2.97	3.61	4.00	4.28	4.49	4.67	4.82	4.96	5.07	5.17	5.27
	.01	4.07	4.70	5.09	5.38	5.60	5.79	5.94	6.08	6.20	6.31	6.41
20	.05	2.95	3.58	3.96	4.23	4.45	4.62	4.77	4.90	5.01	5.11	5.20
	.01	4.02	4.64	5.02	5.29	5.51	5.69	5.84	5.97	6.09	6.19	6.29
24	.05	2.92	3.53	3.90	4.17	4.37	4.54	4.68	4.81	4.92	5.01	5.10
	.01	3.96	4.54	4.91	5.17	5.37	5.54	5.69	5.81	5.92	6.02	6.11
30	.05	2.89	3.49	3.84	4.10	4.30	4.46	4.60	4.72	4.83	4.92	5.00
	.01	3.89	4.45	4.80	5.05	5.24	5.40	5.54	5.56	5.76	5.85	5.93
40	.05	2.86	3.44	3.79	4.04	4.23	4.39	4.52	4.63	4.74	4.82	4.91
	.01	3.82	4.37	4.70	4.93	5.11	5.27	5.39	5.50	5.60	5.69	5.77
60	.05	2.83	3.40	3.74	3.98	4.16	4.31	4.44	4.55	4.65	4.73	4.81
	.01	3.76	4.28	4.60	4.82	4.99	5.13	5.25	5.36	5.45	5.53	5.60
120	.05	2.80	3.36	3.69	3.92	4.10	4.24	4.36	4.48	4.56	4.64	4.72
	.01	3.70	4.20	4.50	4.71	4.87	5.01	5.12	5.21	5.30	5.38	5.44
∞	.05	2.77	3.31	3.63	3.86	4.03	4.17	4.29	4.39	4.47	4.55	4.62
	.01	3.64	4.12	4.40	4.60	4.76	4.88	4.99	5.08	5.16	5.23	5.29

From B. J. Winer, *Statistical Principles in Experimental Design,* McGraw-Hill, 1962; abridged from H. L. Harter, D. S. Clemm, and E. H. Guthrie, "The probability integrals of the range and of the studentized range," WADC Tech. Rep., 58–484, Vol. 2, 1959, Wright Air Development Center, Table II.2, pp. 243–281. Reproduced by permission of the McGraw-Hill Companies, Inc.

TABLE C.7

Critical Values of Chi Square: The χ^2-Tables

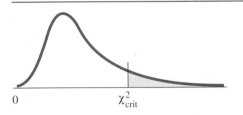

0   χ^2_{crit}

| | *Alpha Level* | |
df	$\alpha = .05$	$\alpha = .01$
1	3.84	6.64
2	5.99	9.21
3	7.81	11.34
4	9.49	13.28
5	11.07	15.09
6	12.59	16.81
7	14.07	18.48
8	15.51	20.09
9	16.92	21.67
10	18.31	23.21
11	19.68	24.72
12	21.03	26.22
13	22.36	27.69
14	23.68	29.14
15	25.00	30.58
16	26.30	32.00
17	27.59	33.41
18	28.87	34.80
19	30.14	36.19
20	31.41	37.57
21	32.67	38.93
22	33.92	40.29
23	35.17	41.64
24	36.42	42.98
25	37.65	44.31
26	38.88	45.64
27	40.11	46.96
28	41.34	48.28
29	42.56	49.59
30	43.77	50.89
40	55.76	63.69
50	67.50	76.15
60	79.08	88.38
70	90.53	100.42

TABLE C.8

Critical Values of the Mann–Whitney *U*

To be significant, the U_{obt} must be equal to or *less than* the critical value. (Dashes in the table indicate that no decision is possible.) Critical values for $\alpha = .05$ are **dark numbers** and for $\alpha = .01$ are light numbers.

Two-Tailed Test

n₂ (*Number of Scores in Group 2*)	α	|	|	|	|	n₁ (*Number of Scores in Group 1*)	|	|	|	|
		1	2	3	4	5	6	7	8	9
1	.05	—	—	—	—	—	—	—	—	—
	.01	—	—	—	—	—	—	—	—	—
2	.05	—	—	—	—	—	—	—	**0**	**0**
	.01	—	—	—	—	—	—	—	—	—
3	.05	—	—	—	—	**0**	**1**	**1**	**2**	**2**
	.01	—	—	—	—	—	—	—	—	0
4	.05	—	—	—	**0**	**1**	**2**	**3**	**4**	**4**
	.01	—	—	—	—	—	0	0	1	1
5	.05	—	—	**0**	**1**	**2**	**3**	**5**	**6**	**7**
	.01	—	—	—	—	0	1	1	2	3
6	.05	—	—	**1**	**2**	**3**	**5**	**6**	**8**	**10**
	.01	—	—	—	0	1	2	3	4	5
7	.05	—	—	**1**	**3**	**5**	**6**	**8**	**10**	**12**
	.01	—	—	—	0	1	3	4	6	7
8	.05	—	**0**	**2**	**4**	**6**	**8**	**10**	**13**	**15**
	.01	—	—	—	1	2	4	6	7	9
9	.05	—	**0**	**2**	**4**	**7**	**10**	**12**	**15**	**17**
	.01	—	—	0	1	3	5	7	9	11
10	.05	—	**0**	**3**	**5**	**8**	**11**	**14**	**17**	**20**
	.01	—	—	0	2	4	6	9	11	13
11	.05	—	**0**	**3**	**6**	**9**	**13**	**16**	**19**	**23**
	.01	—	—	0	2	5	7	10	13	16
12	.05	—	**1**	**4**	**7**	**11**	**14**	**18**	**22**	**26**
	.01	—	—	1	3	6	9	12	15	18
13	.05	—	**1**	**4**	**8**	**12**	**16**	**20**	**24**	**28**
	.01	—	—	1	3	7	10	13	17	20
14	.05	—	**1**	**5**	**9**	**13**	**17**	**22**	**26**	**31**
	.01	—	—	1	4	7	11	15	18	22
15	.05	—	**1**	**5**	**10**	**14**	**19**	**24**	**29**	**34**
	.01	—	—	2	5	8	12	16	20	24
16	.05	—	**1**	**6**	**11**	**15**	**21**	**26**	**31**	**37**
	.01	—	—	2	5	9	13	18	22	27
17	.05	—	**2**	**6**	**11**	**17**	**22**	**28**	**34**	**39**
	.01	—	—	2	6	10	15	19	24	29
18	.05	—	**2**	**7**	**12**	**18**	**24**	**30**	**36**	**42**
	.01	—	—	2	6	11	16	21	26	31
19	.05	—	**2**	**7**	**13**	**19**	**25**	**32**	**38**	**45**
	.01	—	0	3	7	12	17	22	28	33
20	.05	—	**2**	**8**	**13**	**20**	**27**	**34**	**41**	**48**
	.01	—	0	3	8	13	18	24	30	36

continued

TABLE C.8 (CONT.)

Critical Values of the Mann–Whitney U

Two-Tailed Test

n_1 *(Number of Scores in Group 1)*

10	11	12	13	14	15	16	17	18	19	20
—	—	—	—	—	—	—	—	—	—	—
—	—	—	—	—	—	—	—	—	—	—
0	**0**	**1**	**1**	**1**	**1**	**1**	**2**	**2**	**2**	**2**
—	—	—	—	—	—	—	—	—	0	0
3	**3**	**4**	**4**	**5**	**5**	**6**	**6**	**7**	**7**	**8**
0	0	1	1	1	2	2	2	2	3	3
5	**6**	**7**	**8**	**9**	**10**	**11**	**11**	**12**	**13**	**13**
2	2	3	3	4	5	5	6	6	7	8
8	**9**	**11**	**12**	**13**	**14**	**15**	**17**	**18**	**19**	**20**
4	5	6	7	7	8	9	10	11	12	13
11	**13**	**14**	**16**	**17**	**19**	**21**	**22**	**24**	**25**	**27**
6	7	9	10	11	12	13	15	16	17	18
14	**16**	**18**	**20**	**22**	**24**	**26**	**28**	**30**	**32**	**34**
9	10	12	13	15	16	18	19	21	22	24
17	**19**	**22**	**24**	**26**	**29**	**31**	**34**	**36**	**38**	**41**
11	13	15	17	18	20	22	24	26	28	30
20	**23**	**26**	**28**	**31**	**34**	**37**	**39**	**42**	**45**	**48**
13	16	18	20	22	24	27	29	31	33	36
23	**26**	**29**	**33**	**36**	**39**	**42**	**45**	**48**	**52**	**55**
16	18	21	24	26	29	31	34	37	39	42
26	**30**	**33**	**37**	**40**	**44**	**47**	**51**	**55**	**58**	**62**
18	21	24	27	30	33	36	39	42	45	48
29	**33**	**37**	**41**	**45**	**49**	**53**	**57**	**61**	**65**	**69**
21	24	27	31	34	37	41	44	47	51	54
33	**37**	**41**	**45**	**50**	**54**	**59**	**63**	**67**	**72**	**76**
24	27	31	34	38	42	45	49	53	56	60
36	**40**	**45**	**50**	**55**	**59**	**64**	**67**	**74**	**78**	**83**
26	30	34	38	42	46	50	54	58	63	67
39	**44**	**49**	**54**	**59**	**64**	**70**	**75**	**80**	**85**	**90**
29	33	37	42	46	51	55	60	64	69	73
42	**47**	**53**	**59**	**64**	**70**	**75**	**81**	**86**	**92**	**98**
31	36	41	45	50	55	60	65	70	74	79
45	**51**	**57**	**63**	**67**	**75**	**81**	**87**	**93**	**99**	**105**
34	39	44	49	54	60	65	70	75	81	86
48	**55**	**61**	**67**	**74**	**80**	**86**	**93**	**99**	**106**	**112**
37	42	47	53	58	64	70	75	81	87	92
52	**58**	**65**	**72**	**78**	**85**	**92**	**99**	**106**	**113**	**119**
39	45	51	56	63	69	74	81	87	93	99
55	**62**	**69**	**76**	**83**	**90**	**98**	**105**	**112**	**119**	**127**
42	48	54	60	67	73	79	86	92	99	105

continued

STATISTICAL TABLES

TABLE C.8 (CONT.)

Critical Values of the Mann–Whitney U

One-Tailed Test

n_2 (Number of Scores in Group 2)	α	\multicolumn{9}{c}{n_1 (Number of Scores in Group 1)}
		1	2	3	4	5	6	7	8	9
1	.05	—	—	—	—	—	—	—	—	—
	.01	—	—	—	—	—	—	—	—	—
2	.05	—	—	—	—	0	0	0	1	1
	.01	—	—	—	—	—	—	—	—	—
3	.05	—	—	0	0	1	2	2	3	3
	.01	—	—	—	—	—	—	0	0	1
4	.05	—	—	0	1	2	3	4	5	6
	.01	—	—	—	—	0	1	1	2	3
5	.05	—	0	1	2	4	5	6	8	9
	.01	—	—	—	0	1	2	3	4	5
6	.05	—	0	2	3	5	7	8	10	12
	.01	—	—	—	1	2	3	4	6	7
7	.05	—	0	2	4	6	8	11	13	15
	.01	—	—	0	1	3	4	6	7	9
8	.05	—	1	3	5	8	10	13	15	18
	.01	—	—	0	2	4	6	7	9	11
9	.05	—	1	3	6	9	12	15	18	21
	.01	—	—	1	3	5	7	9	11	14
10	.05	—	1	4	7	11	14	17	20	24
	.01	—	—	1	3	6	8	11	13	16
11	.05	—	1	5	8	12	16	19	23	27
	.01	—	—	1	4	7	9	12	15	18
12	.05	—	2	5	9	13	17	21	26	30
	.01	—	—	2	5	8	11	14	17	21
13	.05	—	2	6	10	15	19	24	28	33
	.01	—	0	2	5	9	12	16	20	23
14	.05	—	2	7	11	16	21	26	31	36
	.01	—	0	2	6	10	13	17	22	26
15	.05	—	3	7	12	18	23	28	33	39
	.01	—	0	3	7	11	15	19	24	28
16	.05	—	3	8	14	19	25	30	36	42
	.01	—	0	3	7	12	16	21	26	31
17	.05	—	3	9	15	20	26	33	39	45
	.01	—	0	4	8	13	18	23	28	33
18	.05	—	4	9	16	22	28	35	41	48
	.01	—	0	4	9	14	19	24	30	36
19	.05	0	4	10	17	23	30	37	44	51
	.01	—	1	4	9	15	20	26	32	38
20	.05	0	4	11	18	25	32	39	47	54
	.01	—	1	5	10	16	22	28	34	40

continued

TABLE C.8 (CONT.)

Critical Values of the Mann–Whitney U

One-Tailed Test

n_1 *(Number of scores in Group 1)*

10	11	12	13	14	15	16	17	18	19	20
—	—	—	—	—	—	—	—	—	**0**	**0**
—	—	—	—	—	—	—	—	—	—	—
1	**1**	**2**	**2**	**2**	**3**	**3**	**3**	**4**	**4**	**4**
—	—	—	0	0	0	0	0	0	1	1
4	**5**	**5**	**6**	**7**	**7**	**8**	**9**	**9**	**10**	**11**
1	1	2	2	2	3	3	4	4	4	5
7	**8**	**9**	**10**	**11**	**12**	**14**	**15**	**16**	**17**	**18**
3	4	5	5	6	7	7	8	9	9	10
11	**12**	**13**	**15**	**16**	**18**	**19**	**20**	**22**	**23**	**25**
6	7	8	9	10	11	12	13	14	15	16
14	**16**	**17**	**19**	**21**	**23**	**25**	**26**	**28**	**30**	**32**
8	9	11	12	13	15	16	18	19	20	22
17	**19**	**21**	**24**	**26**	**28**	**30**	**33**	**35**	**37**	**39**
11	12	14	16	17	19	21	23	24	26	28
20	**23**	**26**	**28**	**31**	**33**	**36**	**39**	**41**	**44**	**47**
13	15	17	20	22	24	26	28	30	32	34
24	**27**	**30**	**33**	**36**	**39**	**42**	**45**	**48**	**51**	**54**
16	18	21	23	26	28	31	33	36	38	40
27	**31**	**34**	**37**	**41**	**44**	**48**	**51**	**55**	**58**	**62**
19	22	24	27	30	33	36	38	41	44	47
31	**34**	**38**	**42**	**46**	**50**	**54**	**57**	**61**	**65**	**69**
22	25	28	31	34	37	41	44	47	50	53
34	**38**	**42**	**47**	**51**	**55**	**60**	**64**	**68**	**72**	**77**
24	28	31	35	38	42	46	49	53	56	60
37	**42**	**47**	**51**	**56**	**61**	**65**	**70**	**75**	**80**	**84**
27	31	35	39	43	47	51	55	59	63	67
41	**46**	**51**	**56**	**61**	**66**	**71**	**77**	**82**	**87**	**92**
30	34	38	43	47	51	56	60	65	69	73
44	**50**	**55**	**61**	**66**	**72**	**77**	**83**	**88**	**94**	**100**
33	37	42	47	51	56	61	66	70	75	80
48	**54**	**60**	**65**	**71**	**77**	**83**	**89**	**95**	**101**	**107**
36	41	46	51	56	61	66	71	76	82	87
51	**57**	**64**	**70**	**77**	**83**	**89**	**96**	**102**	**109**	**115**
38	44	49	55	60	66	71	77	82	88	93
55	**61**	**68**	**75**	**82**	**88**	**95**	**102**	**109**	**116**	**123**
41	47	53	59	65	70	76	82	88	94	100
58	**65**	**72**	**80**	**87**	**94**	**101**	**109**	**116**	**123**	**130**
44	50	56	63	69	75	82	88	94	101	107
62	**69**	**77**	**84**	**92**	**100**	**107**	**115**	**123**	**130**	**138**
47	53	60	67	73	80	87	93	100	107	114

From the *Bulletin of the Institute of Educational Research,* 1, No. 2, Indiana University, with permission of the publishers.

STATISTICAL TABLES

TABLE C.9

Critical Values of the Wilcoxon T

To be significant, the T_{obt} must be equal to or *less than* the critical value. (Dashes in the table indicate that no decision is possible.) In the table N is the number of nonzero differences that occurred when T_{obt} was calculated.

		Two-Tailed Test			
N	$\alpha = .05$	$\alpha = .01$	**N**	$\alpha = .05$	$\alpha = .01$
5	—	—	28	116	91
6	0	—	29	126	100
7	2	—	30	137	109
8	3	0	31	147	118
9	5	1	32	159	128
10	8	3	33	170	138
11	10	5	34	182	148
12	13	7	35	195	159
13	17	9	36	208	171
14	21	12	37	221	182
15	25	15	38	235	194
16	29	19	39	249	207
17	34	23	40	264	220
18	40	27	41	279	233
19	46	32	42	294	247
20	52	37	43	310	261
21	58	42	44	327	276
22	65	48	45	343	291
23	73	54	46	361	307
24	81	61	47	378	322
25	89	68	48	396	339
26	98	75	49	415	355
27	107	83	50	434	373

continued

TABLE C.9 (CONT.)

Critical Values of the Wilcoxon T

| | One-Tailed Test | | | | |
N	$\alpha = .05$	$\alpha = .01$	N	$\alpha = .05$	$\alpha = .01$
5	0	—	28	130	101
6	2	—	29	140	110
7	3	0	30	151	120
8	5	1	31	163	130
9	8	3	32	175	140
10	10	5	33	187	151
11	13	7	34	200	162
12	17	9	35	213	173
13	21	12	36	227	185
14	25	15	37	241	198
15	30	19	38	256	211
16	35	23	39	271	224
17	41	27	40	286	238
18	47	32	41	302	252
19	53	37	42	319	266
20	60	43	43	336	281
21	67	49	44	353	296
22	75	55	45	371	312
23	83	62	46	389	328
24	91	69	47	407	345
25	100	76	48	426	362
26	110	84	49	446	379
27	119	92	50	466	397

From F. Wilcoxon and R. A. Wilcox, *Some Rapid Approximate Statistical Procedures,* New York: Lederle Laboratories, 1964. Reproduced with the permission of the American Cyanamid Company.

STATISTICAL TABLES

D Answers to Odd-Numbered Questions

Chapter 1

1. To conduct research and to understand the research of others
3. (a) To two more decimal places than were in the original scores.
   (b) If the number in the third decimal place is 5 or greater, round up the number in the second decimal place. If the number in the third decimal place is less than 5, round down by not changing the number in the second decimal place.
5. Perform squaring and taking a square root first, then multiplication and division, and then addition and subtraction.
7. It is the "dot" placed on a graph when plotting a pair of X and Y scores.
9. A proportion is a decimal indicating a fraction of the total. To transform a number to a proportion, divide the number by the total.
11. (a) $5/15 = .33$ (b) $10/50 = .20$
    (c) $1/1000 = .001$
13. (a) 33%   (b) 20%   (c) .1%
15. (a) 13.75   (b) 10.04   (c) 10.05   (d) .08
    (e) 1.00
17. $Q = (8 + -2)(64 + 4) = (6)(68) = 408$
19. $D = (-3.25)(3) = -9.75$
21. (a) $(.60)40 = 24; (.60)35 = 21; (.60)60 = 36$
    (b) $(.60)135 = 81$
    (c) $115/135 = 0.85$; multiplied by 100 is 85%
23. (a) Space the labels to reflect the actual distance between the scores.
    (b) So that they don't give a misleading impression.

Chapter 2

1. (a) The large groups of individuals (or scores) to which we think a law of nature applies.
   (b) A subset of the population that represents or stands in for the population.
   (c) We assume that the relationship found in a sample reflects the relationship found in the population.
   (d) All relevant individuals in the world, in nature.
3. It is the consistency with which one or close to one Y score is paired with each X.
5. The design of the study and the scale of measurement used.
7. The independent variable is the overall variable the researcher is interested in; the conditions are the specific amounts or categories of the independent variable under which participants are tested.
9. To discover relationships between variables, which may reflect how nature operates.
11. (a) A statistic describes a characteristic of a sample of scores. A parameter describes a characteristic of a population of scores.
    (b) Statistics use letters from the English alphabet. Parameters use letters from the Greek alphabet.
13. The problem is that a statistical analysis cannot prove anything.
15. His sample may not be representative of all college students. Perhaps he selected those few students who prefer carrot juice.
17. Ratio scales provide the most specific information, interval scales provide less specific information, ordinal scales provide even less specific information, and nominal scales provide the least specific information.
19. Samples A (Y scores increase) and D (Y scores increase then decrease).
21. Studies A and C. In each, as the scores on one variable change, the scores on the other variable change in a consistent fashion.
23. Because each relationship suggests that in nature, as the amount of X changes, Y also changes.

25.

Variable	Qualitative or Quantitative	Continuous, Discrete or Dichotomous	Type of Measurement Scale
gender	qualitative	dichotomous	nominal
academic major	qualitative	discrete	nominal
number of minutes before and after an event	quantitative	continuous	interval
restaurant ratings (best, next best, etc.)	quantitative	discrete	ordinal
speed	quantitative	continuous	ratio
dollars in your pocket	quantitative	discrete	ratio
change in weight	quantitative	continuous	interval
checking account balance	quantitative	discrete	interval
reaction time	quantitative	continuous	ratio
letter grades	quantitative	discrete	ordinal
clothing size	quantitative	discrete	ordinal
registered voter	qualitative	dichotomous	nominal
therapeutic approach	qualitative	discrete	nominal
schizophrenia type	qualitative	discrete	nominal
work absences	quantitative	discrete	ratio
words recalled	quantitative	discrete	ratio

Chapter 3

1. (a) N is the number of scores in a sample.
   (b) f is frequency, the number of times a score occurs.
   (c) $rel.\,f$ is relative frequency, the proportion of time a score occurs.
   (d) cf is cumulative frequency, the number of times scores at or below a score occur.

3. (a) A histogram has a bar above each score; a polygon has datapoints above the scores that are connected by straight lines.
   (b) Histograms are used with a few different interval or ratio scores, polygons are used with a wide range of interval/ratio scores.

5. (a) Relative frequency (the proportion of time a score occurs) may be easier to interpret than simple frequency (the number of times a score occurs).
   (b) Percentile (the percent of scores at or below a score) may be easier to interpret than cumulative frequency (the number of scores at or below a score).

7. A negatively skewed distribution has only one tail at the extreme low scores; a positively skewed distribution has only one tail at the extreme high scores.

9. The graph showed the relationship where, as scores on the X variable change, scores on the Y variable change. A frequency distribution shows the relationship where, as X scores change, their frequency (shown on Y) changes.

11. It means that the score is either a high or low extreme score that occurs relatively infrequently.

13. (a) The middle IQ score has the highest frequency in a symmetric distribution; the higher and lower scores have lower frequencies, and the highest and lowest scores have a relatively very low frequency.
   (b) The agility scores form a symmetric distribution containing two distinct "humps" where there are two scores that occur more frequently than the surrounding scores.
   (c) The memory scores form an asymmetric distribution in which there are some very infrequent, extremely low scores, but there are not correspondingly infrequent high scores.

15. It indicates that the test was difficult for the class, because most often the scores are low or middle scores, and seldom are there high scores.

17. (a) 35% of the sample scored at or below the score.
   (b) The score occurred 40% of the time.
   (c) It is one of the highest and least frequent scores.
   (d) It is one of the lowest and least frequent scores.
   (e) 50 participants had either your score or a score below it.
   (f) 60% of the area under the curve and thus 60% of the distribution is to the left of (below) your score.

19. (a) 70, 72, 60, 85, 45
   (b) Because .20 of the area under the curve is to the left of 60, it's at the 20th percentile.
   (c) With .50 of the area under the curve to the left of 70, .50 of the sample is below 70.
   (d) With .50 of the area under the curve below 70, and .20 of the area under the curve below 60, then .50 − .20 = .30. of the area under the curve is between 60 and 70.
   (e) .20

(f) With .30 of the scores between 70 and 80, and .50 of the area under the curve below 70, a total of .30 + .50 = .80 of scores are below 80, so it's at the 80th percentile.

21.

Score	f	rel. f	cf
53	1	.06	18
52	3	.17	17
51	2	.11	14
50	5	.28	12
49	4	.22	7
48	0	.00	3
47	3	.17	3

23.

Score	f	rel. f	cf
16	5	.33	15
15	1	.07	10
14	0	.00	9
13	2	.13	9
12	3	.20	7
11	4	.27	4

25. (a) Nominal: scores are names of categories.
    (b) Ordinal: scores indicate rank order, no zero, adjacent scores not equal distances.
    (c) Interval: scores indicate amounts, zero is not none, negative numbers allowed.
    (d) Ratio: scores indicate amounts, zero is none, negative numbers not allowed.

27. (a) Bar graph; this is a nominal (categorical) variable.
    (b) Polygon; we will have many different ratio scores.
    (c) Histogram; we will have only 8 different ratio scores.
    (d) Bar graph; this is an ordinal variable.

29. (a) Multiply .60(100) = 60% of the time.
    (b) We expect .60 of 50, which is .60(50) = 30.

Chapter 4

1. It indicates where on a variable most scores tend to be located.
3. The mode is the most frequently occurring score, used with nominal scores.
5. The mean is the average score, the mathematical center of a distribution, used with symmetrical distributions of interval or ratio scores.
7. Because here the mean is not near most of the scores.
9. Deviations convey (1) whether a score is above or below the mean and (2) how far the score is from the mean.
11. (a) $X - \overline{X}$
    (b) $\Sigma(X - \overline{X})$

(c) The total distance that scores are above the mean equals the total distance that other scores are below the mean.

13. (a) $\Sigma X = 638, N = 11, \overline{X} = 58$
    (b) The mode is 58.

15. (a) Mean
    (b) Median (these ratio scores are skewed)
    (c) Mode (this is a nominal variable)
    (d) Median (this is an ordinal variable)

17. (a) The person with −5; it is farthest below the mean.
    (b) The person with −5; it is in the tail where the lowest-frequency scores occur.
    (c) The person with 0; this score equals the mean, which is the highest-frequency score.
    (d) The person with +3; it is farthest above the mean.

19. Mean errors do not change until there has been 5 hours of sleep deprivation. Mean errors then increase as a function of increasing sleep deprivation.

21. She is correct *unless* the variable is something on which it is undesirable to have a high score. Then, being below the mean with a negative deviation is better.

23. (a) The means for conditions 1, 2, and 3 are 15, 12, and 9, respectively.
    (b)

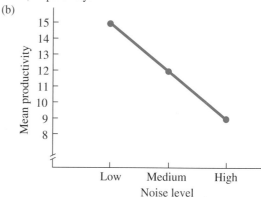

    (c)

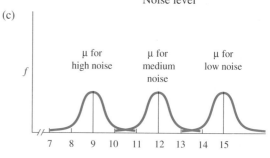

    (d) Apparently the relationship is that as noise level increases, the typical productivity score decreases from around 15 to around 12 to around 9.

25. (a) It is the variable that supposedly influences a behavior; it is manipulated by the researcher.
    (b) It reflects the behavior that is influenced by the independent variable; it measures participants' behavior.

27. (a) The independent variable.
   (b) The dependent variable.
   (c) Produce a line graph when the independent variable is an interval or ratio scale, a bar graph when the independent variable is nominal or ordinal.

29. (a) Line graph; income on Y axis, age on X axis; find median income per age group—income is skewed.
   (b) Bar graph; positive votes on Y axis, presence or absence of a wildlife refuge on X axis; find mean number of votes, if normally distributed.
   (c) Line graph; running speed on Y axis, amount of carbohydrates consumed on X axis; find mean running speed, if normally distributed.
   (d) Bar graph; alcohol abuse on Y axis, ethnic group on X axis; find mean rate of alcohol abuse per group, if normally distributed.

Chapter 5

1. It is needed for a complete description of the data, indicating how spread out scores are and how accurately the mean summarizes them.
3. (a) The range is the distance between the highest and lowest scores in a distribution.
   (b) Because it includes only the most extreme and often least-frequent scores, so it does not summarize most of the differences in a distribution.
   (c) With nominal or ordinal scores or with interval/ratio scores that cannot be accurately described by other measures.
5. (a) Variance is the average of the squared deviations around the mean.
   (b) Variance equals the squared standard deviation, and the standard deviation equals the square root of the variance.
7. Because a sample value too often tends to be smaller than the population value. The unbiased estimates of the population involve the quantity $N - 1$, resulting in a slightly larger estimate.
9. (a) The lower score $= \overline{X} - 1S_X$ and the upper score $= \overline{X} + 1S_X$.
   (b) The lower score $= \mu - 1\sigma_X$ and the upper score $= \mu + 1\sigma_X$.
   (c) Use $\overline{X}$ to estimate μ, then the lower score $= \mu - 1s_X$ and the upper score $= \mu + 1s_X$.
11. (a) Range $= 8 - 0 = 8$, so the scores spanned 8 different scores.
   (b) $\Sigma X = 41$, $\Sigma X^2 = 231$, $N = 10$, so $S_X^2 = (231 - 168.1)/10 = 6.29$: The average squared deviation of creativity scores from the mean of 4.10 is 6.29.
   (c) $S_X = \sqrt{6.29} = 2.51$: The "average deviation" of creativity scores from the mean of 4.10 is 2.51.
13. (a) With $\overline{X} = 4.1$ and $S_x = 2.51$, the scores are $4.1 - 2.51 = 1.59$, and $4.1 + 2.51 = 6.61$.

   (b) The portion of the normal curve between these scores is 68%, so $(.68)(1000) = 680$.
   (c) Below 1.59 is about 16% of a normal distribution, so $(.16)(1000) = 160$.
15. (a) Because the sample tends to be normally distributed, the population should be normal too.
   (b) Because $\overline{X} - 1297/17 = 76.29$, we would estimate the μ to be 76.29.
   (c) The estimated population variance is $(99{,}223 - 98{,}953.47)/16 = 16.85$.
   (d) The estimated standard deviation is $\sqrt{16.85} = 4.10$.
   (e) Between 72.19 (76.29 − 4.10) and 80.39 (76.29 + 4.10)
17. (a) Guchi. Because his standard deviation is larger, his scores are spread out around the mean, so he tends to be a more inconsistent student.
   (b) Pluto, because his scores are closer to the mean of 60, so it more accurately describes all of his scores.
   (c) Pluto, because we predict each will score at his mean score, and Pluto's individual scores tend to be closer to his mean than Guchi's are to his mean.
   (d) Guchi, because his scores vary more widely above and below 60.
19. (a) Compute the mean and sample standard deviation in each condition.
   (b) Changing conditions A, B, C changes dependent scores from around 11.00 to 32.75 to 48.00, respectively.
   (c) The S_X for the three conditions are .71, 1.09, and .71, respectively. These seem small, showing little spread, so participants scored consistently in each condition.
   (d) Yes.
21. (a) Study A has a relatively narrow/skinny distribution, and Study B has a wide/fat distribution.
   (b) In A, about 68% of scores will be between $35(40 - 5)$ and $45(40 + 5)$; in B, 68% will be between $30(40 - 10)$ and $50(40 + 10)$.
23. (a) For conditions 1, 2, and 3, we'd expect μs of about 13.33, 8.33, and 5.67, respectively.
   (b) Somewhat inconsistently, because based on s_X we'd expect a σ_X of 4.51, 2.52, and 3.06, respectively.
25. The shape of the distribution, a measure of central tendency and a measure of variability.
27. (a) The flat line graph indicates that all conditions produced close to the same mean, but a wide variety of different scores was found throughout the conditions.
   (b) The mean for men was 14 and their standard deviation was 3.
   (c) The researcher found $\overline{X} = 14$ and is using it to estimate μ, and is estimating σ_X by using the sample data to compute s_X.

29. (a) A bar graph; rain/no rain on X; mean laughter time on Y.
    (b) A bar graph; divorced/not divorced on X; mean weight change on Y.
    (c) A bar graph; alcoholic/not alcoholic on X; median income on Y.
    (d) A line graph; amount paid on X; mean number of ideas on Y.
    (e) A line graph; number of siblings on X; mean vocabulary size on Y.
    (f) A bar graph; type of school on X; median income rank on Y.

Chapter 6

1. (a) A z-score indicates the distance, measured in standard deviation units, that a score is above or below the mean.
   (b) z-scores can be used to interpret scores from any normal distribution of interval or ratio scores.
3. It is the distribution that results after transforming a distribution of raw scores into z-scores.
5. (a) It is our model of the perfect normal z-distribution.
   (b) It is used as a model of any normal distribution of raw scores after being transformed to z-scores.
   (c) The raw scores should be at least approximately normally distributed, they should be from a continuous interval or ratio variable, and the sample should be relatively large.
7. (a) That it is normally distributed, that its μ equals the μ of the raw score population, and that its standard deviation (the standard error of the mean) equals the raw score population's standard deviation divided by the square root of N.
   (b) Because it indicates the characteristics of any sampling distribution, without our having to actually measure all possible sample means.
9. (a) Convert the raw score to z, use z with the z-tables to find the proportion of the area under the appropriate part of the normal curve, and that proportion is the rel. f, or use it to determine percentile.
   (b) In column B or C of the z-tables, find the specified rel. f or the rel. f converted from the percentile, identify the corresponding z at the proportion, transform the z into its raw score, and that score is the cutoff score.
   (c) Compute the standard error of the mean, transform the sample mean into a z-score, follow the steps in part (a) above.
11. (a) Small. This will give him a large positive z-score, placing him at the top of his class.
    (b) Large. Then he will have a small negative z and be relatively close to the mean.
13. $\Sigma X = 103$, $\Sigma X^2 = 931$, and $N = 12$, so $S_X = 1.98$ and $\overline{X} = 8.58$.
    (a) For $X = 10$, $z = (10 - 8.58)/1.98 = +.72$.
    (b) For $X = 6$, $z = (6 - 8.58)/1.98 = -1.30$.

15. (a) $z = +1.0$ (b) $z = -2.8$
    (c) $z = -.70$ (d) $z = -2.0$
17. (a) .4706   (b) .0107
    (c) $.3944 + .4970 = .8914$
    (d) $.0250 + .0250 = .05$
19. From the z-table the 25th percentile is at approximately $z = -.67$. The cutoff score is then $X = (-.67)(10) + 75 = 68.3$.
21. To make the salaries comparable, compute z. For City A, $z = (47,000 - 65,000)/15,000 = -1.2$. For City B, $z = (70,000 - 85,000)/20,000 = -.75$. City B is the better offer, because her income will be closer to (less below) the average cost in that city.
23. (a) $z = (60 - 56)/8 = .50$, so .1915 of the curve is between 60 and 56, plus .50 of the curve below the mean gives a total of .6915 or 69.15% of the curve is expected to be below 60.
    (b) $z = (54 - 56)/8 = -.25$, so .0987 of the curve is between 54 and 56, plus .50 of the curve that is above the mean for a total of .5987 or 59.87% scoring above 54.
    (c) The approximate upper .20 of the curve (.2005) is above $z = +.84$ so the corresponding raw score is $X = (+.84)(8) + 56 = 62.72$.
25. (a) This is a rather infrequent mean.
    (b) The sampling distribution of means.
    (c) All other means that might occur in this situation.
27. (a) With normally distributed, interval, or ratio scores.
    (b) With normally distributed, interval, or ratio scores.
29. (a) $.40 (500) = 200$
    (b) We are asking for its frequency.
    (c) $35/50 = .70$
    (d) $.70(100) = 70\%$

Chapter 7

1. (a) In experiments the researcher manipulates one variable and measures participants on another variable; in correlational studies the researcher measures participants on two variables.
   (b) In experiments the researcher computes the mean of the dependent (Y) scores for each condition of the independent variable (each X); in correlational studies the researcher examines the relationship over all $X-Y$ pairs by computing a correlation coefficient.
3. You don't necessarily know which variable occurred first, nor have you controlled other variables that might cause scores to change.
5. (a) A scatterplot is a graph of the individual data points formed from a set of $X-Y$ pairs.
   (b) It is a data point that lies outside of the general pattern in the scatterplot, because of an extreme X or Y score.
   (c) A regression line is the summary straight line drawn through the center of the scatterplot.

7. (a) As the X scores increase, the Y scores tend to increase.
   (b) As the X scores increase, the Y scores tend to decrease.
   (c) As the X scores increase, the Y scores do not only increase or only decrease.

9. (a) Particular Y scores are not consistently paired with particular X scores.
   (b) The variability in Y at each X is equal to the overall variability in all Y scores in the data.
   (c) The Y scores are not close to the regression line.
   (d) Knowing X does not improve accuracy in predicting Y.

11. (a) That the X variable forms a relatively strong relationship with Y (r is relatively large), so by knowing someone's X we come close to knowing his or her Y.
    (b) That the X variable forms a relatively weak relationship with Y (r is relatively small), so by knowing someone's X we have only a general idea if he or she has a low or high Y score.

13. He is drawing the causal inference that more people cause fewer bears, but it may be the number of hunters, or the amount of pesticides used, or the noise level associated with more people, etc.

15. (a) With $r = -.40$, the scatterplot is skinnier.
    (b) With $r = -.40$, there is less variability in Y at each X.
    (c) With $r = -.40$, the Y scores hug the regression line more closely.
    (d) No. He thought a positive r was better than a negative r. Consider the absolute value.

17. Disagree. Exceptionally smart people will produce a *restricted range* of IQ scores and grade averages. With an unrestricted range, the r would be larger.

19. Compute r. $\Sigma X = 38$, $\Sigma X^2 = 212$, $(\Sigma X)^2 = 1444$, $\Sigma Y = 68$, $\Sigma Y^2 = 552$, $(\Sigma Y)^2 = 4624$, $\Sigma Y = 317$, and $N = 9$. $r = (2853 - 2584)/\sqrt{(464)(344)} = +.67$. This is a strong positive linear relationship, so a nurse's burnout score will allow reasonably accurate prediction of her absenteeism.

21. Compute r_S: $\Sigma D^2 = 312$; $r = 1 - (1872/990) = -.89$. This is a very strong negative relationship, so that the most dominant consistently weigh the most, and the less dominant weigh less.

23. (a) Neither variable.
    (b) Impossible to determine: being creative may cause intelligence, or being intelligent may cause creativity, or maybe something else (e.g., genes) causes both.
    (c) Also impossible to determine.
    (d) We do not call either variable the independent variable in a correlational coefficient.

25. (a) Correlational, Pearson r.
    (b) Experiment, with age as a (quasi-) independent variable.

(c) Correlational, Pearson r.
(d) Correlational, Spearman r_S after creating ranks for number of visitors to correlate with attractiveness rankings.
(e) Experiment.

Chapter 8

1. It is the line that summarizes a scatterplot by, on average, passing through the center of the Y scores at each X.

3. Y' is the predicted Y score for a given X, computed from the regression equation.

5. (a) The Y intercept is the value of Y when the regression line crosses the Y axis.
   (b) The slope indicates the direction and degree that the regression line is slanted.

7. (a) The standard error of the estimate.
   (b) It is a standard deviation, indicating the "average" amount that the Y scores deviate from their corresponding values of Y'.
   (c) It indicates the "average" amount that the actual scores differ from the predicted Y' scores, so it is the "average" error.

9. $S_{Y'}$ is inversely related to the absolute value of r. Because a smaller $S_{Y'}$ indicates the Y scores are closer to the regression line (and Y') at each X, which is what happens with a stronger relationship (a larger r).

11. (a) r^2 is the coefficient of determination, or the proportion of variance in Y that is accounted for by the relationship with X.
    (b) r^2 indicates the proportional improvement in accuracy when using the relationship with X to predict Y scores, compared to using the overall mean of Y to predict Y scores.

13. (a) Foofy; because r is positive.
    (b) Linear regression.
    (c) "Average" error is the $S_{Y'} = 3.90$.
    (d) $r^2 = (+.44)^2 = .19$, so our error is 19% smaller.
    (e) The proportion of variance accounted for (the coefficient of determination).

15. The researcher measured more than one X variable and then correlated them with one Y variable, and used the Xs to predict Y.

17. (a) He should use multiple correlation and multiple regression, simultaneously considering the hours people study and their test speed when he predicts error scores.
    (b) With a multiple R of $+.67$, $R^2 = .45$: He will be 45% more accurate at predicting error scores when considering hours studied and test speed than if these predictors are not considered.

19. (a) Compute r: $\Sigma X = 45$, $\Sigma X^2 = 259$, $(\Sigma X)^2 = 2025$, $\Sigma Y = 89$, $\Sigma Y^2 = 887$, $(\Sigma Y)^2 = 7921$, $\Sigma XY = 460$, and $N = 10$, so $r = (4600 - 4005)/\sqrt{(565)(949)} = +.81$.

(b) $b = (4600 - 4005)/565 = +1.05$ and $a = 8.9 - (1.05)(4.5) = 4.18$, so $Y' = (+1.05)X + 4.18$.

(c) Using the regression equation, for people with an attraction score of 9, the predicted anxiety score is $Y'(+1.05)9 + 4.18 = 13.63$.

(d) Compute $S_{Y'}$: $S_Y = 3.081$, so $S_{Y'} = (3.081)\sqrt{1 - .81^2} = 1.81$. The "average error" is 1.81 when using Y' to predict anxiety scores.

21. (a) $b = (2853 - 2584)/(1908 - 1444) = .58$; $a = 7.556 - (.58)(4.222) = 5.11$, so $Y' = .58X + 5.11$

(b) $S_{Y'} = 2.061\sqrt{1 - .67^2} = 1.53$

(c) $Y' = (.58)4 + 5.11 = 7.43$

(d) $r^2 = (.67)^2 = .45$; very useful, providing a 45% improvement in the accuracy of predictions

23. (a) The size of the correlation coefficient indirectly indicates this, but the standard error of the estimate most directly communicates how much better (or worse) than predicted she's likely to perform.

(b) Not very useful: Squaring a small correlation coefficient produces a small proportion of variance accounted for.

25. (a) Yes, by each condition.

(b) The $\overline{X}$ and S_X per condition.

27. (a) The independent (X) variable.

(b) The dependent (Y) variable.

(c) The mean dependent score of their condition.

(d) The variance (or standard deviation) of that condition.

29. (a) It is relatively strong.

(b) There is a high degree of consistency.

(c) The variability is small.

(d) The data points are close to the line,

(e) r is relatively large.

(f) We can predict Y scores reasonably accurately.

Chapter 9

1. (a) It is our expectation or confidence in the event.

(b) The relative frequency of the event in the population.

3. (a) Sampling with replacement is replacing the individuals or events from a sample back into the population before another sample is selected.

(b) Sampling without replacement is not replacing the individuals or events from a sample before another is selected.

(c) Over successive samples, sampling without replacement increases the probability of an event, because there are fewer events that can occur; with replacement, each probability remains constant.

5. It indicates that by chance, we've selected too many high or low scores so that our sample is unrepresentative. Then the mean does not equal the μ that it represents.

7. It indicates whether or not the sample's z-score (and the sample $\overline{X}$) lies in the region of rejection.

9. The p of a hurricane is $160/2005 = .80$. The uncle is looking at an unrepresentative sample over the past 13 years. Poindexter uses the gambler's fallacy, failing to realize that p is based on the long run, and so in the next few years there may not be a hurricane.

11. She had sampling error, obtaining an unrepresentative sample that contained a majority of Poindexter supporters, but the majority in the population were Dorcas supporters.

13. (a) $+1.96$.

(b) Yes: $\sigma_{\overline{X}} = 1.28$, so $z = (72.1 - 75)/1.28 = -2.27$,

(c) No, we should reject that the sample represents this population.

(d) Because such a sample and mean is unlikely to occur if we are representing this population.

15. (a) $+1.645$

(b) Yes: $\sigma_{\overline{X}} = 2.19$, so $z = (36.8 - 33)/2.19 = 1.74$.

(c) Such a sample and mean is unlikely to occur in this population.

(d) Reject that the sample represents this population.

17. $\sigma_{\overline{X}} = 1.521$, so $z = (34 - 28)/1.521 = +3.945$. No, this z-score is beyond the critical value of ± 1.96, so this sample is unlikely to be representing this population.

19. No, the gender of one baby is independent of another, and a boy now is no more likely than at any other time, with $p = .5$.

21. (a) For Fred's sample, $\mu = 26$, and for Ethel's, $\mu = 18$.

(b) The population with $\mu = 26$ is most likely to produce a sample with $\overline{X} = 26$, and the population with $\mu = 18$ is most likely to produce a sample with $\overline{X} = 18$.

23. (a) The $\overline{X} = 321/9 = 35.67$; $\sigma_{\overline{X}} = 5/\sqrt{9}, = 1.67$. Then $z = (35.67 - 30)/1.67 = +3.40$. With a critical value of ± 1.96, conclude that the football players do not represent this population.

(b) Football players, as represented by your sample, form a population different from non-football players, having a μ of about 35.67.

25. (a) $z = (27 - 43)/8 = -2.0$, $rel. f = .0228$

(b) $z = (51 - 43)/8 = +1.0$, $rel. f = .1587$

(c) $z = (42 - 43)/8 = -.13$, $z = (44 - 43)/8 = +.13$, $rel. f = .0517 + .0517 = .1034$

(d) $z = (33 - 43)/8 = -1.25$, $z = (49 - 43)/8 = +.75$, $rel. f = .1056 + .2266 = .3322$

(e) For each the probability equals the above $rel. f$.

27. (a) $\sigma_{\overline{X}} = 18/\sqrt{40} = 2.846$; $z = (46 - 50)/2.8 = -1.41$; $rel. f = .0793$.

(b) $p = .0793$

Chapter 10

1. A sample may (1) poorly represent one population because of sampling error, or (2) represent some other population.

3. α stands for the criterion probability; it determines the size of the region of rejection and the theoretical probability of a Type I error.

5. They describe the predicted relationship that may or may not be demonstrated in an experiment.

7. (a) Use a one-tailed test when predicting the direction the scores will change.
   (b) Use a two-tailed test when predicting a relationship but not the direction that scores will change.

9. (a) Power is the probability of not making a Type II error.
   (b) So we can detect relationships when they exist and thus learn something about nature.
   (c) When results are not significant, we worry if we missed a real relationship.
   (d) In a one-tailed test the critical value is smaller than in a two-tailed test; so the obtained value is more likely to be significant.

11. (a) We will demonstrate that changing the independent variable from a week other than finals week to finals week increases the dependent variable of amount of pizza consumed; we will not demonstrate an increase.
    (b) We will demonstrate that changing the independent variable from not performing breathing exercises to performing them changes the dependent variable of blood pressure; we will not demonstrate a change.
    (c) We will demonstrate that changing the independent variable by increasing hormone levels changes the dependent variable of pain sensitivity; we will not demonstrate a change.
    (d) We will demonstrate that changing the independent variable by increasing amount of light will decrease the dependent variable of frequency of dreams; we will not demonstrate a decrease.

13. (a) A two-tailed test because we do not predict the direction that scores will change.
    (b) H_0: $\mu = 50$, H_a: $\mu \neq 50$
    (c) $\sigma_{\bar{X}} = 12/\sqrt{49} = 1.71$;
        $z_{obt} = (54.63 - 50)/1.71 = +2.71$
    (d) $z_{crit} = \pm 1.96$
    (e) Yes, because z_{obt} is beyond z_{crit}, the results are significant: Changing from the condition of no music to the condition of music results in test scores changing from a μ of 50 to a μ of around 54.63.

15. (a) The probability of a Type I error is $p < .05$. The error would be concluding that music influences scores when really it does not.

(b) By rejecting H_0, there is no chance of making a Type II error. It would be concluding that music does not influence scores when really it does.

17. She is incorrect about Type I errors, because the total size of the region of rejection (which is α) is the same regardless of whether a one- or two-tailed test is used; α is also the probability of making a Type I error, so it is equally likely using either type of test.

19. (a) She is correct; that is what $p < .0001$ indicates.
    (b) She is incorrect. In both studies the researchers decided the results were unlikely to reflect sampling error from the H_0 population; they merely defined unlikely differently.
    (c) The probability of a Type I error is less in Study B.

21. (a) The researcher decided that the difference between the scores for brand X and other brands is too large to have resulted by chance if there wasn't a real difference between them.
    (b) The $p < .44$ indicates an α of .44, so the probability is .44 that the researcher made a Type I error. This p is far too large for us to accept the conclusion.

23. (a) One-tailed.
    (b) H_0: The μ of all art majors is less than or equal to that of engineers who are at 45; H_a: the μ of all art majors is greater than that of engineers who are at 45.
    (c) $\sigma_{\bar{X}} = 14/\sqrt{25} = 2.8$; $z_{obt} = (49 - 45)/2.8 = +1.428$. This is not larger than $z_{crit} = +1.645$; the results are not significant.
    (d) We have no evidence regarding whether, nationwide, the visual memory ability of the two groups differs or not.
    (e) $z = +1.43, p > .05$.

25. (a) The independent variable is manipulated by the researcher; the dependent variable measures participants' scores.
    (b) Dependent variable.
    (c) Interval and ratio scores measure actual amounts, but an interval variable allows negative scores; nominal variables are categorical variables; ordinal scores are the equivalent of 1st, 2nd, etc.
    (d) A normal distribution is bell shaped and symmetrical with two tails; a skewed distribution is asymmetrical with only one pronounced tail.

27. This will produce two normal distributions on the same axis, with the morning's centered around a μ of 40 and the evening's centered around a μ of 60.

29. (a) Because they study the laws of nature and a relationship is the telltale sign of a law at work.
    (b) A real relationship occurs in nature (in the "population"). One produced by sampling error occurs by chance, and only in the sample data.
    (c) Nothing.

Chapter 11

1. (a) The t-test and the z-test.
   (b) Compute z if the standard deviation of the raw score population (σ_X) is known; compute t if σ_X is estimated by s_X.
   (c) That we have one random sample of interval or ratio dependent scores, and the scores are approximately normally distributed.

3. (a) $s_{\bar{X}}$ is the estimated standard error of the mean; $\sigma_{\bar{X}}$ is the true standard error of the mean.
   (b) Both are used as a "standard deviation" to locate a sample mean on the sampling distribution of means.

5. (a) ρ
   (b) ρ_s
   (c) Determine if the coefficient is significant by comparing it to the appropriate critical value; if the coefficient is significant, compute the regression equation and graph it, compute the proportion of variance accounted for, and interpret the relationship.

7. To describe the relationship and interpret it psychologically, sociologically, etc.

9. (a) Power is the probability of rejecting H_0 when it is false (not making a Type II error).
   (b) When a result is not significant.
   (c) Because then we may have made a Type II error.
   (d) When designing a study.

11. (a) $H_0: \mu = 68.5; H_a: \mu \neq 68.5$
    (b) $s_X^2 = 130.5; s_{\bar{X}} = \sqrt{130.5/10} = 3.61$;
       $t_{obt} = (78.5 - 68.5)/3.61 = +2.77$
    (c) With $df = 9$, $t_{crit} = \pm 2.262$.
    (d) Using this book rather than other books produces a significant improvement in exam scores: $t_{obt}(9) = 2.77, p < .05$.
    (e) $(3.61)(-2.262) + 78.5 \leq \mu \leq (3.61)(+2.262) + 78.5 = 70.33 \leq \mu \leq 86.67$

13. (a) $H_0: \mu = 50$ $H_a: \mu \neq 50$
    (b) $t_{obt} = (53.25 - 50)/8.44 = +.39$
    (c) For $df = 7$, $t_{crit} = \pm 2.365$
    (d) $t(7) = +.39, p > .05$, so the results are not significant, so do not compute the confidence interval.
    (e) She has no evidence that the arguments change people's attitudes toward this issue.

15. Disagree. Everything Poindexter said was meaningless, because he failed to first perform significance testing to eliminate the possibility that his correlation was merely a fluke resulting from sampling error.

17. (a) $H_0: \rho = 0; H_a: \rho \neq 0$
    (b) With $df = 70$, $r_{crit} = \pm .232$.
    (c) $r(70) = +.38, p < .05$ (and even, $< .01$)
    (d) The correlation is significant, so he should conclude that the relationship exists in the population, and he should estimate that ρ is approximately $+.38$.
    (e) The regression equation and r^2 should be computed.

19. $H_0: \mu = 12; H_a: \mu \neq 12; \bar{X} = 8.667, s_X^2 = 4.67; s_{\bar{X}} = \sqrt{4.67/6} = .882; t_{obt} = (8.667 - 12)/.882 = -3.78; df = 5, t_{crit} = \pm 2.571$. The results are significant, so there is evidence of a relationship: Without uniforms, $\mu = 12$; with uniforms, μ is around 8.67. The 95% confidence interval is $(.882)(-2.571) + 8.67 \leq \mu \leq (.882)(2.571) + 8.67$, so μ with uniforms is between 6.40 and 10.94.

21. (a) $t(42) = +6.72, p < .05$
    (b) $t(5) = -1.72, p > .05$

23. The df of 80 is .33 of the distance between the df at 60 and 120, so the target t_{crit} is .33 of the distance from 2.000 to 1.980: $2.000 - 1.980 = .020$, so $(.020)(.33) = .0066$, and thus $2.000 - .0066$ equals the target t_{crit} of 1.993.

25. (a) We must first believe that it is a real relationship, which is what significant indicates.
    (b) Significant means only that we believe that the relationship exists in the population; a significant relationship is unimportant if it accounts for little of the variance.

27. The z- and t-tests are used to compare the μ of a population when exposed to one condition of the independent variable, to the scores in a sample that is exposed to a different condition. A correlation coefficient is used when we have measured the X and Y scores of a sample.

29. (a) They are equally significant.
    (b) Study B's is larger.
    (c) Study B's is smaller, further into the tail of the sampling distribution.
    (d) The probability of a Type I error is .031 in A, but less than .001 in B.

31. (a) Compute the mean and standard deviation for each set of scores, compute the Pearson r between the scores and test if it is significant (two-tailed). If so, perform linear regression. The size of r^2 (and r) is the key.
    (b) We have the mean and (estimated) standard deviation, so perform the two-tailed, one-sample t-test, comparing the $\bar{X} = 97.4$ to $\mu = 106$. Whether t_{obt} is significant is the key. Then compute the confidence interval for the μ represented by the sample.
    (c) Along with the mean, compute the standard deviation for this year's team. Perform the one-tailed z-test, comparing $\bar{X} = 77.6$ to $\mu = 71.1$. Whether z_{obt} is significant is the key.

Chapter 12

1. (a) The independent-samples t-test and the related-samples t-test.
   (b) Whether the scientist created independent samples or related samples.

3. (a) Each score in one sample is paired with a score in the other sample by matching pairs of participants or by repeatedly measuring the same participants in all conditions.

(b) The scores are interval or ratio scores; the populations are normal and have homogeneous variance.

5. (a) $s_{\bar{X}_1 - \bar{X}_2}$ is the standard error of the difference—the "standard deviation" of the sampling distribution of differences between means from independent samples.

   (b) $s_{\bar{D}}$ is the standard error of the mean difference, the "standard deviation" of the sampling distribution of $\bar{D}$ from related samples.

   (c) n is the number of scores in each condition; N is the number of scores in the experiment.

7. It indicates a range of values of μ_D, one of which $\bar{D}$ is likely to represent.

9. (a) It indicates the size of the influence that the independent variable had on dependent scores.

   (b) It measures effect size using the magnitude of the difference between the means of the conditions.

   (c) It indicates the proportion of variance in the dependent scores accounted for by changing the conditions of the independent variable in the experiment, indicating how consistently close to the mean of each condition the scores are.

11. She should graph the results, compute the appropriate confidence interval, and compute the effect size.

13. (a) H_0: $\mu_1 - \mu_2 = 0$; H_a: $\mu_1 - \mu_2 \neq 0$

    (b) $s_{pool}^2 = 23.695$; $s_{\bar{X}_1 - \bar{X}_2} = 1.78$;
    $t_{obt} = (43 - 39)/1.78 = +2.25$

    (c) With $df = (15 - 1) + (15 - 1) = 28$, $t_{crit} = \pm 2.048$.

    (d) The results are significant: In the population, hot baths (with μ about 43) produce different relaxation scores than cold baths (with μ about 39).

    (e) $(1.78)(-2.0480) + 4 \le \mu_1 - \mu_2 \le (1.78)$
    $(+2.048) + 4 = .35 \le \mu_1 - \mu_2 \le 7.65$

    (f) $d = (43 - 39)/\sqrt{23.695} = .82$;
    $r_{pb}^2 = (2.25)^2/[(2.25)^2 + 28] = .15$. This is a moderate to large effect.

    (g) Label the X axis as bath temperature; label the Y axis as mean relaxation score; plot the data point for cold baths at a Y of 39 and for hot baths at a Y of 43; connect the data points with a straight line.

15. (a) H_0: $\mu_D = 0$; H_a: $\mu_D \neq 0$

    (b) $t_{obt} = (2.63 - 0)/.75 = +3.51$

    (c) With $df = 7$, $t_{crit} = \pm 2.365$, so people exposed to much sunshine exhibit a significantly higher well-being score than when exposed to less sunshine.

    (d) $.86 \le \mu_D \le 4.40$

    (e) The $\bar{X}$ of 15.5, the $\bar{X}$ of 18.13.

    (f) $r_{pb}^2 = (3.51)^2/[(3.51)^2 + 7] = .64$; they are on average about 64% more accurate.

    (g) By accounting for 64% of the variance, these results are very important.

17. (a) H_0: $\mu_D \le 0$; H_a: $\mu_D > 0$

    (b) $\bar{D} = 1.2$, $s_D^2 = 1.289$, $s_{\bar{D}} = .359$;
    $t_{obt} = (1.2 - 0)/.359 = +3.34$

    (c) With $df = 9$, $t_{crit} = +1.833$

(d) The results are significant. In the population, children exhibit more aggressive acts after watching the show (with μ about 3.9), than they do before the show (with μ about 2.7).

(e) $(.359)(-2.262) + 1.2 \le \mu_D \le (.359)$
$(+2.262) + 1.2 = .39 \le \mu_D \le 2.01$

(f) $d = 1.2/\sqrt{1.289} = 1.14$; a relatively large difference.

19. You cannot test the same people first when they're males and then again when they're females.

21. (a) Two-tailed.

    (b) H_0: $\mu_1 - \mu_2 = 0$, H_a: $\mu_1 - \mu_2 \neq 0$

    (c) $\bar{X}_1 = 11.5$, $s_1^2 = 4.72$; $\bar{X}_2 = 14.1$, $s_2^2 = 5.86$,
    $s_{pool}^2 = 5.29$, $s_{\bar{X}_1 - \bar{X}_2} = 1.03$, $t_{obt} = (11.5 - 14.1)/1.03$
    $= -2.52$ With $df = 18$, $t_{crit} = \pm 2.101$, so t_{obt} is significant.

    (d) $(1.03)(-2.101) + -2.6 \le \mu_1 - \mu_2 \le (1.03)$
    $(+2.101) + -2.6 = -4.76 \le \mu_1 - \mu_2 \le -.44$

    (e) Police who've taken this course are more successful at solving disputes than police who have not taken it. The μ for the police with the course is around 14.1, and the μ for police without the course is around 11.5. The absolute difference between these μs will be between 4.76 and .44.

    (f) $d = 1.13$; $r_{ph}^2 = .26$; taking the course is important.

23. The independent variable is manipulated by the researcher to create the conditions in which participants are tested; presumably this will cause a behavior to change. The dependent variable measures the behavior of participants that is to be changed.

25. (a) When the dependent variable is measured using normally distributed, interval, or ratio scores, and has homogeneity of variance.

    (b) z-test, performed in a one-sample experiment when μ and σ_X under one condition are known; one sample t-test, performed in a one-sample experiment when μ under one condition is known but σ_X must be estimated using s_X; independent samples t-test, performed when independent samples of participants are tested under two conditions; related samples t-test, performed when participants are tested under two conditions of the independent variable, with either matching pairs of participants or with repeated measures of the same participants.

27. (a) To predict Y scores by knowing someone's X score.

    (b) We predict the mean of a condition for participants in that condition.

    (c) The independent variable.

    (d) When the predicted mean score is close to most participants' actual score.

29. A confidence interval *always* uses the two-tailed t_{crit}.

Chapter 13

1. (a) Analysis of variance.
   (b) A study that contains one independent variable.
   (c) An independent variable.
   (d) A condition of the independent variable.
   (e) Another name for a level.
   (f) All samples are independent.
   (g) All samples are related, either through a repeated-measures or matched-samples design.
3. Because the hypotheses require more than two levels, or because it's easy to obtain additional information.
5. (a) It is the probability of making a Type I error after comparing all possible pairs of means in an experiment.
   (b) Multiple *t*-tests result in an experiment-wise error rate larger than alpha, but performing ANOVA and then post hoc tests keeps the experiment-wise error rate equal to alpha.
7. (a) When F_{obt} is significant and k is greater than 2. The F_{obt} indicates only that two or more sample means differ significantly; post hoc tests determine which levels differ significantly.
   (b) When F_{obt} is not significant or when $k = 2$.
9. The effect size, as the proportion of variance in dependent scores accounted for by changing the levels of the independent variable.
11. The researcher measured participants on more than one dependent variable in each condition of the independent variable.
13. (a) $H_0: \mu_1 = \mu_2 = \mu_3 = \mu_4$
    (b) H_a: not all μs are equal.
    (c) H_0 is that a relationship is not represented; H_a is that one is.
15. (a) The MS_{bn} is less than the MS_{wn}; either term is a poor estimate of σ^2_{error} and H_0 is assumed to be true.
    (b) He made a computational error—F_{obt} cannot be a negative number.
17. (a) This is a pretest–posttest design, comparing the weights of one group before and after they dieted.
    (b) A within-subjects design.
19. (a)

Source	Sum of Squares	df	Mean Square	F
Between	134.800	3	44.933	17.117
Within	42.000	16	2.625	
Total	176.800	19		

(b) With $df = 3$ and 16, $F_{crit} = 3.24$, so F_{obt} is significant, $p < .05$.
(c) For $k = 4$ and $df_{wn} = 16$, $q_k = 4.05$, so $HSD = (4.05)(\sqrt{2.625/5}) = 2.93$: $\bar{X}_4 = 4.4$, $\bar{X}_6 = 10.8$, $\bar{X}_8 = 9.40$, $\bar{X}_{10} = 5.8$. Only ages 4 and 10, and ages 6 and 8 do not differ significantly.

(d) Because $\eta^2 = 134.8/176.8 = .76$, this relationship accounts for 76% of the variance, so it's a very important relationship.
(e) Label the X axis as the factor of age and the Y axis as the mean creativity score. Plot the mean score for each condition, and connect adjacent data points with straight lines.
21. (a) A Type I error: rejecting a true H_0, thus claiming the factor influences behavior when really it does not.
    (b) It produces scientific "facts" that are wrong.
23. (a) By testing the same participants under all conditions or by testing different participants who are matched on a variable.
    (b) The related-samples *t*-test and the within-subjects ANOVA.
    (c) By selecting different and unmatched participants for each condition.
    (d) The independent-samples *t*-test and the between-subjects ANOVA.
25. (a) Both measure effect size, indicating the proportion of variance accounted for.
    (b) The improvement in predicting scores when using our level means.
27. (a) It is manipulated by the researcher to create the conditions; it is presumably the cause of a behavior.
    (b) They are the different conditions created by the researcher under which participants are tested.
    (c) It measures the behavior of participants that is "caused" by the factor.
29. (a) Levels: statistics grades; dependent variable (DV): math phobia scores.
    (b) Levels: recall vs. recognition. DV: memory score.
    (c) Levels: 1, 3, 5, or 7 weeks. DV: aggressiveness.
    (d) Predictor: First exam grade. Criterion: final exam grade.
    (e) Levels: male vs. female. DV: number who voted.
    (f) Levels: pilot vs. navigator. DV: reaction time.

Chapter 14

1. To examine the interaction between two independent variables, or for the efficiency of simultaneously studying multiple factors.
3. A main effect mean is based on scores in a level of one factor while collapsing across the other factor. A cell mean is the mean of scores from a particular combination of a level of factor A with a level of factor B.
5. (a) A confounded comparison involves two cells that differ along more than one factor. It occurs with cells that are diagonally positioned in a study's diagram.
   (b) An unconfounded comparison involves two cells that differ along only one factor. It occurs with means within the same column or within the same row of a diagram.

(c) Because we cannot determine which factor produced the difference.

7. (a) H_0 is that the μs represented by the level means from factor A are all equal; H_a is that not all μs are equal.

(b) H_0 is that the μs represented by the level means from factor B are all equal; H_a is that they are not all equal.

(c) H_0 is that the μs represented by the cell means do not form an interaction; H_a is that they do form an interaction.

9. (a) 4×3;　(b) 30;　(c) 40;　(d) 10

11. *Study 1:* For A, means are 7 and 9; for B, means are 3 and 13. Apparently there are effects for A and B but not for $A \times B$.

　　Study 2: For A, means are 7.5 and 7.5; for B, means are 7.5 and 7.5. There is no effect for A or B but there is an effect for $A \times B$.

　　Study 3: For A, means are 8 and 8; for B, means are 11 and 5. There is no effect for A, but there are effects for B and $A \times B$.

13. Perform Tukey's post hoc comparisons on each main effect and the interaction, graph each main effect and interaction and compute its η^2; where appropriate, compute confidence intervals for the μ represented by a cell or level mean.

15. Only the main effect for difficulty level is significant.

17. (a) For low reward $\overline{X} = 8$; for medium $\overline{X} = 10$; and for high $\overline{X} = 12$. It appears that as reward increases, performance increases.

(b) For low practice $\overline{X} = 7$; for medium $\overline{X} = 8$; for high $\overline{X} = 15$. It appears that increasing practice increases performance.

(c) Yes: How the scores change with increasing reward depends on the level of practice, and vice versa.

(d) By comparing the three means within each column and the three means within each row.

19. (a)

Source	Sum of Squares	df	Mean Square	F
Factor A	7.20	1	7.20	1.19
Factor B	115.20	1	115.20	19.04
Interaction	105.80	1	105.80	17.49
Within groups	96.80	16	6.05	
Total	325.00			

For each factor, $df = 1$ and 16, so $F_{crit} = 4.49$: factor B and the interaction are significant, $p < .05$.

(b) For factor A, $\overline{X}_1 = 8.9$, $\overline{X}_2 = 10.1$; for factor B, $\overline{X}_1 = 11.9$, $\overline{X}_2 = 7.1$; for the interaction, $\overline{X}_{A_1B_1} = 9.0$, $\overline{X}_{A_1B_2} = 8.8$, $\overline{X}_{A_2B_1} = 14.8$, $\overline{X}_{A_2B_2} = 5.4$.

(c) Because factor A is not significant and factor B contains only two levels, such tests are unnecessary.

For A $\times$ B, *adjusted k* = 3, so $q_k = 3.65$, $HSD = (3.65)(\sqrt{6.05/5} = 4.02$; the only significant differences are between males and females tested by a male, and between females tested by a male and females tested by a female.

(d) Conclude that a relationship exists between gender and test scores when testing is done by a male, and that male versus female experimenters produce a relationship when testing females, $p < .05$.

(e) For B, $\eta^2 = 115.2/325 = .35$; for A $\times$ B, $\eta^2 = 105.8/325 = .33$.

21.

Source	Sum of Squares	df	Mean Square	F
A: Fantasy	42.467	2	21.233	13.134
B: Meditation	.833	1	.833	.515
A $\times$ B: Interaction	141.267	2	70.633	43.691
Within	38.800	24	1.617	
Total	223.367	29		

The main effect of fantasy and the interaction are significant. For the main effect of fantasy, $\overline{X}_{A1} = 6.9$, $\overline{X}_{A2} = 4.8$, $\overline{X}_{A3} = 7.6$; $HSD = 3.53(\sqrt{1.617/10}) = 1.419$; only low versus high do not differ significantly ($\eta^2 = 42.467/223.367 = 19$). For the interaction, with meditation the cell means for the low-, medium-, and high-fantasy levels are 4, 6.4, and 9.4; with no meditation the means are 9.8, 3.2, and 5.8; $HSD = 4.17(\sqrt{1.617/5}) = 2.37$; all unconfounded comparisons are significant ($\eta^2 = 14.267/223.367 = .63$.) Conclude that: with meditation each increase in fantasy increases relaxation; without meditation, low fantasy produces the highest relaxation, which drops drastically with medium fantasy and then increases slightly with high fantasy.

23. That the relationship in the sample data was not a chance pattern; that it reflects a relationship in the population (in nature).

25. Are there related samples (because of matching or repeated measures) or are there independent samples?

27. (a) The researcher selects one group of participants and measures their scores on X and Y.

(b) Are both normally distributed, interval/ratio variables?

(c) Pearson r.

(d) Spearman r_S.

29. (a) The researcher manipulates two variables, for each creating at least two or more groups that differ on a variable or are tested in different situations.

(b) Is it a normally distributed, interval or ratio variable with homogeneous variance.

(c) The between-subjects, within-subjects or mixed design ANOVA.

Chapter 15

1. Both types of procedures test whether, due to sampling error, the data poorly represent the absence of the predicted relationship in the population.

3. (a) Nominal or ordinal scores.
   (b) They form nonnormal distributions, or they do not have homogeneous variance, so they are transformed to ranks.
   (c) Transform them to ranks.

5. (a) When categorizing participants on one variable.
   (b) When categorizing participants on two variables.

7. (a) That the sample frequencies are unlikely to represent the frequencies in the population described by H_0.
   (b) That category membership on one variable depends on or is correlated with category membership on the other variable.

9. (a) It is the correlation coefficient between the two variables in a significant 2×2 chi square design.
   (b) ϕ^2 indicates the improvement in predicting participants' category membership on one variable by knowing their category membership on the other variable.
   (c) C is the correlation coefficient between the two variables in a significant two-way chi square that is not a 2×2 design.
   (d) C^2 indicates the improvement in predicting participants' category membership on one variable by knowing their category membership on the other variable.

11. (a) The one-way χ^2.
    (b) H_0: The elderly population is 30% Republican, 55% Democrat, and 15% other; H_a: Affiliations in the elderly population are not distributed this way.
    (c) For Republicans, $f_e = (.30)(100) = 30$; for Democrats, $f_e = (.55)(100) = 55$; and for others, $f_e = (.15)(100) = 15$.
    (d) $\chi^2_{obt} = 4.80 + 1.47 + .60 = 6.87$
    (e) For $df = 2$, $\chi^2_{crit} = 5.99$, so the results are significant: Party membership in the population of senior citizens is different from party membership in the general population, and it is distributed as in our samples, $p < .05$.

13. (a) The frequency with which students *dislike* each professor also must be included.
    (b) She can perform a separate one-way χ^2 on the data for each professor to test for a difference between the frequency for "like" and "dislike," or she can perform a two-way χ^2 to determine if whether students like or dislike one professor is correlated with whether they like or dislike the other professor.

15. (a) The two-way χ^2.
    (b) H_0: Gender and political party affiliation are independent in the population; H_a: Gender and political party affiliation are dependent in the population.
    (c) For males, Republican $f_e = (75)(57)/155 = 27.58$, Democrat $f_e = (75)(66)/155 = 31.94$, and other $f_e = (75)(32)/155 = 15.48$. For females, Republican $f_e = (80)(57)/155 = 29.42$, Democrat $f_e = (80)(66)/155 = 34.06$ and other $f_e = (80)(32)/155 = 16.52$.
    (d) $\chi^2_{obt} = 3.33 + 3.83 + .14 + 3.12 + 3.59 + .13 = 14.14$
    (e) With $df = 2$, $\chi^2_{crit} = 5.99$, so the results are significant: In the population, frequency of political party affiliation depends on gender, $p < .05$.
    (f) $C = \sqrt{14.14/(155 + 14.14\,)} = .29$, indicating a somewhat consistent relationship.

17. $\chi^2 (2, N = 40) = 13.31, p < .05$

19. (a) Yes. Because for the Mann–Whitney test, $U_1 = 32$ and $U_2 = 4$; therefore, $U_{obt} = 4$. $U_{crit} = 5$. The two groups of ranks differ significantly, as do the original maturity scores, $p < .05$.
    (b) Return to the raw scores: For students who have not taken statistics, $\overline{X} = 41.67$, so you would expect μ to be around 41.67. For statistics students, $\overline{X} = 69.67$, so you would expect their μ to be around 69.67.

21. (a) She should use the Kruskal–Wallis H test, because this is a nonparametric one-way, between-subjects design.
    (b) She should assign ranks to the 15 scores, assigning a 1 to the lowest score in the study, a 2 to the next lowest, and so on.
    (c) She should perform the post hoc comparisons: For each possible pair of conditions, she should rerank the scores and perform the rank sums test.
    (d) She will determine which types of patients have significantly different improvement ratings.

23. (a) The design involved a within-subjects factor with two conditions.
    (b) The raw scores were ordinal scores.
    (c) That the ranks in one group were significantly higher or lower than those in the other group.

25. For the independent variable, the number of conditions we have, the number of independent variables we have, and whether independent or related samples were tested. For the dependent variable, whether it requires a parametric or nonparametric procedure.

Appendix A: Additional Statistical Formulas

Section A.1: Creating Grouped Frequency Distributions

1.

Score	f	rel. f	cf
53	1	.06	18
52	3	.17	17
51	2	.11	14
50	5	.28	12
49	4	.22	7
48	0	.00	3
47	3	.17	3

3.

Score	f	rel. f	cf
128–131	1	.04	28
124–127	2	.07	27
120–123	6	.21	25
116–119	8	.29	19
112–115	4	.14	11
108–111	3	.11	7
104–107	2	.07	4
100–103	2	.07	2

Section A.2: Performing Linear Interpolation

1. The target z-score is between $z = .670$ at .2514 of the curve and $z = .680$ at .2483. With .2500 at .0014/.0031 of the distance between .2514 and .2483, the corresponding z-score is .00452 above .67, at .67452.

3. The df of 50 is bracketed by $df = 40$ with $t_{crit} = 2.021$, and $df = 60$ with $t_{crit} = 2.000$. Because 50 is at .5 of the distance between 40 and 60, the target t_{crit} is .5 of the .021 between the brackets, which is 2.0105.

Section A.3: The One-Way, Within-Subjects Analysis of Variance

1. (a) It tells you that the researcher tracked participants' weight gain at different times and, at each, measured their mood.

(b) On some occasions when participants' weight increased, their mood significantly decreased.

3. (a) $H_0: \mu_1 = \mu_2 = \mu_3$; H_a: Not all μs are equal.

(b) $SS_{tot} = 477 - 392.04$; $SS_A = 445.125 - 392.04$; and $SS_{subs} = 1205/3 - 392.04$

Source	Sum of Squares	df	Mean Square	F
Subjects	9.63	7		
Factor A	53.08	2	26.54	16.69
A × Subjects	22.25	14	1.59	
Total	84.96	23		

(c) With $df_A = 2$ and $df_{A \times subs} = 14$, the F_{crit} is 3.74. The F_{obt} is significant.

(d) The $q_k = 3.70$ and $HSD = 1.65$. The means for 0, 1, and 2 hours are 2.13, 4.25, and 5.75, respectively. Significant differences occurred between 0 and 1 hour and between 0 and 2 hours, but not between 1 and 2 hours.

(e) Eta squared $(\eta^2) = 53.08/84.96 = .62$.

(f) The variable of amount of practice is important in determining performance scores, but although 1 or 2 hours of practice significantly improved performance compared to no practice, 2 hours was not significantly better than 1 hour.

GLOSSARY

Alpha The Greek letter α, which symbolizes the criterion, the size of the region of rejection of a sampling distribution, and the theoretical probability of making a Type I error

Alternative hypothesis The statistical hypothesis describing the population parameters that the sample data represent if the predicted relationship does exist; symbolized by H_a

Analysis of variance The parametric procedure for determining whether significant differences exist in an experiment containing two or more sample means; abbreviated ANOVA

ANOVA Abbreviation of analysis of variance

As a function of A way to describe a relationship using the format "changes in Y as a function of changes in X"

Bar graph A graph in which a free-standing vertical bar is centered over each score on the X axis; used with nominal or ordinal scores

Beta The Greek letter β, which symbolizes the theoretical probability of making a Type II error

Between-subjects ANOVA The type of ANOVA that is performed when a study involves between-subjects factors

Between-subjects factor An independent variable that is studied using independent samples in all conditions

Biased estimator A formula for a sample's variability that involves dividing by N that is biased toward underestimating the corresponding population variability

Bimodal distribution A symmetrical frequency polygon with two distinct humps where there are relatively high-frequency scores and with center scores that have the same frequency

Bivariate normal distribution An assumption of the Pearson correlation coefficient that the Y scores at each X form a normal distribution and the X scores at each Y form a normal distribution.

Cell In a two-way ANOVA, the combination of one level of one factor with one level of the other factor

Central limit theorem A statistical principle that defines the mean, standard deviation, and shape of a theoretical sampling distribution

χ^2-distribution The sampling distribution of all possible values of χ^2 that occur when the samples represent the distribution of frequencies described by the null hypothesis

Chi square procedure The nonparametric inferential procedure for testing whether the frequencies of category membership in the sample represent the predicted frequencies in the population

Class interval The name for each group of scores in a grouped frequency distribution

Coefficient of alienation The proportion of variance not accounted for by a relationship; computed by subtracting the squared correlation coefficient from 1

Coefficient of determination The proportion of variance accounted for by a relationship; computed by squaring the correlation coefficient

Cohen's d A measure of effect size that reflects the magnitude of the difference between the means of conditions

Collapsing In a two-way ANOVA, averaging together all scores from all levels of one factor in order to calculate the main effect means for the other factor

Complete factorial design A two-way ANOVA design in which all levels of one factor are combined with all levels of the other factor

Condition An amount or category of the independent variable that creates the specific situation under which subjects' scores on the dependent variable are measured

Confidence interval for a single μ A range of values of μ, one of which is likely to be represented by the sample mean

Confidence interval for μ_D A range of values of μ_D, one of which is likely to be represented by the sample mean ($\overline{D}$) in a related-samples t-test

453

Confidence interval for the difference between two μs A range of differences between two population μs, one of which is likely to be represented by the difference between two sample means

Confounded comparison In a two-way ANOVA, a comparison of two cells that differ along more than one factor

Contingency coefficient The statistic that describes the strength of the relationship in a two-way chi square when there are more than two categories for either variable; symbolized by C

Continuous scale A measurement scale that allows for fractional amounts of the variable being measured

Correlation coefficient A number that describes the type and the strength of the relationship present in a set of data

Correlational study A procedure in which participants' scores on two variables are measured, without manipulation of either variable, to determine whether they form a relationship

Criterion probability The probability that defines whether a sample is too unlikely to have occurred by chance and thus is unrepresentative of a particular population

Criterion variable The variable in a relationship whose unknown scores are predicted through use of the known scores on the predictor variable

Critical value The value of the sample statistic that marks the edge of the region of rejection in a sampling distribution; values that fall beyond it lie in the region of rejection

Cumulative frequency The frequency of the scores at or below a particular score; symbolized by cf

Curvilinear relationship See *Nonlinear relationship*

Data point A dot plotted on a graph to represent a pair of X and Y scores

Degree of association See *Strength of a relationship*

Degrees of freedom The number of scores in a sample that are free to vary, and thus the number that is used to calculate an estimate of the population variability; symbolized by df

Dependent events Events for which the probability of one is influenced by the occurrence of the other

Dependent variable In an experiment, the variable that is measured under each condition of the independent variable

Descriptive statistics Procedures for organizing and summarizing data so that the important characteristics can be described and communicated

Design The way in which a study is laid out so as to demonstrate a relationship

Deviation The distance that separates a score from the mean and thus indicates how much the score differs from the mean

Dichotomous variable A discrete variable that has only two possible amounts or categories

Discrete scale A measurement scale that allows for measurement only in whole amounts

Distribution An organized set of data

Effect size A measure of the amount of influence that changing the conditions of the independent variable had on dependent scores; may be computed as the proportion of variance accounted for

Error variance The inherent variability within a population, estimated in ANOVA by the mean square within groups

Estimated population standard deviation The unbiased estimate of the population standard deviation, calculated from sample data using degrees of freedom $(N - 1)$; symbolized by s_X

Estimated population variance The unbiased estimate of the population variance, calculated from sample data using degrees of freedom $(N - 1)$; symbolized by s_X^2

Estimated standard error of the mean An estimate of the standard deviation of the sampling distribution of means, used in calculating the one-sample t-test; symbolized by $s_{\bar{X}}$

Eta squared The proportion of variance in the dependent variable that is accounted for by changing the levels of a factor, and thus a measurement of effect size; symbolized by η^2

Expected frequency In chi square, the frequency expected in a category if the sample data perfectly represent the distribution of frequencies in the population described by the null hypothesis; symbolized by f_e

Experiment A research procedure in which one variable is actively changed or manipulated, the scores on another variable are measured, and all other variables are kept constant, to determine whether a relationship exists

Experimental hypotheses Two statements made before a study is begun, describing the predicted relationship that may or may not be demonstrated by the study

Experiment-wise error rate The probability of making a Type I error when comparing all means in an experiment

Factor In ANOVA, an independent variable

F-distribution The sampling distribution of all possible values of F that occur when the null hypothesis is true and all conditions represent one population μ

Fisher's protected *t*-test The post hoc procedure performed with ANOVA to compare means from a factor in which all levels do not have equal ns

F-ratio In ANOVA, the ratio of the mean square between groups to the mean square within groups

Frequency The number of times each score occurs within a set of data; also called simple frequency; symbolized by f

Frequency polygon A graph that shows interval or ratio scores (X axis) and their frequencies (Y axis), using data points connected by straight lines

Friedman χ^2 test The nonparametric version of the one-way, repeated-measures ANOVA for ranked scores

Goodness-of-fit test A name for the one-way chi square, because it tests how "good" the "fit" is between the data and H_0

Grouped distribution A distribution formed by combining different scores to make small groups whose total frequencies, relative frequencies, or cumulative frequencies can then be manageably reported

Heteroscedasticity An unequal spread of Y scores around the regression line (that is, around the values of Y')

Histogram A graph similar to a bar graph but with adjacent bars touching, used to plot the frequency distribution of a small range of interval or ratio scores

Homogeneity of variance A characteristic of data describing populations represented by samples in a study that have the same variance

Homoscedasticity An equal spread of Y scores around the regression line and around the values of Y'

Incomplete factorial design A two-way ANOVA design in which not all levels of the two factors are combined

Independent events Events for which the probability of one is not influenced by the occurrence of the other

Independent samples Samples created by selecting each participant for one sample, without regard to the participants selected for any other sample

Independent-samples t-test The t-test used with data from two independent samples

Independent variable In an experiment, a variable that is changed or manipulated by the experimenter; a variable hypothesized to cause a change in the dependent variable

Individual differences Variations in individuals' traits, backgrounds, genetic makeup, etc., that influence their behavior in a given situation and thus the strength of a relationship

Inferential statistics Procedures for determining whether sample data represent a particular relationship in the population

Interval estimation A way to estimate a population parameter by describing an interval within which the population parameter is expected to fall

Interval scale A measurement scale in which each score indicates an actual amount and there is an equal unit of measurement between consecutive scores, but in which zero is simply another point on the scale (not zero amount)

Interval size The number of values spanned by each class interval in a grouped frequency distribution

Kruskal–Wallis H test The nonparametric version of the one-way, between-subjects ANOVA for ranked scores

Level In ANOVA, each condition of the factor (independent variable); also called treatment

Linear regression The procedure for describing the best-fitting straight line that summarizes a linear relationship

Linear regression equation The equation that defines the straight line summarizing a linear relationship by describing the value of Y' at each X

Linear regression line The straight line that summarizes the scatterplot of a linear relationship by, on average, passing through the center of all Y scores

Linear relationship A correlation between the X scores and Y scores in a set of data in which the Y scores tend to change in only one direction as the X scores increase, forming a slanted straight regression line on a scatterplot

Line graph A graph of an experiment when the independent variable is an interval or ratio variable; plotted by connecting the data points with straight lines

Main effect In a two-way ANOVA, the effect on the dependent scores of changing the levels of one factor while ignoring (collapsing over) the other factor

Main effect mean The overall mean of one level of a factor while ignoring (collapsing over) the influence of the other factor

Mann–Whitney U test The nonparametric version of the independent-samples t-test for ranked scores when n is less than or equal to 20

Margin of error Expressing the expected error when estimating a population parameter as plus or minus some amount

Matched-samples design An experiment in which each participant in one sample is matched on an extraneous variable with a participant in the other sample

Mean The score located at the mathematical center of a distribution

Mean square In ANOVA, an estimated population variance, symbolized by MS

Mean square between groups In ANOVA, the variability in scores that occurs between the levels in a factor or the cells in an interaction

Mean square within groups In ANOVA, the variability in scores that occurs in the conditions, or cells; also known as the error term

Measure of central tendency A score that summarizes the location of a distribution on a variable by indicating where the center of the distribution tends to be located

Measures of variability Measures that summarize the extent to which scores in a distribution differ from one another

Median The score located at the 50th percentile; symbolized by Mdn

Mode The most frequently occurring score in a sample

Multiple correlation coefficient The correlation that describes the relationship between multiple predictor (X) variables and one criterion (Y) variable

Multiple regression equation The procedure for simultaneously using multiple predictor (X) variables to predict scores on one criterion (Y) variable

Multivariate statistics Procedures applied to a study that measures two or more dependent variables

Negative linear relationship A linear relationship in which the Y scores tend to decrease as the X scores increase

Negatively skewed distribution A frequency polygon with low-frequency, extreme low scores but without corresponding low-frequency, extreme high ones, so that its only pronounced tail is in the direction of the lower scores

Nemenyi's procedure The post hoc procedure performed with the Friedman χ^2 test

Nominal scale A measurement scale in which each score is used simply for identification and does not indicate an amount

Nonlinear relationship A relationship in which the Y scores change their direction of change as the X scores change; also called a curvilinear relationship

Nonparametric statistics Inferential procedures that do not require stringent assumptions about the parameters of the raw score population represented by the sample data; usually used with scores most appropriately described by the median or the mode

Nonsignificant Describes results that are considered likely to result from chance sampling error when the predicted relationship does not exist; it indicates failure to reject the null hypothesis

Normal curve The symmetric, bell-shaped curve produced by graphing a normal distribution

Normal distribution A set of scores in which the middle score has the highest frequency, and proceeding toward higher or lower scores the frequencies at first decrease slightly but then decrease drastically, with the highest and lowest scores having very low frequency

Null hypothesis The statistical hypothesis describing the population parameters that the sample data represent if the predicted relationship does not exist; symbolized by H_0

Observed frequency In chi square, the frequency with which participants fall into a category of a variable; symbolized by f_0

One-sample t-test The parametric procedure for a one-sample experiment when the standard deviation of the raw score population must be estimated

One-tailed test The test used to evaluate a statistical hypothesis that predicts that scores will only increase or only decrease

One-way ANOVA The analysis of variance performed when an experiment has only one independent variable

One-way chi square The chi square procedure for testing whether the sample frequencies of category membership on one variable represent the predicted distribution of frequencies in the population

Ordinal scale A measurement scale in which scores indicate rank order

Outlier A data point that lies outside of the general pattern in a scatterplot; created by an unusual X or Y score

Parameter See *Population parameter*

Parametric statistics Inferential procedures that require certain assumptions about the parameters of the raw score population represented by the sample data; usually used with scores most appropriately described by the mean

Participants The individuals who are measured in a sample; also called *subjects*

Pearson correlation coefficient The correlation coefficient that describes the linear relationship between two interval or ratio variables; symbolized by r

Percent A proportion multiplied times 100

Percentile The percentage of all scores in the sample that are at or below a particular score

Phi coefficient The statistic that describes the strength of the relationship in a two-way chi square when there are only two categories for each variable; symbolized by ϕ

Point-biserial correlation coefficient The correlation coefficient that describes the linear relationship between scores from one continuous interval or ratio variable and one dichotomous variable; symbolized by r_{pb}

Point estimation A way to estimate a population parameter by describing a point on the variable at which the population parameter is expected to fall

Pooled variance The weighted average of the sample variances in a two-sample experiment; symbolized by s^2_{pool}

Population The infinitely large group of all possible scores that would be obtained if the behavior of every individual of interest in a particular situation could be measured

Population parameter A number that describes a characteristic of a population of scores, symbolized by a letter from the Greek alphabet; also called a parameter

Population standard deviation The square root of the population variance, or the square root of the average squared deviation of scores around the population mean; symbolized by σ_X

Population variance The average squared deviation of scores around the population mean; symbolized by σ^2_X

Positive linear relationship A linear relationship in which the Y scores tend to increase as the X scores increase

Positively skewed distribution A frequency polygon with low-frequency, extreme high scores but without corresponding low-frequency, extreme low ones, so that its only pronounced tail is in the direction of the higher scores

Post hoc comparisons In ANOVA, statistical procedures used to compare all possible pairs of sample means in a significant effect, to determine which means differ significantly from each other

Power The probability that a statistical test will detect a true relationship and allow the rejection of a false null hypothesis

Predicted *Y* score   In linear regression, the best prediction of the *Y* scores at a particular *X*, based on the linear relationship summarized by the regression line; symbolized by Y'

Predictor variable   The variable from which known scores in a relationship are used to predict unknown scores on another variable

Probability   A mathematical statement indicating the likelihood that an event will occur when a particular population is randomly sampled; symbolized by p

Probability distribution   The probability of every event in a population, derived from the relative frequency of every event in that population

Proportion   A decimal number between 0 and 1 that indicates a fraction of a total

Proportion of the area under the curve   The proportion of the total area beneath the normal curve at certain scores, which represents the relative frequency of those scores

Proportion of variance accounted for   The proportion of the error in predicting scores that is eliminated when, instead of using the mean of *Y*, we use the relationship with the *X* variable to predict *Y* scores; the proportional improvement in predicting *Y* scores thus achieved

Qualitative variable   A variable that reflects a quality or category

Quantitative variable   A variable that reflects a quantity or amount

Random sampling   A method of selecting samples so that all members of the population have the same chance of being selected for a sample

Range   The distance between the highest and lowest scores in a set of data

Rank sums test   The nonparametric version of the independent-samples *t*-test for ranked scores when *n* is greater than 20; also, the post hoc procedure performed with the Kruskal–Wallis *H* test

Ratio scale   A measurement scale in which each score indicates an actual amount, there is an equal unit of measurement, and there is a true zero

Rectangular distribution   A symmetric frequency polygon shaped like a rectangle; it has no discernible tails because its extreme scores do not have relatively low frequencies

Region of rejection   That portion of a sampling distribution containing values considered too unlikely to occur by chance, found in the tail or tails of the distribution

Regression line   The line drawn through the long dimension of a scatterplot that best fits the center of the scatterplot, thereby visually summarizing the scatterplot and indicating the type of relationship that is present

Related samples   Samples created by matching each participant in one sample with a participant in the other sample or by repeatedly measuring the same participant under all conditions; also called dependent samples

Related-samples *t*-test   The *t*-test used with data from two related (dependent) samples

Relationship   A correlation between two variables whereby a change in one variable is accompanied by a consistent change in the other

Relative frequency   The proportion of time a score occurs in a distribution, equal to the proportion of the total number of scores that is made up by the score's simple frequency; symbolized by $rel.\ f$

Relative frequency distribution   A distribution of scores, organized to show the proportion of time each score occurs in a set of data

Relative standing   A description of a particular score derived from a systematic evaluation of the score using the characteristics of the sample or population in which it occurs

Repeated-measures design   A related-samples design in which the same participants are measured repeatedly under all conditions of an independent variable

Representative sample   A sample whose characteristics accurately reflect those of the population

Restriction of range   In correlation, improper limitation of the range of scores obtained on one or both variables, leading to an underestimate of the strength of the relationship between the two variables

Sample   A relatively small subset of a population, intended to represent the population; a subset of the complete group of scores found in any particular situation

Sample standard deviation   The square root of the sample variance or the square root of the average squared deviation of sample scores around the sample mean; symbolized by S_X

Sample statistic   A number that describes a characteristic of a sample of scores, symbolized by a letter from the English alphabet; also called a statistic

Sample variance   The average squared deviation of a sample of scores around the sample mean; symbolized by S_X^2

Sampling distribution of differences between the means   A frequency distribution showing all possible differences between two means that occur when two independent samples of a particular size are drawn from the population of scores described by the null hypothesis

Sampling distribution of mean differences   A frequency distribution showing all possible mean differences that occur when the difference scores from two related samples of a particular size are drawn from the population of difference scores described by the null hypothesis

Sampling distribution of means   A frequency distribution showing all possible sample means that occur when samples of a particular size are drawn from the raw score population described by the null hypothesis

Sampling distribution of *r*   A frequency distribution showing all possible values of r that occur when samples are drawn from a population in which ρ is zero

Sampling distribution of r_S   A frequency distribution showing all possible values of r_S that occur when samples are drawn from a population in which ρ_S is zero

Sampling error   The difference, due to random chance, between a sample statistic and the population parameter it represents

Sampling with replacement   A sampling procedure in which previously selected individuals or events are returned to the population before any additional samples are selected

Sampling without replacement   A sampling procedure in which previously selected individuals or events are not returned to the population before additional samples are selected

Scatterplot   A graph of the individual data points from a set of X–Y pairs

Significant   Describes results that are too unlikely to accept as resulting from chance sampling error when the predicted relationship does not exist; it indicates rejection of the null hypothesis

Simple frequency   The number of times that a score occurs in data

Simple frequency distribution   A distribution of scores, organized to show the number of times each score occurs in a set of data

Skewed distribution   A frequency polygon similar in shape to a normal distribution except that it is not symmetrical and it has only one pronounced tail

Slope   A number that indicates how much a linear regression line slants and in which direction it slants; used in computing predicted Y scores; symbolized by b

Spearman rank-order correlation coefficient   The correlation coefficient that describes the linear relationship between pairs of ranked scores; symbolized by r_S

Squared sum of X   A result calculated by adding all scores and then squaring their sum; symbolized by $(\Sigma X)^2$

Standard error of the difference   The estimated standard deviation of the sampling distribution of differences between the means of independent samples in a two-sample experiment; symbolized by $s_{\overline{X}_1 - \overline{X}_2}$

Standard error of the estimate   A standard deviation indicating the amount that the actual Y scores in a sample differ from, or are spread out around, their corresponding Y' scores; symbolized by $S_{Y'}$

Standard error of the mean   The standard deviation of the sampling distribution of means; used in the z-test (symbolized by $\sigma_{\overline{X}}$) and estimated in the one-sample t-test (symbolized by $s_{\overline{X}}$)

Standard error of the mean difference   The standard deviation of the sampling distribution of mean differences between related samples in a two-sample experiment; symbolized by $s_{\overline{D}}$

Standard normal curve   A theoretical perfect normal curve, which serves as a model of the perfect normal z-distribution

Standard scores   See z-score

Statistic   See *Sample statistic*

Statistical hypotheses   Two statements (H_0 and H_a) that describe the population parameters the sample statistics will represent if the predicted relationship exists or does not exist

Statistical notation   The standardized code for the mathematical operations performed in formulas and for the answers obtained

Strength of a relationship   The extent to which one value of Y within a relationship is consistently associated with one and only one value of X; also called the degree of association

Sum of squares   The sum of the squared deviations of a set of scores around the mean of those scores

Sum of the deviations around the mean   The sum of all differences between the scores and the mean; symbolized as $\Sigma(X - \overline{X})$

Sum of the squared Xs   A result calculated by squaring each score in a sample and adding the squared scores; symbolized by ΣX^2

Sum of X   The sum of the scores in a sample; symbolized by ΣX

Tail (of a distribution)   The far-left or far-right portion of a frequency polygon, containing the relatively low-frequency, extreme scores

t-distribution   The sampling distribution of all possible values of t that occur when samples of a particular size represent the raw score population(s) described by the null hypothesis

Test of independence   A name for the two-way chi square, because it tests whether the frequencies in the categories of one variable are independent of the categories of the other variable

Total area under the curve   The area beneath the normal curve, which represents the total frequency of all scores

Transformation   A systematic mathematical procedure for converting a set of scores into a different but equivalent set of scores

Treatments   The conditions of the independent variable; also called levels

Treatment effect   The result of changing the conditions of an independent variable so that different populations of scores having different μs are produced

Treatment variance   In ANOVA, the variability between scores from different populations that would be created by the different levels of a factor

t-test for independent samples   The parametric procedure used for significance testing of sample means from two independent samples

t-test for related samples   The parametric procedure used for significance testing of sample means from two related (dependent) samples

Tukey's *HSD* multiple comparisons test   The post hoc procedure performed with ANOVA to compare means from a factor in which all levels have equal n

Two-tailed test   The test used to evaluate a statistical hypothesis that predicts a relationship, but not whether scores will increase or decrease

Two-way ANOVA   The parametric inferential procedure performed when an experiment contains two independent variables

Two-way, between-subjects ANOVA   The parametric inferential procedure performed when both factors are between-subjects factors

Two-way chi square   The chi square procedure for testing whether, in the population, frequency of category membership on one variable is independent of frequency of category membership on the other variable

Two-way interaction effect   The effect produced by manipulating two independent variables such that the influence of changing the levels of one factor depends on which level of the other factor is present

Two-way, mixed-design ANOVA   The parametric inferential procedure performed when the design involves one within-subjects factor and one between-subjects factor

Two-way, within-subjects ANOVA   The parametric inferential procedure performed when both factors are within-subjects factors

Type I error   A statistical decision-making error in which a large amount of sampling error causes rejection of the null hypothesis when the null hypothesis is true (that is, when the predicted relationship does not exist)

Type II error   A statistical decision-making error in which the closeness of the sample statistic to the population parameter described by the null hypothesis causes the null hypothesis to be retained when it is false (that is, when the predicted relationship does exist)

Type of relationship   The form of the correlation between the X scores and the Y scores in a set of data, determined by the overall direction in which the Y scores change as the X scores change

Unbiased estimator   A formula for a sample's variability that involves dividing by $N - 1$ that equally often under- and over-estimates the corresponding population variability (See *Biased estimator*)

Unconfounded comparison   In a two-way ANOVA, a comparison between two cells that differ along only one factor

Ungrouped distribution   A distribution that shows information about each score individually (See *Grouped distribution*)

Unimodal distribution   A distribution whose frequency polygon has only one hump and thus has only one score qualifying as the mode

Univariate statistics   Procedures applied to a study that measures only one dependent variable

Variable   Anything that, when measured, can produce two or more different scores

Variance   A measure of the variability of the scores in a set of data, computed as the average of the squared deviations of the scores around the mean

Variance of Y scores around Y'   In regression, the average squared deviation between the actual Y scores and corresponding predicted Y' scores, symbolized by $S_{Y'}$

Wilcoxon T test   The nonparametric version of the related-samples t-test for ranked scores

Within-subjects ANOVA   The type of ANOVA performed when a study involves within-subjects factors

Within-subjects factor   The type of factor created when an independent variable is studied using related samples in all conditions because participants are either matched or repeatedly measured

Y intercept   The value of Y at the point where the linear regression line intercepts the Y axis; used in computing predicted Y scores; symbolized by a

z-distribution   The distribution of z-scores produced by transforming all raw scores in a distribution into z-scores

z-score   The statistic that describes the location of a raw score in terms of its distance from the mean when measured in standard deviation units; symbolized by z; also known as a standard score because it allows comparison of scores on different kinds of variables by equating, or standardizing, the distributions

z-test   The parametric procedure used to test the null hypothesis for a single-sample experiment when the true standard deviation of the raw score population is known

INDEX

A

a, 164. *See also* Y intercept
Absolute value, 109
Accounting for variance, 103
Accuracy, correlation coefficients and, 144
Alpha (α)
  error and, 225–226
  one-sample *t*-test and, 235, 240
  published research and, 224, 246–247
  SPSS and, 255
  two-tailed test and, 216
Alternative hypothesis (H_a)
  hypothesis testing and, 211–213
  one-way chi square and, 356
  power and, 228–229
  Type I error: rejecting H_0 when H_0 is true, 224–227
  Type I/Type II error comparison and, 228–229
  Type II error: retaining H_0 when H_0 is false, 227–228
American Psychological Association (APA)
  publication rules, 54
  and use of mean, 79
Analysis of variance. *See* One-way ANOVA
ANOVA. *See* One-way ANOVA; Two-way ANOVA
Apparent limits, 380
Area under the curve, 91
"Around", quantification of, 88–89
As a function of
  definition of, 19
  and drawing conclusions from experiments, 24
Assumptions, 208
Average of deviations, 88

B

b. *See* Slope
Bar graph
  chi square procedures and, 362
  of experiments, 76–77

frequency distribution graphing and, 38–41
  relative frequency and, 49
Beta (β), 227
Between-subjects ANOVA, 312, 350
Between-subjects factor, 291–292
Biased estimators, 96–97
Bimodal distribution, 44–46, 62–63

C

Causality/causal relationship, 26–27
Cell, 320–321, 324–325
Cell means, 324, 338
Central limit theorem, 125–126
Central tendency
  definition/importance of, 60–61
  summation and, 60–61
Chi square procedures (x^2)
  and choosing nonparametric procedure, 364–365
  critical values of, 431
  definition of, 352
  distribution, 355
  formula for, 354–355, 378
  Kruskal-Wallis H test and, 369–370
  and logic of nonparametric procedures for ranked data, 363–364
  Mann-Whitney U test and, 365–366
  one-way chi square, 352–357
  published research and, 362–363
  rank sums test and, 366–367
  tied rank resolution and, 364
  two-way chi square, 357–362
  Wilcoxon T test and, 368–369
Class interval, 379
Coefficient of alienation, 177
Coefficient of determination, 177
Coefficients. *See* Correlation coefficients
Cohen's *d*, 280–282, 289
Collapsing, 321
Complete factorial design, 320–321
Computational formulas
  definition of, 85
  for estimated population standard deviation, 99, 108

for estimated population variance, 98
for Pearson correlation coefficient, 148–149, 159
for proportion of variance accounted for, 176–177
for proportion of variance not accounted for, 176–177
for sample standard deviation, 94, 108
for sample variance, 93, 108
for Spearman rank-order correlation coefficient, 152–153, 159
of standard error of the estimate, 170–171
Computers in statistics, 3. *See also* SPSS
Condition, independent variables and, 23–24
Confidence, 187
Confidence interval
  for difference between two μs, 289
  for μ, 317
  for μD, 277
  for single μ, 234, 243–245, 268–269, 289, 310
Confounded comparison, 339–340
Constant (K)
  mathematical constants, 95
  standard deviation and, 95
  transformations and, 69
Contingency coefficient (C), 361–362, 378
Continuous scales, 28–29, 40
Correlation coefficients. *See also* Pearson correlation coefficient
  characteristics of, 138
  concepts of, 136–137
  drawing conclusions from, 137–138
  importance of, 136
  and interpretation of Pearson *r* (correlation coefficients), 250–251
  multiple, 179
  and one-tailed tests of Pearson *r* (correlation coefficients), 251
  Pearson correlation coefficient, 147–150
  and restriction of range problem, 153–154
  scatterplots and, 138–139
  and significance testing in one-sample *t*-test, 402–404

Spearman rank-order, 151–153
and statistics in published research,
154–155
and sum of the cross products, 135–136
and testing the Pearson r (correlation
coefficients), 247–250
zero association and, 146
Correlational studies
mean/central tendencies and, 73
relationships and, 25–27
Criterion probability, 198
Criterion variable, 162–163
Critical value of z (z_{crit})
comparing to obtained z, 217–218
nonsignificant results and, 219
region of rejection and, 215–216
significant results and, 218
Critical values, 198–201, 436–437
of chi square (x_2 tables), 431
comparing z_{obt} to, 217–218
F-tables, 426–428
interpolating, 383–384
of Mann-Whitney U test, 432–435
and performing z-test, 216
r-tables, 424
r_S-tables, 425
t-tables, 423
Cumulative frequency (cf)
computing, 51–52
formula for, 59
grouped frequency distribution and,
54–55
Curvilinear relationships, 141–142

D

Data, definition of, 1–2
Data point
definition of, 8–9
and line graphs of experiments, 75–76
and reading graphs, 19–20
Definitional formula, 85, 89
Degrees of freedom (df)
ANOVA and, 295
computing $df_{bn}/df_{wn}/df_{tot}$, 302–303
and confidence interval for μD, 277
definition of, 97–98, 97–98
between groups for factor A, 388
one-sample t-test and, 238–240, 242
one-way chi square and, 356
relationships and, 361–362
two-way ANOVA and, 349
two-way chi square and, 360
Dependent events, 189
Dependent variables
definition of, 23–24
experiments and, 73–74
inferential statistics and, 209
multivariate statistics and, 312
and performing z-test, 215
two-sample t-test and, 272
two-way ANOVA and, 321
Descriptive statistics, 20–21
Design of study. *See also* Two-way
ANOVA
correlational studies, 25–27
experiments, 22–25

Deviation, around the mean, 69–72
df. See Degrees of freedom (df)
Dichotomous variable, 29
Differences between conditions (SS_{bn}),
311, 317
Differences (total) between all scores in
experiment (SS_{tot}), 311, 317
Direction from the mean, 69–70
Discrete scales, 28–30, 76–77
Distance as difference between scores, 61
Distance from the mean, 69–70
Distributions, 36. *See also* Frequency
distributions

E

Effect size, 280–283
Ellipse, 140
Empirical, 1
Error
estimated standard error of the mean,
236
and experiment summarization, 73–74
population representation and, 194–196
power and, 228–229
in predicted Y score, 168–169
and predicting scores with mean, 71–72
random sampling and, 193–194
standard error of the estimate, 170–171
standard error of the estimate and,
173–174
standard error of the mean, 126–127
Type I error: rejecting H_0 when H_0 is true,
224–227
Type I/Type II error comparison, 228–229
Type II error: retaining H_0 when H_0 is
false, 227–228
variability and, 102
Error variance ($μ^2_{error}$), 298, 385
Estimated population standard
deviation, 97, 99
Estimated population variance, 97–99, 236
Estimated population variance of the
difference scores, 274
Estimated standard error of the mean, 236
Estimation
and computing confidence interval for
single μ, 243–245
interval/point, 243
and predicting scores with mean, 71–72
Eta squared, 311, 317
Experiment-wise error rate, 293
Experimental hypotheses, 209–210
Experiments. *See also* Research
dependent variables and summarization
of, 73–74
drawing conclusions from, 24–25
graphing, 75–77
relationships and, 22–25
steps of statistical analysis of, 78–79

F

F. See also One-way ANOVA; Two-way
ANOVA
ANOVA and, 293–294, 349
interpreting within-subject, 390

F-distribution, 304–305
F-ratio
and comparing mean squares, 296–298
one-way, within-subject analysis of
variance and, 385–386
theoretical components of, 298–299
F-tables, 426–428
Factor. *See* Independent variables
Fisher's protected t-test, 307
F_{obt}, computing, 300–304
Formulas. *See also* Computational formulas
basic concepts of, 5–6
for chi square, 354–355, 378
for Cohen's d, 280–282, 289
computing $df_{bn}/df_{wn}/df_{tot}$, 302–303
for confidence interval for difference
between two μs, 289
for confidence interval for μ, 317
for confidence interval for single μ,
244–245, 259, 289, 310
for contingency coefficient (C), 362,
378
and critical values of Pearson r, 259
and critical values of Spearman r_S, 259
for degrees of freedom between groups
for factor A, 388
for deviation, 69
for estimated standard error of the mean,
236
for eta squared, 317, 350
for expected frequency in cell of two-way
chi square (f_c), 359–360
for F_{obt}, 350
for grouped frequency distribution,
379–381
for HSD, 317
for independent-samples t-test, 289
for independent-samples t_{obt}, 266
for linear regression equation, 163–164,
184
for mean, 66
for median estimation, 83
for one-sample t-test, 236–237, 259
for one-way, within-subject analysis of
variance, 386–390
Pearson, 159
for percentile for score with known cf, 59
for phi coefficient (φ), 362, 378
for pooled variance, 265
for population mean, 73
for proportion of variance accounted
for, 184
for proportion of variance accounted for
(r^2_{pb}), 283
for proportion of variance not accounted
for, 184
for protected t-test, 317
for range, 87, 108
for related-samples t-test, 275, 289
for relative frequency ($rel.f$), 59
for estimated population variance of the
difference scores, 274
for sample mean, 66, 83
for sample variance, 89
for slope of linear regression line,
164–166, 184
Spearman rank-order, 152–153, 159

for standard error of estimate, 184
for standard error of the difference $(s_{X_1-X_2})$, 265 for standard error of the mean difference, 275
sum of squares between groups (SS_{bn}), 301
sum of squares within groups (SS_{wn}), 302
summary of two-way ANOVA, 349
for transformation, 206
for transforming raw score in a population into z-score, 113, 134
for transforming raw score in a sample into z-score, 112, 134
for transforming sample mean into z-score, 127–128, 134, 206
for transforming z-score in population into raw score, 114, 134
for transforming z-score in sample into raw score, 113–114, 134
for true standard error of the mean, 126–127, 134
for Tukey's HSD multiple comparisons test, 308
Tukey's HSD post hoc comparisons test and, 350
for two-way ANOVA degrees of freedom, 350
for two-way ANOVA mean square, 350
two-way between subjects ANOVA, 350
for variance of Y scores around Y', 184
for Y intercept of linear regression line, 184
z-test, 216–217, 233
Frequency distributions
APA publication rules and, 53–54
bimodal distribution and, 44–46
cumulative frequency and, 51–52
formulas for creation of, 379–381
grouped/ungrouped, 54–55
importance of, 37
labeling of, 45–46
and measure of central tendency, 62
normal distribution and, 42–43
percentile and, 51–53
rectangular distribution and, 45
relative frequency and, 47–51
simple, 37–42
skewed distribution and, 43–44, 46
Frequency (f), 36–37, 54–55
Friedman χ^2 test, 371–372
Function (as a function of)
definition of, 19
and drawing conclusions from experiments, 24

G
Gambler's fallacy, 187
Goodness-of-fit test, 353
Graphs
bar graphs, 38–41, 49, 76–77, 362
creation of, 7–9
of experiments, 75–77
of grouped distributions, 380–381
histogram and, 38, 40–41
one-way ANOVA and, 311

polygon and, 38, 40–41
relationships and, 18–20
relative frequency and, 48–49
of simple frequency distributions, 38–42
two-sample t-test, 279–280
Greater than ($>$), 207
Greater than or equal to ($\geq$), 207
Grouped distributions, 54–55, 379–381

H
H_0 (null hypothesis), 212–213, 235, 237–238, 240–242, 250–251
H_a (alternative hypothesis), 211–213, 241–242
Heteroscedasticity, 172–173
Histogram
frequency distribution graphing and, 38, 40–41
relative frequency and, 49
Homogeneity of variance, 262, 321
Homoscedasticity, 171–172
HSD (honestly significant differences). *See also* Tukey's HSD multiple comparisons test
formulas for, 317
Tukey's HSD multiple comparisons test, 307–308
two-way ANOVA and, 337–340
Hypothesis testing. *See also* Statistical hypotheses
alternative hypothesis and, 211–212
experimental hypotheses, 209–210
H_0 and, 218. *See also* Null hypothesis (H_0)
and importance of z-test, 208
nonsignificant results and, 219–220
null hypothesis and, 212–213. *See also* Null hypothesis (H_0)
one-sample experiment design and, 210–211
one-tailed testing and, 221–223
and performing z-test, 215–218
published research and, 224
related-samples t-test and, 272–273
significant results and, 218–219
statistical hypotheses creation and, 211–213
statistical hypotheses logic and, 213–214

I
Incomplete factorial design, 320
Independent events, 188–189, 262
Independent samples, 262, 321
Independent-samples t-test
and computing standard error of the difference, 265–266
computing t_{obt}, 265, 289
and confidence interval for difference between two μs, 268–269
and estimating population variance, 264–265
formulas for, 265, 289
interpreting, 267–268
one-tailed testing and, 269–270

requirements for, 262
sampling distribution for, 263–264
statistical hypotheses for, 262–263
when to perform, 284
Independent variables
definition of, 22–24
effect size and, 280
Individual differences, 17
Inferential statistics
and decisions regarding representative samples, 202–203
definition of, 21, 186
mode/mean/median comparisons and, 67–69
nonparametric statistics as, 351
research role of, 208–209, 229–230
setting up procedures for, 209–213
two-way ANOVA as, 318
z-test and, 208
Inflection points, 91
Interaction effect
definition of, 339
post hoc comparisons on, 339–340
two-way ANOVA and, 319, 324–326
Interval estimation, 243
Interval scale, 28–29
simple frequency histograms and, 40–41
Interval size, 379
Inverted U-shaped relationship, 141

K
k (number of levels in a factor), 290
Known true population variability, 96
Kruskal-Wallis H test, 369–372

L
Labeling, of frequency distributions, 45–46
Less than ($<$, 207
Less than or equal to ($\leq$), 207
Level, definition of, 23, 290
Line graphs, 75–76
Linear interpolation, 381–384
Linear regression
importance of, 161
linear regression line and, 161–163
predicted Y score, 160
statistics in published research of, 179
Linear regression equation (Y'), 163–164, 167, 184
Linear regression line, 161–163
formula for Y intercept of, 166–167
Linear relationships, 139–141
Locations and scores, 61

M
Main effect mean, 321
Main effect of factor A, 321–323
Main effect of factor B, 323
Mann-Whitney U test, 365–366, 432–435
Margin of error, 243
Matched-samples design, 271–272
Mathematical constants, 95. *See also* Constant (K)

Mathematical operations. *See also* Formulas
  identification of, 5
  order of, 5–6
Mean
  compared with median/mode, 67–69
  definition/computation of, 65–66
  deviation around the, 69–72
  estimated standard error of the mean, 236
  and measures of variability, 85
  population, 72–73
  population mean (μ), 72–73
  for predicted Y scores, 160
  predicting scores with, 71–72
  sample mean and, 79
  standard error of the, 126–127
  uses of, 66–67
  z-scores and, 112
Mean difference, 273–279
Mean square between groups (MS_{bn})
  computing, 303
  one-way ANOVA and, 295–296, 299–300
Mean square within groups, 349, 385–386, 388
Mean square within groups (MS_{wn})
  computing, 303
  one-way ANOVA and, 295, 299–300
Measure of central tendency, 61–62
Measurement scales, 27–30
Measures of variability
  $(\Sigma X)^2$, 84–85
  ΣX^2, 85
  definition/importance of, 85–86
  descriptive/inferential measures of variability, 101
  estimated population standard deviation, 97, 99
  estimated population variance, 97–99
  mathematical constants and, 95
  population variance/standard deviation, 95–99
  published research and, 103–104
  range, 87
  sample standard deviation, 89–92, 94
  sample variance, 88–89, 94
  standard deviation, 87–90
  and strength of relationship of variability, 101–102
  variability and errors in prediction, 102
  variance/standard deviation application to research, 100–103
Median
  compared with mode/mean, 67–69
  definition of, 63–64
  uses of, 65
Mode
  compared with median/mean, 67–69
  definition of, 62
  uses of, 63
mu (μ). *See* Population mean (μ)
Multiple correlation coefficient, 179
Multiple regression equation, 179
Multivariate analysis of variance (MANOVA), 312
Multivariate statistics, 312

N
n (total number of scores in condition), 260, 337
N (total number of scores in study), 37–38, 260
Negative linear relationship, 141
Negatively skewed distribution, 44, 46, 68
Nemenyi's procedure, 365, 372
Nominal scale
  and bar graphs of experiments, 76–77
  characteristics of, 27, 29
  simple frequency bar graphs and, 39–41
Nonlinear relationships, 141–142
Nonnormal distribution, 43–46
Nonparametric procedures
  and choosing nonparametric procedure, 364–365
  Friedman χ^2 test and, 371–372
  Kruskal-Wallis H test and, 369–370
  and logic of nonparametric procedures for ranked data, 363–364
  Mann-Whitney U test and, 365–366
  rank sums test and, 366–367
  tied rank resolution and, 364
  Wilcoxon T test and, 368–369
Nonparametric statistics. *See also* Chi square procedures (x^2)
  definition of, 209, 351
  importance of, 351–352
Nonsignificant results, hypothesis testing and, 219–220
Nonsymmetrical. *See* Skewed distribution
Normal curve/distribution. *See also* z-scores
  and area under the curve, 91–92
  and computing percentiles, 52–53
  and measure of central tendency, 86
  median location in, 64
  relative frequency and, 49–51
  simple frequency distributions and, 42–43, 46
  standard deviation rule and, 94
Not equal to ($\neq$), 207
Null hypothesis (H_0)
  ANOVA and, 293
  F-ratio and, 298–299
  hypothesis testing and, 212–213
  and interaction effects, 326
  and main effect of factor A, 322–323
  nonsignificant results and, 220
  one-way chi square and, 353–354, 356
  power and, 228–229
  Type I error: rejecting H_0 when H_0 is true, 224–227
  Type I/Type II error comparison and, 228–229
  Type II error: retaining H_0 when H_0 is false, 227–228
Number of levels in a factor (k), 290

O
Observed frequency, 353, 355–356, 358–360
One-sample t-test
  and computing confidence interval for single μ, 243–245
  definition of, 234
  formula for, 236–237
  importance of, 234–235
  and interpretation of Pearson r (correlation coefficients), 250–251
  interpreting, 240–241
  one-tailed t-test, 241–242
  and one-tailed tests of Pearson r (correlation coefficients), 251
  performing, 235–237
  published research and, 246–247
  and significance testing of correlation coefficients, 402–404
  t-distribution/degrees of freedom and, 238–240
  and testing the Pearson r (correlation coefficients), 247–250
  and using t-tables, 240, 242
  when to perform, 234
One-tailed test
  for decreasing scores, 223
  definition of, 210
  for increasing scores, 221–222
  r-tables, 424
  relationships and, 229
  Type I error and, 225
One-way ANOVA
  and computing F_{obt}, 300–304
  and confidence interval for single μ, 310
  controlling experiment-wise error rate, 292–293
  Fisher's protected t-test and, 307
  format for summary table of, 317
  graphing, 311
  importance of, 291
  and interpreting F_{obt} with F-distribution, 304–305
  and logic of F-ratio, 296–298
  and mean square between groups, 295–296, 299–300
  and mean square within groups, 295, 299–300
  order of operations in, 293–294
  overview of, 291–292
  post hoc comparisons and, 306–308
  published research and, 312
  statistical hypotheses in, 293
  summary of performance steps for, 309–310
  and theoretical components of F-ratio, 298–299, 298–299
  Tukey's HSD multiple comparisons test and, 307–308
  within-subject analysis of variance, 384–390
One-way chi square
  computing, 354–355
  definition of, 352–353
  hypotheses/assumptions of, 353–354
  interpreting, 355–356
  testing other hypotheses with, 356
Ordinal scale
  about, 27–29
  and bar graphs of experiments, 76–77
  simple frequency bar graphs and, 39–41
Outlier, 138–139
Over the long run, 186–187
Overestimation, 71–72

P

Parameters, definition of, 22
Parametric statistics
    analysis of variance, 292
    definition of, 209
    two-sample *t*-test and, 262
Participants (subjects), 384
    definition of, 13
PASW. *See* SPSS
Pearson correlation coefficient (ρ) or (r)
    about, 147–148
    computational formula for, 148–149, 159
    definition of, 147–148
    interpreting, 250–251
    linear regression and, 164
    one-tailed testing and, 251
    r-tables, 424
    sampling distribution of, 249–250
Pearson Product Moment Correlation
        Coefficient, 147
Percentile
    as cumulative frequency, 51–53
    for score, 59
Percents, definition of, 7
Perfect association, 143–144
Phi coefficient (ϕ), 361–362, 378
Plotting. *See* Graphs
Plus or minus ($\pm$), 109
Point estimation, 243
Polygon
    frequency distribution graphing and, 38,
        40–41
    relative frequency and, 49
Pooled variance, 265
Population mean (μ), 72–73. *See also*
        Mean
Population standard deviation (σX), 95–99
Population variance, 95–99
Populations
    about, 13–14
    definition of, 186
    descriptive/inferential statistics and, 21
    inferring relationship in, 78–79
    parameters and, 22
    sampling error and, 194–196
    two-sample *t*-test and, 261–262, 272
Positive linear relationship, 140–141
Positively skewed distribution, 44, 46
    median location in, 64
    mode/mean/median comparisons and, 68
Post hoc comparisons, 293–294, 336–337,
        339–340
Power. *See also* Effect size
    error and, 228–229
    maximizing in statistical tests, 253–255
    sufficient scores for, 235
    Type II error and, 241, 253
Predicted *Y* score
    computing, 167
    error description in, 168–169
    linear regression and, 160
Predictions
    correlation coefficients and accuracy of,
        144–146
    descriptive statistics and, 21
    standard error of the estimate and,
        173–174

variability and errors in, 102
    Y score, 160, 167
Predictor variable, 162–163
Pretest/posttest, 272
Probability distribution, creation of, 188
Probability (p)
    creation of probability distribution, 188
    definition of, 186
    and factors affecting probability of event,
        188–189
    importance of, 186–187
    logic of, 186–187
    odds/chance definitions, 185
    p as, 186
    power and, 228–229, 235, 241, 253–255
    random sampling and, 193–194
    representative samples and, 193–194,
        197–200, 202–203
    sampling distribution of means and,
        191
    from standard normal curve (individual
        scores), 190
    from standard normal curve (sample
        means), 191
Proportion of total area under the curve,
        50–51
Proportion of variance accounted for (r_{pb}^2)
    effect size and, 282–283
    predicted *Y* scores and, 174–177, 184
Proportion of variance not accounted for
        ($1 - r^2$)
    predicted *Y* scores and, 177–178, 184
Proportions
    cumulative frequency/percentile and, 52
    definition of, 7
    finding unknown, 383
    relative frequency and, 47–51
    z-tables, 419–422
Protected *t*-test
    formula for, 317
    formulas for, 317
Published research. *See* Research

Q

q_k. *See* Range (q_k)
Quasi-independent variables, 23

R

r-tables, 424
Random sample
    definition of, 14
    probability and, 186
    sampling error and, 193–194
Range (q_k)
    as measure of variability, 87
    one-way ANOVA and, 429–430
Rank sums test, 366–367
Ratio scale, 28–29, 40–41
Raw score, transformed into *z*-score, 112
r_{crit}, 235
Real limits, 380
Rectangular distribution, 45
Region of rejection, 197–198, 216
Regression line, 140, 161–163
Regular frequency distribution, 38

Related-samples *t*-test
    as method to calculate two-sample
        t-test, 262
    computing, 274–275
    computing confidence interval for, 277
    formula for, 275, 289
    interpreting, 275–276
    one-tailed testing and, 277–278
    requirements for, 271–272
    statistical hypotheses for, 272–273,
        273–274
    when to perform, 284
Relationships
    as associations, 15–17
    causal, 26–27
    cumulative frequency/percentile and,
        51–53
    definition of, 15
    graphing, 18–20
    and inferences in population, 78–79
    linear, 139–141
    nonexistence of, 17–18
    nonlinear, 141–142
    strength of, 16–17, 142–146
    two-sample *t*-test and, 261–262
    in two-way chi square, 361–362
    types of, 139–142
    variability and strength of, 101–102
Relative degree of consistency, 143–144
Relative frequency, 47–51, 187
Relative frequency distribution, 48–49
Relative frequency (*rel.f*), 47–51, 54–55, 59
Relative standing, 110
Repeated-measures design, 271–272
Representative samples
    definition of, 14
    making intelligent decisions regarding,
        202–203
    random sampling and, 193–194
    region of rejection and, 197–198
    and standard error of the mean,
        199–200
Research. *See also* Central tendency
    alpha (α) and, 246–247
    APA rules, 53
    causality and, 26–27
    chi square procedures and, 362–363
    correlation coefficients and, 154–155
    correlational studies and, 25–26
    descriptive/inferential statistics and,
        20–21
    and drawing conclusions from
        experiments, 24–25
    experiments and, 22–25
    hypothesis testing and, 224
    linear regression and, 179
    logic of, 12–13
    mean and, 79
    measurement scales/scores and, 27–29
    and measuring variables for data
        acquisition, 14–15
    one-sample *t*-test and, 246–247
    one-way ANOVA and, 312
    parameters and, 21–22
    relationships and, 15–18
    role of inferential statistics, 208–209
    samples/populations and, 13–14

and steps of statistical analysis of experiments, 78–79
two-sample *t*-test and, 284
use of terms in, 30–31
variability and, 103–104
variance/standard deviation application to, 100–103
Research design, 209
Restriction of range problem, 153–154
Rounding numbers, 6

S

Sample, definition of, 13–14, 186
Sample mean
formula for, 83
mean, 79
relative frequency of, 128–129
and sampling distribution of means, 124–128
Sample standard deviation (S_X), 89–92, 94, 160
Sample variance, 88–89, 94, 160
Sampling distribution of differences between means, 263–264
Sampling distribution of mean differences, 274
Sampling distribution of means
and critical value identification, 198–200
and performing *z*-test, 215–218
probability and, 191
setting up, 197–198, 200–201
and standard error of the mean, 126–127
z-scores and, 124–125
Sampling distribution of r_S, 252–253
Sampling error. *See also* Error
population representation and, 194–196
random sampling and, 193–194
and role of inferential statistics, 208
Sampling with replacement, 189
Sampling without replacement, 189
Scale of measurement, 209
Scales
characteristics of, 27–30
and measure of central tendency, 62
Scatterplots. *See also* Correlation coefficients
correlation coefficients and, 138–139, 143–146
described by H_0, 248
two-sample *t*-test and, 280
Scores
as locations, 61
characteristics of, 27–30
dependent variables and, 23–24
deviation of, 69–72
grouped frequency distribution and, 379–381
Mann-Whitney *U* test and, 432–435
median as, 63–64
percentile for *cf*, 59
predicting with mean, 71–72
and sum of deviations around the mean, 70–71
ΣX^2. *See* Sum of the squared *X*s $(\Sigma X)^2$

Significance testing. *See* One-sample *t*-test; *z*-test
Significant results
definition of, 218
hypothesis testing and, 218–219
Type I/Type II errors and, 228–230
Simple frequency distribution, 37–42, 37–42
Skewed distribution
about, 43–44, 46
median location in, 64
mode/mean/median comparisons and, 67–69
Slope
linear regression equation and, 163–164, 184
of regression line, 164–166
Spearman rank-order correlation coefficient (ρs)
about, 151–153
linear regression equation and, 159
r_S-tables, 425
testing, 252–253
SPSS
central tendency/variability/*z*-scores and, 397–399
chi square procedures and, 373, 411–414
and computing percentiles, 52
correlation coefficients and, 155, 399–401
entering data, 392–394
frequency distributions/percentiles and, 395–397
Friedman χ^2 test and, 415
hypothesis testing and, 255
Kruskal-Wallis *H* test and, 415
linear regression and, 180
linear regression equation and, 399–401
Mann-Whitney *U* test and, 414
measures of variability and, 104
mode/mean/median comparisons and, 79
one-sample *t*-test and, 255
one-sample *t*-test and significance testing of correlation coefficients, 402–404
one-way ANOVA and, 312–313
one-way, between-subjects ANOVA and, 407–409
one-way, within-subjects ANOVA and, 416–417
and transformation into *z*-scores, 130
two-sample *t*-test and, 284, 404–407
two-way, between-subjects ANOVA and, 409–411
Wilcoxon *T* test and, 415
$(\Sigma X)^2$. *See* Squared sum of *X* $(\Sigma X)^2$
Squared point-biserial correlation coefficient, 282–283
Squared sum of *X* $(\Sigma X)^2$, 84–85
Standard deviation
mathematical constants and, 95
and performing *z*-test, 215
population variance/standard deviation, 95–99
sample standard deviation (S_X), 90
of sampling distribution of means (standard error of the mean), 126–127
variance and, 87–90
z-scores and, 110–112

Standard error of the difference, 265–266
Standard error of the estimate, 170–174, 184
Standard error of the mean difference, 275
Standard error of the mean ($\sigma\overline{X}$)
formula for, 126–127, 191
and performing *z*-test, 216
and population representation, 199–200
Standard normal curve
and relative frequency of sample means, 128–129
z-tables, 419–422
Statistical hypotheses
for independent-samples *t*-test, 262–263
one-sample *t*-test and, 235–237
in one-way ANOVA, 293
for related-samples *t*-test, 272–273
and testing the Pearson *r* (correlation coefficients), 248
and testing the Spearman r_S (correlation coefficients), 252–253
Statistical notation, 4–6
Statistical tables
of chi square (x_2 tables), 431
F-tables, 426–428
Mann-Whitney *U* test, 432–435
q_k values, 429–430
r-tables, 424
r_S-tables, 425
t-tables, 423
Wilcoxon *T* test and, 436–437
z-tables, 419–422
Statistics
basic standard notation, 4–5
definitions of, 1–4
descriptive, 20–21
importance of, 1–4
inferential, 21
Strength of relationship, 16–17. *See also* Relationships
linear/nonlinear variables, 142–146
standard error of the estimate and, 173–174
Subscripts, 85
Sum of *X* (ΣX), 60–61
Sum of deviations around the mean, 70–71, 70–71
Sum of squares between groups (SS_{bn}), 301
Sum of squares (SS_{tot}), 301
Sum of squares within groups (SS_{wn}), 302, 349
Sum of the cross products, 136
Sum of the squared $X_2(\Sigma X)^2$, 84–85
Summation (Σ), 60–61
Symmetrical distribution, 66–67

T

t-distribution, 238–240
t-tables, 240, 242
t-tables, 423
t-test. *See* One-sample *t*-test
Tables. *See* Statistical tables
Tails of distribution, 43–44
t_{crit}, 235, 238–242, 244–245
Test of independence, 358
Theoretical probability distribution, 188

Tied ranks, 151
t_{obt}, 235–238, 240–242
   and independent-samples t-test, 266–267
   and related-samples t-test, 275
Total area under the curve, 50–51
Total number of scores in condition (n), 260, 337
Total number of scores in study (N), 37–38, 260
Transformation
   definition of, 6–7
   mean and, 69
Treatment, 23, 290
Treatment effect, 290
Treatment variance (μ^2_{treat}), 298–299
Tukey's HSD multiple comparisons test, 307–308
Tukey's HSD post hoc comparisons test, 350
Two-sample t-test. *See also* Independent-samples t-test; Related-samples t-test
   graphing, 279–280
   importance of, 261
   measuring effect size in, 280–283
   published research and, 283
   relationships and, 261–262
Two-tailed test
   and creation of statistical hypotheses, 248
   definition of, 210
   r-tables, 424
   relationships and, 229
   sampling distribution and, 215–216
   uses of, 223
Two-way analysis of variance (ANOVA). *See* Two-way ANOVA
Two-way ANOVA
   and computing degrees of freedom, 332
   computing (degrees of freedom), 332
   and computing each F, 334
   and computing F_{obt}, 327–328
   and computing mean square between groups for factor A, 332–333
   and computing mean square between groups for factor B, 333
   and computing mean square between groups for interaction, 333
   and computing mean square within groups, 333
   computing (mean squares), 332–335
   computing (overview), 327–328
   and computing sum of squares between groups for factor A, 330
   and computing sum of squares between groups for factor B, 330
   and computing sum of squares between groups for interaction, 331
   and computing sum of squares within groups, 331–332
   and computing sums and means, 328–329
   computing (sums and means), 328–329
   computing (sums of squares), 329–332
   and computing total sum of squares, 329
   and computing total sum of squares between groups, 331
   eta squared effect size and, 341–343
   and graphing interaction effect, 337–339, 337–339, 338–339

and graphing/post hoc comparisons with main effects, 336–337, 336–340
   importance of, 319
   and interaction effects of A/B factors, 323
   interpreting, 340–341
   and interpreting each F, 335–336
   and interpreting F, 335
   and main effect of factor A, 321–323
   and main effect of factor B, 323
   overview of, 319–327
   and post hoc comparisons on interaction effects, 339–340
   summary of performance steps for, 343–344
   summary table of, 349
   two-way between-subjects ANOVA, 318
   two-way mixed-design ANOVA, 318
   two-way within-subjects ANOVA, 318
Two-way between-subjects ANOVA, 318
Two-way chi square
   computing, 359–360
   definition of, 357–359
   formula for, 359–360
Two-way interaction effect, 326
Two-way mixed-design ANOVA, 318
Two-way within-subjects ANOVA, 318
Type I error. *See also* Error
Type I error: rejecting H_0 when H_0 is true, 224–229, 240–241, 246–247, 250, 292–293
Type II error: retaining H_0 when H_0 is false, 227–229, 241, 253–254. *See also* Error
Types of relationships, 139–142

U
Unbiased estimators, 97
Unconfounded comparison, 339–340
Underestimation
   biased estimators and, 96–97
   and predicting scores with mean, 71–72
Ungrouped distributions, 54
Unimodal distribution, 62–63, 66–67
Univariate statistics, 312
Unrepresentative samples, 14

V
Variability
   correlation coefficients and, 144–145
   descriptive/inferential measures of, 101
   and strength of relationship, 101–102
Variables
   definition/use of, 14–15
   dependent, 23–24, 73–74
   dichotomous, 29
   independent, 22–23
   predictor/criterion, 162–163
   quasi-independent, 23
   and relationship with interval variables, 150
Variance
   ANOVA and, 299–300

computational formula for, 94
   homogeneity of, 262
   one-way, within-subject analysis of, 384–390
   pooled, 265
   and proportion of variance accounted for, 174–175
   standard deviation and, 87–89
   variance of Y scores around Y', 169–170, 184

W
Wilcoxon T test, 368–369, 436–437
Within-subjects factor, 291

X
X axis
   and "as a function of" changes, 19, 24
   bar graphs and, 39
   graph creation and, 7–9
   normal distribution and, 42–43
   relationships and, 15–16, 18–20

Y
Y axis
   and "as a function of" changes, 19, 24
   graph creation and, 7–9
   normal distribution and, 42–43
   relationships and, 15–16, 18–20
Y intercept
   about, 163–164
   formula for, 184
   linear regression equation and, 163–164
Y prime (Y'), 160, 164

Z
z-distribution
   definition of, 115–116
   and probability from standard normal curve (individual scores), 190
   and relative frequency of sample means, 128–129
z obtained (z_{obt})
   comparing to critical values, 217–218
   formula for, 216
   nonsignificant results and, 219–220
   published research and, 224
   significant results and, 218
z-scores
   comparing different variables using, 116–117
   critical values and, 198–199
   definition of, 110–112
   in determining probability of particular sample means, 191–192
   determining relative frequency of raw scores with, 117–120
   formula for, 127–128
   formula for transforming raw score in a population into, 113
   formula for transforming raw score in a sample into, 112

importance of, 110
linear interpretation of, 381–384
sample mean ($\overline{X}$) and, 124–128
for transforming sample mean into
   z-score, 206
for transforming z-score in population
   into raw score, 114
for transforming z-score in sample into
   raw score, 113–114
and use of z-table, 120–122

z-table, 120–122, 419–422
z-test. *See also* Critical value of $z(z_{crit})$; z
   obtained (z_{obt})
   and computing confidence interval, 245
   and computing t_{obt}, 236
   formula for, 216–217, 233
   importance of, 208
   inferential statistics and, 208
   one-tailed testing and, 223, 229
   performing, 215–218

power and, 228-229
summary of, 220
Type I error and, 228
Type II error and, 228
when to perform, 234
z_{crit}. *See* Critical value of z ($z$crit)
Zero, as sum of deviations around the mean,
   70–71
Zero association, 146